Nanotechnology in Fuel Cells

Nanotechnology in Fuel Cells

Edited by

HUAIHE SONG
State Key Laboratory of Chemical Resource Engineering, College of Materials Science and Engineering, Beijing University of Chemical Technology, Beijing, China

TUAN ANH NGUYEN
Institute for Tropical Technology, Vietnam, Academy of Science and Technology, Hanoi, Vietnam

GHULAM YASIN
Institute for Advanced Study, College of Physics and Optoelectronic Engineering, Shenzhen University, Shenzhen, Guangdong, China

Elsevier
Radarweg 29, PO Box 211, 1000 AE Amsterdam, Netherlands
The Boulevard, Langford Lane, Kidlington, Oxford OX5 1GB, United Kingdom
50 Hampshire Street, 5th Floor, Cambridge, MA 02139, United States

Copyright © 2022 Elsevier Inc. All rights reserved.

No part of this publication may be reproduced or transmitted in any form or by any means, electronic or mechanical, including photocopying, recording, or any information storage and retrieval system, without permission in writing from the publisher. Details on how to seek permission, further information about the Publisher's permissions policies and our arrangements with organizations such as the Copyright Clearance Center and the Copyright Licensing Agency, can be found at our website: www.elsevier.com/permissions.

This book and the individual contributions contained in it are protected under copyright by the Publisher (other than as may be noted herein).

Notices
Knowledge and best practice in this field are constantly changing. As new research and experience broaden our understanding, changes in research methods, professional practices, or medical treatment may become necessary.

Practitioners and researchers must always rely on their own experience and knowledge in evaluating and using any information, methods, compounds, or experiments described herein. In using such information or methods they should be mindful of their own safety and the safety of others, including parties for whom they have a professional responsibility.

To the fullest extent of the law, neither the Publisher nor the authors, contributors, or editors, assume any liability for any injury and/or damage to persons or property as a matter of products liability, negligence or otherwise, or from any use or operation of any methods, products, instructions, or ideas contained in the material herein.

British Library Cataloguing-in-Publication Data
A catalogue record for this book is available from the British Library

Library of Congress Cataloging-in-Publication Data
A catalog record for this book is available from the Library of Congress

ISBN: 978-0-323-85727-7

For Information on all Elsevier publications
visit our website at https://www.elsevier.com/books-and-journals

Publisher: Matthew Deans
Acquisitions Editor: Simon Holt
Editorial Project Manager: Gabriela D. Capille
Production Project Manager: Debasish Ghosh
Cover Designer: Victoria Pearson

Typeset by MPS Limited, Chennai, India

Contents

List of contributors   xiii
Foreword   xvii

Section 1   Basic principles

1. Nanotechnology-based fuel cells: an introduction   3
Nirali H. Gondaliya

1.1 Introduction	3
1.2 Fuel cells	5
1.3 Nanotechnology and fuel cells	6
1.4 Current research on nanofuel cells	7
1.5 Conclusion	8
References	8

2. Microfluidic concept-based fuel cells   11
Biswajit S. De, Neeraj Khare, Anastasia Elias and Suddhasatwa Basu

2.1 Introduction	11
2.2 Theory	13
2.3 Fabrication and design of microfluidic fuel cells	16
2.4 Performance evaluation of microfluidic fuel cells	21
2.5 Perspective and conclusions	35
References	36

3. Alcohol fuel cell on-a-chip   41
Cauê A. Martins

3.1 Introduction	41
3.2 Fuel cell on-a-chip	42
3.3 General view of direct alcohol microfluidic fuel cells	45
3.4 Testing direct alcohol microfluidic fuel cells	50
3.5 Nanotechnology in direct alcohol microfluidic fuel cells	54
3.5.1 Nanoparticle anodes and cathodes for methanol microfluidic fuel cells	55
3.5.2 Nanoparticle anodes and cathodes for ethanol microfluidic fuel cells	56

	3.5.3	Nanoparticle anodes and cathodes for ethylene glycol microfluidic fuel cells	60
	3.5.4	Nanoparticle anodes and cathodes for glycerol microfluidic fuel cells	61
3.6	Perspectives and future engagements		65
Acknowledgments			67
Appendix			**68**
References			72

4. General aspects in the modeling of fuel cells: from conventional fuel cells to nano fuel cells 77
Pablo A. García-Salaberri

4.1	Introduction		77
	4.1.1	Proton exchange membrane fuel cells (PEMFCs)	80
4.2	Numerical modeling		88
	4.2.1	Macroscopic continuum modeling	88
	4.2.2	Pore-scale modeling	90
	4.2.3	Hybrid continuum/pore-scale modeling	92
4.3	PEMFC macroscopic modeling		94
	4.3.1	Assumptions	97
	4.3.2	Conservations equations	98
	4.3.3	Source terms	104
	4.3.4	Boundary conditions	110
Acknowledgments			113
References			113

5. Mathematical modeling for fuel cells 123
Abdalla M. Abdalla, Mohamed K. Dawood, Shahzad Hossain, Mohamed El-Sabahy, Basem E. Elnaghi, Shabana P.S. Shaikh and Abul K. Azad

5.1	Introduction		123
5.2	Fuel cells simulation and modeling		124
5.3	Process design and mathematical formulation		124
	5.3.1	Solid oxide fuel cells	125
	5.3.2	Molten carbonate fuel cells	126
	5.3.3	Proton exchange membrane fuel cells	128
	5.3.4	Direct alkaline fuel cell	131
	5.3.5	Microbial fuel cells	132
5.4	Key challenges of mathematical modeling		134
5.5	Conclusion		135
References			135

Section 2 Nanostructures and nanomaterials for fuel cells

6. Nanostructures and nanomaterials in microbial fuel cells — 139
Saranya Narayanasamy and Jayapriya Jayaprakash

6.1	Introduction	139
6.2	Nanostructured materials in microbial fuel cells	140
	6.2.1 Nanostructures as electrode materials	141
6.3	Nanocomposite materials	148
	6.3.1 Anode materials	149
	6.3.2 Cathode materials	156
	6.3.3 Membranes	159
6.4	Conclusion	162
	Acknowledgment	163
	References	163

7. Metal-organic frameworks for fuel cell technologies — 173
Muhammad Rizwan Sulaiman and Ram K. Gupta

7.1	Introduction	173
7.2	Structure of metal-organic frameworks	176
	7.2.1 Secondary building units	178
	7.2.2 Open metal sites	179
	7.2.3 Pores	179
	7.2.4 Functional groups	180
	7.2.5 Development of porous structure	181
	7.2.6 Design of metal-organic framework derivative	183
7.3	Metal-organic frameworks for fuel cells applications	183
	7.3.1 Proton-conducting metal-organic frameworks	184
7.4	Metal-organic frameworks as oxygen reduction reaction catalyst	189
7.5	Conclusion	196
	References	196

8. Advanced carbon-based nanostructured materials for fuel cells — 201
Muhammad Rizwan Sulaiman and Ram K. Gupta

8.1	Introduction	201
8.2	Important oxygen reduction reaction characterization notions	204
	8.2.1 Onset potential (E_{ons})	204
	8.2.2 Current density	205
	8.2.3 Tafel slope	205
	8.2.4 Electron transfer number and HO_2^- percentage	205

	8.2.5	Turnover frequency		206
8.3	Approaches to synthesize nanocarbons			206
	8.3.1	Zero-dimensional carbonaceous materials		206
	8.3.2	One-dimensional carbonaceous materials		211
	8.3.3	Graphene-based two-dimensional carbonaceous materials		216
	8.3.4	Three-dimensional carbon materials		220
8.4	Conclusion			224
References				224

9. Covalent organic framework-based materials as electrocatalysts for fuel cells — 229

Anuj Kumar, Shashank Sundriyal, Tribani Boruah, Charu Goyal, Sonali Gautam, Dipak Kumar Das and Tuan Anh Nguyen

9.1	Introduction		229
	9.1.1	Motivations and scope	229
	9.1.2	Fuel cell and its chemistry	230
	9.1.3	Chemistry of covalent organic frameworks	232
	9.1.4	Interlayer stacking in COFs	235
9.2	Recent advancements in COF-based electrocatalysts for ORR		235
	9.2.1	Pristine metal-free COFs	236
	9.2.2	Macrocycle-incorporated COFs	237
	9.2.3	COF-derived SAC-based materials	239
	9.2.4	Nanocarbon-supported COFs	241
9.3	Designing approaches of COF-based electrocatalysts		242
	9.3.1	Geometric orientation-based approaches	242
	9.3.2	Different bonding-based approaches	244
9.4	Concluding remarks		245
	9.4.1	Challenges	246
	9.4.2	Prospects and research directions	246
References			246

10. Nano-inks for fuel cells — 251

Fahimeh Hooriabad Saboor and Tuan Anh Nguyen

10.1	Introduction		251
10.2	Ink rheological parameters and influencing factors		253
10.3	Recent progress in nano-ink preparation and utilization in fuel cell application		254
	10.3.1	The effect of solvents and additives	254
	10.3.2	The effect of the dispersion method	256
	10.3.3	The effect of the ink composition	256
10.4	Conclusion		258
References			258

11. Nanomaterial and nanocatalysts in microbial fuel cells — 261
Sumisha Anappara, Karthick Senthilkumar and Haribabu Krishnan

- 11.1 Introduction — 261
- 11.2 Application of nanomaterial structures in anodes — 263
  - 11.2.1 Carbon-based nanomaterials — 263
  - 11.2.2 Nanoconducting polymer — 266
  - 11.2.3 Metal-based nanomaterials — 268
- 11.3 Application of nanomaterial structures in cathodes — 270
  - 11.3.1 Graphene-based nanomaterial — 271
  - 11.3.2 Carbon nanotubes and nanofibers — 272
  - 11.3.3 Transition metal oxide nanomaterials — 273
  - 11.3.4 Conducting polymer nanomaterials — 274
- 11.4 Ion-exchange membranes — 275
  - 11.4.1 Perfluorinated membranes — 276
  - 11.4.2 Sulfonated membranes — 276
  - 11.4.3 Nonfluorinated and nonsulfonated membranes — 277
- 11.5 Conclusion — 279
- References — 279

12. Nanomembranes in fuel cells — 285
Yunfeng Zhang

- 12.1 The introduction of nanomembranes in fuel cells — 285
  - 12.1.1 Description of nanomembranes — 285
  - 12.1.2 Proton transport channels — 288
- 12.2 Polymer-based nanomembranes — 295
  - 12.2.1 Polymer membranes with tailored main chains — 295
  - 12.2.2 Polymer membranes with functional side chains — 299
  - 12.2.3 Polymer membranes with special groups — 304
  - 12.2.4 Cross-linking membranes — 310
- 12.3 Hybrid membranes containing nanofillers — 316
  - 12.3.1 Effect of nanofillers — 316
  - 12.3.2 Nanoparticles — 317
  - 12.3.3 Nanocarbon materials — 323
- References — 338

13. Shape-controlled metal nanoparticles for fuel cells applications — 349
Ajit Behera

- 13.1 Introduction — 349
- 13.2 What is shape-controlled catalyst — 350
- 13.3 Platinum catalyst — 350

13.4	Fe, Co, and Ni catalysts	353
13.5	Various functionality of shape-selected nanoparticles	354
	13.5.1 Activity enhancement	354
	13.5.2 Selectivity enhancement	355
	13.5.3 Optical behavior	356
13.6	Summary	357
	References	357

14. Fuel cells recycling 361
Ajit Behera

14.1	Introduction to fuel cells and recycling	361
14.2	Nanomaterials used in fuel cells	362
	14.2.1 Nanomaterials in the polymer electrolyte membrane or proton-exchange membrane	363
	14.2.2 Nanomaterials in the molten carbonate fuel cells	364
	14.2.3 Materials in phosphoric acid fuel cell	365
	14.2.4 Materials in solid oxide fuel cells	365
14.3	Recycling processes	366
	14.3.1 Hydrometallurgical process	367
	14.3.2 Pyrohydrometallurgical process	367
	14.3.3 Hydrothermal process	368
	14.3.4 Selective electrochemical dissolution	368
	14.3.5 Transient dissolution through potential alteration	369
	14.3.6 Membrane and Pt-recovery acid process	369
	14.3.7 Alcohol solvent process	370
14.4	Microbial fuel cell recycling mechanism	370
14.5	Summary	371
	References	371

15. Micro/nanostructures for biofilm establishment in microbial fuel cells 375
Linbin Hu, Jun Li, Qian Fu, Liang Zhang, Xun Zhu and Qiang Liao

15.1	Introduction	375
15.2	Nanostructure for promoting electrochemical active bacteria/electrode interaction	377
	15.2.1 Carbon nanomaterials decoration	377
	15.2.2 Conductive polymer coating	383
	15.2.3 Other types of modifications	387
15.3	Microstructure for increasing specific surface area	388
	15.3.1 Electrodes with random microstructure	388
	15.3.2 Electrode materials with regular microstructure	395
15.4	Outlook	401

Acknowledgments		402
References		402

16. Nanomaterials in biofuel cells — **411**

Sangeetha Dharmalingam, Vaidhegi Kugarajah and John Solomon

16.1	Introduction	411
16.2	Types of biofuel cell	412
	16.2.1 Enzymatic fuel cell	413
	16.2.2 Microbial fuel cell	413
16.3	Application of nanomaterials in biofuel cells	414
	16.3.1 Nanomaterials in electrodes	415
	16.3.2 Nanomaterials in membranes	426
16.4	Conclusion	434
References		435

Index — *445*

List of contributors

Abdalla M. Abdalla
Mechanical Engineering Department, Faculty of Engineering, Suez Canal University, Ismailia, Egypt

Sumisha Anappara
Department of Chemical Engineering, National Institute of Technology Calicut, Kozhikode, India

Abul K. Azad
Faculty of Integrated Technologies, Universiti Brunei Darussalam, Bandar Seri Begawen, Brunei

Suddhasatwa Basu
Indian Institute of Technology, New Delhi, India; CSIR—Institute of Minerals & Materials Technology (IMMT), Bhubaneswar, India; University of Alberta, Edmonton, AB, Canada

Ajit Behera
Department of Metallurgical & Materials Engineering, National Institute of Technology, Rourkela, Rourkela, India

Tribani Boruah
Northeast Hill University (NEHU), Umshing Mawkynroh, Shillong, India

Mohamed K. Dawood
Mechanical Engineering Department, Faculty of Engineering, Suez Canal University, Ismailia, Egypt

Biswajit S. De
Indian Institute of Technology, New Delhi, India

Sangeetha Dharmalingam
Department of Mechanical Engineering, Anna University, Chennai, India

Anastasia Elias
Indian Institute of Technology, New Delhi, India

Basem E. Elnaghi
Electrical Engineering Department, Faculty of Engineering, Suez Canal University, Ismailia, Egypt

Mohamed El-Sabahy
Mechanical Engineering Department, Faculty of Engineering, Suez Canal University, Ismailia, Egypt

Qian Fu
Key Laboratory of Low-grade Energy Utilization Technologies and Systems, Chongqing University, Ministry of Education, Chongqing, China; Institute of Engineering Thermophysics, School of Energy and Power Engineering, Chongqing University, Chongqing, China

Pablo A. García-Salaberri
Department of Thermal and Fluids Engineering, University Carlos III of Madrid, Leganés, Spain

Sonali Gautam
Department of Chemistry, GLA University, Mathura, India

Nirali H. Gondaliya
Department of Physics, SVM Institute of Technology, Bharuch, India

Charu Goyal
Department of Chemistry, GLA University, Mathura, India

Ram K. Gupta
Department of Chemistry, Kansas Polymer Research Center, Pittsburg State University, Pittsburg, KS, United States

Shahzad Hossain
Bangladesh Atomic Energy Commission, Dakha, Bangladesh

Linbin Hu
Key Laboratory of Low-grade Energy Utilization Technologies and Systems, Chongqing University, Ministry of Education, Chongqing, China; Institute of Engineering Thermophysics, School of Energy and Power Engineering, Chongqing University, Chongqing, China

Jayapriya Jayaprakash
Department of Applied Science and Technology, Alagappa College of Technology, Anna University, Chennai, India

Neeraj Khare
Indian Institute of Technology, New Delhi, India

Haribabu Krishnan
Department of Chemical Engineering, National Institute of Technology Calicut, Kozhikode, India

Vaidhegi Kugarajah
Department of Mechanical Engineering, Anna University, Chennai, India

Anuj Kumar
Department of Chemistry, GLA University, Mathura, India

Dipak Kumar Das
Department of Chemistry, GLA University, Mathura, India

Jun Li
Key Laboratory of Low-grade Energy Utilization Technologies and Systems, Chongqing University, Ministry of Education, Chongqing, China; Institute of Engineering Thermophysics, School of Energy and Power Engineering, Chongqing University, Chongqing, China

Qiang Liao
Key Laboratory of Low-grade Energy Utilization Technologies and Systems, Chongqing University, Ministry of Education, Chongqing, China; Institute of Engineering Thermophysics, School of Energy and Power Engineering, Chongqing University, Chongqing, China

Cauê A. Martins
Institute of Physics, Federal University of Mato Grosso do Sul, Campo Grande, Brazil

Saranya Narayanasamy
Department of Applied Science and Technology, Alagappa College of Technology, Anna University, Chennai, India

Tuan Anh Nguyen
Institute for Tropical Technology, Vietnam Academy of Science and Technology, Hanoi, Vietnam

Fahimeh Hooriabad Saboor
Department of Chemical Engineering, University of Mohaghegh Ardabili, Ardabil, Iran

Karthick Senthilkumar
Department of Chemical Engineering, National Institute of Technology Calicut, Kozhikode, India

Shabana P.S. Shaikh
Department of Physics, SP Pune University, Pune, India

John Solomon
Department of Mechanical Engineering, Anna University, Chennai, India

Muhammad Rizwan Sulaiman
Department of Chemistry, Kansas Polymer Research Center, Pittsburg State University, Pittsburg, KS, United States; Department of Plastic Engineering, Kansas Technology Center, Pittsburg State University, Pittsburg, KS, United States

Shashank Sundriyal
Advanced Carbon Products Department, CSIR-National Physical Laboratory, New Delhi, India

Liang Zhang
Key Laboratory of Low-grade Energy Utilization Technologies and Systems, Chongqing University, Ministry of Education, Chongqing, China; Institute of Engineering Thermophysics, School of Energy and Power Engineering, Chongqing University, Chongqing, China

Yunfeng Zhang
Sustainable Energy Laboratory, China University of Geosciences Wuhan, Wuhan, China

Xun Zhu
Key Laboratory of Low-grade Energy Utilization Technologies and Systems, Chongqing University, Ministry of Education, Chongqing, China; Institute of Engineering Thermophysics, School of Energy and Power Engineering, Chongqing University, Chongqing, China

Foreword

Most of us are aware of the finite nature of coal, natural gas, and petroleum, which by itself is generating a great impact upon society, as we know it. In addition, much has been said about the impact on climate change resulting from the continued and intensive use (or even abusive use) of these materials. Of course, alternative approaches using other energy sources have been invented in the last decades and exhaustively deepened within time by intensive research work, looking for innovative devices using renewable and sustainable sources of energy.

Fuel cells are one of such routes, for a more clean and sustainable environment/society. Considering their high efficiency, I share the vision that fuel cells may turn out game changers in mobile, stationary, and portable applications in a near future. There are mainly two areas, in my opinion, driving forward fuel cells in this quest for efficient and clean energy sources. One relates to the overall materials employed, which are critical for the fuel cell operation. Another one relates to the conceptual design of the cell itself, which may operate under flow or stationary modes, some with portable features and/or using microfluidics. Without a doubt, nanotechnology is the main field feeding these two topics and this deserves being highlighted. Controlling the overall events at a nanoscale level generates a wide range of opportunities to enhance the performance of these cells, as well as to reduce their cost and environmental impact.

Fuel cells are also being very recently employed in highly innovative applications. This includes their interface with biosensing devices Carneiro et al. [1], a work developed in the context of the project Symbiotic, funded by the European Commission, in FET-Open call under Horizon 2020 (https://cordis.europa.eu/project/id/665046). In brief, a methanol fuel cell generates energy that depends on the concentration of a cancer biomarker present, basically taking advantage of many changes occurring at a nanoscale level, both in the biosensors and in fuel cells.

Thus it is by far obvious that the progression of clean energies from fuel cells walks side-by-side with nanotechnology, as wisely deepened in this book. *Nanotechnology in Fuel Cells* focuses on basic knowledge around the fuel cell concepts to complex modeling, while also including a detailed use of different nanomaterials. Overall, I consider this book a valuable source of information to researchers working in the field of fuel cells, being also timely for professors in the same field.

Goreti Sales
BioMark Sensor Research/UC, ISEP Team, Chemical Engineering Department, Faculty of Sciences and Technology, University of Coimbra, Coimbra, Portugal

Reference

[1] L.P.T. Carneiro, N.S. Ferreira, A.P.M. Tavares, A.M.F.R. Pinto, A. Mendes, M.G.F. Sales, A passive direct methanol fuel cell as transducer of an electrochemical sensor, applied to the detection of carcinoembryonic antigen, Biosensors and Bioelectronics 175 (2021) 112877. Available from: https://doi.org/10.1016/j.bios.2020.112877.

SECTION 1:
BASIC PRINCIPLES

CHAPTER 1

Nanotechnology-based fuel cells: an introduction

Nirali H. Gondaliya
Department of Physics, SVM Institute of Technology, Bharuch, India

1.1 Introduction

Since one and half century ago research started in the field of fuel cells (FCs) and since then there is no looking back. In 1800 William Nicholson and Anthony Carlisle described the process of using electricity to break water into hydrogen and oxygen [1], then Sir William Grove in 1839 discovered that it may be possible to generate electricity by reversing the electrolysis of water and he develop cell using platinum as electrode which was immersed in nitric acid and a zinc electrode in zinc sulfate to generate about 12 A of current at about 1.8 V. After his name this cell was termed as "Grove cell," and beside this, he was also known as the "Father of the Fuel Cells" [2]. After few decades in 1889, two researchers, Charles Langer and Ludwig Mond, officially coined the term "fuel cells" using air and coal gas [3]. While, the founder of physical chemistry Friedrich Wilhelm Ostwad provided much of the theoretical understanding of how FC operates and in 1893, he experimentally determined the interconnected roles of the various components of the FCs: electrodes, electrolyte, oxidizing and reducing agents, anions, and cations [4]. It was in early 1900s that many attempts were made to develop FCs that could convert coal or carbon into electricity, but the advent of the IC engine temporarily quashed any hopes of further development of the fledgling technology. Almost after four decades, Francis Bacon developed an inexpensive alternative to the catalysts used by Mond and Langer. It was perhaps the first successful FC device in 1932, with a hydrogen−oxygen cell using alkaline electrolytes [5] and nickel electrodes. Their research resulted in the invention of gas-diffusion electrodes in which the fuel gas on one side is effectively kept in controlled contact with an aqueous electrolyte on the other side [6].

By the mid-1990s, O. K. Davtyan of the Soviet Union had published the results of experimental work on solid electrolytes, his major focus was on increase in conductivity and mechanical strength while its major drawback was its short life rating [7]. By the late 1950s, research into solid oxide technology, that is, solid oxide fuel cells (SOFCs) began to accelerate and the need for highly efficient and stable power supplies for space satellites and manned spacecraft created exciting new opportunities. The initial support of the National Aeronautics and Space Administration (NASA) provided the catalyst needed to launch the commercial FCs industry in the United States (US). NASA started developing FC generators for manned space missions, which resulted in the invention of the proton-exchange membrane fuel cell (PEMFC) by Willard Thomas Grubb, later refined by Leonard Niedrach through the use of platinum as a catalyst on the membranes [8,9] Thereafter, in cooperation with NASA the Grubb-Niedrach, FC was further developed and eventually used in the Gemini space program of the mid-1960s [8]. This cooperation of NASA accelerated the urge of better and better development in FCs technology, which resulted in to the development of a 1.5 kW Alkaline fuel cell (AFC) for use in the Apollo space missions by International Fuel Cells, which was able to provide power and clean water for an entire space mission [8]. This mission was in collaboration between government and industry players, emphasizing a continuum toward possible triple-helix dynamics [10]. After that there was no looking back, NASA has been addressing aerospace challenges, dramatically advancing FC technologies for aerospace applications with the aid of industry and universities, which indicates a move toward triple-helix dynamics [10]. This act of NASA attracted attention of scientific fraternity and in 1992 altogether 21 university and research centers were created as result of partnerships between minority institutions and NASA [11]. Hence this was the starting period of collaboration between industry, government, and universities.

After successful entry of FCs into space technology, it gained attention in the field of military applications in the Soviet Union, where they used it for onboard power for a submarine and later for the Soviet manned space program also [12]. This led to substantial investment of government agencies and companies for further development and refinement of the technology, as the energy produced by these FCs yields numerous advantages [13]. A massive thrust was observed in commercialization of FC technology and many leading companies like Siemens and Sulzer (Europe) [14], Westinghouse Electrical Cooperation (USA) [15], and Fuji Electric Corporate Research and Development, Ltd and Tokyo Electric Power Co. (Japan) [16] were involved. In the late 1990s, first 100 kW SOFC field unit (Westinghouse SOFC technology) was put into operation in the Netherlands and many companies like ELSAM a Danish Production Companies, Dutch Energy Distribution Companies, EnergieNed and the Dutch Subsidiser NOVEM cooperate [17] were few name who were tried their luck in production of FC. Another company Global Thermoelectric's Fuel Cell Division used FC designed by Julich Research Institute in Germany which was a very good example of collaboration of research institute and industry. Ceramatec-Advanced Ionic Technologies came up with units of 10 kW in capacity. This cell would be used as mobile power generation running on diesel fuel [7]. By the end of 19th century, R and D focus was shifted toward proton-exchange membrane FCs and solid oxide FCs, particularly for stationary applications and by beginning of 20th century development of FCs with energy-efficient and CO_2 savings capability started [18]. The United States Department of Energy announced that a SOFC-microturbine cogeneration unit will be evaluated by the National Fuel Cell Research Centre and Southern California Edison and the FCs were built by Siemens Westinghouse and the microturbine by Northern Research and Engineering Corporation. On the other hand Fritz Haber and Walther H. Nernst in Germany and Edmond Bauer in France experimented with cells using a solid electrolyte. But high costs and low success rate suppressed there interest in continuing their development. While in many other researchers in the world have continued their research and currently, there are several types of FCs under development, each with its own advantages, limitations and potential applications. In 2009 a project develop by Japanese Ene-Farm firm used uninterruptible power supply (UPS) based on FCs and was adopted by North America while Linde North America and Air Products in 2010 developed new hydrogen compression technology Cost optimization of electrolyzer system and performance improvement in the conversion of photosynthetic solar-to-chemical energy [19]. A large number of patents in FCs (57%) was recorded in several places (Japan, Canada, the United States, Taiwan, the United Kingdom, Germany, and Korea), the product of which led to clean energy generation [20]. With increasing demand many countries of the world (2010−19) such as Japan, Germany, Asia, North America, the United States, California, and South Korea contributed in Molten carbonate FCs for stationary applications [20−22].

The most recent development in the field of FCs was teaming up of Doosan Fuel Cell and Korea Shipbuilding and Marine Engineering for the development of an eco-friendly FC for vessels [23]. It plans to invest 72.4 billion won by 2023 to build a power-generation SOFC cell stack manufacturing line and SOFC system assembly line. Hence FCs sector has bright scope in future.

1.2 Fuel cells

A FC is device that converts electromechanical energy to electrical energy with an oxidant and a fuel source (Xia et al., 2015). The difference between the FC and other storage device are: first: in FC liquid reactants or supply of gaseous for the reactions is used;[24] secondly, it is easy to eliminate the reaction products and keep the operation longer [25,26]. A typical FC consists of an electrolyte phase in intimate contact with a porous anode (negative electrode) and a porous cathode (positive electrode). The fuel and oxidant gases flow along the surface of the anode and cathode, respectively, and react electrochemically in the three-phase boundary region established at the gas−electrolyte−electrode interface. Theoretically FCs can produce electrical energy for as long as fuel and oxidant are fed to the porous electrodes, but the degradation or malfunction of some of its components often limits the practical life span of FCs [27].

Various different types of fuels such as, hydrogen, ethanol, methanol, or gaseous fossil fuels like natural gas can be used. Solid or liquid fossil fuels need to be gasified first before they can be used as a fuel. Due to many advantages of the FCs such as no combustion reaction, pollution-free electrical energy, excellent energy efficiency, lightweight, silent operation, and vibration-free operation, it has find its place in various field.

1.3 Nanotechnology and fuel cells

With the development in nanotechnology and their application in FC technology newer methodologies have been development. The penetration of the market dominated by today's bulky batteries is only possible with the paradigm of nanotechnology. The ultimate aim is to minimize and shrink down size of FCs using nanotechnology and microelectromechanical systems [28–32]. Nanotechnology also offers improvements in the proton-exchange membrane. Nano-FC has been one of the potential candidates for portable electronics where indirect hydrogen, direct methanol as well as nanotechnology-based FCs have been of prime focus [33]. Nafion which is similar to Teflon is currently one of the leading membranes for FC, which does not conduct electricity, but conducts positive charges while having a high operating temperature and strong resistance to chemical corrosion [34–36]. Current researches in the field of nanotechnology are focusing on highest yield of possible proton conductivity at high temperature, chemical stability toward oxidation, reduction and hydrolysis, and low cost [37]. Report says that metal phosphate nanoparticles at the boundary between the electrode and the polymer membrane of the FC also enhance the proton conductivity [38–40].

As mentioned in our aforementioned discussion that platinum is one of the most-effective catalysts for FCs, but unfortunately, if platinum is used as an electrocatalyst in FCs for all automotive internal combustion engines, it would deplete the entire global supply of platinum several times [41]. So, now various companies have starting using nanoparticles instead of entire platinum material which have resulted in lowering of cost of FCs. Membrane in FC allows hydrogen ion to pass through the cell but do not allow other atoms or ions such as O_2 to pass through, but now more efficient membranes are created with the help of nanotechnology which reduces the weight of FC considerably as well as it can run for longer time. Nano-FCs are being developed that can be used to replace batteries in handheld devices such as personal digital assistants or laptop computers. Most companies working on this type of FCs are using methanol as a fuel and are calling them DMFCs, which stands for direct methanol FCs. In addition, rather than plugging your device into an electrical outlet and waiting for the battery to recharge, with a DMFC you simply insert a new cartridge of methanol into the device and you are ready to go. In addition to the improvements to catalysts and membranes discussed before, it is necessary to develop a lightweight and safe hydrogen fuel tank to hold the fuel and build a network of refueling stations. To build these tanks, researchers are trying to develop lightweight nanomaterials that will absorb the hydrogen and only release it when needed. The Department of Energy is estimating that widespread usage of hydrogen powered vehicles, as the next-generation vehicles.

1.4 Current research on nanofuel cells

Luigi Osmier and his team members reported that a water-rich ink formulation helps to obtain better gas-phase mass transport properties, as well as increases the H_2/air polymer electrolyte fuel cells performance at high RH, while the mass transport within the cathode catalyst layer pores is more arduous due to presence of liquid water and the effect of ionomer swelling [42].

Lilian and his team through there sensor setup observed that the setup is promising and suitable tool for point-of-care use, capable to screen the carcinoembryonic antigen in very low concentrations with high selectivity and reproducibility [43].

Tim van cleve and team reported the ex situ characterization of catalyst inks (comprising Nafion D2020 ionomer, 30 wt.% PtCo/HSC, and n-PA/H_2O) and electrode layers along with in situ electrochemical testing of the membrane-electrode assembly demonstrated that ink water (and alcohol) content has a significant impact on electrode microstructure and cell performance [44].

Park and Myers reported automated deposition system which is versatile and precise and can be easily employed to create customized electrode structures, for example, electrodes with multiple layers with different compositions to form an electrode with a vertically graded composition. Any individual layer may be patterned in any number of ways limited only by the movement of the X-Y stage. This system enables deposition of uniform platinum group metal-free catalysts onto a membrane with microliter precision and accurate control of loading for high-throughput performance screening using commercially available segmented cell hardware such as the 25 electrode array hardware from Nuvant Systems [45].

Rewatkar and Goel reported first ever, customized catalyst free, printed porous carbon-based cathode structure was established for the origami array-type Al-air FCs [46].

Shaojie Du investigated the rheological characteristics of catalytic layer inks for a PEM FC and the results indicated that the ionomer solution plays a critical role in forming homogeneous colloidal mixtures of Pt/C catalyst in the solvent. The catalyst layer (CL) ink exhibited shear thinning characteristics, and thus it can be categorized as a pseudoplastic fluid. The expected stability of the catalytic layer ink during coating can be accessed via thixotropy measurements. These findings can be helpful for CL design and fabrication in future work [47].

Sobolev et al. reported synthesis and characterization of NiO colloidal ink solution for printing components of solid oxide FCs [48].

1.5 Conclusion

Future commercialization of FCs will have greater use in diversified areas. Portable FCs will replace contemporary bulky battery technology. The cost of FCs will decrease and their efficiency will increase significantly in the future via the use of nanotechnology. For FCs vehicles, recently Roland Gumpert successfully fabricated the world's first electric sports car powered by methanol FCs (Gumpert Nathalie with 800 horsepower, 530 mile range, and 50 mph) [49]. There will be increase in the collaboration between industries and research centers as well as university. Hence FCs have great future ahead.

References

[1] W. Kreuter, H. Hofmann, Electrolysis: the important energy transformer in a world of sustainable energy, International Journal of Hydrogen Energy 23 (1998) 661.
[2] H. Inaba, H. Tagawa, Ceria-based solid electrolytes—review, Solid State Ionics 83 (1996) 1−16.
[3] S. Carlo, A. Catia, E. Benjamin, I. Ioannis, Microbial fuel cells: from fundamentals to applications. A review, Journal of Power Sources 356 (2017) 225−244.
[4] F.M.B. Marques, L.M. Navarro, Performance of double layer electrolyte cells. 2. GCO/YSZ, a case study, Solid State Ionics 100 (1997) 29−38.
[5] L. Carrette, K.A. Friedrich, U. Stimming, Fuel Cells—Fundamentals and Application, Wiley Publication, 2001, pp. 5−39.
[6] <https://www.britannica.com/technology/fuel-cell/Development-of-fuel-cells>.
[7] K. Eguchi, T. Setoguchi, T. Inoue, H. Arai, Electrical-properties of ceria-based oxides and their application to solid oxide fuel-cells, Solid State Ionics 52 (1992) 165−172.
[8] Fuel Cell Today. Fuel cell history. <http://www.fuelcelltoday.com/history>, 2017.
[9] I. Dincer (Ed.), Comprehensive Energy Systems, Elsevier, 2018. ISBN: 978-0-12-814925-6.
[10] NASA, Summary: space applications of hydrogen and fuel cells, 2009 <https://www.nasa.gov/topics/technology/hydrogen/hydrogen_2009.html>.
[11] Mabel Matthews, Sharing partnership success, NASA, 2007, <https://www.nasa.gov/audience/forstudents/postsecondary/features/F_Sharing_Partnership_Success.html>.
[12] <https://fuelcellsworks.com/knowledge/history/>.
[13] M. Jansen, Digging deeper than platinum: South Africa's participation in an emergent fuel cell, M. S. thesis, Queen's University, 2016.
[14] H. Yahiro, Y. Baba, K. Eguchi, H. Arai, High-temperature fuel-cell with ceria-yttria solid electrolyte, Journal of the Electrochemical Society 135 (1988) 2077−2080.
[15] H. Yahiro, Y. Eguchi, K. Eguchi, H. Arai, Oxygen ion conductivity of the ceria samarium oxide system with fluorite structure, Journal of Applied Electrochemistry 18 (1988) 527−531.
[16] H. Yahiro, T. Ohuchi, K. Eguchi, H. Arai, Electrical-properties and microstructure in the system ceria alkaline-earth oxide, Journal of Materials Science 23 (1988) 1036−1041.
[17] J. Faber, C. Geoffroy, A. Roux, A. Sylvestre, P. Abelard, A systematic investigation of the Dc-electrical conductivity of rare-earth doped ceria, Applied Physics A: Materials Science and Processing 49 (1989) 225−232.
[18] Statistics Market, Stationary Fuel Cells—Global Market Outlook (2017−2026), PRNewswire, Telangana, 2020.

[19] Breakthrough Technologies Institute, 2010 Fuel Cells Technologies Market Report, United States Department of Energy—Energy Efficiency and Renewable Energy, Washington, D.C., 2011.
[20] S.S.D. Satyapal, Fuel Cell Technologies Overview, States Energy Advisory Board, Washington, D.C., 2012.
[21] Energy & Environmental Services. Number of fuel cells shipped globally from 2010 to 2019, by application. <https://www.statista.com/statistics/732220/shipments-of-fuel-cellsworldwide-by-application/>, 2020.
[22] V. Bente (Holland Innovation Network China), Overview of hydrogen and fuel cell developments in China, 2019. <https://www.nederlandwereldwijd.nl/binaries/nederlandwereldwijd/documenten/publicaties/2019/03/01/waterstof-in-china/Holland + Innovation + Network + in + China + + Hydrogen + developments. + January + 2019.pdf>, 2020.
[23] <https://www.ship-technology.com/news/doosan-fuel-cell-ksoe-vessels/>, (Last Updated March 19, 2021).
[24] M.F. Ahmer, S. Hameed, Use of organic polymers for energy storage in electrochemical capacitors, Advanced Materials Research 1116 (2015) 202−228.
[25] S.B. Vladimir, Fuel cell problems and solution, The Electrochemical Society Series, second ed., Wiley Publication, 2012. ISBN: 978-1-118-08756-5.
[26] R. Shripad, M. Pradip, Fuel Cell Principles—Design and Analysis, CRC Press; Taylor& Francis Group, 2014. ISBN (13:978-1-4200-8968-4).
[27] J.H. Hirschenhofer, D.B. Stauffer, R.R. Engleman, Fuel Cells: A Handbook, Gilbert/Commonwealth, Inc., Reading, PA, 1994.
[28] K. Lux, K. Rodriuez, Template synthesis of arrays of nano fuel cells, Nanoletters 6 (2006) 288−295.
[29] G. Girishkumar, T.D. Hall, K. Vinodgopal, P.V. Kamat, Single wall carbon nanotube supports for portable direct methanol fuel cells, The Journal of Physical Chemistry B 110 (2006) 107−114.
[30] R.S. Jayashree, L. Gancs, E.R. Choban, A. Primak, D. Natarajan, L.J. Markoski, et al., Air-breathing laminar flow-based microfluidic fuel cell, Journal of the American Chemical Society 127 (2005) 16758−16759.
[31] Y. Ding, M. Chen, J. Erlebacher, Metallic mesoporous nanocomposites for electrocatalysis, Journal of the American Chemical Society 126 (2004) 6876−6877.
[32] M. Brochu, B.D. Gauntt, R. Shah, R.E. Loehman, Comparison between micrometer-and nano-scale glass composites for sealing solid oxide fuel cells, Journal of the American Ceramic Society 89 (3) (2006) 810−816.
[33] K.-W. Park, Y.-E. Sung, S. Han, Y. Yun, T. Hyeon, Origin of the enhanced catalytic activity of carbon nanocoil-supported PtRu alloy electrocatalysts, The Journal of Physical Chemistry 108 (2004) 939−944.
[34] C.K. Dyer, Fuel cells and portable electronics, in Symposium on VLSI Circuits, Digest of Technical Papers, 2004, pp. 124−127.
[35] T.K. Manna, S.M. Mahajan, Nanotechnology in the development of photovoltaic cells, International Conference on Clean Electrical Power, ICCEP '07, 2007, pp. 379−386.
[36] C. Wang, M. Waje, X. Wang, J.M. Tang, R.C. Haddon, Yan, Proton exchange membrane fuel cells with carbon nanotube based electrodes, Nano Letters 4 (2004) 345−348.
[37] Fuel Cell Today Industry Review 2013, Johnson Matthey PLC trading, <http://large.stanford.edu/courses/2014/ph240/ukropina1/docs/fct_review_2013.pdf>.
[38] Y. Zeng, Z. Shao, H. Zhang, Z. Wang, S. Hong, H. Yu, et al., Nanostructured ultrathin catalyst layer based on open-walled PtCo bimetallic nanotube arrays for proton exchange membrane fuel cells, Nano Energy 34 (2017) 344−355. Available from: https://doi.org/10.1016/j.nanoen.2017.02.038.
[39] S. Du, Recent advances in electrode design based on one-dimensional nanostructure arrays for proton exchange membrane fuel cell applications, Engineering 7 (1) (2021) 33−49.
[40] G.L. Wang, L.L. Zou, Q.H. Huang, Z.Q. Zou, H. Yang, Multidimensional nanostructured membrane electrode assemblies for proton exchange membrane fuel cell applications, Journal of Materials Chemistry A 7 (2019) 9447−9477.
[41] Department for Transportation. Platinum and hydrogen for fuel cell vehicles. <http://www.dft.gov.uk/stellent/groups/dft_roads/documents/page/dft_roads_024056.hcsp>, 2003.

[42] O. Luigi, W. Guanxiong, C.C. Firat, K. Sunilkumar, P. Jaehyung, M. Samantha, et al., Utilizing ink composition to tune bulk-electrode gas transport, performance, and operational robustness for a Fe−N−C catalyst in polymer electrolyte fuel cell, Nano Energy 75 (2020) 104943.

[43] L.P.T. Carneiro, S. Nadia, A.P.M. Tavares, A.M.F.R. Pinto, A. Mendes, M.G.F. Sales, A passive direct methanol fuel cell as transducer of an electrochemical sensor, applied to the detection of carcinoembryonic antigen, Biosensors and Bioelectronics 175 (2021) 112877.

[44] V.C. Tim, W. Guanxiong, M. Mason, C.F. Cetinbas, K. Nancy, P. Jaehyung, et al., Tailoring electrode microstructure via ink content to enable improved rated power performance for platinum cobalt/high surface area carbon based polymer electrolyte fuel cells, Journal of Power Sources 482 (2021) 228889.

[45] P. Jaehyung, J.M. Deborah, Novel platinum group metal-free catalyst ink deposition system for combinatorial polymer electrolyte fuel cell performance evaluation, Journal of Power Sources 480 (2020) 228801.

[46] P. Rewatkar, G. Sanket, Catalyst-mitigated Arrayed aluminum-air origami fuel cell with ink-jet printed custom-porosity cathode, Energy 224 (2021) 120017.

[47] S. Du, W. Li, H. Wu, P.-Y.A. Chuang, M. Pan, P.-C. Sui, Effects of ionomer and dispersion methods on rheological behavior of proton exchange membrane fuel cell catalyst layer ink, International Journal of Hydrogen Energy 45 (2020) 29430−29441.

[48] S. Alexander, S. Paz, B. Konstantin, Synthesis and characterization of NiO colloidal ink solution for printing components of solid oxide fuel cells anodes, Ceramics International 46 (2020) 25260−25265.

[49] <https://www.rolandgumpert.com/en/>.

CHAPTER 2

Microfluidic concept-based fuel cells

Biswajit S. De[1], Neeraj Khare[1], Anastasia Elias[1] and Suddhasatwa Basu[1,2,3]
[1]Indian Institute of Technology, New Delhi, India
[2]CSIR—Institute of Minerals & Materials Technology (IMMT), Bhubaneswar, India
[3]University of Alberta, Edmonton, AB, Canada

2.1 Introduction

The development of portable, wearable, and wireless microfluidic lab-on-chip devices is restricted by the lack of personalized micropower sources to provide on-demand power to integrated micro/nanomechanical, biomedical, and chemical sensors. The power demand for integrated miniaturized electronics stimulated focusing on photovoltaic, thermoelectric, and piezoelectric devices [1,2] for harvesting micro-energy. Still, these power sources do not deliver a continuous power supply. The cost of the power source dominates the overall cost of many mass-produced electronics. Solid-state lithium-ion batteries are currently used as a device-level power source for wireless and portable electronics. The battery chemistry of lithium derivatives is favorable because of the high cell voltage and energy density, but limited resources and high lithium cost hamper the growth. A small and customized fuel cell can potentially offer a lower price and higher energy density than a lithium-ion battery.

Fuel cells generally consist of a stacked structure in which two electrodes are sandwiched to membrane electrode assembly (MEA). The fuel and oxidant are circulated through each half-cell, powering an electrochemical reaction. In these devices, each subsystems of energy conversion and energy storage, may be optimized independently, allowing significant stride towards improved functionality and performance. However, in practice the miniaturization of the fuel cell has numerous technical challenges. For instance, in polymer membrane fuel cells, fuel storage units are bulky. In addition, the crossover of reactants and pH dependence for selective ion conductivity across the membrane separating the anode and cathode limits the fuel cell performance [3]. The laminated structure renders the design of membrane electrode assembly (MEA) inadequate for integration into miniaturized devices [4]. The MEA of fuel cell accounts for 24% cost of production, and the conductivity of the membrane is limited by the humidity and operating temperature. The fuel cell requires additional components such as a gas diffusion layer sandwiched between the MEA and flow plates [5–8] for reactants/products transport to and from the reaction sites, heat and electron conduction, with minimum ohmic, fluidic, activation, thermal, and interfacial losses [1–17].

The future of energy technology requires high resiliency, manufacturability, and lightweight [9], which can be attained by scaling down to microscale. Shrinking the characteristic dimensions to microscale improves the many important intensive properties. For example, both the concentration and temperature gradient are reduced in microscaled devices owing to increased mass and heat transfer achieved by short radial diffusion and narrow residence time. A control over the flow in the microchannel is administered because of the prevalence of laminar regime. The high surface-to-volume ratio enables the surface phenomena dictated processes to attain a fast reaction equilibrium [10,11]. Microfluidics provides a consistent framework for the development of laminar flow-based membrane-less microfluidic fuel cell (MFC). Microfluidics offers its advantages in the MFC by downsizing the characteristic dimensions (interelectrode distance, channel depth, length, width, and depth) to microns, which results in an upsizing of fluid intrinsic property (viscosity) and rightsizing the electrochemical performance (power density, and current density) (Fig. 2.1).

MFCs can act as power sources capable of being integrated to and operating in a microfluidic chip framework [12,13]. One prominent advantage of MFC is the economic aspect; fewer components are required in MFC operation than in conventional membrane-based fuel cells. The reaction sites, microelectrode structure, reactant delivery are enclosed a single MFC chip. The viscous flow in MFCs allows

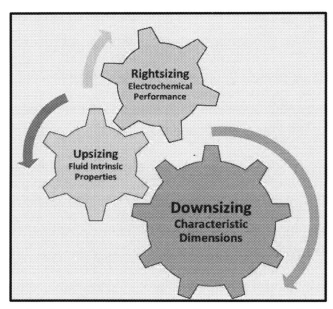

Figure 2.1 The benefits of microfluidics in the fuel cell system can be realized by downsizing the characteristic dimension leading to an upsizing of the intrinsic fluid properties and rightsizing the fuel cell's electrochemical performance.

the membrane-less operation. The catholyte-anolyte co-flow interface behaves as a virtual membrane for the transfer of ions across the electrode. The MFC's membrane-less and room-temperature operation eliminates the necessity of auxiliary cooling, humidification, and water management system. MFCs have increased reactant utilization and a lower ohmic loss because of the microscale interelectrode distance and the elimination of the separation membrane. Facile reactant, product, and ion charge transport in the MFC facilitates a high power density in the MFC. The high surface-to-volume ratio microscale improves the kinetics of reaction in MFC. The MFC can be fabricated with ease by employing well-established techniques available for microfabrication and micromachining.

The performance gain presented by microfluidics in fuel cells has motivated many research groups to exploit the advantages of microsystems. In this chapter, we review the recent advances in MFC technology. The design and fabrication methodology adopted for manufacturing of microchannels and microelectrodes provides an overview of the fabrication process. The performance evaluation concerning the choice of fuel-oxidant, electrolyte flow rate, and catalyst which influences the MFCs characteristics, are discussed.

2.2 Theory

Microfluidics is the manipulation of fluid in microstructures with at least one characteristic dimension in micron-scale [18]. The fluid flow in a typical microfluidic device is characterized by laminar flow predicted by a low Reynolds number. A low Reynolds number signifies that surface forces are more relevant than body forces, corresponding to the prevalence of viscous effects over inertial effects. The momentum conservation Navier-Stokes equations in 3D can be determined for the velocity field $\bar{u}$ of an incompressible Newtonian fluid for microfluidic laminar [19,20]:

$$\rho\left(\frac{\partial \bar{u}}{\partial t} + \bar{u}.\nabla\bar{u}\right) = -\nabla p + \mu \nabla^2 \bar{u} + \bar{f} \qquad (2.1)$$

where $p, \bar{f}, \rho$, and u are the pressure force, body force, density, and viscosity, respectively. The non-linear convective terms for a low Reynolds number can be neglected which results in a linear Stokes flow:

$$\rho\frac{\partial \bar{u}}{\partial t} = -\nabla p + \mu^2 \bar{u} + \bar{f} \qquad (2.2)$$

A pressure driven parabolic velocity flow profile is obtained, and the fluid flow for mass conservation obeys a continuity equation represented as:

$$\frac{\partial \rho}{\partial t} + \nabla(\rho\bar{u}) = 0 \qquad (2.3)$$

For an incompressible fluid flow with constant density, the above equation to

$$\nabla \bar{u} = 0 \tag{2.4}$$

The high surface-to-volume ratio offered by microfluidics and the characteristic length are inversely related. The heterogenous electrochemical reactions in the MFC benefits from the high surface-to-volume ratio of the microscale systems. More power is generated using a small volume of fuels and oxidants. However, there is a trade-off between the microchannel dimension and the parasitic load to pump the fluid. The MFC's net power must account for the power required to pump the reactants in the microchannel. The pump load increases as the channel dimension shrinks, and the power obtained from the MFC diminishes. For a pressure-driven laminar flow, the pressure drop is given by:

$$\nabla p = \frac{32\mu L U}{D_h^2} \tag{2.5}$$

Here, L is the channel length, U is mean velocity, and D_h is the hydraulic diameter. The power to pump fluid (W) can be evaluated by multiplying the flow rate, Q, to the above equation as:

$$W = Q \times \nabla p = \frac{32\mu L U Q}{D_h^2} \tag{2.6}$$

The reactant flow rate has a massive implication on the overall power density of the MFC. Flow optimization is required to keep the pumping power minimum to achieve a maximum net power output. The pumping power obtained from the equation (2.6) applies to a straight channel with a fully developed flow and neglects entry and exit effect losses at the inlet and the outlet of the channel.

A microfluidic system a high Peclet number, $Pe = UD_h/D$, signifying that the species transport by advection dominates over transport by diffusion. Thereforethe rate of diffusive transport in the direction normal to flow is much lower than convective transport transverse to the direction of flow. The viscous flow inhibits convective mixing in the microchannel as the fuel and oxidant move towards the outlet. The catholyte and anolyte streams adhere to their respective sides without requiring a membrane. This controlled phenomenon is exploited in the MFC at the fluid-fluid interface to provide functionality. The parallel co-flow of two laminar fluid streams with similar viscosity and density inside a microchannel gives rises to an interface. The interface behaves as a virtual membrane in the membrane-less MFC for the species transport across the electrodes. The species diffusion normal to the direction of the flow at the liquid-liquid interface at the channel creates a thin region solution containing both anolyte (fuel) and catholyte (oxidant). This zone is called the inter-diffusion zone or the mixing region. An hourglass-shaped mixing width is obtained across the channel cross-section because of the maximized diffusion at the channel wall and

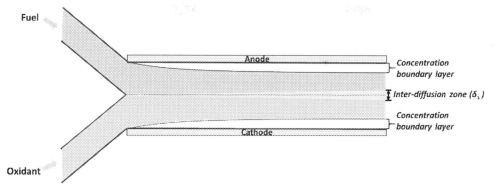

Figure 2.2 The performance of microfluidic fuel cells is hampered by the formation of inter-diffusion zone (δ_x) at the channel center and the concentration boundary layer on the electrode surface.

minimum diffusion in the channel middle. For a given constant fluid velocity and microchannel dimension, the mixing zone is confined to a narrow width at the interface and increases with the downstream position (z) along with the flow (Fig. 2.2). For a co-flow of two laminar streams in a microchannel, the inter-diffusion zone with a maximum width (δ_x) is given as [14].

$$\delta_x \propto \left(\frac{DHz}{U}\right)^{1/3} \qquad (2.7)$$

where, H and D represents the height of the channel and diffusion coefficient, respectively.

The mass transfer of species in the MFC is influenced by the inter-diffusion zone occurring at the channel center as represented by equation (2.7). The diffusivity of the species and the time for which the species were in contact influences the degree of mixing. The fuel crossover arising from the diffusion can be averted by operating the MFC in a regime with a Peclet number in the range of $10^3 - 10^4$. The width of the inter-diffusion zone determines the desired electrode positioning in the microchannel of MFC to prevent the fuel-oxidant crossover. The anode and the cathode must be adequately separated from the anolyte-catholyte liquid interface, considering the operating flow rates and channel geometry.

Both the ohmic resistance and the fuel utilization of the MFC are influenced by the electrode's orientation and position. High ionic conductivity is required between the electrodes to provide a conductive path with low ohmic resistance for ionic charge transport. A supporting electrolyte with high mobility having hydronium or hydroxide ions is normally used to attain high conductivity. The highly mobile constituents of the supporting electrolyte redistributes quickly to shield the effects due to electric double-layers in the channel. This stabilized the perturbation to the colaminar flow arising from the electromigration of the species from fuel and oxidant. The ohmic resistance (R_Ω) for the transport of the ions in the channel is represented as:

$$R_\Omega = \frac{d_e}{\sigma\ A_c} \qquad (2.8)$$

Here, d_e is distance between the electrodes for the charge transfer, A_c is the charge transfer cross-sectional area, and σ is the ionic conductivity. Equation (2.8) suggests the employment of closely spaced electrodes and a supporting electrolyte with high ionic conductivity. However, equation (2.7) governs the minimum interelectrode spacing based on the interdiffusion layer's thickness. Higher ohmic resistance is realized in MFCs than in conventional devices as the interelectrode distance is higher that in the membrane-based device where the electrodes are sandwiched tightly in the membrane-electrode assembly (MEA). To mitigate the higher ohmic resistance in the MFC, a high concentration of supporting electrolyte is employed. The laminar co-flow in the membrane-less microchannel allows the choice of two different streams with pH bias configured independently to achieve open-circuit potential and cell voltage and improve the reaction kinetics.

The electrochemical reaction rate in the MFC is influenced by the formation of the concentration boundary layer occurring on the leading edge of the electrode, which depends on the channel geometry, electrode arrangement, and flow rate. The rate of electrochemical reactions tends to be faster than the diffusion rates as the concentration of fuel and oxidant at the electrode surface reaches zero [21]. A concentration boundary layer on the electrode surface is formed due to the fast depletion rate of reactants on the electrodes which impedes the reactants from reaching the active surface the electrode, thereby lowering the MFC performance (Fig. 2.2). A thin concentration boundary layer over the electrodes is required so that the concentration gradient can drive the reactant through the concentration boundary layer [22]. Operating the MFC at a higher inlet flow rate replenishes the depleted layer and pushes the concentration boundary layer towards the electrodes. This improves the electrochemical reaction kinetics of the MFC because of the thinner depletion layer. However, the upper limit of the inlet velocity is limited by the hydrodynamic instability induced in the flow at higher flow rates which causes an oscillating motion in the electrolyte streams, and the electrolyte touch contrary electrode. Generally, the MFC's performance is controlled by combining the ohmic resistance, mass transport, and kinetics of electrochemical reaction.

2.3 Fabrication and design of microfluidic fuel cells

The fabrication techniques established previously for MEMS and microfluidic lab-on-chip analytical applications have been widely adopted for the fabrication of ere adopted to fabricate the MFCsC. MFCs ares typically comprised of a microelectrodes pair encompassed in a microchannel with tight support to prevent any loss through leakages.

Standard lithography and soft lithography techniques that are commonly used in microfluidics have also been widely used for the rapid prototyping of the microchannels of the MFCs [15,16]. Polydimethylsiloxane (PDMS) is a transparent, hydrophobic polymer ubiquitous in the fabrication of microfluidic devices. This elastomer is easily patterned by mixing a prepolymer with a crosslinker and then coating this mixture onto a from the mold with raised microstructures; when peeled from the mold a negative imprint of the lines is transferred into the PDMS. This layer can then be and which is consecutively sealed using bonded to a solid substrate patterned with the microelectrodes [17,18]. PDMS is widely used as it is compatible with most solvents and electrolytes used in the MFC. The soft lithography technique for microchannel fabrication is described in brief. A silicon wafer or glass substrate is pretreated and cleaned to remove any organic residue on the surface. The pristine substrate is spin-coated with photoresist. The thickness of the photoresist defines the microchannel height, which is controlled by the duration of spin-coating and rotational speed depending on the property of the photoresist. The photoresist coated substrate is soft baked to stabilize the coating. A CAD software is used to draft the design of the microchannel which is printed to a photomask. The photomask is aligned over the substrate and subjected to UV light using a mask-aligner optical lithography system. The photomask allows selective exposure of UV light to transfer the design of the microchannel to the photoresist. The substrate is subjected to a post-exposure bake followed by immersion in a developer and rinsing with water to remove the unexposed photoresist. The positive pattern is obtained from the remaining polymerised photoresist which is used as a reusable master of the microchannel. The pre-polymer of PDMS and curing agent mixed thoroughly, degassed in vacuum, and poured on the master template followed by curing on a hot plate. The polymer solidifies and the negative channel design is transferred to the PDMS which is carefully peeled from the master and cut to the desired shape. The PDMS microchannel structure is sealed with either glass or PDMS, reversibly or irreversibly. Irreversible bonding is done by plasma treating the surfaces of each component, and then pressing the components together.

Martins et al. [19,20] adopted the soft lithography technique for the fabrication of a glycerol/bleach-based MFC (Fig. 2.3A). As descried above, the negative design was transferred to the PDMS by a positive template of the microchannel. The microchannel and the flow-through electrodes were sealed irreversibly to a glass capping window by exposing the substrates to a plasma, and then pressing them together. The microfluidic cell has provision for the fuel and oxidant entry through two inlets, which results in a colaminar flow of the electrolyte in the microchannel. Jindal et al. [21] fabricated a Y-shaped PDMS microchannel (Fig. 2.3B) to introduce fuel and oxidant to test a direct formic acid MFC performance. The microchannel was defined on the master template

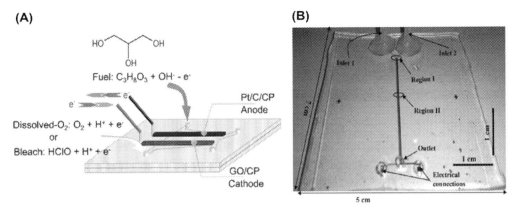

Figure 2.3 (A) Glycerol/bleach microfluidic fuel cell with flow-through electrodes and (B) Formic acid microfluidic fuel cell with electrospun CNx nanofiber catalyst. The devices were fabricated by standard lithography and soft lithography techniques. *(A) Reproduced from C.A. Martins, P. Pei, M. Tellis, O.A. Ibrahim, E. Kjeang, Graphene-oxide-modified metal-free cathodes for glycerol/bleach microfluidic fuel cells 8 (2020) 8286–93 with permission from the ACS. (B) Reproduced from A. Jindal, S. Basu, N. Chauhan, T. Ukai, D.S. Kumar, K.T. Samudhyatha, Application of electrospun CNx nanofibers as cathode in microfluidic fuel cell, Journal of Power Sources 342 (2017) 165–174 with permission from Elsevier.*

using SU8 photoresist on a silicon wafer and selectively exposed to UV by performing lithography. The thickness of the Y-shaped microchannel was set by the photoresist layer thickness deposited by spin-coating on the silicon substrate. The microchannel allowed a co-laminar flow of the fuel and oxidant, limiting the mixing of the reactants to a thin layer between the fuel and oxidant streams [22]. The provision for inlet and outlet of the fuel and oxidant and making electrical connections were made by punching holes through the PDMS. The catalyst was patterned on a glass slide by a lift-off technique. The glass slide was coated with a positive photoresist and selectively exposed by UV lithography to define the area of the microelectrodes. The catalyst was deposited on the patterned glass slide, and lift-off was performed to remove the excess photoresist and catalyst using a developer, thereby defining the electrode area on the glass slide. The Y-shaped channel was aligned and soft-bonded with the electrodes patterned glass substrate such that the anode and cathode are enclosed within the microchannel width.

Polymethyl methacrylate (PMMA) is another polymeric material employed frequently in MFCs. Xu et al. [23] designed a T-shaped counter-flow direct formic acid MFC with with electrodes placed in the two feeding channels (Fig. 2.4A). The counter-flow design was made on a PMMA sheet using CO_2 laser micro-machining. T-shaped grooves with windows for exposing the catalyst surface was machined on another PMMA sheet. Two PMMA sheets exposing the catalyst surface was machined on another PMMA sheet. Two PMMA sheets encompassed the channel at the top and bottom. The electrodes were

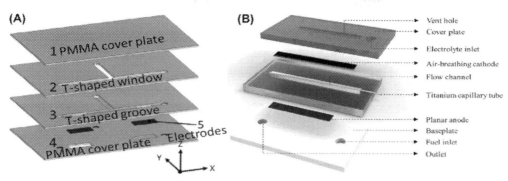

Figure 2.4 Cell architectures of (A) air-breathing counter-flow formic acid microfluidic fuel cell, and (B) air-breathing immersed fuel micro-jet microfluidic fuel cell. *(A) Reproduced from H. Xu, H. Zhang, H. Wang, D.Y.C. Leung, L. Zhang, J. Cao, et al. Counter-flow formic acid microfluidic fuel cell with high fuel utilization exceeding 90%, Applied Energy 160 (2015): 930–936 [23] with permission from Elsevier]. (B) reproduced from Y. Zhou, B. Zhang, X. Zhu, D.D. Ye, R. Chen, T. Zhang, et al. Enhancing fuel transport in air-breathing microfluidic fuel cells by immersed fuel micro-jet, Journal of Power Sources 445 (2020) 227326, with permission from Elsevier.*

placed on the bottom sheet with holes for the inlet and outlet of the electrolyte. The cathode was exposed to air through a window on the bottom PMMA sheet to ensure sufficient oxygen supply. Precise control over the speed and laser power enable the laser cutter to be a useful micro-machining MFC tool. Zhou et al. [24] proposed an air-breathing MFC with fuel micro-jet to enhance fuel transport and replenish the boundary layer concentration of the fuel. The schematic of the cell is depicted in Figure 4B. The cell architecture was fabricated on PMMA plates by a laser cutter. A parallel flow of anolyte and catholyte in a rectangular flow channel was used to introduce the electrolyte in the MFC. A titanium capillary tube with a micro-opening was installed at the catholyte—anolyte interface of the rectangular microchannel center. The capillary micro-opening located midway along the microchannel length fabricated by laser micro-machining through which an additional fuel was jetted. The rectangular channel and the titanium capillary tube were enclosed with a planar anode and air-breathing cathode supported by the cover plate with inlet and outlet ports of the electrolytes. The cathode side cover plate had vent holes to allow air diffusion rendering the MFC air-breathing.

MFC fabrication using inexpensive Computer Numerical Control (CNC) and cutting plotter technique for micro-machining has also been demonstrated. Ortiz-Ortega et al. [25] demonstrated an air-breathing direct formic acid nanofluidic fuel cell with through electrodes (Fig. 2.5A). The channel for electrolyte flow was patterned and cut on a silicone elastomer film using a cutting plotter. The electrodes were placed on the channel polymer film and sandwiched between two PMMA support plates fabricated by CNC. The top plate included two inlets and a window for breathing air, and the bottom plate had one outlet for the reaction by-product. A similar methodology using the CNC and cutting plotter was adopted by Abrego et al. [9] for the fabrication of air-breathing V-shaped channel geometry for direct methanol MFC with flow-through electrodes (Fig. 2.5)

Advances in 3D printing technology have led to the development of rapid prototyping of microfluidic devices. As described above, the manufacturing of microfluidic devices involves multiple steps (piranha cleaning, spin-coating, heating, developing, curing) and several pieces of equipment such as a UV lithography system, spin coater, and a plasma chamber, where all the processes are performed in a cleanroom. On the contrary, 3D printing is a relatively low-cost process, offering a fast and reproducible device. Guima et al. [26] presented a glycerol MFC which was 3D-printed using fused deposition modeling (FDM) in which a thermoplastic filament is heated through a nozzle and dispensed on to a substrate which undergoes relative motion in the x-y direction with respect to the nozzle, allowing three dimensional structures to be printed. The schematic representation of the 3D MFC is shown in Fig. 2.5B. The cell was designed in CAD software and transferred to the 3D

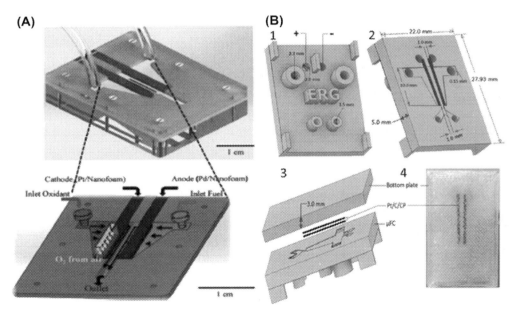

Figure 2.5 (A) Formic acid flow through nanofluidic cell fabricated by CNC and cutting plotter. (B) Glycerol microfluidic fuel cell fabricated by 3D printing, 1 and 2: microfluidic channel made of polylactic acid with porous electrodes, 3: the assembly of the microfluidic fuel cell with transparent PETG back cover, 4: view from the bottom of the cell. *(A) Reproduced from E. Ortiz-Ortega, M.A. Goulet, J.W. Lee, M. Guerra-Balcázar, N. Arjona, E. Kjeang, et al. A nanofluidic direct formic acid fuel cell with a combined flow-through and air-breathing electrode for high performance, Lab Chip 14 (2014) 4596–4598, with permission from the RSC]. (B) Reproduced from K.E. Guima, P.H.L. Coelho, M.A. G. Trindade, M.A.G. Trindade, C.A. Martins, 3D-Printed glycerol microfluidic fuel cell, Lab Chip 20 (2020) 2057–2061, with permission from the RSC].*

printer. Polylactic acid (PLA) was used to print the microfluidic channel (Fig. 2.5B 1 and 2) for the colaminar flow of the electrolytes, and polyethylene terephthalate glycol (PETG) was used as a transparent cover (Fig. 2.5B3). Prior to the device assembly, the electrodes were positioned in their corresponding position on the PLA microchannel, and the substrate was then attached to PETG cover by polyvinyl chloride (PVC) glue. The electrical connections to the electrodes were made using aluminum wires glued by silver conductive epoxy. The cell contains two ports (inlet and outlet), and holes for electrical contact. The estimated cost of the 3D-printed MFC quoted was USD 1.85 per cell which was close that of PDMS-based MFC [18]. In addition to enabling rapid manufacturing of microfluidic devices with traditional, planar designs (parallel channels in which fluids flow parallel to the plane of the substrate), 3D printing is also enabling the implementation of some other less conventional systems designs in which the fluids flow through stacks of materials. The present scope of 3D printing technology for MFCs application is limited by the minimum achievable resolution of the printed structure in the X, Y, and Z axes. The surface finish of the 3D printed devices may cause perturbation in flow of the liquid electrolytes. The development in the material used for 3D printing will help to adopt reactive reactants for the MFC application.

2.4 Performance evaluation of microfluidic fuel cells

Evaluating the performance is imperative to apprehend the feasibility of the MFC operation. The kinetics and mass transfer in the MFC are dependent on the fuel-oxidant pair, electrolyte flow rate, and the catalyst choice, directly impacting the MFC performance (Fig. 2.6). Some recently reported MFCs, along with their characteristic features and key findings are summarized in Table 2.1.

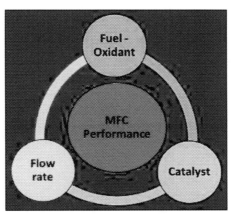

Figure 2.6 Fuel-oxidant pair, electrolyte flow rate, and catalyst are the key influencers of the microfluidic fuel cell performance.

Table 2.1 A few recent reports of the microfluidic fuel cell with their characteristics and key findings.

Authors	Fuel	Oxidant	Flow rate	Anode	Cathode	Max. current density	Max. power density	Features of the microfluidic fuel cell
Ortiz-Ortega et al. [25]	0.5 M H_2SO_4 + 0.5 M HCOOH	Air + 0.5 M H_2SO_4 bubbled with pure O_2	100 µL min^{-1}	Pd/C (20 wt%) oncarbon nanofoam	Pt/C (30 wt%) oncarbon nanofoam	500 mA cm^{-2}	100 mW cm^{-2}	Air-breathing and flow-through 3D nanoporous electrodes
Xu et al. [27]	1 M HCOOH + 0.5–3 M H_2SO_4	0.5–3 M H_2SO_4	1–20 µL min^{-1}	Pt black loaded carbon paper (3 mg cm^{-2})	Pt black loaded carbon paper (3 mg cm^{-2})	13.8–30 mA cm^{-2}	2.1–3.63 mW cm^{-2}	T-shaped, counter-flow, 91.4% fuel utilization
Jindal et al. [21]	2.1 M HCOOH + 0.5 M H_2SO_4	0.144 M $KMnO_4$ + 0.5 M H_2SO_4	50–500 µL min^{-1}	Pt thin film (50 nm)	CNx nanofiber	9.79 mA cm^{-2}	3.43 mW cm^{-2}	Double Y-shaped, electrospun CNx nanofiber
Deng et al. [28]	0.5 M HCOOH + 1 M H_2SO_4 (acidic), 0.5 M HCOONa + 2 M KOH (alkaline)	Air + 1 M H_2SO_4 (acidic)Air + 2 M KOH (alkaline)	300 µL min^{-1}	Pd (5 mg cm^{-2}) on graphite rods	Pt (3.5 mg cm^{-2}) on carbon paper	225.94 mA cm^{-3} (acidic) 384.1 mA cm^{-3} (alkaline)	50.30 mW cm^{-3} (acidic) 65.4 mW cm^{-3} (acidic)	Scale-up strategy for MFC using electrodes array, air-breathing, alkaline and acidic media operations
Zhou et al. [24]	0.5 M HCOONa + 2 M KOH	Air + 2 M KOH	200 µL min^{-1}	Pd (5 mg cm^{-2}) on carbon paper	Pt black (3 mg cm^{-2}) on carbon paper	~316 mA cm^{-3}	119.3 mW cm^{-3}	Air-breathing cathode, immersed capillary micro-jet to replenish fuel at anode
Wang et al. [29]	Neat CH_3OH vapor	Air + 3 M KOH	50 µL min^{-1}	PtRu (10 mg cm^{-2})	Pt/C (2 mg cm^{-2})	~290 mA cm^{-2}	55.4 mW cm^{-2}	Vapor fed methanol MFC
Abrego-Martinez et al. [30]	5 M CH_3OH + 0.5 M KOH	Air + 5 M CH_3OH + 0.5 M KOH	10 mL hr^{-1}	Ag/Pt/CP	Pt/CP	25 mA cm^{-2}	1.2 mW cm^{-2}	Mixed-reactant, air-breathing, 2-cell stack operation
Kwok et al. [31]	CH_3OH + H_2 + 2 M NaOH	Air + 2 M NaOH	100 µL min^{-1}	PtRu (10 mg cm^{-2})	Pt/C (2 mg cm^{-2})	307.2 mA cm^{-2}	133.1 mW cm^{-2}	Dual fuel-breathing anode, H_2 fuel generated from CH_3OH by photocatalysis, air-breathing cathode

Martins et al. [32]	1 M glycerol + 1 M KOH	Clorox (ClO$^-$, 3–8%) in 2 M KOHClorox (ClO$^-$, 3–8%) in 1 M H$_2$SO$_4$	100 µL min^{-1}	Pt/C nanoparticles on carbon paper	Pt/C nanoparticles on carbon paper	337 mA cm^{-2} (all alkaline) 638 mA cm^{-2} (mixed media)	71.2 mW cm^{-2} (all alkaline) 315 mW cm^{-2} (mixed media)	Mixed media system attained OCV 1.9–2.0 V. Performance comparable to conventional AEM glycerol fuel cells.
Hashemi et al. [33]	H$_2$ gas (4–16 mL h^{-1}) + 1 M H$_2$SO$_4$	O$_2$ gas (2 mL h^{-1}) + 1 M H$_2$SO$_4$	1 mL h^{-1}	Pt thin film	Pt thin film	26.6 mA cm^{-2}	~ 3.8 mW cm^{-2}	Two-phase flow of gas-electrolyte in bridge-shaped MFC
Liu et al. [10]	2 M HCOONa + 2 M KOH	0.5 H$_2$O$_2$ + 1 M H$_2$SO$_4$	9.32 ± 0.65 mL h^{-1} (anolyte) 9.85 ± 1.26 mL h^{-1} (catholyte)	Pd nanoparticles loaded on 3D-herringbone-like structure carbon fiber	Pd nanoparticles loaded on 3D-herringbone-like structure carbon fiber	83.3 mA cm^{-2}	29.9 mW cm^{-2}	Carbon fiber braided into 3D herringbone structure, functions as both flow channel and electrode, pump-free MFC
Guima et al. [26]	1 M glycerol + 1 M KOH	Sodium hypochlorite (6%) in 1 M H$_2$SO$_4$	100–900 µL min^{-1}	Pt/C/CP	Pt/C/CP	67.7–275.2 mA cm^{-2}	48.3–310.7 mW cm^{-2}	3D-printed MFC, low manufacturing cost

Ortiz-Ortega et al. [25] demonstrated the performance of an air-breathing formic acid MFC (Fig. 2.5A). The performance of the device was evaluated based on the concentration of fuel, electrolyte flow rate, and anode catalyst Pd loading. The formic acid (HCOOH) concentration was varied from 0.1–3 M, and the highest current density was 70 mW cm^{-2} obtained for 3 M formic acid concentration. The limiting behavior of the reactant was analyzed by studying the effect of flow rate (100–200 μL minute^{-1}). The concentration of formic acid was kept constant at 0.5 M to avoid electrode poisoning. At 200 μL minute^{-1} anolyte and 100 μL minute^{-1} catholyte flow rate, a power density of 68 mW cm^{-2} was obtained. The air-breathing configuration contributed to enhanced oxygen transport in the catholyte of the MFC. The Pd catalyst loading on the carbon nanofoam at the anode was changed from 0.15–0.6 mg. The open-circuit voltage (OCV) of 0.85 V and a maximum power density of 100 mW cm^{-2} was attained for 0.6 mg Pd loading. The performance enhancement was due to the carbon nanofoam with high surface area and implementation of the air-breathing cathode with dissolved oxygen in the catholyte. A comparison between an open system (air-breathing) and a closed system (without air-breathing) was made to determine the contribution of oxygen at the cathode from the air and dissolved oxygen. The current density for the open system (320 mA cm^{-2}) was 1.7 times higher than that with the closed system (188 mA cm^{-2}). The open system attained a maximum power density of 70 mW cm^{-2} and the closed system resulted in 52 mW cm^{-2}. The enhanced performance of the air-breathing over dissolved oxygen cathode in the fuel cell was expected owing to the higher oxygen concentration and diffusivity in the air than that in an aqueous solution [34]. The power density reported in that work for the colaminar flow formic acid MFE was substantially higher than the previously reported literature [34–39], which resulted from merging air-breathing cathode with flow-through porous 3D electrodes.

To overcome kinetic limitations observed in other devices, Xu et al. [27] proposed a T-shaped formic acid MFC with the anolyte and catholyte counter-flow to enhance fuel utilization The electrodes were arranged in the two feeding channels (Fig. 2.7A) instead of positioning them face-to-face (Fig. 2.7B) in traditional MFC with the co-laminar flow. This resulted in a significant restriction in fuel crossover to the cathode because the convective transport at the cathode was opposite to diffusive transport. This was contrary to the co-laminar scheme, where the fuel crossover from the anolyte to catholyte was determined by diffusion with convective transport orthogonal to diffusive transport. This was contrary to the co-laminar scheme, where the fuel crossover from the anolyte to catholyte was determined by diffusion with convective transport orthogonal to diffusive transport. In their device, the catholyte consisted of 1 M HCOOH supported by 0.5–3 M sulfuric acid (H$_2$SO$_4$), and the anolyte was 0.5–3 M H$_2$SO$_4$ with flow rates 1–20 μL minute^{-1}. The anode and cathode consisted of Pt black (3 mg cm^{-2}) loaded on carbon paper.

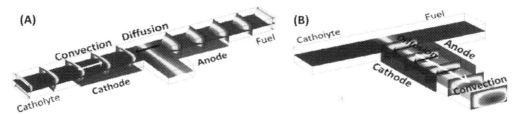

Figure 2.7 Formic acid MFC with (A) counter-flow of fuel and oxidant with the electrodes in the feeding channel, and (B) coflow of fuel and oxidant with electrodes placed face-to-face. *Reproduced from H. Xu, H. Zhang, H. Wang, D.Y.C. Leung, L. Zhang, J. Cao, et al. Counter-flow formic acid microfluidic fuel cell with high fuel utilization exceeding 90% q 160 (2015) 930–6, with permission from Elsevier.*

The cathode was exposed to air to ensure a sufficient supply of oxygen. The ohmic resistance because of the side–by–side arrangement of the electrodes in the counter-flow scheme was 31.6 Ω cm^2, 44 times higher than that in the co-flow face-to-face arrangement. This was owing to the large species transport pathway in the counter-flow electrode arrangement. A high concentration of the supporting electrolyte was adopted to compensate for the large ohmic resistance. The current density achieved in the counter-flow arrangement was 30 mA cm^{-2} and a peak power density of 3.63 mW cm^{-2} for the electrolyte flow rate 1 minute^{-1} with 1 M HCOOH in 3 M H$_2$SO$_4$. y adopting a counter-flow of fuel and oxidant at 1 minute^{-1} 91.4% fuel utilization MFC. Reduced energy consumption and improved energy efficiency due to the low flow rate in the counter-flow MFC were additional benefits.

To address catalyst poisoning arising from fuel crossover, Jindal et al. [21] synthesized carbon nitride (CNx) nanofibers by electrospinning for use as a cathode in a device in which a Pt thin film (50 nm) was used as the anode (Fig. 2.3B). Formic acid and potassium permanganate (KMnO$_4$) were used as fuel and oxidant, respectively, H$_2$SO$_4$ as a supporting electrolyte. The CNx nanofiber was inactive towards formic acid electrooxidation resulting in tolerance to fuel crossover occurring from anolyte to catholyte and resistant to CO poisoning. Stable operation and non-varying peak current density was observed for the CNx nanofiber in a strong KMnO$_4$ oxidative condition. The fuel and oxidant flow rates were optimized (50–500 μL minute^{-1}) to adjust the interfacial width of the mixing region in the MFC. At 300 μL minute^{-1} flow rate, the interfacial width of the fuel-oxidant interface was 86 μm, which was less than the interelectrode distance (100 μm) that gave rise to a stable interface. The maximum power density increased (2.44 to 3.43 mW cm^{-2}) with the reduction in electrolyte flow rate (500 to 300 μL minute^{-1}). A lower OCV of 0.8 V was observed at 500 μL minute^{-1} as compared to 1.1 V at 300 μL minute^{-1}. The reduced OCV at a high flow rate was attributed to fuel crossover along the microchannel length.

Deng et al. [28] demonstrated a scale-up strategy for MFC by adopting an array arrangement for electrodes (Fig. 2.8A–C). The air-breathing cathode consisted of Pt

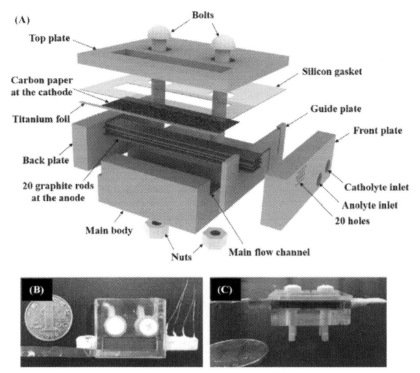

Figure 2.8 Scaled-up array arrangement of formic acid MFC (A) schematic representation of the cell, (B), and (C) actual snapshots of the assembled cells. *Reproduced from B. Deng, D. Ye, B. Zhang, X. Zhu, R. Chen, Q. Liao, Current density distribution in air-breathing microfluidic fuel cells with an array of graphite rod anodes, International Journal of Hydrogen Energy (2020), with permission from Elsevier.*

black (3.5 mg cm^{-2}) catalyst ink spray-coated on carbon paper. A staggered pattern distributed in four layers of 20 graphite rods electrodeposited with Pd (5 mg cm^{-2}) was used as a cathode. The performance of the MFC was evaluated in both acidic and alkaline media with an electrolyte 300 μL minute^{-1} flow rate. The acidic media was 0.5 M HCOOH in 1 M H$_2$SO$_4$ as anolyte and 1 M H$_2$SO$_4$ as catholyte. For alkaline media, 0.5 M sodium formate (HCOONa) in 2 M KOH as the anolyte, and 2 M KOH as the catholyte. The maximum current density and the peak power density were 384.1 mA cm^{-3} and 65.4 mW cm^{-3}, respectively, in the alkaline media. The current density and the power density were 1.7 times and 1.3 times lower, respectively, in acidic electrolyte than in alkaline media. An OCV of 1.05 V and 0.98 V was obtained for the MFC under alkaline and acidic electrolyte, respectively. The lower performance of the MFC in the acidic media resulted from the higher activation loss at low current density than in the alkaline media. At high current density, mass transfer loss was observed in acidic media due to high resistance to transfer proton from anode to cathode arising from the CO_2 generation and accumulation at the anode. Significant variation in the current density distribution in the acidic media was observed owing to the CO_2 bubble generation. Stable current density distribution for the alkaline electrolyte was obtained in the alkaline media due to the absence of CO_2 bubbles. The study provides a direction for the scale-up of the MFCs.

One limitation observed in air-breathing MFCs is that a fuel concentration boundary layer arising at the anode can limit the cell performance and significantly hinder fuel transport (Fig. 2.9A). Zhou et al. [24] proposed that the fuel concentration at the

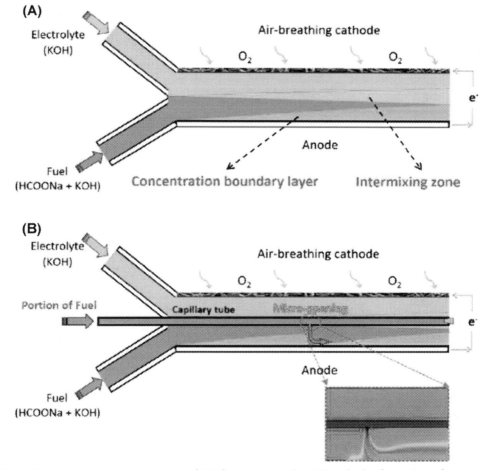

Figure 2.9 Schematic representation of (A) flow-over mode MFC with the formation of a concentration boundary layer along the length of the microchannel, and (B) micro-jet mode MFC with fuel replenishment. *Reproduced from Y. Zhou, B. Zhang, X. Zhu, D.D. Ye, R. Chen, T. Zhang, et al. Enhancing fuel transport in air-breathing microfluidic fuel cells by immersed fuel micro-jet, Journal of Power Sources 445 (2020) 227326, with permission from Elsevier.*

anode can be replenished actively by jetting fresh fuel perpendicular to the anode through a capillary with a micro-jet (Fig. 2.9B). This enhanced fuel transport and replenished the boundary layer concentration of fuel. Unlike multiple inlets/outlets proposed network described previously to overcome the limitation of the concentration boundary layer in MFCs [11,40], the micro-jet system offers easy integration into a rectangular flow channel. The MFC was air-breathing, having 2 M KOH as catholyte and 0.5 M HCOONa and 2 M KOH as anolyte. Carbon paper was employed as a substrate for a planar anode loaded with 5 mg cm^{-2} of Pd, and an air-breathing cathode spray-coated with Pt black (3 mg cm^{-2}). The MFC performance was compared between a micro-jet and flow-over mode. Mass transport limitation in the flow-over mode arising from the concentration boundary layer was observed. The fuel from the micro-jet and the anolyte was introduced at 100 μL minute^{-1}. The micro-jet case showed a better performance where the fuel transfers to the boundary layer mitigated the mass transport limitation. The maximum power density in the micro-jet case (83.1 mW cm^{-3}) was 25% higher than in the flow-over mode. A maximum current density of ~316 mA cm^{-3} (calculated) and an OCP of 1.04 V was exhibited for the micro-jet case. The positioning of the micro-jet at a distance from the inlet in the channel played a vital role in replenishing the fuel and eliminating the fuel concentration boundary layer. Positioning the micro-jet at a location having a relatively thick concentration boundary layer along the downstream was required to achieve effective fuel replenishment. The maximum power density for micro-jet location 10 mm from the inlet was 83.1 mW cm^{-3}, which was 19.6% and 25% higher than the micro-jet location at 5 mm case and flow-over mode, respectively. The immersed fuel micro-jet contributed to a 40.9% improvement in the cell performance under optimal conditions in contrast to the flow-over mode.

The fuel utilization in the MFC can be improved by using a volatile fuel in the vapor phase. Wang et al. [29] proposed methanol vapor as fuel MFC to achieve better fuel utilization than the MFC with liquid methanol. The schematic of the vapor feed methanol MFC is illustrated in Fig. 2.10. In the vapor feed methanol MFC, a fuel reservoir containing neat liquid methanol was stored near the fuel cell. Methanol vapor evaporates from the liquid surface in the reservoir and diffuses to the anode catalyst layer. The liquid catholyte (3 M KOH) was circulated at the cathode at a 50 μL minute^{-1} flow rate. The anode and cathode catalysts comprised of Pt-Ru (10 mg cm^{-2}) and Pt/C (2 mg cm^{-2}), respectively. A vertical configuration of the channel was chosen as the horizontal configuration led to the crossover of low-density methanol to the cathode, thereby lowering the performance. The maximum power density was 25.3 mW cm^{-2} for 10 M methanol solution, which increased to 31.6 mW cm^{-2} for neat methanol solution. This was owing to the lower mass transfer resistance in neat methanol. The methanol crossover at high concentration was evaded using a short electrode (3 mm) and vertical MFC configuration. The fuel boundary layer depletion was eliminated using

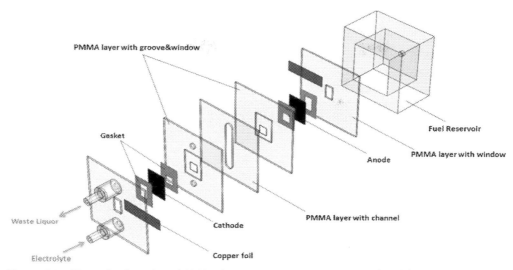

Figure 2.10 Vapor feed methanol MFC schematic representation. *Reproduced from Y. Wang, D.Y.C. Leung, J. Xuan, H. Wang, A vapor feed methanol microfluidic fuel cell with high fuel and energy efficiency, Applied Energy 147 (2015) 456–465, with permission from Elsevier.*

vapor methanol, and no benefits of a higher electrolyte flow rate were observed. This was contrary to liquid methanol MFC, or any liquid fuel feed MFC, where a higher flow rate is suggested to replenish the fuel over the catalyst surface. A substantial effect on the methanol vapor feed MFC performance was observed with the increase in the catholyte KOH concentration (1–3 M). A maximum power density of 55.4 mW cm^{-2} and maximum current density of 290 mA cm^{-2} with 3 M KOH was achieved, which was higher than that with 1 M KOH (31.6 mW cm^{-2}, 204 mA cm^{-2}). The reduced ionic transfer resistance and improved methanol oxidation reaction at high electrolyte concentration was the rationale for the enhanced performance. The vapor feed methanol MFC performance was better than the liquid methanol counterpart because the liquid methanol MFC suffers from low fuel utilization, fuel crossover, and wastage. The fuel and energy efficiencies were 1.35% and 0.33%, respectively, for liquid feed MFC, and that for vapor feed was 40% and 9.4%, respectively. The estimated efficiencies are competitive with several PEM-based vapor feed direct methanol fuel cells [41–49].

A two-cell stack air-breathing MFC design was proposed by Abrego-Martínez et al. [30]. The fuel and oxidant mixed-reactant consisting of 5 M methanol in 0.5 M KOH delivered concurrently to the cathode and anode. The electrolyte stream at 10 mL hour^{-1} was supplied to fill the channel with a 40 μL solution containing the reactant mixture. The anode was made of Pt film on carbon paper (Pt/CP), and the cathode comprised a film of Pt and Ag on carbon paper (Pt/Ag/CP). Ag is inactive towards the methanol oxidation reaction, and the layer of Ag on Pt inhibits the methanol molecules from reaching the Pt surface. The Pt plays the role of oxygen reduction reaction catalyst where the oxygen molecule permeates through the porous Ag layer and reacts on the Pt surface. The strategic design of the methanol tolerant Ag/Pt/CP cathode catalyst enabled the mixed-reactant operation of the MFC [50]. The two-cell stack configuration of the MFC is depicted in Fig. 2.11. The electrodes in the hourglass-shaped MFC were placed perpendicular to the inlet flow direction. The interelectrode distance was 3 mm, and Cell A and Cell B were distanced 12 mm apart. The MFC generates a maximum power density of 1.2 mW cm^{-2} and a current density of 25 mA cm^{-2} for 5 M methanol concentration in the reactant mixture. The two-cell MFC stack performance was evaluated, Cell A and Cell B individually, and Cell A and Cell B in series. The OCV and maximum current density for individual cells A and B were 0.445 V and 21.1 mA cm^{-2}, and 0.449 V and 24.3 mA cm^{-2}, respectively. However, the maximum power density of cell B (1.42 mW cm^{-2}) was 34% more than cell A (1.06 mW cm^{-2}). The improved distribution of the mixed-reactant stream in cell B arising from the bridge joining the two cells attributed to the improved performance. The OCV of 0.890 V was achieved for the cells A and B connected in series, and the delivered power output by the cell stack was 75% higher than the individual cells. A stable operation for 10 hours was exhibited by the stack denoting the durable operation of the catalyst under a concentrated methanol environment. The total power of the MFC stack can be improved by merely combining cells in series with the pair of electrodes arranged perpendicular to the flow (Fig. 2.11).

Kwok et al. [31] demonstrated MFC operation with an air-breathing cathode and a dual fuel-breathing anode (Fig. 2.12). PtRu on carbon paper (10 mg cm^{-2}) was used as the fuel-breathing anode, and Pt/C (2 mg cm^{-2}) was the air-breathing cathode. The MFC was supplied with 2 M NaOHas electrolyte at a flow rate of

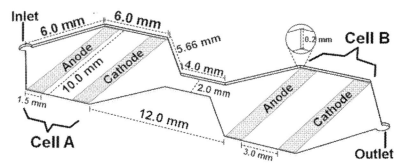

Figure 2.11 A two-cell stack air-breathing mixed-reactant MFC with the electrodes arranged perpendicular to the electrolyte flow. *Reproduced from J.C. Abrego-Martínez, A. Moreno-Zuria, F.M. Cuevas-Muñiz, L.G. Arriaga, S. Sun, M. Mohamedi, Design, fabrication and performance of a mixed-reactant membrane-less micro direct methanol fuel cell stack, Journal of Power Sources 371 (2017) 10–17, with permission from Elsevier.*

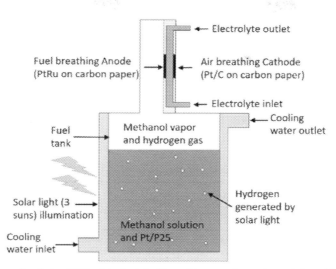

Figure 2.12 Schematic of the MFC with dual fuel-breathing anode and an air-breathing cathode. Dual fuel vapor was generated by utilizing the incident solar light by photocatalysis. *Reproduced from Y.H. Kwok, Y. Wang, M. Wu, F. Li, Y. Zhang, H. Zhang, et al. A dual fuel microfluidic fuel cell utilizing solar energy and methanol, Journal of Power Sources 409 (2019) 58–65, with permission from Elsevier.*

100 μL minute^{-1}. As depicted in, the dual fuel supplied to the anode comprised of methanol vapor and H_2 from a fuel tank with liquid methanol attached to the MFC. The H_2 was produced from methanol in the fuel tank by photocatalysis using Pt/P25 in sunlight. Hydrogen oxidation reaction (HOR) and methanol oxidation reaction (MOR) occurred simultaneously at the fuel-breathing anode to utilize the generated H_2 and the methanol vapor. The H_2 generation rate from methanol in the fuel tank under light illumination (3000 W m^{-2}) was 1.76 mL minute^{-1} g^{-1} $_{catalyst}$ using Pt/P25 photocatalyst with 1% loading of Pt (1%Pt/P25). An optimized 1:1 ratio of methanol and water in the fuel tank was used to generate a dual fuel mixture. The excess of methanol led to carbon monoxide generation, which poison the Pt catalyst in the MFC. The stability of the 1%Pt/P25 photocatalyst exhibited increased stability in the 1:1 water/methanol mixture with an H_2 generation rate of 1.91 mL min^{-1} g^{-1} $_{catalyst}$. 100 mg of 1%Pt/P25 photocatalyst was used in 100 mL of 1:1 water/methanol mixture to produce H_2 at 0.19 sccm for optimum operation of the MFC anode. The performance of the MFC was initially tested by simulating the dual fuel by externally controlling the H_2 input rate (0.25 sccm) and vaporized methanol from the fuel tank containing a 1:1 water/methanol mixture. Methanol is volatile and gets easily vaporized in an ambient environment. Controlling the concentration of methanol in the mixture was imperative as the MOR is a sluggish reaction compared to HOR. Excess methanol vapor in the fuel stream causes competition between the methanol and hydrogen molecules to occupy the active oxidation sites on the anode resulting in a

declined MFC performance. A maximum current density of 197.1 mA cm^{-2} and a maximum power density of 79.9 mW cm^{-2} was obtained for the simulated run of the MFC. The dual fuel mixture from the photo-reforming of methanol in the fuel tank boosted the performance of the MFC significantly attaining a 133.1 mW cm^{-2} maximum power density and 307.2 mA cm^{-2} maximum current density with the OCV reaching up to 1 V. The improvement in the performance of the MFC with the photocatalytic process was due to intermediate products (formaldehyde, formate, and formic acid) formed during the oxidation of methanol to CO_2 in a multistep process [48]. The MFC also utilized intermediate products as useful fuels.

Reports on direct glycerol MFC [51–55] indicated that the devices generally suffered from low power density due to the partial oxidation of glycerol to render carbonyl compounds, compromising the MFC performance by catalyst poisoning. The direct glycerol MFC was circumvented using bleach catholyte, as demonstrated by Martins et al. [32]. The MFC was operated at all-alkaline (anolyte: 1 M glycerol in 1 M KOH, catholyte: bleach (ClO$^-$, 3–8%) in 2 M KOH), and mixed media (anolyte: 1 M glycerol in 1 M KOH, catholyte: bleach (ClO$^-$, 3–8%) in 1 M H_2SO_4) electrolyte at 100 µL minute^{-1} flow rate. The cathode and anode were made of by Pt/C nanoparticle modified carbon paper to perform bleach electrochemical reduction and glycerol electrochemical oxidation reaction. The cell operated at all-alkaline electrolyte displayed an OCP of 1.0 V with 337 mA cm^{-2} maximum current density and 71.2 mW cm^{-2} maximum power density. The all-alkaline glycerol MFC performance was comparable to that of conventional direct glycerol fuel cells (DGFCs) with anion exchange membranes [56]. However, the all-alkaline cell operation was limited by the deterioration of the peak power density by 40% after 8 successive runs due to the instability arising from alkaline bleach reduction reaction at the cathode. The challenge was circumvented using a mixed media electrolyte configuration, which attained OCV of 1.98–2.0 V. Moreover, the mixed media cell exhibited 638 mA cm^{-2} maximum current density and 315 mW cm^{-2} maximum power density. The stability test resulted in only a 10% drop in the peak power density for successive runs. The glycerol/bleach MFC displayed the highest performance compared to the previously reported glycerol MFC, and the mixed electrolyte glycerol MFC provides power density higher than DGFCs with membranes [57].

The saturation limit and solubility of gas in the liquid electrolyte gas has hindered the use of high energy density fuel such as H_2 gas in the MFC. MFC functioning on a single-phase co-laminar flow of H_2 and O_2 dissolved in liquid electrolyte suffer from low power density, poor fuel utilization, and higher mass transfer limitation [58]. Hashemi et al. [33] developed a two-phase flow of gas-electrolyte MFC to address the challenges arising in the single-phase flow of dissolved fuel and oxidant. The two-phase flow of gas and electrolyte was achieved as the bubbles of H_2 and O_2 deformed to adapt to the morphology of the microchannel to minimize the surface energy.

Fig. 2.13 depicts the schematic diagram of the cross-sectional area of the bridged-shaped microchannel. The stream of H_2 gas (fuel) and O_2 gas (oxidant) were encapsulated by a thin film of ion conducting electrolyte (1 M H_2SO_4) touching the electrode surface (Pt film for both anode and cathode). The H_2 and O_2 diffuse through the very thin electrolyte film separating the gas streams to reach the Pt electrode surface. The proton conduction from cathode to anode took place through the liquid electrolyte which was supplied to the MFC at 1 mL hour^{-1}. The gas bubbles in the two-phase flow behave as fuel reservoirs thereby mitigating the mass transport limitation faced in the single-phase flow caused by low solubility of H_2 and O_2 in the electrolyte. The O_2 was provided to the MFC at 2 mL hour^{-1} constant flow rate. The H_2 flow rate was varied from 4 to 16 mL hour^{-1}, and a steep increase in the maximum current density (7.5 to 26.6 mA cm^{-2}) was observed. The OCV was 1 V, and negligible variation with flow rate was observed. The maximum power density was ~3.8 mW cm^{-2} attained by the two-phase flow MFC, which was an order of magnitude higher than that obtained from single-phase flow. The introduction of controlled gas-electrolyte two-phase flow alleviated the mass transport limitation associated with the single-phase flow.

The fabrication of the MFC microchannels can be a complex multistep process. Liu et al. [10] proposed using carbon fibers (CFs) to simultaneously serve the purpose of electrode and microchannel and named it a dual-functional electrode (Fig. 2.14). The

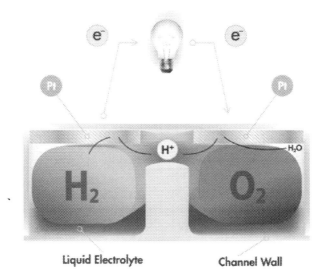

Figure 2.13 Schematic representation of the two-phase flow H_2-O_2 microfluidic fuel cell (MFC) with a bridged-shaped microchannel. *Reproduced from S.M.H. Hashemi, M. Neuenschwander, P. Hadikhani, M.A. Modestino, D. Psaltis, Membrane-less micro fuel cell based on two-phase flow, Journal of Power Sources 348 (2017) 212–218 [33] with permission from Elsevier.*

CF threads were knitted together to form a 3D braided herringbone-like pattern, and a pretreatment rendered it hydrophilic. The MFC was pump-free as the hydrophilic CF transported the electrolyte by capillary action aided by gravitational flow. The pretreated 3D herringbone-like CF surface was electrodeposited with Pd enabling it to function as an electrode. The anolyte was sodium formate (HCOONa) in 2 M KOH, and the catholyte was 0.5 M H_2O_2 in 1 M H_2SO_4. The third stream of 1 M sodium sulfate (Na_2SO_4) solution was supplied at the middle channel to separate the electrolyte flow at the respective electrodes and prevent fuel-oxidant crossover. The anolyte and catholyte attained a flow rate of 9.32 ± 0.65 and 9.85 ± 1.26 mL hour^{-1}, respectively, arising from the capillary effect in the CFs. The concentration of HCOONa in the anolyte was varied from 1–6 M, and the MFC's performance was determined. The MFC has a stable OCP of 1.44 V, and negligible variation was observed with varying fuel concentration. The maximum power density with 2 M HCOONa was 29.9 mW cm^{-2}, which was 39.7% higher than that with 1 M HCOONa due to improved fuel mass transport. However, increasing the concentration further to 6 M HCOONa reduced the peak power density to 25.5 mW cm^{-2}. A similar trend was followed by the maximum current density depicting the best performance with 2 M HCOONa to be 83.3 mA cm^{-2}. The high HCOONa concentration hinders the OH^- ion adsorption on the catalyst active sites, impeding the formate oxidation kinetics at the anode. The higher concentrations of HCOONa increase the viscosity, resulting in decreased anolyte flow rate and slower product removal, thereby affecting the MFC performance. The catholyte flow rate was unaffected by changing the concentration of the anolyte because of the flow in the middle channel, which averted the fuel crossover.

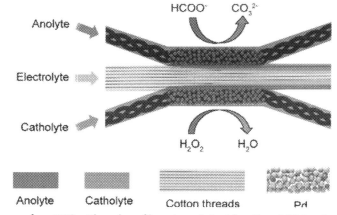

Figure 2.14 A pump-free MFC with carbon fibers based dual-functional 3D herringbone electrodes. *Reproduced from Z. Liu, D. Ye, R. Chen, B. Zhang, X. Zhu, Q. Liao, A dual-functional three-dimensional herringbone-like electrode for a membrane-less microfluidic fuel cell, Journal of Power Sources (2019) 438, with permission from Elsevier.*

The dual-functional electrode exhibited 50 minutes of stable operation with the continuous supply of fuel and oxidant. The performance of the MFC with 3D herringbone-like structured dual-functional electrodes surpassed the performance of the previously reported pump-free MFCs [59–64].

3D printing technology for the rapid and low-cost manufacturing of the MFCs is desirable compared to the conventional multistep microfabrication techniques. Guima et al. [26] manufactured a 3D-printed direct glycerol MFC with flow-through porous electrodes (Fig. 2.4B), achieving high power density at low cost. Pt/C (60%) on carbon paper (Pt/C/CP) was used as anode and cathode. The mixed-media configuration was adopted, having anolyte of 1 M glycerol in 1 M KOH and sodium hypochlorite (6%) in 1 M H_2SO_4 as the catholyte. An OCP of 1.8 V was achieved by the MFC using the mixed media condition. The performance of the MFC was estimated by changing the flow rate of the electrolyte in the range of 100–900 µL minute^{-1}. At low flow rates (100–300 µL minute^{-1}), the performance was limited by mass transport occurring due to oxidant starvation and depletion of H_3O^+ to suffice the electron supply. The mass transport limitation was averted by high electrolyte flow rates (500–900 µL minute^{-1}). A steep increase in the maximum power density (48.3–193 mW cm^{-2}) was noted, increasing the electrolyte flow rate (100–900 µL minute^{-1}). A maximum power density of 310.7 mA cm^{-2} was observed for the MFC operated at 500 µL minute^{-1} electrolyte flow rate. The 3D-printed direct glycerol MFC depicted a performance comparable with previously reported literature [20,53–55,65], attributed to the mixed-media operations and flow -through porous media.

2.5 Perspective and conclusions

Microfluidic electrochemical energy conversion has the potential of being a disruptive technology in energy systems of the future. In the past decades, MFCs have been extensively studied to develop a portable devices providing high power density. Several studies concerning the MFCs focus on improving the performance to make suitable for portable power applications and replace batteries. The desired channel geometry and flow scheme are adopted for microfabrication of the MFC to incorporate the choice of fuel-oxidant and catalysts, while withstanding the required flow rate to tune the operation to achieve maximum performance. The capital cost of performing the electrochemical conversion in the MFCs is substantially lower than for conventional fuel cells. The low capital cost associated with the MFCs is due to the condensing of the conventional fuel cells to micron-scale, where the intensive properties are greatly enhanced. The typical requirements of the carefully monitoring and controlling operating conditions such as humidity, pH, and temperature in the fuel cells are averted in the MFC due to membrane-less operations and fewer components. The flexibility in choosing the fuel-oxidant followed by advancement in resilient catalyst to withstand the reactants in the MFCs paves the way for more opportunities in developing commercial MFCs. The scale-up of MFCs is required to reap the advantages of the microsystem to the desired energy demand. Advancement in the microfabrication techniques such as 3D printing pushes the research boundary for the rapid and reproducible MFC production subjected to scaling up by parallelizing microchannels.

The critical factor to make MFC suitable alternative to compete with existing power source is by attaining high performance in terms of power density and high fuel utilization. The chapter emphasized the challenges and opportunities in MFC system. The formation of a concentration boundary layer, low fuel utilization, and slow reaction kinetics can be averted by implementing strategic design and flow scheme in the microchannels. MFC performance was evaluated in the light of the fuel-oxidant, electrolyte flow rate, and catalysts [66–87]. The advantages offered by the MFCs will play a pivotal role as more renewable technology penetrates the energy market. To that end, improving the MFC's performance and efficiency is necessary, and undoubtedly, it will continue to occupy an essential space amongst the research community.

References

[1] G. Zhou, L. Huang, W. Li, Z. Zhu, Harvesting ambient environmental energy for wireless sensor networks: a survey, Journal of Sensors (2014) 2014.
[2] B. Dziadak, è. Makowski, A. Michalski, Survey of energy harvesting systems for wireless sensor networks in environmental monitoring, Metrology and Measurement Systems 23 (2016) 495–512.
[3] Z. Yang, J. Zhang, M.C.W. Kintner-Meyer, X. Lu, D. Choi, J.P. Lemmon, et al., Electrochemical energy storage for green grid, Chemical Reviews 111 (2011) 3577–3613.
[4] J.D. Morse, Micro-fuel cell power sources, International Journal of Energy Research 31 (2007) 576–602.
[5] Y. Wang, K.S. Chen, J. Mishler, S.C. Cho, X.C. Adroher, A review of polymer electrolyte membrane fuel cells: technology, applications, and needs on fundamental research, Applied Energy 88 (2011) 981–1007.
[6] M.S. Ismail, K.J. Hughes, D.B. Ingham, L. Ma, M. Pourkashanian, Effects of anisotropic permeability and electrical conductivity of gas diffusion layers on the performance of proton exchange membrane fuel cells, Applied Energy 95 (2012) 50–63.
[7] J. Cho, J. Park, H. Oh, K. Min, E. Lee, J.-Y. Jyoung, Analysis of the transient response and durability characteristics of a proton exchange membrane fuel cell with different micro-porous layer penetration thicknesses, Applied Energy 111 (2013) 300–309.
[8] X. Zhou, J. Qiao, L. Yang, J. Zhang, A review of graphene-based nanostructural materials for both catalyst supports and metal-free catalysts in PEM fuel cell oxygen reduction reactions, Advanced Energy Materials 4 (2014) 1301523.
[9] J.C. Abrego-Martínez, A. Moreno-Zuria, Y. Wang, F.M. Cuevas-Muñiz, L.G. Arriaga, S. Sun, et al., Fabrication and evaluation of passive alkaline membraneless microfluidic DMFC, International Journal of Hydrogen Energy 42 (2017) 21969–21975.
[10] Z. Liu, D. Ye, R. Chen, B. Zhang, X. Zhu, Q. Liao, A dual-functional three-dimensional herringbone-like electrode for a membraneless microfluidic fuel cell, Journal of Power Sources (2019) 438.
[11] S.K. Yoon, G.W. Fichtl, P.J.A. Kenis, Active control of the depletion boundary layers in microfluidic electrochemical reactors, Lab Chip 6 (2006) 1516–1524.

[12] J.L. Cohen, D.A. Westly, A. Pechenik, H.D. Abruña, Fabrication and preliminary testing of a planar membraneless microchannel fuel cell, Journal of Power Sources 139 (2005) 96−105.
[13] N.Da Mota, D.A. Finkelstein, J.D. Kirtland, C.A. Rodriguez, A.D. Stroock, H.D. Abruña, Membraneless, room-temperature, direct borohydride/cerium fuel cell with power density of over 0.25 W/cm^2, Journal of the American Chemical Society 134 (2012) 6076−6079.
[14] R.F. Ismagilov, A.D. Stroock, P.J.A. Kenis, G. Whitesides, H.A. Stone, Experimental and theoretical scaling laws for transverse diffusive broadening in two-phase laminar flows in microchannels, Applied Physics Letters 76 (2000) 2376−2378.
[15] D.C. Duffy, J.C. McDonald, O.J.A. Schueller, G.M. Whitesides, Rapid prototyping of microfluidic systems in poly (dimethylsiloxane), Analytical Chemistry 70 (1998) 4974−4984.
[16] J.C. Love, J.R. Anderson, G.M. Whitesides, Fabrication of three-dimensional microfluidic systems by soft lithography, MRS Bulletin 26 (2001) 523−528.
[17] B.S. De, A. Singh, A.L. Elias, N. Khare, S. Basu, Electrochemical neutralization energy-assisted membrane-less microfluidic reactor for water electrolysis, Sustainable Energy & Fuels 4 (2020) 6234−6244.
[18] E. Kjeang, R. Michel, D.A. Harrington, N. Djilali, D. Sinton, A microfluidic fuel cell with flow-through porous electrodes, Journal of the American Chemical Society 130 (2008) 4000−4006.
[19] C.A. Martins, P. Pei, M. Tellis, O.A. Ibrahim, E. Kjeang, Graphene-oxide-modified metal-free cathodes for glycerol/bleach micro fluidic fuel cells 8 (2020) 8286−93.
[20] C.A. Martins, O.A. Ibrahim, P. Pei, E. Kjeang, Towards a fuel-flexible direct alcohol microfluidic fuel cell with flow-through porous electrodes: assessment of methanol, ethylene glycol and glycerol fuels, Electrochim Acta 271 (2018) 537−543.
[21] A. Jindal, S. Basu, N. Chauhan, T. Ukai, D.S. Kumar, K.T. Samudhyatha, Application of electrospun CNxnanofibers as cathode in microfluidic fuel cell, Journal of Power Sources 342 (2017) 165−174.
[22] S.A.M. Shaegh, N.-T. Nguyen, S.H. Chan, A review on membraneless laminar flow-based fuel cells, International Journal of Hydrogen Energy 36 (2011) 5675−5694.
[23] H. Xu, H. Zhang, H. Wang, D.Y.C. Leung, L. Zhang, J. Cao, et al., Counter-flow formic acid microfluidic fuel cell with high fuel utilization exceeding 90%, Applied Energy 160 (2015) 930−936.
[24] Y. Zhou, B. Zhang, X. Zhu, D.D. Ye, R. Chen, T. Zhang, et al., Enhancing fuel transport in air-breathing microfluidic fuel cells by immersed fuel micro-jet, Journal of Power Sources 445 (2020) 227326.
[25] E. Ortiz-Ortega, M.A. Goulet, J.W. Lee, M. Guerra-Balcázar, N. Arjona, E. Kjeang, et al., A nanofluidic direct formic acid fuel cell with a combined flow-through and air-breathing electrode for high performance, Lab Chip 14 (2014) 4596−4598.
[26] K.E. Guima, P.H.L. Coelho, M.A.G. Trindade, M.A.G. Trindade, C.A. Martins, 3D-Printed glycerol microfluidic fuel cell, Lab Chip 20 (2020) 2057−2061.
[27] H. Xu, H. Zhang, H. Wang, D.Y.C. Leung, L. Zhang, J. Cao, et al. Counter-flow formic acid microfluidic fuel cell with high fuel utilization exceeding 90% 160 (2015) 930−936.
[28] B. Deng, D. Ye, B. Zhang, X. Zhu, R. Chen, Q. Liao, Current density distribution in air-breathing microfluidic fuel cells with an array of graphite rod anodes, International Journal of Hydrogen Energy (2020).
[29] Y. Wang, D.Y.C. Leung, J. Xuan, H. Wang, A vapor feed methanol microfluidic fuel cell with high fuel and energy efficiency, Applied Energy 147 (2015) 456−465.
[30] J.C. Abrego-Martínez, A. Moreno-Zuria, F.M. Cuevas-Muñiz, L.G. Arriaga, S. Sun, M. Mohamedi, Design, fabrication and performance of a mixed-reactant membraneless micro direct methanol fuel cell stack, Journal of Power Sources 371 (2017) 10−17.
[31] Y.H. Kwok, Y. Wang, M. Wu, F. Li, Y. Zhang, H. Zhang, et al., A dual fuel microfluidic fuel cell utilizing solar energy and methanol, Journal of Power Sources 409 (2019) 58−65.
[32] C.A. Martins, O.A. Ibrahim, P. Pei, E. Kjeang, "Bleaching" glycerol in a microfluidic fuel cell to produce high power density at minimal cost, Chemical Communications 54 (2017) 192−195.

[33] S.M.H. Hashemi, M. Neuenschwander, P. Hadikhani, M.A. Modestino, D. Psaltis, Membrane-less micro fuel cell based on two-phase flow, Journal of Power Sources 348 (2017) 212−218.
[34] R.S. Jayashree, L. Gancs, E.R. Choban, A. Primak, D. Natarajan, L.J. Markoski, et al., Air-breathing laminar flow-based microfluidic fuel cell, Journal of the American Chemical Society 127 (2005) 16758−16759.
[35] X. Zhu, B. Zhang, D. Ye, J. Li, Q. Liao, Air-breathing direct formic acid micro fluidic fuel cell with an array of cylinder anodes 247 (2014) 346−353.
[36] F.R. Brushett, W.-P. Zhou, R.S. Jayashree, P.J.A. Kenis, Alkaline microfluidic hydrogen-oxygen fuel cell as a cathode characterization platform, Journal of the Electrochemical Society 156 (2009) B565−B571.
[37] R.S. Jayashree, S.K. Yoon, F.R. Brushett, P.O. Lopez-Montesinos, D. Natarajan, L.J. Markoski, et al., On the performance of membraneless laminar flow-based fuel cells, Journal of Power Sources 195 (2010) 3569−3578.
[38] S.A.M. Shaegh, N.-T. Nguyen, S.H. Chan, W. Zhou, Air-breathing membraneless laminar flow-based fuel cell with flow-through anode, International Journal of Hydrogen Energy 37 (2012) 3466−3476.
[39] S.A.M. Shaegh, N.-T. Nguyen, S.H. Chan, Air-breathing microfluidic fuel cell with fuel reservoir, Journal of Power Sources 209 (2012) 312−317.
[40] J. Phirani, S. Basu, Analyses of fuel utilization in microfluidic fuel cell, Journal of Power Sources 175 (2008) 261−265.
[41] L. Feng, J. Zhang, W. Cai, W. Xing, C. Liu, Single passive direct methanol fuel cell supplied with pure methanol, Journal of Power Sources 196 (2011) 2750−2753.
[42] S. Eccarius, F. Krause, K. Beard, C. Agert, Passively operated vapor-fed direct methanol fuel cells for portable applications, Journal of Power Sources 182 (2008) 565−579.
[43] G. Jewett, Z. Guo, A. Faghri, Performance characteristics of a vapor feed passive miniature direct methanol fuel cell, International Journal of Heat and Mass Transfer 52 (2009) 4573−4583.
[44] C. Xu, A. Faghri, X. Li, Development of a high performance passive vapor-feed DMFC fed with neat methanol, Journal of the Electrochemical Society 157 (2010) B1109.
[45] I. Chang, S. Ha, J. Kim, J. Lee, S.W. Cha, Performance evaluation of passive direct methanol fuel cell with methanol vapour supplied through a flow channel, Journal of Power Sources 184 (2008) 9−15.
[46] Z. Guo, A. Faghri, Vapor feed direct methanol fuel cells with passive thermal-fluids management system, Journal of Power Sources 167 (2007) 378−390.
[47] X. Li, A. Faghri, Effect of the cathode open ratios on the water management of a passive vapor-feed direct methanol fuel cell fed with neat methanol, Journal of Power Sources 196 (2011) 6318−6324.
[48] X. Li, A. Faghri, Development of a direct methanol fuel cell stack fed with pure methanol, International Journal of Hydrogen Energy 37 (2012) 14549−14556.
[49] C. Xu, A. Faghri, X. Li, Improving the water management and cell performance for the passive vapor-feed DMFC fed with neat methanol, International Journal of Hydrogen Energy 36 (2011) 8468−8477.
[50] J.C. Abrego-Martínez, Y. Wang, J. Ledesma-García, F.M. Cuevas-Muñiz, L.G. Arriaga, M. Mohamedi, A pulsed laser synthesis of nanostructured bi-layer platinum-silver catalyst for methanol-tolerant oxygen reduction reaction, International Journal of Hydrogen Energy 42 (2017) 28056−28062.
[51] S. Miao, S. He, M. Liang, G. Lin, B. Cai, O.G. Schmidt, Microtubular fuel cell with ultrahigh power output per footprint, Advanced Materials 29 (2017) 1607046.
[52] A.S. Hollinger, P.J.A. Kenis, Manufacturing all-polymer laminar flow-based fuel cells, Journal of Power Sources 240 (2013) 486−493.
[53] A. Dector, F.M. Cuevas-Muñiz, M. Guerra-Balcázar, L.A. Godínez, J. Ledesma-García, L.G. Arriaga, Glycerol oxidation in a microfluidic fuel cell using Pd/C and Pd/MWCNT anodes electrodes, International Journal of Hydrogen Energy 38 (2013) 12617−12622.

[54] J. Maya-Cornejo, M. Guerra-Balcázar, N. Arjona, L. Álvarez-Contreras, F.J.R. Valadez, M.P. Gurrola, et al., Electrooxidation of crude glycerol as waste from biodiesel in a nanofluidic fuel cell using Cu@ Pd/C and Cu@ Pt/C, Fuel 183 (2016) 195—205.

[55] J. Maya-Cornejo, E. Ortiz-Ortega, L. Álvarez-Contreras, N. Arjona, M. Guerra-Balcázar, J. Ledesma-García, et al., Copper—palladium core—shell as an anode in a multi-fuel membraneless nanofluidic fuel cell: toward a new era of small energy conversion devices, Chemical Communications 51 (2015) 2536—2539.

[56] Z. Zhang, L. Xin, W. Li, Supported gold nanoparticles as anode catalyst for anion-exchange membrane-direct glycerol fuel cell (AEM-DGFC), International Journal of Hydrogen Energy 37 (2012) 9393—9401.

[57] N. Benipal, J. Qi, J.C. Gentile, W. Li, Direct glycerol fuel cell with polytetrafluoroethylene (PTFE) thin film separator, Renewable Energy 105 (2017) 647—655.

[58] J.L. Cohen, D.J. Volpe, D.A. Westly, A. Pechenik, H.D. Abruña, A dual electrolyte H_2/O_2 planar membraneless microchannel fuel cell system with open circuit potentials in excess of 1.4 V, Langmuir 21 (2005) 3544—3550.

[59] J.P. Esquivel, F.J. Del Campo, J.L.G. De La Fuente, S. Rojas, N. Sabate, Microfluidic fuel cells on paper: meeting the power needs of next generation lateral flow devices, Energy Environment Science 7 (2014) 1744—1749.

[60] T.S. Copenhaver, K.H. Purohit, K. Domalaon, L. Pham, B.J. Burgess, N. Manorothkul, et al., A microfluidic direct formate fuel cell on paper, Electrophoresis 36 (2015) 1825—1829.

[61] S. Lal, V.M. Janardhanan, M. Deepa, A. Sagar, K.C. Sahu, Low cost environmentally benign porous paper based fuel cells for micro-nano systems, Journal of the Electrochemical Society 162 (2015) F1402.

[62] V. Galvan, K. Domalaon, C. Tang, S. Sotez, A. Mendez, M. Jalali-Heravi, et al., An improved alkaline direct formate paper microfluidic fuel cell, Electrophoresis 37 (2016) 504—510.

[63] K. Domalaon, C. Tang, A. Mendez, F. Bernal, K. Purohit, L. Pham, et al., Fabric-based alkaline direct formate microfluidic fuel cells, Electrophoresis 38 (2017) 1224—1231.

[64] Z. Liu, D. Ye, R. Chen, B. Zhang, X. Zhu, J. Li, et al., A woven thread-based microfluidic fuel cell with graphite rod electrodes, International Journal of Hydrogen Energy 43 (2018) 22467—22473.

[65] C.A. Martins, O.A. Ibrahim, P. Pei, E. Kjeang, In situ decoration of metallic catalysts in flow-through electrodes: application of Fe/Pt/C for glycerol oxidation in a microfluidic fuel cell, Electrochim Acta 305 (2019) 47—55.

[66] W. Xu, K. Scott, S. Basu, Performance of a high temperature polymer electrolyte membrane water electrolyser, Journal of Power Sources 196 (2011) 8918—8924.

[67] J. Tayal, B. Rawat, S. Basu, Effect of addition of rhenium to Pt-based anode catalysts in electrooxidation of ethanol in direct ethanol PEM fuel cell, International Journal of Hydrogen Energy 37 (2012) 4597—4605.

[68] S.P.S. Badwal, S. Giddey, A. Kulkarni, J. Goel, S. Basu, Direct ethanol fuel cells for transport and stationary applications—a comprehensive review, Applied Energy 145 (2015) 80—103.

[69] S. Basu, Fuel Cell Science and Technology, Springer, 2007.

[70] B. Hua, N. Yan, M. Li, Y. Sun, Y. Zhang, J. Li, et al., Anode-engineered protonic ceramic fuel cell with excellent performance and fuel compatibility, Advanced Materials 28 (2016) 8922—8926.

[71] Y. Feng, J. Luo, K.T. Chuang, Conversion of propane to propylene in a proton-conducting solid oxide fuel cell, Fuel 86 (2007) 123—128.

[72] J. Mo, Z. Kang, G. Yang, S.T. Retterer, D.A. Cullen, J. Todd, Thin liquid/gas diffusion layers for high—efficiency hydrogen production from water splitting, Applied Energy 177 (2016) 817—822.

[73] J. Mo, R.R. Dehoff, W.H. Peter, T.J. Toops, J.B. Green Jr, F.-Y. Zhang, Additive manufacturing of liquid/gas diffusion layers for low-cost and high-efficiency hydrogen production, International Journal of Hydrogen Energy 41 (2016) 3128—3135.

[74] K. Eom, E. Cho, J. Jang, H.-J. Kim, T.-H. Lim, B.K. Hong, et al., Optimization of GDLs for high-performance PEMFC employing stainless steel bipolar plates, International Journal of Hydrogen Energy 38 (2013) 6249—6260.

[75] M. Prasanna, H.Y. Ha, E. Cho, S.-A. Hong, I.-H. Oh, Influence of cathode gas diffusion media on the performance of the PEMFCs, Journal of Power Sources 131 (2004) 147–154.
[76] F.-Y. Zhang, S.G. Advani, A.K. Prasad, Performance of a metallic gas diffusion layer for PEM fuel cells, Journal of Power Sources 176 (2008) 293–298.
[77] E.C. Kumbur, K.V. Sharp, M.M. Mench, On the effectiveness of Leverett approach for describing the water transport in fuel cell diffusion media, Journal of Power Sources 168 (2007) 356–368.
[78] A. Higier, H. Liu, Optimization of PEM fuel cell flow field via local current density measurement, International Journal of Hydrogen Energy 35 (2010) 2144–2150.
[79] T. Zhou, H. Liu, Effects of the electrical resistances of the GDL in a PEM fuel cell, Journal of Power Sources 161 (2006) 444–453.
[80] W.K. Epting, J. Gelb, S. Litster, Resolving the three-dimensional microstructure of polymer electrolyte fuel cell electrodes using nanometer-scale X-ray computed tomography, Advanced Functional Materials 22 (2012) 555–560.
[81] J. Park, H. Oh, Lee Y. Il, K. Min, E. Lee, J.-Y. Jyoung, Effect of the pore size variation in the substrate of the gas diffusion layer on water management and fuel cell performance, Applied Energy 171 (2016) 200–212.
[82] M.P. Manahan, M.C. Hatzell, E.C. Kumbur, M.M. Mench, Laser perforated fuel cell diffusion media. Part I: related changes in performance and water content, Journal of Power Sources 196 (2011) 5573–5582.
[83] G.M. Whitesides, The origins and the future of microfluidics, Nature 442 (2006) 368–373.
[84] N.-T. Nguyen, S.T. Wereley, Fundamentals and Applications of Microfluidics, Artech House Inc, Boston, MA, 2002, pp. 354–369.
[85] M. Gad-el-Hak, Liquids: the holy grail of microfluidic modeling, Physics Fluids 17 (2005) 100612.
[86] A. Bazylak, D. Sinton, N. Djilali, Improved fuel utilization in microfluidic fuel cells: a computational study 143 (2005) 57–66.
[87] E.R. Choban, L.J. Markoski, A. Wieckowski, P.J.A. Kenis, Microfluidic fuel cell based on laminar flow, Journal of Power Sources 128 (2004) 54–60.

CHAPTER 3

Alcohol fuel cell on-a-chip

Cauê A. Martins
Institute of Physics, Federal University of Mato Grosso do Sul, Campo Grande, Brazil

3.1 Introduction

The previous chapters made clear the applicabilities of fuel cells in clean energy conversion. This chapter aims to provide a general overview of miniaturized fuel cells fed by alcohol, including basics of functioning, anodic reactions, involving alcohols, nanoparticle (NP) used as anodes, some relevant achievements to shed a light on ongoing and emerging devices, challenges still needing to be addressed, and an outlook for future researchers and industrials. This chapter is not a schematic review because it reports only a fraction of the works regarding miniaturized alcohol fuel cells. A revision and a deep discussion about general liquid microfluidic fuel cells (μFCs) were made by Oliveira et al. [1].

The use of renewable power sources is pivotal to transit toward a sustainable lifestyle, reached when the demands for energy does not impact environmental issues. Among the alternatives, fuel cells are finding their way to the market. As discussed in previous chapters, fuel cells convert chemical energy into electric energy through a coupled and continuous redox reaction [2,3]. Fuel cells are versatile and allow the use of different configurations since fuels, oxidants, electrodes, and membranes. Besides the different configuration, they operate basically in the same way, fuel is electrooxidized at the anode while an oxidant is reduced at the cathode. As a result, electrons flow through an external circuit while ions go through an internal membrane.

The commercial fuel cell is fed by hydrogen and oxygen, producing clean energy, heat, and water [2]. However, hydrogen production from water electrolysis is an expensive process, so new fuels have been tested for direct liquid fuel cells, such as alcohols [1,4]. Methanol (MeOH), ethanol (EtOH), ethylene glycol (EgOH), and glycerol (GlOH) are conceivable fuels for Direct Alcohol Fuel Cells (DAFCs). In DAFCs, the alcohol is oxidized at the anode leading to electron harvesting, so the more advanced is the electrooxidation the more current will flow. Complete methanol, ethanol, ethylene glycol, and glycerol electrooxidation reactions are shown in Eqs. (3.1)–(3.4). In all cases, CO_2 is chemically converted to carbonate in an alkaline medium, as following: $CO_2 + 2OH^- \rightarrow CO_3^{2-} + H_2O$.

$$CH_3(OH) + 6OH^- \rightarrow CO_2 + 5H_2O + 6e^- \quad (3.1)$$

$$C_2H_5(OH) + 12\,OH^- \rightarrow 2CO_2 + 9H_2O + 12\,e^- \quad (3.2)$$

$$C_2H_4(OH)_2 + 10OH^- \rightarrow 2CO_2 + 8H_2O + 10e^- \quad (3.3)$$

$$C_3H_5(OH)_3 + 14OH^- \rightarrow 3CO_2 + 11H_2O + 14e^- \quad (3.4)$$

These alcohols can be renewably obtained from biomass and show different characteristics. Methanol has only one hydrated carbon, which makes it easier to be oxidized, but is highly toxic. Ethanol is not toxic but is mainly converted into acetic acid at room temperature [5]. Ethylene glycol and glycerol show two and three hydrated carbon in the main chain, respectively, which makes it difficult for complete electrooxidation. The commercially available Pt/C NPs show limited ability to cleavage the C—C bonds of these alcohols, which induces incomplete electrooxidation, fewer electrons, and more partially oxidized compounds in the system [5—7]. Moreover, the membrane that separates the anode from the cathode augments the ohmic loss to the already low power density under room temperature, which may turn its application unfeasible. Therefore there is an impending need to overcome these challenges for electrochemical energy conversion based on widely available biomass-derived alcohol.

3.2 Fuel cell on-a-chip

A cost-effective way of using biomass-derived alcohols to feed electrochemical energy converters rose with the development of μFCs or laminar flow-based fuel cells [8,9]. These miniaturized devices with few square centimeters [8] work basically by introducing two liquids into a single microfluidic channel, in which the opposite sides are the anode and the cathode. Dissolved oxygen is one of the most used oxidants since is available in the atmosphere, ideal catholyte for most microfluidic application. The other liquid is the anolyte, which carries the fuel, such as methanol, ethanol, ethylene glycol, and glycerol. The most important characteristic of a μFC is the absence of an ion-exchange membrane [10], thus it is also called membraneless μFC. This concept is based on the dynamic of the microfluidics [8]. The colaminar flow built inside the microchannel—micro to milliliter—replaces the need for a membrane. The laminar flow separates the streams, permitting ionic transport at the interface between the liquids [8,9]. In this context, the ion-exchange membrane is no longer required.

Different from conventional fuel cells, the reactants, and ions crossover and ohmic polarization related to the use of the membrane are disregarded [9]. Therefore the exclusion of the membrane-ohmic-drop boosts the power density for microfluidic direct alcohol fuel cells (μDAFCs) compared to conventional systems [11]. The most common configuration is the so-called flow-by when the liquids contact the electrodes on the way to the outlet [12], as shown in Fig. 3.1A. This configuration is simple and practical but is highly dependent on the residence time and displays low reactant utilization. Another powerful configuration was developed by Jayashree et al. [13]. These authors built an air-breathing μFC, where the cathode has contact with oxygen from the air, besides being fed by electrolyte [13], as illustrated in Fig. 3.1B. This configuration surpassed drawbacks such as low diffuse and low oxygen concentration. Then, the performance is improved by increased oxygen reduction reaction rate, and current and power densities accordingly [13]. The feasibility of using air-breathing cathodes for μDAFCs will be further discussed [14—16].

Another configuration that changed the magnitude of the output power produced by miniaturized devices is the flow-through porous electrodes, developed by Kjeang et al. [17]. In this configuration, the liquids go through porous electrodes toward the colaminar microchannel, as illustrates in Fig. 3.1C. The interaction between the fuel and oxidant is then maximized, which increases reactant utilization and current and power densities accordingly [17]. This configuration is the most promising for μDAFCs, especially for alcohols with more than one C—C bond, such as glycerol, because the flow-through increases the collision factor and the successive reactions, harvesting high energy [11,18,19]. Another configuration less used but with potential practical application is the dual-pass flow-through cell [9], as shown in Fig. 3.1D. Theoretically, this configuration maximizes the reactant utilization by leading the streams twice through the porous electrodes.

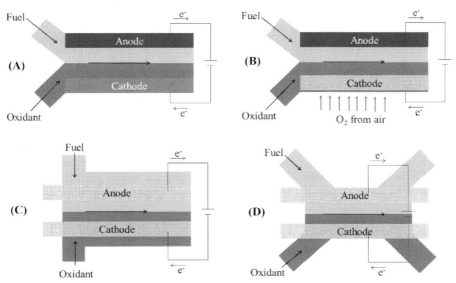

Figure 3.1 The most common architecture used in microfluidic direct alcohol fuel cells. Namely, (A) flow-by, (B) air-breathing, (C) flow-through, and (D) dual-pass flow-through cell.

The absence of a membrane enables the use of solutions with different pHs since the interaction takes place only at the interface [20]. This configuration allows the use of an optimized half-cell reaction since the Nernstian potential is pH-dependent. Namely, it is possible to couple an anodic reaction in an alkaline medium to a cathodic reaction in an acidic medium to enlarge the open-circuit voltage (OCV) of the cell. Therefore these miniaturized fuel cells make biomass-derived suitable for energy conversion, even producing more energy than conventional fuel cells [11]. This concept is promising for low-power devices whether used as a single system [21] or even to power regular macroscale systems whether properly scaled.

The mechanics of μFCs is different from a conventional macrosystem. A mathematical comprehension of these systems is found elsewhere [8], here we lead the reader to the basics. The most important dynamics are a result of the virtual separator of the streams, which requires laminar flow to maintain the ionic transfer steady. The laminar flow regime is characterized by a low Reynolds' number ($R_e < 100$). At this condition, the two streams flow parallel building a colaminar flow [8]. Reynolds' number (R_e) depends on the fluid density (ρ), average velocity (U), hydraulic diameter (D_h), and dynamic viscosity (μ) as depicted in Eq. 3.5. Hence the inertial liquid properties as viscosity and liquid/surface interaction play an important role in microfluidics.

$$R_e = \frac{\rho U D_h}{\mu} \quad (3.5)$$

The width between electrodes and flow rates of the streams are key variables for high performance. The residence time (t_r) of the reactants in the microchannel must not exceed the time for the reactants to reach the opposite electrode side, which is called diffusion time (t_d). On the other hand, the t_r must be long enough to allow surface reaction. The t_r can be estimated using Einstein's relation for Brownian diffusion in one dimension [22], as depicted in Eq. 3.6.

$$t_r = \frac{L}{U} < \frac{W^2}{2D} = \overline{t_d} \quad (3.6)$$

Where L is the channel length, W is the channel width, U is the mean velocity, and D is the diffusion coefficient. The mean velocity is calculated by dividing the flow rate (Q) by both, channel height (H) and width (W). μFCs also depend on Péclet numbers (P_e), calculated as indicated in Eq. 3.7.

$$P_e = \frac{UH}{D} \quad (3.7)$$

The microfluidic systems experience high Pe, showing that the convective flow rate is higher than the diffusive transversal mass transfer [8]. μFCs present a diffusive mixing of reactants only at the thin interface between the streams, with a maximum width (δ_x), as describes Eq. 3.8, for liquids with similar density. Where z is the downstream position [23].

$$\delta_x \propto \left(\frac{DHz}{U}\right)^{1/3} \quad (3.8)$$

Finally, the main advantage of μFCs is being membraneless but there is still ohmic resistance for ionic transport at the interface of the streams (R_{ohm}) [8]. Therefore the architecture of the cell along with the inertial property of ionic conductivity (σ) must be taken into account, as shown in Eq. 3.9. Where d_{elec} is the distance between the electrodes and A_{cs} is the cross-sectional area for charge transfer.

$$R_{ohm} = \frac{d_{elec}}{\sigma \; A_{cs}} \quad (3.9)$$

It seems clear how important is rational engineering of the μFCs, which applies for μDAFCs. Some examples and recommendations are provided along with the equations in the book of E. Kjeang [8]. For instance, a proper cell design in terms of Péclet numbers could be found for a (0.5 × 0.5 × 10) mm³ channel using ~1 μL min⁻¹ for species with a diffusion coefficient of ~10^{-10} m² s⁻¹ [8]. The author also recommended flow rates in the order of μL to mL min⁻¹ as an adequate flow rate window considering the microdynamics [8].

3.3 General view of direct alcohol microfluidic fuel cells

Different research groups have been investigating μDAFCs fed by methanol [16,24–42], ethanol [14,15,43–51], ethylene glycol [52–55], and glycerol [1,11,18,19,56–62]. These works show different cell configurations, electrolytes, oxidants, and thermodynamic properties, but they are all based on the alcohol electrooxidation coupled to an oxidant electroreduction, as illustrated in Fig. 3.2. The easy handling macroparameters shown in the previous section are pivotal to engineer an efficient μDAFC. Furthermore, the catalysts are responsible for the surface reactions. The most common catalyst is Pt/C NPs, which lack ability in C-C cleavage [6,63,64], decreasing the efficiency of the ethanol, ethylene glycol, and glycerol electrooxidation reactions. The limited electron harvest is translated in low current and low power densities accordingly. Moreover, Pt is well known as a powerful adsorbent, which is a disadvantage, since CO is an intermediate of alcohol electrooxidation, and is not electrooxidized (removed) in low potentials, becoming a poisoning reactant. These drawbacks unleashed a wide research field in nanotechnology for miniaturized energy converters, aiming to synthesize low-cost and efficient anodes. It is worth noticing that the same applies to the cathode side.

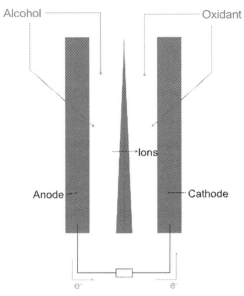

Figure 3.2 Illustrative cross-section of a microchannel showing the interdiffusion zone at the interface of the liquid, featuring electrodes in flow-by or flow-through configuration, as indicated by the red/right and blue/left arrows.

Although this work is not a schematic revision, a representative number of works were collected from the literature and used as a sample of the ongoing researches, as shown in the Appendix. Fig. 3.3 shows a percentual division of μDAFC based on the type of catalyst. Firstly, it is clear that the number of papers regarding microfluidic direct methanol fuel cells (μDMFCs) is bigger than the others, which is the natural order of events since methanol is the simplest among these alcohols, containing only one carbon and then requires less energy to be electrooxidized. Microfluidic direct ethanol fuel cells (μDEtFCs) appear as the second, with a close number of papers to microfluidic direct glycerol fuel cells (μDGFCs). The reason for the interest in ethanol lies in the fact that such alcohol is not toxic neither volatile, different from methanol. Moreover, tropical countries have installed gas stations with ethanol fuel available, making the distribution easy and less expensive. The interest in glycerol is based on the worldwide biodiesel production, which led to a massive glycerol production since the alcohol is a 10% byproduct of the biodiesel fabrication [65]. Therefore glycerol can transit from a "waste" with a very low market price to become a fuel (and a matrix for electrosynthesis [65]), which increases interest and aggregate value to both biodiesel and glycerol. Among these biomass-derived alcohols, ethylene glycol (from microfluidic ethylene glycol fuel cells, μDEgFCs) is the less studied so far, mainly due to the lack of immediate environmental and economic interest; however, it might turn into a powerful fuel in the future because both carbons from the molecule are hydrated, while ethanol contains only one –COH group.

Alcohol fuel cell on-a-chip 47

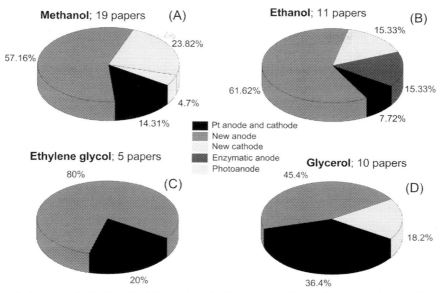

Figure 3.3 Percentual division for (A) methanol, (B) ethanol, (C) ethylene glycol, and (D) glycerol fed microfluidic fuel cells based on the type of anode and cathode. Some works report both new anode and cathode. The number of peer-reviewed papers is indicated in the figure. It is worth noticing that this is a qualitative point of view from a significant sample of the literature, and a schematic revision and statistics were not performed.

In general, the μDAFCs used in this sample of the literature reported the development of new nanomaterials to be used as anodes in the majority, achieving more than 50% of the works regarding methanol, ethanol, and ethylene glycol, as shown in Fig. 3.3A−C. This evidences the demand for new materials and the lack of interest in a new configuration and experimental conditions, which are properly investigated using known commercial NPs, such as Pt/C NPs. The mechanics and chemistry of a μDAFCs can be improved, suggesting that easy handling parameters may build cells more powerful than cells with complex nanomaterials [11,66]. Hence researchers and industrials should gather efforts on both new materials and improved configuration investigation to faster the development in the field. Regarding μDGFC, the works about new anodes are less than 50% and the works using Pt NPs on both catalysts are 36.4%, which suggests an impending interest in architecture, chemistry, and physical parameters more than new nanomaterials.

Most of the 57.2% of the works dealing μDMFC using new anodes (Fig. 3.3A) reports PtRu as the candidate catalyst. The macroanalysis is dubious because it is not possible to know the motivation of the researchers, but I risk to hypothesize that is a consequence of the development and application of PtRu in a conventional fuel cell since most of these authors used commercial PtRu catalysts. The "ripeness" of the technology also led to 23.8% of works, involving new cathodes (e.g., refs. [33,41]), assuming PtRu or Pt NPs as anodes. Also, a photoanode appears to be a candidate of an anode, where the photoexcited material improves electron transfer [25]. Most of the works regarding μDEtFC (Fig. 3.3B) and μDEgFC (Fig. 3.3C) also perform measurements involving anodes different from Pt/C, few shown cells operating with Pt/C as anode and cathode, and some works suggested the use of an enzyme to selective electrooxidize ethanol [44,46]. Only a few papers are found for μDEgFC in this sample of the literature and it seems the investigation of new cathodes is barely found (Fig. 3.3C). The μDGFC technologies are more recent than the others, with the first report in 2013 [61], at least under my knowledge. Hence there are still research groups investigating cells with commercial Pt/C as anode and cathode to optimize performance without adding new and complex material—36.4% of the works (Fig. 3.3D). As expected the focus of the majority of the works is investigating new anodes (45.4%), but almost 20% also investigates new cathodes.

Another major classification can be set for μDAFCs fed by oxygen, oxygen from the air, and new oxidants, shown in Fig. 3.4 for the sample literature sample (Appendix). Oxygen reduction is usually the oxidant cathodic reaction that accepts the electrons harvested at the anode side, along with the ions (e.g., H^+) coming through the stream's interface. The miniaturized systems boosted the anodic reaction by using laminar flow and porous electrode;

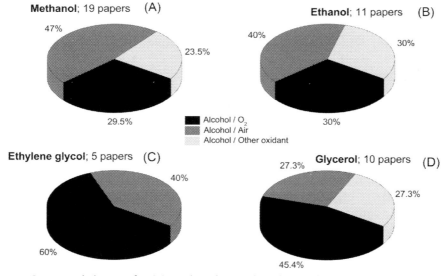

Figure 3.4 Percentual division for (A) methanol, (B) ethanol, (C) ethylene glycol, and (D) glycerol fed microfluidic fuel cells based on the type of oxidant. The number of peer-reviewed papers is indicated in the figure. It is worth noticing that this is a qualitative point of view from a significant sample of the literature, and a schematic revision and statistics were not performed.

therefore extracting high current density from alcohol, eventually achieving 6–14 e$^-$ per molecule of fuel (Eqs. 3.1–3.4). On the cathodic side of the cell, O$_2$ can be reduced to water, but especially on Pt surfaces, the oxidant is reduced to H$_2$O$_2$ consuming 2e$^-$ per molecule (E$_R$° = 0.695 V), shown in Eq. 3.10. Once turned into H$_2$O$_2$, the energy required to convert it into H$_2$O is high, rationalized as E$_R$° = 1.776 V, finally extracting two more electrons (Eq. 3.11). Some catalysts led to a direct conversion of O$_2$ into H$_2$O, involving 4e$^-$ (Eq. 3.12) that requires a theoretical reversible potential of 1.229 V. Thus a complete (or efficient) alcohol electrooxidation may anticipate oxygen starvation—when the cathodic reaction is not enough to collect the electrons and to complete the overall reaction efficiently. This evidences the demand for efficient cathodes and/or alternative oxidants. Among the µDMFCs, only ~30% of the works reported O$_2$ dissolved in the electrolyte as catholyte (Fig. 3.4A). Almost half of the works reported air-breathing cathodes. As commented in Fig. 3.1B (detailed in ref. [13]), the oxygen from the air in addition to O$_2$-dissolved increases the cathodic reaction, making the cathode suitable for µDAFCs [16,30,33,40,42,58]. Such technology was also applied for µDEtFCs [15,45,51,58] (~40%, Fig. 3.4B), µDEgFCs [54,58] (~40%, Fig. 3.4C), and µDGFCs (~30%, Fig. 3.4D).

$$O_2 + 2H^+ + 2e^- \rightarrow H_2O_2 \qquad (3.10)$$

$$H_2O_2 + 2H^+ + 2e^- \rightarrow 2H_2O \qquad (3.11)$$

$$O_2 + 4H^+ + 4e^- \rightarrow 2H_2O \qquad (3.12)$$

Besides the powerful strategy of air-breathing electrodes, some researchers used liquid oxidants to boost the cathodic reaction. This type of investigation appeared in a great part of work dealing with methanol, ethanol, and glycerol (Fig. 3.4). For instance, Arun et al. developed a methanol/perborate µFC [32]. The authors claimed that sodium perborate (NaBO$_3$ · 4H$_2$O) is cheap, does not cause environmental impact, is nontoxic, and largely available [32]. They used sodium perborate dissolved in H$_2$SO$_4$ since perborate in its peroxo salt anionic form is a source of H$_2$O$_2$ in acidic solution [32], as shown in Eq. 3.13. Priya et al. suggested percarbonate as an oxidant for µDEtFCs [43]. This interesting strategy uses another peroxo salt (2Na$_2$CO$_3$·3H$_2$O$_2$) to provide H$_2$O$_2$ for the catholyte (Eq. 3.14). This oxidant is also widely available, low cost, and environmental-friendly [43]. The direct use of H$_2$O$_2$ has also been reported [36,39].

$$[B(OH)_3(O_2H)]^- + H^+ \rightleftharpoons [B(OH)_3] + H_2O_2 \qquad (3.13)$$

$$2N_2CO_3 \cdot 3H_2O_2 \rightleftharpoons 2Na_2CO_3 + 3H_2O_2 \qquad (3.14)$$

Another powerful liquid oxidant firstly applied in formate all-alkaline cells is hypochlorite (ClO⁻), which can be found in a cheap bleach solution [67]. Martins et al. used bleach as a source of ClO⁻ in alkaline media and as HClO in acidic media for μDGFCs [11]. Although the reactions involve few electrons (Eqs. 3.15–3.17), the liquid oxidant worked properly, avoiding oxidant starvation [11]. Guima et al. used a similar strategy in another investigation [56].

$$ClO^- + H_2O + 2e^- \rightleftharpoons Cl^- + 2OH^- \quad (3.15)$$

$$HClO + H^+ + 2e^- \rightleftharpoons Cl^- + H_2O \quad (3.16)$$

$$HClO + H^+ + e^- \rightleftharpoons 1/2 Cl_2 + H_2O \quad (3.17)$$

The few numbers of papers and the potential applicabilities of these technologies make clear that much work is still required to improve performance in terms of catalysts. This research field still demands innovation in terms of catalysts and configuration, to open new classes of μDAFCs.

3.4 Testing direct alcohol microfluidic fuel cells

The most common performance parameters of fuel cells are obtained from a polarization curve that is the cell voltage (in V) versus current density (in mA cm^{-2}), as seen in the previous chapters. The cross-sectional area of the anode is usually used for current normalization in μFCs testing. The polarization curve is usually built from the potential equivalent to the OCV until the short cut voltage (or close to that value). The OCV is the widest cell potential and can be predicted by independent half-cell reactions. Although the μFC testing shows forced-convection mass-transport, stationary half-cell measurements enlight important preliminary interpretation. Caneppele and Martins summarize the experimental steps researchers used to follow to predict the OCV before fuel cell testings [68]. After synthesized, the NPs are ex situ characterized by using physical-chemical techniques. At this point, it is important to mention that the characterized nanomaterials are not necessarily the same material will work as catalysts in the μFC—this is because the NPs experience applied potential and current flow, which may lead to a modification in surface crystallography and surface chemical composition. Hence a good method for in situ characterization is using electrochemical methods [69]. For that the NPs are immobilized on an inert electrode and submitted to electrochemical measurements (e.g., voltammetry) as shown in Fig. 3.5. The voltammogram in the presence of the fuel or the oxidant provides the onset potential (E$_{onset}$) that is related to the Gibbs free energy to form an activated complex and allow further electrons transfers, meaning that the required energy to start the reaction was

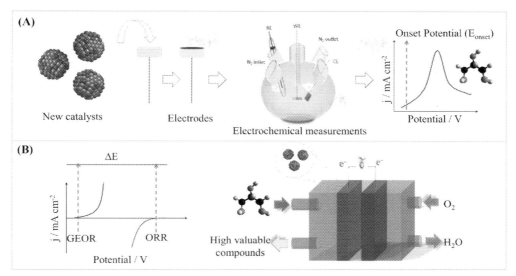

Figure 3.5 Schematic illustration of the processes usually used to test new nanomaterials candidates to be used as catalysts in fuel cells; since (A) half-cell measurements to (B) their applications. *Reprinted from G.L. Caneppele, C.A. Martins, Revisiting glycerol electrooxidation by applying derivative voltammetry: from well-ordered bulk Pt to bimetallic nanoparticles, Journal of Electroanalytical Chemistry 865 (2020) 114139. Copyright (2021), with permission from Elsevier.*

achieved. The difference between E_{onset} from the cathode and the anode gives a theoretical ΔE that may be similar to the OCV (Fig. 3.5).

This set of experiments have been reported in conjunction with the μFC testing (e.g., refs. [11,62]) and assures the reader that the cell is operating properly. Researchers must take into account that the experimental simulation (half-cell and cell testing) are electrolysis tests, where the reaction is a consequence of well-planned experiments. However, the μFC must produce current flow as a consequence of a spontaneous reaction, with negative Gibbs free energy variation ($\Delta G < 0$). Therefore considering $\Delta G = -nFE$, the cell potential (E or ΔE in Fig. 3.5) must be positive, otherwise $\Delta G > 0$; where n is the number of moles of electrons and F is the constant of Faraday. The main point is, whether the authors do not show the half-cell measurements, they must report at least the theoretical $E_R°$. Finally, after built a polarization curve, the profile of the cell can be interpreted.

The product of current density with cell voltage builds the power density curve (mW cm^{-2}) versus current density (mA cm^{-2}), from which the most important parameter is obtained, the maximum power density (P_{max}). P_{max} is the inflection point at the intermediary polarization regime of the cell. This value depends on several variables, such as type of catalysts, temperature, flow rate, conductivity, pH of the electrolytes, reactants concentration, and assembly (single or multiple cells). It is barely possible to rationalize all these variables to normalize the results from the literature for a fair comparison. Hence here I grouped the μDAFCs fed by the different alcohols to see the evolution of maximum power density in time, as shown in Fig. 3.6 (from papers reported in the Appendix).

Lam et al. proposed one of the first membraneless direct methanol fuel cell with conventional size, but replacing the membrane for a gap [66]. Also in 2009, Shen et al. proposed a μFC fed by methanol but using Nafion® strip between the electrodes [39]. In 2012, Gago et al. published a membraneless microfluidic methanol fuel cell built with a Pd anode and PtSn cathode, reaching ∼6 mW cm^{-2} of P$_{max}$ [41]. Along the years the P$_{max}$ of the μDMFCs enhanced exponentially until 2017, with values higher than 200 mW cm^{-2} (Fig. 3.6A). The first breakthrough appeared in 2015, when Wang et al. developed a methanol vapor to feed the anode of the

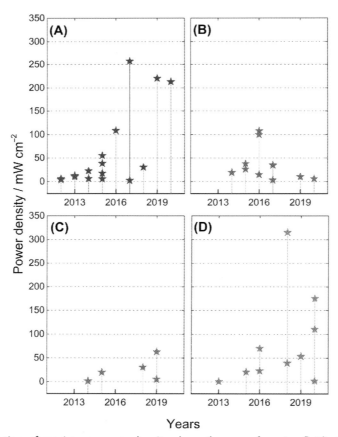

Figure 3.6 Evolution of maximum power density along the years for microfluidic fuel cells fed (A) methanol, (B) ethanol, (C) ethylene glycol, and (D) glycerol.

μDMFC, using PtRu as anode at room temperature, achieving 55.4 mW cm^{-2} [28]. The next benchmark was achieved by Wang and Leung in 2016, using circular stacking μDMFC with PtRu anode to reach 108.7 mW cm^{-2} [24]. In 2017 a new platform of maximum power density was established with the benchmark for a μDMFC to date [38]. Miao et al. developed a microtubular system containing rolled-up microtubes (RUMTs) anodes made of Pt/RuO$_2$, capable of producing ~257 mW cm^{-2} [38]. In 2019 Kwok et al. built a μDMFC fed by methanol vapor and hydrogen with additional solar energy-driven force (three suns) to reach remarkable ~220 mW cm^{-2} [25]. In 2020 a similar strategy was used for the same group to reach 213 mW cm^{-2} [30] (Fig. 3.6A).

It is important to note that the complexity and stability of the systems must be considered in addition to the values of power density. For instance, simple devices with an average power density (compared to the literature) or Pt-free-catalyst-systems are also important even though they do not reach the benchmark values (> 200 mW cm^{-2}) since the cost-effectiveness might be of industrial interest. In 2018 Martins et al. built a methanol/O$_2$ μFC with commercial Pt/C anode and cathode producing 30 mW cm^{-2} simply by using mixed media and flow-through electrode configuration [18]. Moreover, the same system produced similar output power by replacing methanol with ethylene glycol or glycerol [18]. Other reports are important due to partial Pt replacement, such as the work of Arun et al. that used RuNi to replace 50% of the Pt content at the anode side, and still yielded 38.2 mW cm^{-2} [32]. Therefore none parameter can be used alone to choose one system instead of another.

Ethanol electrooxidation faces additional challenges regarding the alcohol electrooxidation, which are the C—C bond and the hydrocarbon. Ethanol is known to electrooxidize mainly to acetic acid in room temperature, which limits electron harvest [5]. Therefore despite the effort to synthesize efficient nanocatalysts (Fig. 3.3B), the output power of μDEtFCs are lower than those from μDMFC, as seen in Fig. 3.6B. In general, the maximum power densities are lower than 50 mW cm^{-2} [43–46,48–51]. López-Rico et al. in two independent works reported the benchmark power density for cells fed by ethanol [15,47]. The authors reported an air-breathing cell built with Pd catalysts on the anode and cathode to produce ~100 mW cm^{-2} [15]. In another breakthrough, the authors designed a mixed-media μDEtFCs with ethanol in KOH feeding a Pd-NiO anode and O$_2$ in H$_2$SO$_4$ feeding a Pt/C cathode to produce 108 mW cm^{-2} [47].

Among the alcohols described here, ethylene glycol arouses less interest so far. Among the few reports, the maximum power densities range from 1.6–5.2 [52,54] to ~20 mW cm^{-2} [58] (Fig. 3.6C), even using complex nanomaterials as anodes. Martins et al. found ~30 mW cm^{-2} as maximum power density using commercial Pt/C as anode and cathode by improving performance in a mixed-media μDEgFC [18]. The benchmark of ~63 mW cm^{-2} was found by Kumar et al. in an all-alkaline μDEgFC [53]. The authors used a new flexible carbon aerogel made of carbon nanotubes (CNTs) with shell-confined Pd-Ni NPs, while Pt/C was used at the cathode. Moreover, their μDEgFC showed stable performance during 60 hours operation [53].

Most of the μDGFC showed power densities ranging 0.3 [61] to 39 mW cm^{-2} [18], as seen in Fig. 3.6D. The use of core-shell Cu@Pd anodes was explored in air-breathing μDGFC, achieving ~20−23 mW cm^{-2} [58,59]. The first system to break the 50 mW cm^{-2} line was reported by Yu et al. in 2016 [57]. These authors used a complex Pt-free cathode made of NiCo$_2$O$_4$/N-graphene and a PtRu/C anode in a glycerol/O$_2$ system to produce >70 mW cm^{-2} at 75ºC [57]. Working at room temperature, Martins et al. used Pt/C as anode and cathode in a mixed-media glycerol/O$_2$ cell to produce 39.5 mW cm^{-2} [18], which was improved to 53.6 mW cm^{-2} by in situ modifying the anode with Fe (Fe/Pt/C/CP anodes) [19]. The benchmark power density of all μFC to date (Fig. 3.6D) was achieved by Martins et al. in 2018 [11]. They allied electrodes with flow-through configuration and mixed-media with a liquid oxidant to produce 315.3 mW cm^{-2} maximum power density [11], which is higher than the previous microtubular cell fed methanol aforementioned [38] and conventional glycerol fuel cell, as discussed by the authors [11]. The flow-through configuration increase reactants collision on the Pt/C/CP electrodes, harvesting electron from glycerol in alkaline, which efficiently reduced HClO from bleach at the cathode side [11]. After this work, the use of HClO as a liquid oxidant in μDGFC has been explored for high power production [56,62]. Guima et al. 3D-printed a μFC and fed the porous Pt/C/CP anode and cathode with glycerol in KOH and HClO, respectively, to produce 175.2 mW cm^{-2} with the first all 3D-printed μFC [56]. Another application in PDMS-cell was made by Martins et al. [62]. These authors in situ modified a CP cathode with graphene oxides (GOs) dispersion to build a metal-free GO/CP cathode. Then the GO/CP metal free cathode with Pt/C anode were fed with HClO and glycerol in alkaline media, respectively, to produce remarkable 110 mW cm^{-2} [62]. This work opened up the possibility of using glycerol and liquid oxidant with metal-free cathode for high performance.

3.5 Nanotechnology in direct alcohol microfluidic fuel cells

At this point, it is clear that the synthesis of new nanomaterials is the key to improve performance whether well designed in conjunction with proper chemistry and engineering of the μFCs. The nanomaterials must be well-characterized, the results must be statistically relevant, and the cost-effectiveness of their production must be worth synthesizing. Filling the literature with complex and time-consuming synthesized nanomaterials with few or none use may confuse researchers about the real aim that should be the transition toward environmental-friendly and powerful systems. A simple but effective nanomaterial should be preferable instead of a complex and nonefficient one. In this context, I comment on some of the nanomaterials that contribute to the knowledge in fundamental surface reaction, to improve performance and/or to reduce the total cost of the systems.

3.5.1 Nanoparticle anodes and cathodes for methanol microfluidic fuel cells

The development and investigation of PtRu-based catalysts for methanol electrooxidation reaction [70−72] led to efficient μDMFC equipped with active and stable commercial PtRu/C anodes. In general, more attention has been spent on the engineering of the system than on the nanomaterials, which rendered significant power density improvements. Therefore researchers have been interested in using commercial PtRu for μDMFC more than synthesizing new materials. However, decreasing the total cost of the devices by partially or completely replacing noble-metals is still worth addressing.

Arun et al. reported the synthesis and characterization of PtRu, PtNi, and PtRuNi dispersed on carbon for the methanol electrooxidation [32]. These authors found that the 3−5 nm $Pt_{50}Ru_{40}Ni_{10}$/C NPs enhanced the reaction in 1.0 mol L^{-1} in an acidic medium. Even though it is still a challenge to rationalize the reason for catalytic improvement, researchers used to summarize as crystallographic and/or electronic effects. Arun et al. attributed the methanol electrooxidation enhancement to the electronic effect on the Pt [32]. Finally, a single all-acid μDMFC built by $Pt_{50}Ru_{40}Ni_{10}$/C (anode) and Pt/C (cathode) fed with methanol and perborate produced 38.2 mW cm^{-2} [32], which is high power considering the low Pt content.

Kwok et al. also gave an important contribution by using advanced NPs dispersed on advanced carbon materials [40]. They synthesized Ru@Pt dispersed on graphene CNT composite aerogel, which was used as an anode in flow-through μDMFC, leading to high performance [40]. In another work, Kowk et al. addressed the development of anode with advanced carbon supports by synthesizing ultrafine Pt NPs dispersed on graphene aerogel [42]. The NPs were 1.5−3.0 nm and well-dispersed on the GOs sheets, leading to >350% increase in specific power density (in terms of mW mg_{Pt}^{-1}) compared with Pt/C [42]. Huo et al. also addressed the advanced carbon supports [36]. These authors found an >300% area augment by using CNTs with Pt as an anode in μDMFC, improving power density [36].

Miao et al. developed a pioneer μDMFC working on nanomaterials and engineer, which rendered the benchmark power density of 257 mW cm^{-2} [38]. Using what the authors called RUMTs nanotechnology, they built a Pt-coated RuO_2 nanomembrane. Using steps of deposition of the double layer RuO_2 on a GeO_2 sacrificial layer and removal, they controlled the diameter of the RUMTs by balancing the strain and membrane thickness. Other steps are cycles of lithography/electron beam to prepare GeO_2, Au, bilayers of Pt-coated RuO_2, and Au outer-wall and inner-wall electrodes.

More information is found in ref. [38]. Finally, the all-acid μDMFC was fed with methanol and O_2 to produce power.

μDMFC experiences methanol crossover, when the fuel crosses the colaminar flow toward the cathode. Therefore methanol tolerant cathodes are required to surpass this issue [33,41]. Gago et al. found that PtS cathode shows only 35% decrease in power density in the presence of methanol (mixed reactants) against 80% for a Pt cathode [41]. Most important, the authors found no changes in performance using $CoSe_2$ cathode [41]. In a similar approach, Ma et al. investigated PtTi and PtSe in air-breathing μDMFC [33]. These authors found a performance loss of 55% and 20% for PtTi and PtSe cathodes respectively in mixed-reactants conditions [33]. These findings opened up the possibility of using single flow configuration in μDMFC.

3.5.2 Nanoparticle anodes and cathodes for ethanol microfluidic fuel cells

Differently from methanol, ethanol has been investigated in several types of complex NP anodes. In 2015 Ponmani et al. investigated binary and ternary electrocatalysts based on Ni, Sn, and Pt for μDEtFC [48], with the highest performance of 37.33 mW cm^{-2} found for $Pt_{80}Sn_{10}Ni_{10}$/C [48]. In 2016 Armenta-González et al. meticulously investigated PdAg/multiwalled carbon nanotubes (MWCNTs) for ethanol electrooxidation by half-cell measurements and in μDEtFC [45]. Although the performance is still lower than previous work [48], they reported the use of a Pt-free anode, which is imperative to reduce the total cost of the cells [45]. These authors used the spray-pyrolisis method to synthesize 4−7 nm PdAg NPs well-dispersed on MWCNTs, with improved electrocatalytic properties, such as onset potential and current density, leading to high cell performance when coupled to oxygen reduction reaction [45]. López-Rico et al. found one of the highest performances reported (99.66 mW cm^{-2}) for a μDEtFC using Pd anodes [15]. Besides the activity of the anode, the authors also attributed such high performance to a possible hydrogen evolution at the cathode side.

Still searching for Pt-free anodes, López-Rico et al. investigated Pd-NiO in acrylic and paper-based μDEtFC [47]. The Pd-NiO was carried out at room temperature and sonication. The authors presented improvement on the activity of the Pd-NiO for ethanol electrooxidation compared with Pd in terms of mA mg^{-1} cm^{-2}. Interestingly, the performance of the all-alkaline paper-based μDEtFC was ~twice higher than that found for the acrylic cell. The authors attributed such enhancement to the recorded open circuit and decrease in ohmic losses. Using mixed-media—alkaline ethanol electrooxidation and acidic oxygen reduction allied with hydrogen reduction—in the acrylic system, these authors reported the benchmark maximum power density of 108 mW cm^{-2} [47]. Fig. 3.7 illustrates

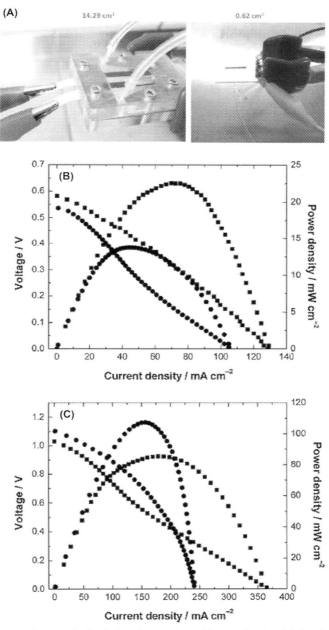

Figure 3.7 (A) Volume of an acrylic (circles) and paper-based (squares) microfluidic ethanol fuel cells and performances in (B) all-alkaline and (C) mixed-media configurations using Pd-NiO/C anode. *Reprinted from C.A. López-Rico, J. Galindo-de-la-Rosa, E. Ortiz-Ortega, L. Álvarez-Contreras, J. Ledesma-García, M. Guerra-Balcázar, et al., High performance of ethanol co-laminar flow fuel cells based on acrylic, paper and Pd-NiO as anodic catalyst, Electrochimica Acta 207 (2016) 164–176. Copyright (2021), with permission from Elsevier.*

their low-volume acrylic and paper-based μDEtFC and summarizes their findings in terms of cell performance [47].

Priya et al. investigated ternary PtSnCe [50] and PtSnW [49] both dispersed on mesoporous carbon (MC). These authors allied the complex anodes to the reduction of perborate on Pt/MC cathodes to reach high performance. They found 34.18 mW cm^{-2} using Pt$_{80}$Sn$_{15}$W$_{05}$/MC [49] and 34.8 mW cm^{-2} with PtSnCe/MC [50]. Although the electronic or crystallographic effects are not yet rationalized, in one of the works [50], the authors argued that the improved activity might be a result of the incorporation of Sn into the Pt crystalline structure and formation of SnO$_2$ phase.

Besides metallic electrocatalysts, enzymatic anodes have been investigated as bionanodes in μDEtFC due to their high specific reactions. Galindo-de-la-Rosa et al. built a multilayer anode based on alcohol dehydrogenase, poly(methylene blue)/tetrabutylammonium bromide/Nafion®, and glutaraldehyde on carbon paper [46]. They assembled such an anode to a Pt/C cathode to work with simulated body fluids, human serum, and blood enriched with ethanol. The optimized system rendered 1.035 V of OCV and 3.154 mW cm^{-2} of power density [46]. In another work, Galindo-de-la-Rosa et al. used alcohol dehydrogenase immobilized on TiO$_2$ nanotubes for a μDEtFC in the absence and presence of UV-light [44]. The authors prepared a catalytic ink containing the enzyme, NAD$^+$, tetrabutylammonium bromide, and Nafion® in phosphate buffer solution (PBS); afterward, they immersed the TiO$_2$ nanotubes to prepare the catalyst. The anode was coupled to a Pt/C cathode and evaluated in N$_2$-saturated ethanol in PBS as fuel and O$_2$-saturated KOH solution as catholyte [44]. Fig. 3.8 shows the performance of the system and illustrates a possible mechanism. The irradiation of UV-light promotes electrons from the valence band of the TiO$_2$ toward the conduction band, forming electron(e$^-$)/holes(h$^+$) pairs. The h$^+$ oxidizes ethanol, harvesting electrons continuously. The alcohol dehydrogenase also may oxidize ethanol, maximizing performance [44]. Although more investigation is still required, the use of photo bioanodes may be an alternative to avoid noble-metal and enhance selective reactions.

Some effort has also been done to develop ethanol-tolerant cathode but is still underexplored. Estrada-Solís et al. synthesized AgPt by pulsed laser deposition and used this material as an ethanol-tolerant cathode in a μDEtFC [51]. These authors investigated oxygen reduction reaction using O$_2$-saturated in alkaline solution for different ethanol concentrations and found that the current density decreases 40% but the onset potential is approximately constant up to 2 mol L^{-1}. The μDEtFC with the AgPt cathode in mixed-reactants condition proves efficient, achieving 0.75 V of OCV and 10 mW cm^{-2} of maximum power density [51].

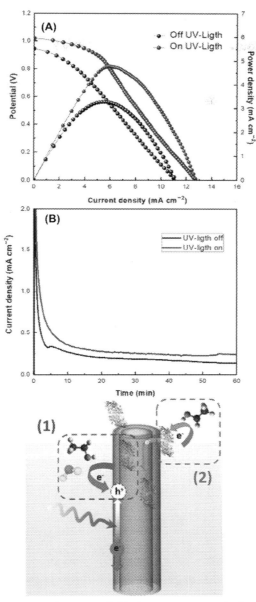

Figure 3.8 (A) Polarization, power density, and (B) chronoamperometric (at 0.02 V) curves at the dark and UV irradiation of a microfluidic ethanol fuel cell equipped with ADH-TNT bioanode and Pt/C cathode. Featuring ethanol electrooxidation mechanism (1) under UV-light irradiation via h$^+$ and (2) catalyzed by the enzyme (*ADH-TNT*, Alcohol dehydrogenase enzyme-TiO$_2$ nanotube). *Reprinted with permission from J. Galindo-de-la-Rosa, A. Álvarez, M.P. Gurrola, J.A. Rodríguez-Morales, G. Oza, L.G. Arriaga, et al., Alcohol dehydrogenase immobilized on TiO2 nanotubes for ethanol microfluidic fuel cells, ACS Sustainable Chemistry & Engineering 8 (29) (2020) 10900–10910. Copyright (2021) American Chemical Society.*

3.5.3 Nanoparticle anodes and cathodes for ethylene glycol microfluidic fuel cells

Although the use of ethylene glycol in μDEgFC is recent with few works, relevant achievements have been reported in terms of nanotechnology. Arjona et al. synthesized semi-spherical AuPd particles dispersed on polyaniline for ethylene glycol electrooxidation [52]. This catalyst was used as the anode in an all-alkaline μDEgFC, but power is still far lower (1.5–1.8 mW cm^{-2} [52]) than μFCs fed by other alcohols (Fig. 3.6), which is nonattractive for a noble-metal anode. In 2019 López-Coronel et al. made the first assessments on the influence of well-ordered NPs on cell performance [54]. These authors synthesized Pd nanocubes and PdAg NPs with a high density of defects (100). Fig. 3.9 shows detailed micrographs of these materials. The half-cell measurements revealed that both well-ordered Pd and PdAg are active for ethylene glycol electrooxidation in 1 mol L^{-1} KOH, and the current density increases by increasing the concentration of the alcohol from 0.1 to 1.0 mol L^{-1}. Moreover, the binary catalyst shifts the onset potential toward lower values, facilitating the reaction [54]. When these anodes were applied in a mixed-media μDEgFC with alkaline anodic and acidic cathodic reaction, the power density reached remarkable 14.44 and 9.66 mW cm^{-2} for Pd nanocubes and well-ordered PdAg, respectively [54]. The authors argued that protons may undergo the colaminar channel reaching the anode and dissolving Ag, which is a hypothesis for the inversion of performance between half-cell and cell [54]. Besides the authors' hypothesis, here it is important to highlight that the diffusion-controlled half-cell measurements provide interesting predictions, but may not match the reality of the forced-convection mass transport of a cell. This work offers the understanding of the surface reaction been influenced by the material—even though it is in terms of electrocatalytic properties. This type of work is imperative to design smart and applicable advanced materials.

Kumar et al. reported flexible carbon aerogels made by N- and S-doped carbon nanotubes (NSCNTs) shell with the core composed of PdNi, synthesized by chemical vapor deposition and used as an anode for ethylene glycol electrooxidation [53]. The synthesis involves steps of fixing metal ions onto raw cotton fibers, reduction of the ions via a hydrothermal path in presence of thiourea to form conductive carbon fiber aerogels [53]. The organized advanced material is composed of the metallic bimetallic core with the NSCNTs shell. This outstanding nanodesigned material was used as the anode while Pt/C/CP was used as the cathode in a μDEgFC to produce the highest power density reported to date, 62.8 mW cm^{-2}.

Despite the few but relevant works regarding new anodes, the literature lacks reports about the advanced materials used as cathodes for μDEgFC. Since ethylene glycol is still underexplored, this new research field requires more studies in the designing of new anodes and cathodes.

Alcohol fuel cell on-a-chip 61

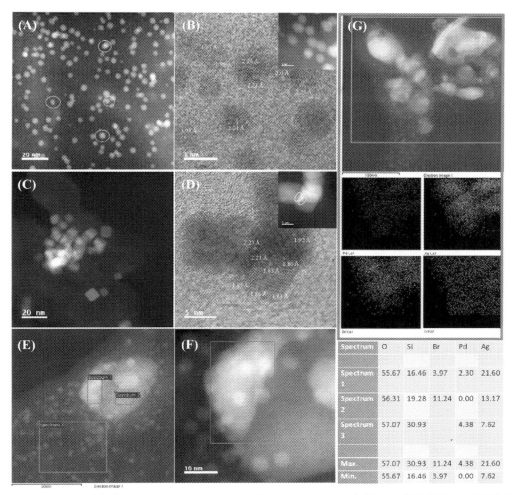

Figure 3.9 Dark-field high resolution transmission electron microscopy (HRTEM) of (A) PdAg and (C) Pd nanocubes. Bright-field HRTEM of (B) PdAg and (D) Pd nanocubes. [(E) and (F)] TEM images of PdAg and (G) EDX elemental mapping of PdAg. (*EDX*, Energy dispersive X-ray; *TEM*, transmission electron microscopy). *Reprinted from A. López-Coronel, E. Ortiz-Ortega, L.J. Torres-Pacheco, M. Guerra-Balcázar, L.G. Arriaga, L. Álvarez-Contreras, et al., High performance of Pd and PdAg with well-defined facets in direct ethylene glycol microfluidic fuel cells, Electrochimica Acta 320 (2019) 134622. Copyright (2021), with permission from Elsevier.*

3.5.4 Nanoparticle anodes and cathodes for glycerol microfluidic fuel cells

The increase in biodiesel fabrication boosted the investigation of glycerol electrooxidation, including its use in μDGFC. In 2013 Dector et al. built μDGFCs with Pd/C and Pd/MWCNTs as anodes [61]. This initial system produced low power density in the order of 0.7 mW cm^{-2} [61]. In 2015 Maya-Conejo et al. published a pioneer work regarding the use of several alcohols, including glycerol, in an air-breathing μDAFC [58]. These authors equipped the μDAFC with Cu@Pd core-shell anode. This work proved that it is possible to work with different fuels in the same cell [58]. The same team explored the use of Cu@Pd and Cu@Pt for μDGFC fed by crude glycerol [59]. Both copper-based materials worked properly using crude glycerol to produce power in the order of ∼17−23 mW cm^{-2} [59]. This work showed the direct use of glycerol byproduct of biodiesel fabrication into an energy conversion. At least under my concern, the other works from the literature regarding μDGFC report the use of pure glycerol acquired as a chemical reactant.

Martins et al. used electrodes in flow-through configuration and mixed-media for a series of papers [11,18,19,62]. Firstly, they used Pt/C/CP porous anode and cathode in a glycerol/O$_2$ μFC [18]. They found 39.5 mW cm^{-2} and 1.23 V of OCV using 0.05 mol L^{-1} glycerol, which is just slightly higher than the performance found when the same system was fed by 3.0 mol L^{-1} methanol (30.0 mW cm^{-2}) and 1.5 mol L^{-1} ethylene glycol (30.3 mW cm^{-2}); they built a fuel-flexible μDAFC [18]. Next, the authors developed a method for in situ modification of μFC [19]. They applied a simple idea firstly used in flow injection analysis for analytical applications [73], where one syringe pumps metallic precursors (containing cations M$^+$) through one electrode under applied potential and the other syringe pumps water to establish the colamir flow. The applied potential reduces M$^+$ to M^0 on the Pt/C previously deposited on carbon paper. As a proof-of-concept, Martins et al. decorated Pt/C/CP anodes with different amounts of Fe and used it for glycerol/O$_2$ μFC [19]. Thus the technique allows controlling the amount of metal deposited by calculating the charge involved in the reduction reaction, as illustrated in Fig. 3.10. As can be seen, the faster the flow rate, the lower the residence time; therefore the charge involved in the Fe reduction on the Pt/C is lower. Using this technique, the authors added very small amounts of Fe at the anode active sites, ranging from 8.6 to 64.2 ng [19]. Most interesting, the electrode is modified exactly where it is needed, increasing the fuel/catalyst interaction, which led to 53.6 mW cm^{-2} of maximum power density [19].

These previous papers proved the potential of using glycerol as fuel to produce high power, and the same team boosted the performance toward the benchmark power density of all single μFCs fed by alcohol using HClO (from bleach) as liquid oxidant [11]. Using bleach as a source of ClO$^-$ (in alkaline medium) or HClO (in acidic medium), Martins et al. built a flow-through μDGFC with Pt/C/CP as anode and cathode. The found 1.0 V of OCV and a remarkable 71.2 mW cm^{-2} of maximum power density for an all-alkaline μDGFC. Finally, they used glycerol electrooxidation in alkaline medium and HClO reduction to reach 1.97 V of OCV and 315.3 mW cm^{-2} [11]. Fig. 3.11 shows the polarization and power density curves. This

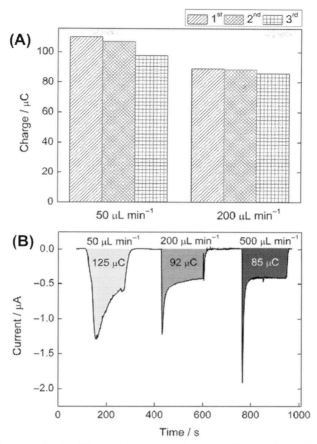

Figure 3.10 (A) Charges obtained from three successive cathodic transients due to the reduction of Fe^{3+} on a Pt/C/CP anode at 50 and 200 µL min^{-1} and (B) the baselines corrected amperograms for the same reduction at the flow rates indicated in the figure. *Reprinted from C.A. Martins, O.A. Ibrahim, P. Pei, E. Kjeang, In situ decoration of metallic catalysts in flow-through electrodes: application of Fe/Pt/C for glycerol oxidation in a microfluidic fuel cell, Electrochimica Acta 305 (2019) 47–55. Copyright (2021), with permission from Elsevier.*

work showed that basic chemistry and engineering should be considered along with the development of new nanomaterials since the benchmark power density was reached with commercial and simple Pt/C/CP electrodes.

Such work [11] inspired Guima et al. to develop the first 3D-printed µFC [56]. These authors used fused deposition modeling as an affordable 3D-printing technique to built µFC with a 0.015 cm^2 cross-sectional area, which showed stable colaminar flow. Fig. 3.12 illustrates the 3D-printed µFC [56]. They used Pt/C/CP as both anode

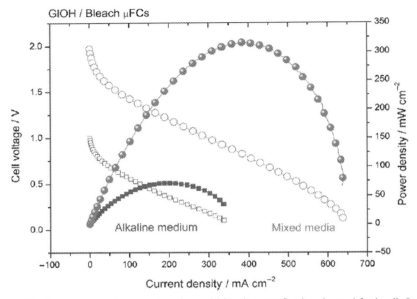

Figure 3.11 All-alkaline and mixed-media glycerol/bleach microfluidic glycerol fuel cell. *Reproduced from C.A. Martins, O.A. Ibrahim, P. Pei, E. Kjeang, "Bleaching" glycerol in a microfluidic fuel cell to produce high power density at minimal cost, Chemical Communications 54 (2018) 192–195 with the permission from The Royal Society of Chemistry.*

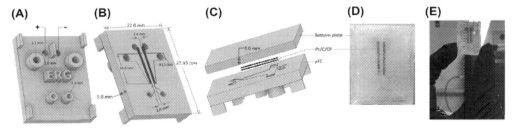

Figure 3.12 Illustrative scheme of the 3D-printed glycerol microfluidic fuel cell. Featuring the [(A) and (B)] microfluidic channel, (C) the assembly profile, (D) real picture of the assembled cell, and (E) a real picture of the cell in comparison to a human hand. *Reproduced from K.-E. Guima, P.-H.L. Coelho, M.A.G. Trindade, C.A. Martins, 3D-printed glycerol microfluidic fuel cell, Lab on a Chip 20 (12) (2020) 2057–2061 with the permission from The Royal Society of Chemistry.*

and cathode and irreversibly sealed the cell using a polyvinyl chloride glue. The cell was fed by glycerol in an alkaline medium and HClO to produce 175 mW cm^{-2} and 1.8 V of OCV [56]. That is the second higher power density reported to date fed by glycerol (Fig. 3.6D), it is lower than that reported by Martins et al. [11]. The authors attributed the lower performance to the hydrophobicity of the polylactic acid (PLA) thermoplastic. Hence the polydimethylsiloxane (PDMS) systems are still better than printed µFC, but the easy, reproducible, and inexpensive 3D-printing technique has risen the attention of the community, since the PLA surface (or other thermoplastic) may be further modified to induce hydrophilicity. Guima et al. have already improved the 3D-printed cell to allow changing of the porous electrodes, without enduringly sealing the cell [74].

Advances have been made to develop new cathodes for µDGFC [57,62]. Yu et al. developed a complex but nonnoble-metal cathode for oxygen reduction reaction coupled to different fuel electrooxidation [57]. They showed that µDGFC built with $NiCo_2O_4$ NPs dispersed on nitrogen-doped graphene as cathode and PtRu/C as anode reached >70 mW cm^{-2} at 75°C. At least to my knowledge, this is one of the first noble-metal-free cathode used in a µDGFC that produces high power density [57].

Martins et al. also made progress looking for metal-free cathodes for µDGFC [62]. For the first time, they found that carbon paper itself is active for HClO (from bleach) reduction reaction; thus they built a PDMS-based flow-through mixed media glycerol/HClO system with Pt/C/CP as anode and CP as the cathode. Such µDGFC showed remarkable 83 mW cm^{-2} at 0.474 and 1.68 V as OCV [62]. This result is a turning point regarding the development of µDGFC, since the CP, which was used just like the electric conductor and support, showed itself as an efficient catalyst for this liquid oxidant. These authors reported at the same work an in situ modification of the cathode, simply by flowing GOs through the CP cathode while water was pumped at the anode to build a colaminar flow [62]. The GO homogeneously wrapped the carbon fibers of the CP as "blankets," as illustrated in Fig. 3.13. As in all in situ techniques, the modification takes place exactly where the reactants will further undergo. Therefore the use of the GO/CP cathode for the HClO reduction reaction in the µDGFC enhanced performance to 110 mW cm^{-2} at 0.506 V and with 1.72 V of OCV. Hence the new challenge of building metal-free (or noble-metal-free) seems to be easily reached whether using liquid oxidants.

3.6 Perspectives and future engagements

Considering that alcohols may be sustainably obtained from biomass and that their complete electrooxidation produces CO_2 that may be recycled into the photosynthesis of their sources, methanol, ethanol, ethylene glycol, and glycerol are economic and environmentally powerful fuels. The increase in application requires future engagement of the field to produce active and stable anodes and cathodes. Most importantly, the *industrial gap* must be fulfilled, meaning that the barrier between laboratory and industry must be surpassed, which is achieved by enduring a more realistic scenario in the laboratory and increasing the interest of the industry by making this technology more affordable. Here I point out my outlook for possible future researches.

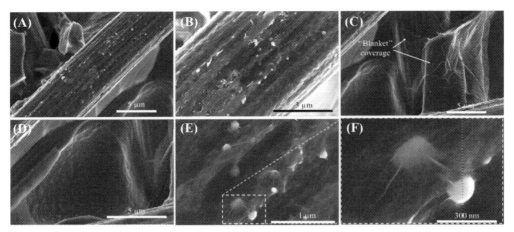

Figure 3.13 (A-F) Micrographs of the "blanket-like" in situ GO-modified carbon paper (*GO*, Graphene oxide). *Reprinted with permission from C.A. Martins, P. Pei, M. Tellis, O.A. Ibrahim, E. Kjeang, Graphene-oxide-modified metal-free cathodes for glycerol/bleach microfluidic fuel cells, ACS Applied Nano Materials 3 (8) (2020) 8286–8293. Copyright (2021) American Chemical Society.*

From a fundamental perspective, the development of active and stable anodes for μDAFCs still requires new nanomaterials. Therefore new bi(multi)metallic or nonmetallic NPs are an open research field, as has been for decades. Moreover, the use of well-ordered NPs for miniaturized systems is still unexplored. However, researchers must be careful in this area since there is no standard protocol for testing these materials in half-cell measurements. As seen in this chapter, the researcher simply chooses to use one or another way of normalized current. For instance, the researchers just normalize output current by geometric area, electroactive area, mass, mass of the main metal, or any combination among these. This lack of method may lead the reader to a misunderstanding. Therefore I encourage the researchers to explain in detail the method they are using to normalize current and the reason for doing so.

Furthermore, the aims of the scientific reports need more connections between fundamental (half-cell) and applied measurements. It seems researchers are completely ignoring the basics to have their work published at any cost. For instance, identifying electrocatalytic properties at stationary experiments is highly relevant, and among several benefits, it helps to understand the surface reaction and the *initial* electrochemical behavior of those materials. However, it is unlikely to directly correlate that result with fuel cell applications—at least those at flow configuration. Even the way of immobilization of the material on a gas difusion layer can change the results completely. Even though the thermodynamics may be similar, if not the same, other variables are not considered in stationary conditions, such as materials carrying, bubbles, etc.

Another issue is the engineering of the μDAFCs. This chapter shows how is possible to reach high performance by using commercially available materials, so the use of a new nanomaterial *must be* advantageous somehow—in terms of stability, cost, etc. Otherwise, there is no sense (so far) for using such materials. For instance, the literature is plenty of systems with complex materials containing expensive noble metals that show less than 10 mW cm^{-2} when assembled in a μDAFCs. The question is: Why would a researcher or industrial choose to spend time and investment in using these materials? If there is no advantage there are limited reasons to publish it. Gains in performance, cost, reaction understanding, application, reactants source, and engineering, should be the focus of the field.

Important specific challenges can be addressed to improve performance. The demand for liquid oxidants may be achieved by using mixed-media technology. Since membraneless system allows independent pH solution in each stream, it may work using aqueous and organic (or ionic liquids) for anolyte and catholyte. Of course, there are additional challenges to make it feasible, but it could be even an alternative to use CO_2 reduction reaction in organic liquids assembled with a fuel electrooxidation, which would ally energy conversion to CO_2 mitigation.

Among the biomass-derived alcohols, ethanol is especially interesting for tropical countries because they already have the distribution system, working full time to provide the fuel for automobiles. Therefore ethanol with additives (such as aromatics and antioxidants) is already available. Glycerol is another choice for the future because of multiple factors: (1) is widely available where biodiesel is produced, (2) has low market price, (3) produces high power in miniaturized devices, (4) and can be turned into carbonyl compounds of industrial interest. Regardless, the fuel used, the μDAFCs can be scaled out—they can be reproduced and assembled to maximize power, which is also an unexplored field.

The μDAFCs technologies are recent and versatile, which led to an increase in the number of publications and patents, so the field is still open for technological improvement and the study of market availability. The use of alcohol seems to be an interesting strategy for energy conversion in low- and high-powered devices. The gather of the electrocatalysis community and engineers is pivotal to build an environmentally friendly feasible product.

Acknowledgments

The author acknowledges Conselho Nacional de Desenvolvimento Cientifíico e Tecnologíco (CNPq, grants # 305691/ 2018−0 and 454516/2014−2), Fundação de Apoio ao Desenvolvimento do Ensino, Ciência e Tecnologia do Estado de Mato Grosso do Sul (FUNDECT, grants #045/2018, #026/2015, and #099/2016), Coordenação de Aperfeiçoamento de Pessoal de Níve Superior (CAPES), Financiadora de Estudos e Projetos (FINEP), and the financial support provided by the CAPESPrInt funding program (#88887.353061/2019−00 and #88887.311920/2018−00).

Appendix Literature data collected until November 2020 regarding methanol, ethanol, ethylene glycol, and glycerol microfluidic fuel cells. Some values are approximated from graphs of the papers. This is not a schematic revision.

Observation	Fuel/oxidant; anode–cathode	OCV/V	Peak power density/ mW cm^{-2}	Date of publication	Country/director	Citations (Google Scholar)	References
Circular six-cell vapor-feed microfluidic fuel cell stack (in series); 60ºC with 3M KOH as electrolyte	Methanol vapor/air; PtRu-Pt/C	5–5.5	108.7	2016	Hong Kong (Dennis Y. C. Leung)	13	[24]
Additional hydrogen from methanol pohotocatalysis; Pt/P25 for methanol reform; 3 suns	Water + methanol + hydrogen/air; PtRu-Pt/C	>0.9	220	2019	Hong Kong (Dennis Y. C. Leung)	11	[25]
Room temperature; 3 M KOH as electrolyte	Methanol vapor/KOH; PtRu-Pt/C	>0.9	55.4	2015	China/United Kingdom (Huizhi Wang)	25	[28]
Additional hydrogen from methanol pohotocatalysis; Pt/P25 for methanol reform	Water + methanol + hydrogen/air; PtRu-Pt/C	>0.9	213	2020	Hong Kong (Dennis Y. C. Leung)	–	[30]
Passive and mixed-reactant 2-cell stack	Mixed 5M MeOH in KOH/O$_2$; Pt/CP-Ag/Pt/CP	0.89	3.33 mW mg$_{Pt}^{-1}$	2017	Canada (Mohamed Mohamedi)	13	[31]
All-acid	Methanol/sodium perborate; Pt$_{50}$Ru$_{40}$Ni$_{10}$/C-Pt/C	>0.8	38.2	2015	India (B. Muthukumaran)	14	[32]
Mixed-reactant; air-breathing	Methanol/air; PtRu/C-Pt$_x$Ti$_y$/C	0.91	6.0	2014	France (Nicolas Alonso-Vante)	8	[33]
Passive; air-breathing	Mixed 5M MeOH in KOH/air; Pt/CP-Ag/Pt/CP	>0.5	2.4	2017	Canada (Mohamed Mohamedi)	9	[16]
Kapton fuel cell fabrication	Methanol in H$_2$SO$_4$/O$_2$; PtRu/C-Pt/C	>0.65	10	2013	United States (P. J. A. Kenis)	20	[35]
All-acid	Methanol/H$_2$O$_2$; Pt/CNTs anode	0.79	5.7	2015	China (Fengchun Sun)	7	[36]
Ladder-shaped microchannel	Methanol/H$_2$O$_2$; PtRu-Pt	0.97	12.24	2013	China (Hui Yang)	15	[37]
Microtubular system; all-acid	Methanol/O$_2$; Pt/RuO$_2$/rolled-up-microtubes-RuO$_2$	>0.6	257	2017	Germany (Oliver G. Schmidt)	7	[38]

Report of the PDMS-based cell	Methanol/H$_2$O$_2$; PtRu-Pt/C	>0.45	3.0	2009	Switzerland (M. A. M. Gijs)	20	[39]
Orthogonal flow microfluidic fuel cell; air-breathing cathodes	Methanol/air; Ru@Pt/G/CNTs-Pt/C	>0.8	16.35 (10.15 mW mg^{-1})	2017	Hong Kong (Dennis Y. C. Leung)	5	[40]
Methanol tolerant cathodes	Methanol/O$_2$; Pd-Pt$_x$S$_y$	>0.8	4	2012	France (Nicolas Alonso-Vante)	46	[41]
Air-breathing	Methanol/air; Pt/G-Pt/C	>0.65	~10 mW mg$_{Pt}^{-1}$	2017	Hong Kong (Dennis Y. C. Leung)	35	[42]
Air-breathing; all-alkaline; individual or mixed alcohols; methanol, ethanol; ethylene glycol, and glycerol	Methanol/air; Cu@Pd anode	0.61	17.1	2015	Mexico (L. G. Arriaga)	34	[58]
Mixed-media; methanol, ethylene glycol, and glycerol investigation	Methanol in KOH/O$_2$ in H$_2$SO$_4$; Pt/C-Pt/C	1.24	30.0	2018	Canada (Erik Kjeang)	24	[18]
All-alkaline; mixed media	Ethanol in KOH/sodium percarbonate in H$_2$SO$_4$; Pt-Pt	1.43	18.96	2014	India (B. Muthukumaran)	4	[43]
Enzymatic anode	Ethanol/O$_2$; ADH-TNTbioanode-Pt/C	>0.9	5.93	2020	Mexico (J. Ledesma-García)	—	[44]
Air-breathing; 3M Ethanol; all-alkaline	Ethanol/air; PdAg/MWCNTs-Pt/C	>0.9	14.5	2016	Mexico (L. G. Arriaga)	18	[45]
Air-breathing	Ethanol/air; Pd-Pd	>1.0	99.66	2016	Mexico (Noé Arjona)	6	[15]
Enzymatic anode; 2-cell stack	Alcohol dehydrogenase/CP-Pt/C	>1.0	3.154	2017	Mexico (L. G. Arriaga)	20	[46]
Mixed-media; room temperature	Ethanol in KOH/O$_2$ in H$_2$SO$_4$; Pd-NiO-Pt/C	1.11	108	2016	Mexico (Noé Arjona)	17	[47]
Pt, Pt$_{90}$Sn$_{10}$, Pt$_{80}$Ni$_{20}$ and Pt$_{80}$Sn$_{10}$Ni$_{10}$ anodes; all-acid; room temperature	Ethanol/sodium percarbonate; Pt$_{80}$Sn$_{10}$Ni$_{10}$-Pt/C	0.71	37.77	2015	India (B. Muthukumaran)	4	[48]
Pt, Pt$_{85}$Sn$_{20}$, Pt$_{90}$W$_{10}$, and Pt$_{80}$Sn$_{15}$W$_{05}$ dispersed on mesoporous carbon (MC)	Ethanol;Pt$_{80}$Sn$_{15}$W$_{05}$/MC anode	0.783	34.18	2017	India (B. Muthukumaran)	—	[49]

Pt-Sn and Pt-Sn-Ce dispersed on mesoporous carbon (MC); room temperature; all-acid	Ethanol/sodium perborate; PtSnCe/MC-Pt/MC	0.703	34.8	2017	India (B. Muthukumaran)	2	[50]
Mixed-reactant; air-breathing	Ethanol/O$_2$; PtRu-AgPt	0.75	10	2019	Canada (Mohamed Mohamedi)	3	[51]
Air-breathing; all-alkaline; individual or mixed alcohols; methanol, ethanol; ethylene glycol; and glycerol	Ethanol/air; Cu@Pd anode	0.67	25.75	2015	Mexico (L. G. Arriaga)	34	[58]
All-alkaline	Ethylene glycol/O$_2$; AuPd/polyaniline anode	0.53	1.6	2014	Mexico (L. G. Arriaga)	24	[52]
All-alkaline; flexible and 3D carbon aerogels (CAs) composed of carbon nanotubes (CNTs) with carbon shell-confined binary palladium-nickel (Pd$_x$–Ni$_y$) nanocatalysts on carbon fibers (Pd$_x$–Ni$_y$/NSCNT/CA)	Ethylene glycol/O$_2$; Pd$_{52}$Ni$_{48}$/NSCNT/CA-Pt/C	~0.9	62.8	2019	United States (Arumugam Manthiram)	28	[53]
All-alkaline; air-breathing; well-ordered (100) PdAg nanoparticles	Ethylene glycol/air; Pd$_{30}$Ag$_{70}$-Pt/C	0.71	5.2	2019	Mexico (Noé Arjona)	4	[54]
Air-breathing; all-alkaline; individual or mixed alcohols; methanol, ethanol; ethylene glycol; and glycerol	Ethylene glycol/air; Cu@Pd anode	0.65	19.95	2015	Mexico (L. G. Arriaga)	34	[58]
Mixed-media; methanol, ethylene glycol, and glycerol investigation	Ethylene glycol in KOH/O$_2$ in H$_2$SO$_4$; Pt/C-Pt/C	1.3	30.3	2018	Canada (Erik Kjeang)	24	[18]

Notes	Configuration	Col3	Col4	Year	Country (Author)	Col7	Ref
All-alkaline; probably the first work about glycerol µFC	Glycerol/O₂; Pd/MWCNTs-Pt/C	0.55	0.7	2013	Mexico (L. G. Arriaga)	59	[61]
Air-breathing; all-alkaline; individual or mixed alcohols; methanol, ethanol, ethylene glycol and glycerol	Glycerol/air Cu@Pd anode	0.62	20.43	2015	Mexico (L. G. Arriaga)	34	[58]
Air-breathing; all-alkaline	Glycerol/air; Cu@Pd-Pt/C	0.79	23.16	2016	Mexico (L. G. Arriaga)	23	[59]
All-alkaline; 75°C	Glycerol/O₂; PtRu/C-NiCo₂O₄/N-graphene	>0.8	>70	2016	United States (Arumugam Manthiram)	13	[57]
Mixed-media	Glycerol in KOH/HClO; Pt/C-Pt/C	1.9	315.3	2018	Canada (Erik Kjeang)	13	[11]
Mixed-media; methanol, ethylene glycol, and glycerol investigation	Glycerol in KOH/O₂ in H₂SO₄; Pt/C-Pt/C	1.3	39.5	2018	Canada (Erik Kjeang)	24	[18]
Mixed-media; in situ electrode modification	Glycerol in KOH/O₂ in H₂SO₄; Fe-decorated Pt/C-Pt/C	1.3	53.6	2019	Canada (Erik Kjeang)	5	[19]
Mixed-media; 3D-printed cell; 500 µL min⁻¹	Glycerol in KOH/HClO; Pt/C-Pt/C	1.8	175.2	2020	Brazil (Cauê A. Martins)	1	[56]
Air-breathing; all-alkaline; 307K	Glycerol/air; Pd-Pt/C-Pt/C	0.88	1.6	2020	India (Hiralal Pramanik)	--	[60]
Mixed-media; in situ electrode modification	Glycerol/HClO; Pt/C-GO	1.72	110.0	2020	Canada (Erik Kjeang)	--	[62]

References

[1] R. Oliveira, J. Santander, R. Rego, Overview of direct liquid oxidation fuel cells and its application as micro-fuel cells, in: F.J. Rodríguez-Varela, T.W. Napporn (Eds.), Advanced Electrocatalysts for Low-Temperature Fuel Cells, Springer International Publishing, Cham, 2018, pp. 129–174. Available from: https://doi.org/10.1007/978-3-319-99019-4_4.

[2] Y. Wang, K.S. Chen, J. Mishler, S.C. Cho, X.C. Adroher, A review of polymer electrolyte membrane fuel cells: technology, applications, and needs on fundamental research, Applied Energy 88 (2011) 981–1007. Available from: https://doi.org/10.1016/j.apenergy.2010.09.030.

[3] J. Wu, X.Z. Yuan, J.J. Martin, H. Wang, J. Zhang, J. Shen, et al., A review of PEM fuel cell durability: degradation mechanisms and mitigation strategies, Journal of Power Sources 184 (2008) 104–119. Available from: https://doi.org/10.1016/j.jpowsour.2008.06.006.

[4] U.B. Demirci, Direct liquid-feed fuel cells: thermodynamic and environmental concerns, Journal of Power Sources 169 (2007) 239–246. Available from: https://doi.org/10.1016/j.jpowsour.2007.03.050.

[5] M.J. Giz, G.A. Camara, The ethanol electrooxidation reaction at Pt (1 1 1): the effect of ethanol concentration, Journal of Electroanalytical Chemistry 625 (2009) 117–122. Available from: https://doi.org/10.1016/j.jelechem.2008.10.017.

[6] C.A. Martins, M.J. Giz, G.A. Camara, Generation of carbon dioxide from glycerol: evidences of massive production on polycrystalline platinum, Electrochimica Acta 56 (2011) 4549–4553. Available from: https://doi.org/10.1016/j.electacta.2011.02.076.

[7] R.G.C.S. Reis, C.A. Martins, G.A. Camara, The Electrooxidation of 2-propanol: an example of an alternative way to look at in situ FTIR data, Electrocatalysis 1 (2010) 116–121. Available from: https://doi.org/10.1007/s12678-010-0018-x.

[8] E. Kjeang, Microfluidic Fuel Cells and Batteries, Springer International Publishing, 2014. Available from: http://www.springer.com/gp/book/9783319063454 (accessed November 4, 2017).

[9] E. Kjeang, N. Djilali, D. Sinton, Microfluidic fuel cells: a review, Journal of Power Sources 186 (2009) 353–369. Available from: https://doi.org/10.1016/j.jpowsour.2008.10.011.

[10] R. Ferrigno, A.D. Stroock, T.D. Clark, M. Mayer, G.M. Whitesides, Membraneless vanadium redox fuel cell using laminar flow, Journal of the American Chemical Society 124 (2002) 12930–12931. Available from: https://doi.org/10.1021/ja020812q.

[11] C.A. Martins, O.A. Ibrahim, P. Pei, E. Kjeang, "Bleaching" glycerol in a microfluidic fuel cell to produce high power density at minimal cost, Chemical Communications 54 (2018) 192–195. Available from: https://doi.org/10.1039/C7CC08190A.

[12] M.-A. Goulet, E. Kjeang, Co-laminar flow cells for electrochemical energy conversion, Journal of Power Sources 260 (2014) 186–196. Available from: https://doi.org/10.1016/j.jpowsour.2014.03.009.

[13] R.S. Jayashree, L. Gancs, E.R. Choban, A. Primak, D. Natarajan, L.J. Markoski, et al., Air-breathing laminar flow-based microfluidic fuel cell, Journal of the American Chemical Society 127 (2005) 16758–16759. Available from: https://doi.org/10.1021/ja054599k.

[14] Q. Yi, L. Sun, H. Chu, Y. Zhang, M. Tang, X. Xiao, et al., A membraneless self-breathing direct alcohol fuel cell, Scientia Sinica Chimica 43 (2013) 1346–1353. Available from: https://doi.org/10.1360/032012-519.

[15] C.A. López-Rico, J. Galindo-de-la-Rosa, L. Álvarez-Contreras, J. Ledesma-García, M. Guerra-Balcázar, L.G. Arriaga, et al., Direct ethanol membraneless nanofluidic fuel cell with high performance, ChemistrySelect 1 (2016) 3054–3062. Available from: https://doi.org/10.1002/slct.201600364.

[16] J.C. Abrego-Martínez, A. Moreno-Zuria, Y. Wang, F.M. Cuevas-Muñiz, L.G. Arriaga, S. Sun, et al., Fabrication and evaluation of passive alkaline membraneless microfluidic DMFC, International Journal of Hydrogen Energy 42 (34) (2017) 21969–21975. Available from: https://doi.org/10.1016/j.ijhydene.2017.07.120.

[17] E. Kjeang, R. Michel, D.A. Harrington, N. Djilali, D. Sinton, A microfluidic fuel cell with flow-through porous electrodes, Journal of the American Chemical Society 130 (2008) 4000–4006. Available from: https://doi.org/10.1021/ja078248c.

[18] C.A. Martins, O.A. Ibrahim, P. Pei, E. Kjeang, Towards a fuel-flexible direct alcohol microfluidic fuel cell with flow-through porous electrodes: assessment of methanol, ethylene glycol and glycerol

[19] C.A. Martins, O.A. Ibrahim, P. Pei, E. Kjeang, In situ decoration of metallic catalysts in flow-through electrodes: application of Fe/Pt/C for glycerol oxidation in a microfluidic fuel cell, Electrochimica Acta 305 (2019) 47−55. Available from: https://doi.org/10.1016/j.electacta.2019.03.018.
[20] J.L. Cohen, D.J. Volpe, D.A. Westly, A. Pechenik, H.D. Abruña, A. Dual, Electrolyte H_2/O_2 planar membraneless microchannel fuel cell system with open circuit potentials in excess of 1.4V, Langmuir 21 (2005) 3544−3550. Available from: https://doi.org/10.1021/la0479307.
[21] J.P. Esquivel, P. Alday, O.A. Ibrahim, B. Fernández, E. Kjeang, N. Sabaté, A metal-free and biotically degradable battery for portable single-use applications, Advanced Energy Materials 7 (18) (2017) 1700275. Available from: https://doi.org/10.1002/aenm.201700275.
[22] A. Einstein, Investigations on the Theory of the Brownian Movement, Courier Corporation, Mineola, NY, 1956.
[23] R.F. Ismagilov, A.D. Stroock, P.J.A. Kenis, G. Whitesides, H.A. Stone, Experimental and theoretical scaling laws for transverse diffusive broadening in two-phase laminar flows in microchannels, Applied Physics Letters 76 (2000) 2376−2378. Available from: https://doi.org/10.1063/1.126351.
[24] Y. Wang, D.Y.C. Leung, A circular stacking strategy for microfluidic fuel cells with volatile methanol fuel, Applied Energy 184 (2016) 659−669. Available from: https://doi.org/10.1016/j.apenergy.2016.11.019.
[25] Y.H. Kwok, Y. Wang, M. Wu, F. Li, Y. Zhang, H. Zhang, et al., A dual fuel microfluidic fuel cell utilizing solar energy and methanol, Journal of Power Sources 409 (2019) 58−65. Available from: https://doi.org/10.1016/j.jpowsour.2018.10.095.
[26] J. Massing, N. van der Schoot, C.J. Kähler, C. Cierpka, A fast start up system for microfluidic direct methanol fuel cells, International Journal of Hydrogen Energy 44 (2019) 26517−26529. Available from: https://doi.org/10.1016/j.ijhydene.2019.08.107.
[27] J.D. Morse, R.S. Upadhye, R.T. Graff, C. Spadaccini, H.G. Park, E.K. Hart, A MEMS-based reformed methanol fuel cell for portable power, Journal of Micromechanics and Microengineering 17 (2007) S237−S242. Available from: https://doi.org/10.1088/0960-1317/17/9/S05.
[28] Y. Wang, D.Y.C. Leung, J. Xuan, H. Wang, A vapor feed methanol microfluidic fuel cell with high fuel and energy efficiency, Applied Energy 147 (2015) 456−465. Available from: https://doi.org/10.1016/j.apenergy.2015.03.028.
[29] A. Arun, M. Gowdhamamoorthi, S. Kiruthika, B. Muthukumaran, Analysis of membraneless methanol fuel cell using percarbonate as an oxidant, Journal of the Electrochemical Society 161 (2014) F311. Available from: https://doi.org/10.1149/2.067403jes.
[30] Y.H. Kwok, Y. Wang, Y. Zhang, H. Zhang, F. Li, W. Pan, et al., Boosting cell performance and fuel utilization efficiency in a solar assisted methanol microfluidic fuel cell, International Journal of Hydrogen Energy 45 (2020) 21796−21807. Available from: https://doi.org/10.1016/j.ijhydene.2020.05.163.
[31] J.C. Abrego-Martínez, A. Moreno-Zuria, F.M. Cuevas-Muñiz, L.G. Arriaga, S. Sun, M. Mohamedi, Design, fabrication and performance of a mixed-reactant membraneless micro direct methanol fuel cell stack, Journal of Power Sources 371 (2017) 10−17. Available from: https://doi.org/10.1016/j.jpowsour.2017.10.026.
[32] A. Arun, M. Gowdhamamoorthi, K. Ponmani, S. Kiruthika, B. Muthukumaran, Electrochemical characterization of Pt−Ru−Ni/C anode electrocatalyst for methanol electrooxidation in membraneless fuel cells, RSC Advances 5 (2015) 49643−49650. Available from: https://doi.org/10.1039/C5RA04958J.
[33] J. Ma, A. Habrioux, C. Morais, N. Alonso-Vante, Electronic modification of Pt via Ti and Se as tolerant cathodes in air-breathing methanol microfluidic fuel cells, Physical Chemistry Chemical Physics 16 (2014) 13820−13826. Available from: https://doi.org/10.1039/C3CP54564D.
[34] J.C. Abrego-Martínez, L.H. Mendoza-Huizar, J. Ledesma-Garcia, L.G. Arriaga, F.M. Cuevas-Muñiz, Highly methanol tolerant cathode based on PtAg for use in microfluidic fuel cell, Journal of Physics: Conference Series 660 (2015) 012079. Available from: https://doi.org/10.1088/1742-6596/660/1/012079.

[35] A.S. Hollinger, P.J.A. Kenis, Manufacturing all-polymer laminar flow-based fuel cells, Journal of Power Sources 240 (2013) 486−493. Available from: https://doi.org/10.1016/j.jpowsour.2013.04.053.

[36] W. Huo, H. He, F. Sun, Microfluidic direct methanol fuel cell by electrophoretic deposition of platinum/carbon nanotubes on electrode surface, International Journal of Energy Research 39 (2015) 1430−1436. Available from: https://doi.org/10.1002/er.3354.

[37] W. Huo, Y. Zhou, H. Zhang, Z. Zou, H. Yang, Microfluidic direct methanol fuel cell with ladder-shaped microchannel for decreased methanol crossover, International Journal of Electrochemical Science 8 (2013) 4827−4838.

[38] S. Miao, S. He, M. Liang, G. Lin, B. Cai, O.G. Schmidt, Microtubular fuel cell with ultrahigh power output per footprint, Advanced Materials 29 (34) (2017) 1607046. Available from: https://doi.org/10.1002/adma.201607046.

[39] M. Shen, S. Walter, M.A.M. Gijs, Monolithic micro-direct methanol fuel cell in polydimethylsiloxane with microfluidic channel-integrated Nafion strip, Journal of Power Sources 193 (2009) 761−765. Available from: https://doi.org/10.1016/j.jpowsour.2009.03.055.

[40] Y.H. Kwok, Y.F. Wang, A.C.H. Tsang, D.Y.C. Leung, Ru@Pt core shell nanoparticle on graphene carbon nanotube composite aerogel as a flow through anode for direct methanol microfluidic fuel cell, Energy Procedia 142 (2017) 1522−1527. Available from: https://doi.org/10.1016/j.egypro.2017.12.602.

[41] A.S. Gago, Y. Gochi-Ponce, Y.-J. Feng, J.P. Esquivel, N. Sabaté, J. Santander, et al., Tolerant chalcogenide cathodes of membraneless micro fuel cells, ChemSusChem 5 (2012) 1488−1494. Available from: https://doi.org/10.1002/cssc.201200009.

[42] Y.H. Kwok, A.C.H. Tsang, Y. Wang, D.Y.C. Leung, Ultra-fine Pt nanoparticles on graphene aerogel as a porous electrode with high stability for microfluidic methanol fuel cell, Journal of Power Sources 349 (2017) 75−83. Available from: https://doi.org/10.1016/j.jpowsour.2017.03.030.

[43] M. Priya, A. Arun, M. Elumalai, S. Kiruthika, B. Muthukumaran, A development of ethanol/percarbonate membraneless fuel cell, Advances in Physical Chemistry 2014 (2014) 1−8. Available from: https://doi.org/10.1155/2014/862691.

[44] J. Galindo-de-la-Rosa, A. Álvarez, M.P. Gurrola, J.A. Rodríguez-Morales, G. Oza, L.G. Arriaga, et al., Alcohol dehydrogenase immobilized on TiO$_2$ nanotubes for ethanol microfluidic fuel cells, ACS Sustainable Chemistry & Engineering 8 (29) (2020) 10900−10910. Available from: https://doi.org/10.1021/acssuschemeng.0c03219.

[45] A.J. Armenta-González, R. Carrera-Cerritos, A. Moreno-Zuria, L. Álvarez-Contreras, J. Ledesma-García, F.M. Cuevas-Muñiz, et al., An improved ethanol microfluidic fuel cell based on a PdAg/MWCNT catalyst synthesized by the reverse micelles method, Fuel 167 (2016) 240−247. Available from: https://doi.org/10.1016/j.fuel.2015.11.057.

[46] J. Galindo-de-la-Rosa, N. Arjona, A. Moreno-Zuria, E. Ortiz-Ortega, M. Guerra-Balcázar, J. Ledesma-García, et al., Evaluation of single and stack membraneless enzymatic fuel cells based on ethanol in simulated body fluids, Biosensors and Bioelectronics 92 (2017) 117−124. Available from: https://doi.org/10.1016/j.bios.2017.02.010.

[47] C.A. López-Rico, J. Galindo-de-la-Rosa, E. Ortiz-Ortega, L. Álvarez-Contreras, J. Ledesma-García, M. Guerra-Balcázar, et al., High performance of ethanol co-laminar flow fuel cells based on acrylic, paper and Pd-NiO as anodic catalyst, Electrochimica Acta 207 (2016) 164−176. Available from: https://doi.org/10.1016/j.electacta.2016.05.002.

[48] K. Ponmani, S. Kiruthika, B. Muthukumaran, Investigation of nanometals (Ni and Sn) in platinum-based ternary electrocatalysts for ethanol electro-oxidation in membraneless fuel cells, Journal of Electrochemical Science and Technology 6 (2015) 95−105. Available from: https://doi.org/10.33961/JECST.2015.6.3.95.

[49] M. Priya, S. Kiruthika, B. Muthukumaran, Mesoporous carbon supported Pt-Sn-W ternary electrocatalyst for membraneless ethanol fuel cell, Journal of Materials and Environmental Sciences 8 (2) (2017) 410−419.

[50] M. Priya, S. Kiruthika, B. Muthukumaran, Synthesis and characterization of Pt−Sn−Ce/MC ternary catalysts for ethanol oxidation in membraneless fuel cells, Ionics 23 (2017) 1209−1218. Available from: https://doi.org/10.1007/s11581-016-1940-6.

[51] M.J. Estrada-Solís, J.C. Abrego-Martínez, A. Moreno-Zuria, L.G. Arriaga, S. Sun, F.M. Cuevas-Muñiz, et al., Use of a bilayer platinum-silver cathode to selectively perform the oxygen reduction reaction in a high concentration mixed-reactant microfluidic direct ethanol fuel cell, International Journal of Hydrogen Energy 44 (33) (2019) 18372−18381. Available from: https://doi.org/10.1016/j.ijhydene.2019.05.078.

[52] N. Arjona, A. Palacios, A. Moreno-Zuria, M. Guerra-Balcázar, J. Ledesma-García, L.G. Arriaga, AuPd/polyaniline as the anode in an ethylene glycol microfluidic fuel cell operated at room temperature, Chemical Communications 50 (2014) 8151−8153. Available from: https://doi.org/10.1039/C4CC03288H.

[53] T.R. Kumar, G.G. Kumar, A. Manthiram, Biomass-derived 3D carbon aerogel with carbon shell-confined binary metallic nanoparticles in CNTs as an efficient electrocatalyst for microfluidic direct ethylene glycol fuel cells, Advanced Energy Materials 9 (16) (2019) 1803238. Available from: https://doi.org/10.1002/aenm.201803238.

[54] A. López-Coronel, E. Ortiz-Ortega, L.J. Torres-Pacheco, M. Guerra-Balcázar, L.G. Arriaga, L. Álvarez-Contreras, et al., High performance of Pd and PdAg with well-defined facets in direct ethylene glycol microfluidic fuel cells, Electrochimica Acta 320 (2019) 134622. Available from: https://doi.org/10.1016/j.electacta.2019.134622.

[55] L. An, R. Chen, Recent progress in alkaline direct ethylene glycol fuel cells for sustainable energy production, Journal of Power Sources 329 (2016) 484−501. Available from: https://doi.org/10.1016/j.jpowsour.2016.08.105.

[56] K.-E. Guima, P.-H.L. Coelho, M.A.G. Trindade, C.A. Martins, 3D-printed glycerol microfluidic fuel cell, Lab on a Chip 20 (12) (2020) 2057−2061. Available from: https://doi.org/10.1039/D0LC00351D.

[57] X. Yu, E.J. Pascual, J.C. Wauson, A. Manthiram, A membraneless alkaline direct liquid fuel cell (DLFC) platform developed with a catalyst-selective strategy, Journal of Power Sources 331 (2016) 340−347. Available from: https://doi.org/10.1016/j.jpowsour.2016.09.077.

[58] J. Maya-Cornejo, E. Ortiz-Ortega, L. Álvarez-Contreras, N. Arjona, M. Guerra-Balcázar, J. Ledesma-García, et al., Copper−palladium core−shell as an anode in a multi-fuel membraneless nanofluidic fuel cell: toward a new era of small energy conversion devices, Chemical Communications 51 (2015) 2536−2539. Available from: https://doi.org/10.1039/C4CC08529A.

[59] J. Maya-Cornejo, M. Guerra-Balcázar, N. Arjona, L. Álvarez-Contreras, F.J. Rodríguez Valadez, M.P. Gurrola, et al., Electrooxidation of crude glycerol as waste from biodiesel in a nanofluidic fuel cell using Cu@Pd/C and Cu@Pt/C, Fuel 183 (2016) 195−205. Available from: https://doi.org/10.1016/j.fuel.2016.06.075.

[60] D. Panjiara, H. Pramanik, Electrooxidation study of glycerol on synthesized anode electrocatalysts Pd/C and Pd-Pt/C in a Y-shaped membraneless air-breathing microfluidic fuel cell for power generation, Ionics 26 (2020) 2435−2452. Available from: https://doi.org/10.1007/s11581-019-03385-8.

[61] A. Dector, F.M. Cuevas-Muñiz, M. Guerra-Balcázar, L.A. Godínez, J. Ledesma-García, L.G. Arriaga, Glycerol oxidation in a microfluidic fuel cell using Pd/C and Pd/MWCNT anodes electrodes, International Journal of Hydrogen Energy 38 (2013) 12617−12622. Available from: https://doi.org/10.1016/j.ijhydene.2012.12.030.

[62] C.A. Martins, P. Pei, M. Tellis, O.A. Ibrahim, E. Kjeang, Graphene-oxide-modified metal-free cathodes for glycerol/bleach microfluidic fuel cells, ACS Applied Nano Materials 3 (8) (2020) 8286−8293. Available from: https://doi.org/10.1021/acsanm.0c01722.

[63] J.F. Gomes, C.A. Martins, M.J. Giz, G. Tremiliosi-Filho, G.A. Camara, Insights into the adsorption and electro-oxidation of glycerol: self-inhibition and concentration effects, Journal of Catalysis 301 (2013) 154−161. Available from: https://doi.org/10.1016/j.jcat.2013.02.007.

[64] C. Busó-Rogero, J. Solla-Gullón, F.J. Vidal-Iglesias, E. Herrero, J.M. Feliu, Adatom modified shape-controlled platinum nanoparticles towards ethanol oxidation, Electrochimica Acta 196 (2016) 270−279. Available from: https://doi.org/10.1016/j.electacta.2016.02.171.

[65] C.A. Martins, P.S. Fernández, G.A. Camara, Alternative uses for biodiesel byproduct: glycerol as source of energy and high valuable chemicals, Increased Biodiesel Efficiency, Springer, Cham, 2018, pp. 159−186. Available from: https://doi.org/10.1007/978-3-319-73552-8_7.

[66] A. Lam, D.P. Wilkinson, J. Zhang, Control of variable power conditions for a membraneless direct methanol fuel cell, Journal of Power Sources 194 (2009) 991–996. Available from: https://doi.org/10.1016/j.jpowsour.2009.05.038.

[67] E. Kjeang, R. Michel, D.A. Harrington, D. Sinton, N. Djilali, An alkaline microfluidic fuel cell based on formate and hypochlorite bleach, Electrochimica Acta 54 (2008) 698–705. Available from: https://doi.org/10.1016/j.electacta.2008.07.009.

[68] G.L. Caneppele, C.A. Martins, Revisiting glycerol electrooxidation by applying derivative voltammetry: from well-ordered bulk Pt to bimetallic nanoparticles, Journal of Electroanalytical Chemistry 865 (2020) 114139. Available from: https://doi.org/10.1016/j.jelechem.2020.114139.

[69] F.J. Vidal-Iglesias, R.M. Arán-Ais, J. Solla-Gullón, E. Herrero, J.M. Feliu, Electrochemical characterization of shape-controlled Pt nanoparticles in different supporting electrolytes, ACS Catalysis 2 (2012) 901–910. Available from: https://doi.org/10.1021/cs200681x.

[70] G. Wu, L. Li, B.-Q. Xu, Effect of electrochemical polarization of PtRu/C catalysts on methanol electrooxidation, Electrochimica Acta 50 (1) (2004) 1–10. Available from: https://doi.org/10.1016/j.electacta.2004.07.006.

[71] J.B. Goodenough, A. Hamnett, B.J. Kennedy, R. Manoharan, S.A. Weeks, Methanol oxidation on unsupported and carbon supported Pt + Ru anodes, Journal of Electroanalytical Chemistry and Interfacial Electrochemistry 240 (1–2) (1988) 133–145. Available from: https://doi.org/10.1016/0022-0728(88)80318-8.

[72] H.-Y. Chou, C.-K. Hsieh, M.-C. Tsai, Y.-H. Wei, T.-K. Yeh, C.-H. Tsai, Pulse electrodeposition of Pt and Pt–Ru methanol-oxidation nanocatalysts onto carbon nanotubes in citric acid aqueous solutions, Thin Solid Films 584 (2015) 98–102. Available from: https://doi.org/10.1016/j.tsf.2014.12.016.

[73] J.W. Dieker, W.E. Van Der Linden, Determination of iron(II) and iron(III) by flow injection and amperometric detection with a glassy carbon electrode, Analytica Chimica Acta 114 (1980) 267–274. Available from: https://doi.org/10.1016/S0003-2670(01)84298-7.

[74] K.-E. Guima, L.E. Gomes, J. Alves Fernandes, H. Wender, C.A. Martins, Harvesting energy from an organic pollutant model using a new 3D-printed microfluidic photo fuel cell, ACS Applied Materials & Interfaces 12 (49) (2020) 54563–54572. Available from: https://doi.org/10.1021/acsami.0c14464.

CHAPTER 4

General aspects in the modeling of fuel cells: from conventional fuel cells to nano fuel cells

Pablo A. García-Salaberri
Department of Thermal and Fluids Engineering, University Carlos III of Madrid, Leganés, Spain

4.1 Introduction

Fuel cells are electrochemical devices that convert the chemical energy of a fuel and an oxidizing agent (oxygen) directly into electric energy and heat [1–4]. They differ from batteries in that they produce electricity continuously as long as the reactants are fed into the cell [5]. The reactions are electrochemically catalyzed, so that no high-temperature combustion processes take place in the cell. As a result, fuel cells offer a superior energetic efficiency due to the lack of Carnot cycle limitations, unlike traditional combustion engines. Moreover, they are environmentally friendly power sources, giving low emissions of air pollutants: virtually zero in the case of hydrogen fuel cells (heat and water are the only products), and a low level of carbon dioxide emissions in the case of hydrocarbon- or alcohol-based fuel cells [6–8]. Further benefits include low noise (no moving parts), scalability and long lasting. These characteristics place fuel cells as promising 21st-century energy-conversion devices for stationary, portable and transport applications [6].

The main types of fuel cells available in the market are categorized by their electrolyte in Table 4.1, indicating the main characteristics of each technology (typical catalysts, operating temperature, fuel and electrical efficiency) [1,2,4,9]. Other types of fuel cells not included in the table are liquid-feed direct ethanol fuel cells and microbial fuel cells [4]. Over the last decades, the interest in terms of research and development on fuel cell technology has shifted drastically. In the 1950s–60s, alkaline fuel cells and the first hydrogen polymer electrolyte membrane fuel cells (PEMFCs) were developed during the "space race" [10,11]. In the 1970s–1980s, due to concerns on the surge of oil price and energy shortages, higher attention was paid to large-scale power sources using molten carbonate fuel cells and phosphoric acid fuel cells [12]. In the 1990s, attention turned to transport and small-to-medium stationary applications, leading to a large research activity on low-temperature hydrogen PEMFCs, as well as

Table 4.1 Comparison of main fuel cell technologies.

	PEMFC	DMFC	AFC	PAFC	SOFC	MCFC
Electrolyte type	Polymer membrane	Polymer membrane	Potassium hydroxide solution	Liquid phosphoric acid	Yttria-stabilized zirconia (YSZ)	Molten carbonate salts
Catalyst	Platinum	Platinum and ruthenium	Nickel and silver	Platinum	Nickel-YSZ and lanthanum strontium manganite	Nickel
Fuel	Hydrogen	Methanol	Hydrogen	Hydrogen	Methane	Methane
Operating temperature (°C)	30—90	30—90	50—200	150—200	600—1000	600—700
Electrical efficiency	30%—60%	20%—45%	50%—65%	30%—50%	30%—60%	45%—55%

The electrical efficiencies are estimated based on the low heating value of the fuel. *AFC*, alkaline fuel cell; *DMFC*, direct methanol fuel cell; *MCFC*, molten carbonate fuel cell; *PAFC*, phosphoric acid fuel cell; *PEMFC*, polymer electrolyte membrane fuel cell; *SOFC*, solid oxide fuel cell.

high-temperature solid oxide fuel cells (SOFCs) [13,14]. In addition, the development of liquid-feed direct methanol fuel cells (DMFCs), a variant of PEMFC technology in which the fuel is an aqueous methanol solution, received increasing attention for portable electronic devices [15]. The use of vapor-feed DMFCs has also been explored to mitigate the problem of methanol crossover, although their complexity and energy requirements are higher due to the need to vaporize the liquid fuel and the difficulty to separate the unused methanol vapor from the generated carbon dioxide [16]. Since 2000, improvements in SOFC technology continue toward the development of power units of lower temperature [17]. However, the largest part of the research activity is devoted to polymer electrolyte membrane (PEM) fuel cells, either PEMFCs or liquid-feed DMFCs [6,18,19]. PEM fuel cells operate at significantly lower temperatures than any other type, are smaller in volume and lighter in weight, offer a rapid start-up, a suitable transient response, and use a solid polymer membrane as electrolyte. These unique features make them the perfect option for portable, transport, small stationary, and auxiliary power applications [20]. Hydrogen PEMFCs are the leading candidate to power upcoming fuel cell vehicles (FCVs); world's largest automotive manufacturers (Ford, Mercedes, BMW, Renault, Nissan, Toyota, Honda, etc.) committed to offer FCVs commercially available between 2013 and 2020 [21]. On the other hand, liquid-feed DMFCs are best suited for portable and mobile applications with medium power requirements due to their lower performance and efficiency, but higher energy density, and ease of handling and storage of liquid methanol [15]. As an example, Dynario Toshiba's battery charger was launched in 2009 and Smart Fuel Cells Co. has commercially produced portable and semi-portable liquid-feed DMFCs since the last decade [22]. Nowadays, the global economic recession has negatively affected fuel cell company's prospects, and several companies have gone out of business due to their lack of profitability. Nevertheless, governments around the world still see fuel cells as a clean technology of future economic growth and employment, and further research founding is being invested to solve fundamental technological problems.

The major barriers in the path to PEMFC broad commercialization are cost and lifetime, as well as the development of a hydrogen infrastructure based on renewable energies [4,18,20,23]. In the last decades, a dramatic reduction of the cost of PEMFCs for light duty vehicles has been achieved, varying from $275/kW in 2002 to $55/kW in 2013 (a factor of five) [20]. However, further effort is needed to meet the target of $30/kW projected by the DOE (United States Department of Energy) to be competitive with internal-combustion engines. As for durability, important improvements have also been made in the automotive industry. A durability of 2500 hours with a 10% stack voltage degradation was demonstrated in 2013, although it is still significantly lower than the target of 5000 hours established for 2020 [24]. Cost and durability issues also affect liquid-feed DMFCs [15]. To overcome these hurdles, it is mandatory to decrease the amount of catalyst loading and develop more durable and cheaper materials [25–29]. Furthermore, the performance of fuel cells must be optimized. Proper water and heat management and cold start are critical engineering challenges to increase the performance and durability of PEM fuel cells [20,30–33]. Additional problems to address in liquid-feed DMFCs are the crossover of methanol through the membrane, the slow kinetics of methanol oxidation at the anode, and the inadequate removal of carbon dioxide bubbles from the anode compartment [15,34].

A thorough understanding of the complex transport and electrochemical processes that take place in PEM fuel cells from a combination of detailed experimental and numerical research is therefore necessary. A wide variety of diagnostic and visualization tools are currently available to investigate cell operation and performance [25–29]. These include electrochemical impedance spectroscopy, cyclic voltammetry, direct optical observation, segmented cells, neutron imaging, and X-ray tomography and radiography [25–29]. Important advances have also been achieved in the characterization of the mechanical, transport and physicochemical properties of fuel cell components using both experimental and numerical methods (see, e.g., [35–44]). On the other hand, numerical modeling plays a crucial role in the study of local multiphysics and multiphase phenomena, which are difficult to explore experimentally due to the small dimensions used in PEM fuel cells. Moreover, numerical models are an indispensable predictive tool to optimize cell performance [20]. The recent progress achieved in diagnostic and visualization tools and materials characterization should go hand in hand with the development of more comprehensive mathematical models (see Fig. 4.1).

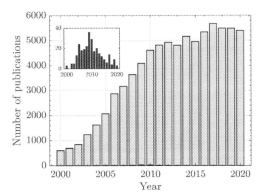

Figure 4.1 Histogram of the number of publications, including the topic "fuel cell" and "nano fuel cell" or "micro fuel cell" (inset) in the period 2000–2020. *Web of Science.*

4.1.1 Proton exchange membrane fuel cells (PEMFCs)

As shown in Fig. 4.2, PEMFCs are made into thin planar structures that provide large active surface areas to volume ratios. The heart of a PEM fuel cell is the so-called membrane electrode assembly (MEA), which is composed of a central PEM, anode and cathode porous catalyst layers (CLs), and two fibrous gas diffusion layers (GDLs). These components are bonded together by hot pressing, and then sandwiched between two bipolar plates (BPPs) for operation [45,46]. The main processes involved in current generation in a PEMFC are as follows:

1. *Reactant transport.* The reactants are supplied to the cell by the channels grooved on the BPPs; the fuel (hydrogen in PEMFCs) is fed to the anode, while the oxidant is fed to the cathode. The reactants flow along the channels parallel to the surface of the MEA and are transported by diffusion and convection to the CLs through the GDLs (see red arrows).
2. *Electrochemical reaction.* Once the reactants reach the CLs, they undergo electrochemical reactions. The fuel oxidation (liberation of electrons, e^-, and protons, H^+) takes place at the anode, while the oxygen reduction (consumption of electrons and protons) takes place at the cathode. The electrochemical reactions that occur in a PEMFC, along with the standard equilibrium potentials versus saturated hydrogen electrode, are summarized below [47].

   Hydrogen oxidation reaction (HOR) at the anode:

   $$H_2 \rightarrow 2H^+ + 2e^- \quad E_a^0 = 0 \text{ V} \tag{4.1}$$

   Oxygen reduction reaction (ORR) at the cathode:

   $$\frac{1}{2}O_2 + 2H^+ + 2e^- \rightarrow H_2O \quad E_c^0 = 1.23 \text{ V}$$

Global reaction:

$$H_2 + \frac{1}{2}O_2 \rightarrow H_2O \quad E^0_{cell} = E^0_c - E^0_a = 1.23 \text{ V}$$

3. *Electron and proton transport.* Electrons generated at the anode are transported across the anode GDL to the ribs of the anode BPP, and then travel through an external circuit to the cathode, thus creating the current output of the cell (green arrows). Protons are transported through the PEM (pink arrow), so that electrons and protons are consumed by the ORR at the cathode.
4. *Product removal.* The by-products generated by the electrochemical reactions (water in PEMFCs) are transported first from the CLs to the channels through the GDL (see violet arrows). Then, they are evacuated from the cell with the reactant flow streams.

Fig. 4.3 illustrates a cross-sectional view of a MEA, indicating the typical dimensions of the components, and the mass, charge and heat transport processes that occur in a PEMFC. The main characteristics and functions of each component are presented below.

Proton exchange, or polymer electrolyte, membrane (PEM). The membrane enables the transport of protons generated at the anode to the cathode CL. In addition, it must provide an effective barrier to avoid the mixing of reactants. PEMs consist of a

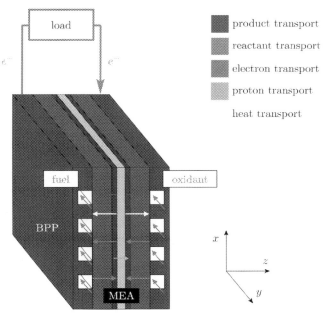

Figure 4.2 Schematic of the principle of operation of a PEMFC, showing the basic components, and the main mass, charge and heat transport processes that occur in a cell. Drawing dimensions are not to scale. *PEMFC*, proton exchange membrane fuel cell.

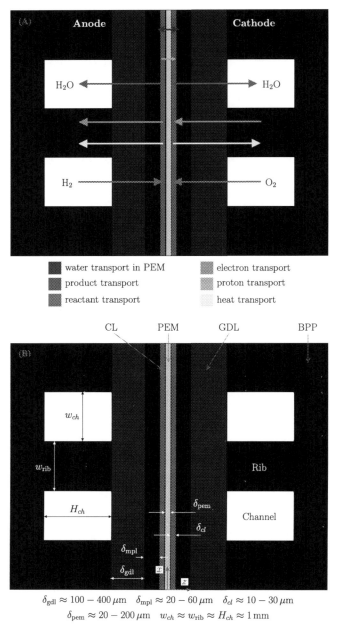

Figure 4.3 Schematic diagram of (A) mass, charge and heat transport processes in a MEA. The net water flux across the PEM can be in both directions in PEMFCs depending on the cell design and operating conditions. (B) Internal components of a PEMFC, showing their typical dimensions. *MEA*, membrane electrode assembly; *PEM*, proton exchange membrane; *PEMFCs*, proton exchange membrane fuel cells.

polytetrafluoroethylene (PTFE) polymer backbone with stabilized perfluorosulfonic acid pendant side chains, and are usually made of commercial Nafion [41]. The PEM has to be well humidified to conduct protons. This basic requirement makes necessary a careful water management and control of the humidification conditions of the gas streams in PEMFCs. A delicate water balance is essential to keep the PEM hydrated, while avoiding water accumulation and blockage of reactant pathways (i.e., flooding) [30,36,37,48]. Water flooding is critical in the cathode compartment where water is generated, leading to a decrease of the cell performance due to limited oxygen transport. Nafion membranes display a rather high impermeability to gaseous species, but they are significantly permeable to liquid water and dissolved species. The fluxes of liquid water through the PEM are mainly driven by three mechanisms: (1) diffusion by species concentration gradient, (2) electroosmotic transport due to the polar attraction between water and protons (proportional to the current density drawn from the cell), and (3) convection by hydraulic pressure gradient. The first two mechanisms (diffusion and electroosmotic drag) are usually dominant in conventional fuel cell designs [49]. Electroosmotic transport is directed from anode to cathode according to the proton flux, while water diffusion is typically directed from cathode to anode (i.e., back diffusion) due to water generation at the cathode [50]. As a result, the net water flux through the PEM (electroosmotic drag + diffusion + convection) can be in both directions in PEMFCs. The thickness of Nafion membranes ranges from 25.4 μm (Nafion 211), passing through 50.8 μm (Nafion 212), 127 μm (Nafion 115), and 183 μm (Nafion 117), up to 254 μm (Nafion 1110) [51,52].

Catalyst layer (CL). CLs are thin porous layers (porosity $\sim 0.3 - 0.4$, thickness $\sim 5 - 40$ μm) adhered to the surface of the PEM, where the electrochemical reactions occur catalyzed by noble metals. Hydrogen oxidation takes place at the anode and oxygen reduction at the cathode. The structure of CLs is composed of a mixture of ionomer and carbon-supported or unsupported catalyst nanoparticles. This ensures the existence of triple boundary active sites needed for the reactions to proceed. Reactants and products are transported through the void space, protons through the ionomer, and electrons through the carbon agglomerates [46,53,54]. In PEMFCs, electrochemical reactions are typically catalyzed by platinum (Pt) [55].

Gas diffusion layer (GDL). The GDL is made of flexible highly porous non-woven carbon paper and felt or woven carbon cloth, and acts as a spacer layer between the BPP and the CL [54]. GDL porosity and thickness range around 0.7−0.9 and 100−400 μm, respectively. The GDL provides several critical functions: (1) a pathway for reactants access and products removal to and from the CLs through its pore volume, (2) electrical and thermal conductivity through its solid fibrous structure, and (3) adequate mechanical support to the PEM and CLs, protecting them from damage and intrusion into the channel when the cell is assembled for operation [56,57]. GDLs are usually treated with a hydrophobic PTFE coating to promote capillary transport of liquid water [30,43,58]. Several GDL manufacturers can be found in the market; the most common carbon-cloth GDLs are those of ELAT, while the most common carbon-paper and carbon-felt GDLs are those of Toray, SIGRACET, Freudenberg and Mitsubishi Rayon Corp. Both types of materials exhibit anisotropic effective transport properties, especially in the case of carbon paper, due to the preferential arrangement of carbon fibers (and pores) in the material plane [57,59−65]. X-ray computed tomography images of various carbon-paper GDLs with different thickness and PTFE content are shown in Fig. 4.4.

Bipolar plate (BPP). The bipolar plates are usually made of metal or graphite to conduct electrons and receive a surface treatment to protect them from corrosion and reduce electrical/thermal contact resistances [45,66]. The channels stamped or machined on the BPPs have a typical cross-section of 1×1 mm^2, and are used to supply the fuel and the oxidant, and to remove the products of the electrochemical reactions. In addition, the BPPs collect the output current of the cell through the ribs or lands (width ~ 1 mm), remove generated heat and provide structural integrity to the cell. Coolant channels are incorporated inside the BPPs to remove the heat released from the reactions and maintain the working temperature as uniform as possible. Cooling needs depend largely on the cell size [67,68]. Fig. 4.5 illustrates the wide variety of flow field

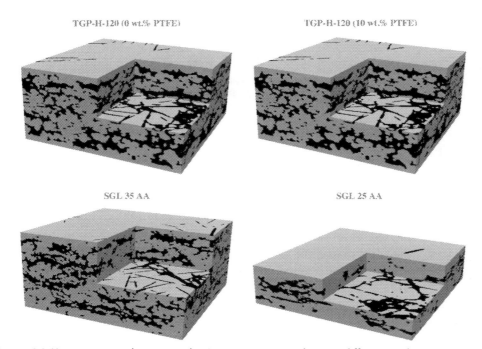

Figure 4.4 X-ray computed tomography images corresponding to different carbon-paper GDLs: untreated Toray TGP-H-120, 10 wt.% PTFE-treated Toray TGP-H-120, Sigracet SGL 35 AA, and Sigracet SGL 25 AA. *GDLs*, gas diffusion layers; *PTFE*, polytetrafluoroethylene. *Courtesy Dr. Iryna V. Zenyuk, National Fuel Cell Research Center, Irvine, CA, United States.*

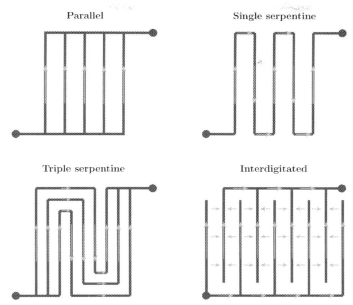

Figure 4.5 Schematic of the main types of flow fields used in active PEMFCs: parallel, single serpentine, triple serpentine, and interdigitated channels. The flow direction is indicated by arrows. *PEMFCs*, proton exchange membrane fuel cells.

designs typically used in active PEMFCs. They include parallel straight channels and serpentine configurations with different number of channel segments. Other widely used design is the interdigitated flow field, which forces under-rib convection in the GDL due to the significant pressure drop existing between adjacent dead-ended channels [69]. Generally speaking, each flow field has its pros and cons: some of them reduce the pressure and concentration drop along the channel, others enhance reactant transport and product removal to/from the CL, etc. A detailed review of conventional and innovative flow fields can be found in [45]. In passive fuel cells, where the reactants are supplied by natural convection and diffusion, the channels are replaced by circular or rectangular holes. Due to their simplicity, passive PEMFCs have received an increasing attention for small portable applications [70–72].

Microporous layer (MPL). In addition to the four basic components discussed earlier (PEM, CLs, GDLs, and BPPs), MEAs used in PEMFCs incorporate a thin MPL applied onto the GDL face adjacent to the CL. MPLs are a fine mixture of carbon powder and hydrophobic PTFE particles, whose porosity and thickness typically range between 0.3–0.6 and 20–60 μm, respectively [73]. The fine pore structure of MPLs significantly affects water management, reducing flooding at high current densities [74–78]. In addition, MPLs reduce electrical and thermal contact resistances at the GDL/CL interface and improve the hydration of the PEM [30,78].

At cell and system levels, PEMFCs require many other components, such as gaskets, end plates, clamps, temperature controllers, flow-rate regulators, pumps, blowers, and power and control electronics. Single cells are stacked, that is, placed in series, to increase the total voltage (equal to the number of cells times the voltage of each cell) and meet specific power requirements. The BPPs provide the electrical connection between adjacent cells in the stack [47,79]. As commented before, the absence of an active feed system in passive PEMFCs makes them simpler, although their performance is significantly lower [72].

The overall performance of a PEMFC is represented by plotting its voltage, V_{cell}, versus current density, I, the so-called polarization curve. The current output (in amperes) is normalized by the cell active area to make results comparable between cells of different size. The power density, P, is obtained as the product of the cell voltage times the current density, $P = V_{cell}I$, and is also plotted against current density, I, to obtain the corresponding power density curve. Examples of polarization and power density curves of a PEMFC are shown in Fig. 4.6, indicating the dominant voltage losses in each region of the polarization curve.

An ideal PEMFC would supply any amount of current while maintaining a constant reversible voltage, $V_{cell} = E_{rev}^0$, as determined by thermodynamics. $E_{rev}^0 \approx 1.2$ V. However, the voltage output of a real PEMFC is lower than the thermodynamically predicted voltage due to inefficiencies in the electrochemical and transport processes involved in current generation. The actual cell voltage, V_{cell}, at a given current density can be expressed as:

$$V_{cell} = E_{rev}^0 - \eta_{mp} - \eta_{act} - \eta_{ohm} - \eta_{mass} \tag{4.2}$$

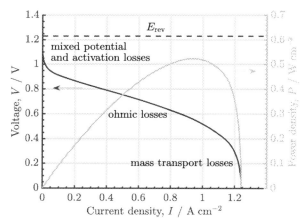

Figure 4.6 Typical polarization and power density curves of a PEMFC. The reversible cell voltage and the predominant voltage losses in each region of the polarization curve are indicated. *PEMFC*, proton exchange membrane fuel cell.

where η_{mp} is the mixed potential loss, η_{act} the activation loss, η_{ohm} the ohmic loss, and η_{mass} the concentration or mass-transport loss. The origin of each term is described below.

Mixed potential losses $\left(\eta_{mp}\right)$ arise due to the reaction of species that cross the membrane (crossover), and are dominant at open-circuit voltage (i.e., $I \approx 0$).

Activation losses $\left(\eta_{act}\right)$ arise due to the finite rate of electrochemical reactions and are more pronounced at low current densities. In PEMFCs, activation losses are mainly caused by the sluggish ORR at the cathode (the overpotential of the HOR is almost negligible).

Ohmic losses $\left(\eta_{ohm}\right)$ are caused by the finite conductivity of cell components to proton and electron transport, and are more pronounced at intermediate current densities. According to Ohm's law, ohmic losses are proportional to the current density, I:

$$\eta_{ohm} = IR_{cell} \tag{4.3}$$

where R_{cell} is the total resistance of the cell, which includes the ionic resistance of the PEM and CLs, R_{proton}, the electronic resistance of electrically conductive solids (CLs, MPLs, GDLs and BPPs), R_{elec}, and the electrical contact resistances at the interfaces between components, R_c:

$$R_{cell} = R_{proton} + R_{elec} + R_c \tag{4.4}$$

Concentration, or mass-transport, losses $\left(\eta_{mass}\right)$ are incurred when insufficient reactants are supplied to the CLs due to the operation of the cell with low feed concentrations and/or when reactant transport is severely hindered by reaction products. These losses are more pronounced at high current densities, when the reactant consumption and product generation rates are high. Mass-transport losses in PEMFCs arise in almost all circumstances at the cathode due to the blockage of oxygen transport by liquid water [36,37,80].

The power output, P, is maximized at intermediate cell voltages. At open circuit no current is drawn from the cell ($I = 0$), while at maximum or limiting current density the cell voltage vanishes ($V_{cell} = 0$ V), so the power density is zero. The peak power density is typically reached at the onset of mass-transport limitations, before the voltage drops sharply. In practice, the sum of the mixed potential losses, η_{mp}, the activation losses, η_{act}, and the concentration losses, η_{mass}, are lumped together in the electrochemical overpotentials of the anode and the cathode CLs, η_a and η_c, respectively. As a result, Eq. (4.2) is more often written as

$$V_{cell} = E^0_{rev} - \eta_a - \eta_c - \eta_{ohm} \tag{4.5}$$

where η_a and η_c are a function of the output current density, I, the parasitic current density, I_p, and the species concentration in the CLs, $C_{i,cl}$.

4.2 Numerical modeling

Two main modeling techniques are usually used to study PEM fuel cells: (1) macroscopic continuum modeling, and (2) pore-scale modeling. Both approaches are discussed below, including recent work on (3) hybrid continuum/pore-scale modeling (see Fig. 4.7).

4.2.1 Macroscopic continuum modeling

The most extended, traditional and indispensable modeling approach used by the fuel cell community is macroscopic continuum modeling. Macroscopic models are based on a volume-averaged continuum formulation of mass, momentum, species, charge and energy conservation equations. The model is closed through suitable constitutive relationships, which describe the various effective transport properties of fuel cell components. These include effective diffusivity, permeability, effective electrical and thermal conductivities, capillary pressure curves, phase change rates, electroosmotic drag coefficient of the PEM, etc. [20,79,81–89]. A detailed review of previous PEMFC and DMFC modeling works reported in the literature can be found in [90,91]. Early attempts, at the end of the 20th century, were mostly focused on simplified but fundamental single-phase models that ignored two-phase transport phenomena (i.e., the effect of liquid water in PEMFCs, and

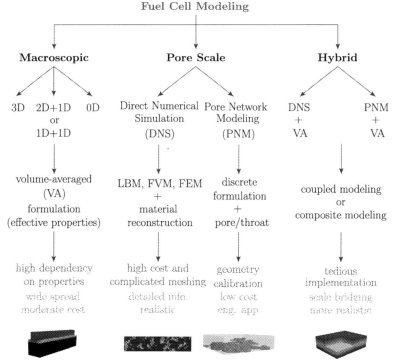

Figure 4.7 Flow chart showing the main types of approaches for modeling fuel cells and components according to their microstructural resolution.

gaseous carbon dioxide/liquid water at the anode/cathode of liquid-feed DMFCs). Significant improvements have been achieved since then. The increased understanding on fuel cell transport and electrochemical processes have led to the development of advanced multiphase computational fluid dynamics (CFD) models in the last decade. Nowadays, a large effort is devoted to the creation of more detailed models that incorporate realistic characterizations of components [83,92]. A wide variety of approaches are adopted in the literature regarding model dimensionality. These include phenomenological 0D models, 1D models accounting for an across-the-channel section, 2D across-the-channel or along-the-channel models, 1D or 2D across-the-channel + 1D along-the-channel models, 3D MEA + 1D along-the-channel models, and fully 3D models [90,91]. The best option with a proper balance between reduced computational cost and high predictive capability is perhaps the 2D MEA + 1D along-the-channel model, although a fully 3D model would be always required for a detailed study of flow fields, especially serpentine or interdigitated channels (see Fig. 4.8).

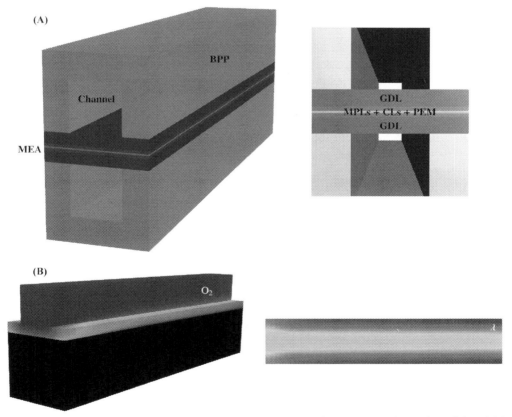

Figure 4.8 Macroscopic continuum modeling. (A) Geometry of a 3D single-channel model and (B) examples of computed distributions of oxygen mass fraction in the cathode and water content in the PEM. *PEM*, proton exchange membrane.

4.2.2 Pore-scale modeling

In tandem to macroscopic continuum modeling, an increasing attention is now devoted to pore-scale modeling. The most outstanding techniques are pore-network modeling (PNM) and direct numerical simulation (DNS), using, e.g., the lattice Boltzmann method (LBM) and the finite volume method (FVM) [35,93]. Both modeling approaches are used to overcome the lack of microstructure resolution inherent to volume-averaged continuum models, and to gain insight into local transport phenomena. Additionally, pore-scale models allow the determination of key effective transport properties that are difficult to measure experimentally due to the small dimensions of MEA components (see, e.g., [36,37,92,94−98] among others). As shown in Fig. 4.9, PNM idealizes the pore space of a material as a network of pore bodies interconnected by throats. The size of the pores and throats and their connectivity are determined according to the morphology of the porous medium [92,95]. Different transport processes can be simulated on the calibrated network, including two-phase, convective and diffusive transport. Two-phase transport is modeled by pore-filling algorithms based on the local capillary resistance of the network, given the dominant role of capillary forces in fuel cell porous components. Convective and

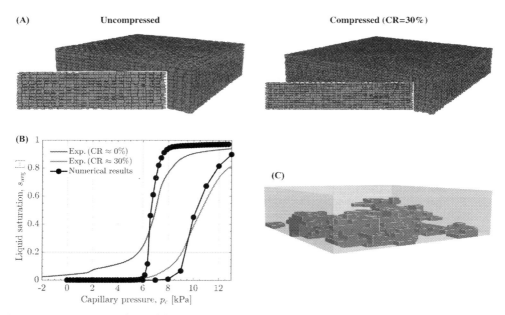

Figure 4.9 Pore network modeling. (A) Pore network representation of an uncompressed and a 30% compressed GDL (10 wt.% PTFE-treated Toray TGP-H-090), (B) computed capillary pressure curve of both samples compared to previous experimental data, and (C) water distribution in the uncompressed sample before water has reached the opposite face from the injection surface. *GDL*, gas diffusion layer; *PTFE*, polytetrafluoroethylene.

diffusive transport are modeled through mass and species conservation equations expressed in terms of discrete resistors [57,59−65]. For example, convection is modeled by solving a discrete mass conservation equation in each pore body

$$q_i = \sum_{j=1}^{n} = g_{i,j}^h (p_j - p_i) = 0 \qquad (4.6)$$

where i and j denote the current and the neighboring pores, n is the number of neighbors, q_i is the total flow through pore i, $g_{i,j}^h$ is the hydraulic conductivity between pores $i-j$ (dependent of the size and length of pores and throats), and p_i and p_j are the pressures in each pore. The total hydraulic conductivity between two adjacent pores is determined as the total conductivity of the flow through half of pore i (pi), the connecting throat (t), and half of pore j (pj), so that the total conductivity of the pore-throat assembly is given by theory for resistors in series as

$$\frac{1}{g_{i,j}^h} = \frac{1}{g_{pi}^h} + \frac{1}{g_t^h} + \frac{1}{g_{pj}^h} \qquad (4.7)$$

where the local hydraulic conductivity is $g^h = KA/L$, with K the permeability, A the cross-sectional area, and L the length of the corresponding pore or throat.

On the other hand, the LBM enables DNS of pore-scale phenomena by solving a space-, time- and velocity-discrete version of the Boltzmann equation [99,100]. The LBM works at the mesoscale, that is, with (fictitious) particle populations. The particle populations stream and collide with each other in specified lattice nodes organized in a rectangular grid. The backbone equation of the LBM can be expressed as

$$\frac{f_\alpha(\mathbf{x}_i + \mathbf{c}_\alpha \Delta t, t + \Delta t) - f_\alpha(\mathbf{x}_i, t)}{\Delta t} = \Omega_\alpha + F_\alpha; \quad \alpha = 0, \ldots, N \qquad (4.8)$$

where $f_\alpha(\mathbf{x}_i, t)$ are the particle distribution functions at location $\mathbf{x}_i$ and time t, $\mathbf{c}_\alpha$ are the discrete lattice velocities, Δt is the time step, Ω_α is the collision operator, F_α is the forcing or external term, and $N+1$ is the number of particle populations. As an example, Fig. 4.10 shows the discrete lattice structure of the D3Q7 model (i.e., 3D with seven velocities) used by García-Salaberri et al. [36,37] to analyze the effective diffusivity of carbon-paper GDLs. The LBM provides an alternative form of solving conservation equations, such as Navier−Stokes or convection-diffusion equations. The relation between the computed particle distribution functions and physically meaningful macroscopic variables (velocity, molar concentration, etc.) is obtained through the Chapmann−Enskog asymptotic analysis of the discretized Boltzmann equation [101,102]. The particulate formulation and local dynamics of the LBM convert this modeling technique in a powerful parallelizable tool to analyze domains with complex boundaries, so that no complicated meshing tools are needed as in traditional CFD techniques. The combination of the LBM

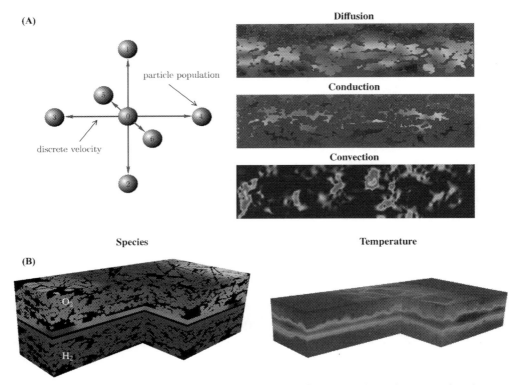

Figure 4.10 Direct numerical simulation. (A) Schematic of a D3Q7 (three-dimensional with seven discrete velocities) lattice structure and cross-sectional view of species concentration (diffusion), electronic potential (conduction) and velocity (convection) fields computed using the lattice Boltzmann method (LBM). (B) Computed pore-scale distributions of species mass fraction (oxygen at the cathode and hydrogen at the anode) and temperature in samples of 10 wt.% PTFE-treated Toray TGP-H-120 and Sigracet SGL 25 AA, respectively, using the finite volume method (FVM). *PTFE*, polytetrafluoroethylene.

with reconstructions of artificial or tomographic images of fuel cell porous media has motivated numerous efforts in the literature to analyze effective transport properties [36,37,42,103–111] and to simulate two-phase transport [112–118].

Comparing both techniques, the simplicity of PNM offers lower computational cost than the LBM. However, the information that can be extracted from DNS is richer, thus making it more appropriate to examine specific transport processes and compute effective properties on real tomography images of porous components.

4.2.3 Hybrid continuum/pore-scale modeling

Currently, there is an increasing need to bridge the gap between volume-averaged and pore-scale models to create more realistic but computationally efficient tools capable of guiding the design of MEAs, forecasting durability and diagnosing operation issues in the most self-predictive way as possible [36,37,42,73,94,96−98,119−130]. This has motivated, for example, the development of hybrid volume-averaged/PNM/DNS models and one-way multiscale models where effective properties extracted from pore-scale simulations are plugged into volume-averaged models (see, e.g., [94,96,117,131]). Inverse modeling techniques, such as global upscaling [132], have also been suggested as an efficient method to transfer local information from pore-scale models to volume-averaged models. Recently, García-Salaberri [92] presented a hybrid model for the analysis of thin porous media in engineering applications, which combined the ease of implementation of macroscopic continuum modeling and the computational power of PNM (see Fig. 4.11). The formulation included a control volume (CV) mesh at the macroscopic scale, which embedded an internal microscopic

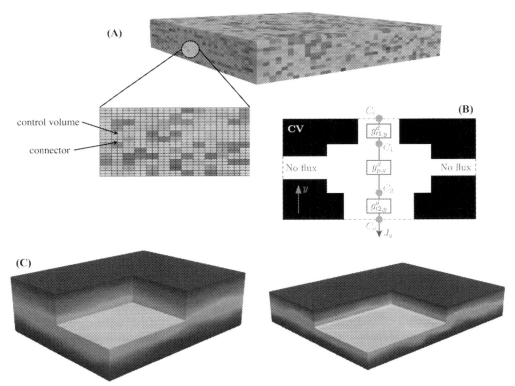

Figure 4.11 Hybrid modeling. (A) spatial pore-size distribution, showing the disposition of the control volumes (CV) and connectors (edges or surfaces between CVs), (B) pore-scale geometry embedded within each CV, and (C) computed species distributions in a sample of 20 wt.% PTFE-treated Toray TGP-H-120 and 5 wt.% PTFE-treated Toray TGP-H-060 using a hybrid continuum-pore network model. *PTFE*, polytetrafluoroethylene.

pore network. The latter was used to extract analytically local anisotropic effective transport properties, which were then mapped into the control volume mesh to resolve macroscopic transport. Thus, for example, the local effective permeability in i-direction, $K_{\text{local},i}^{\text{eff}}$, between two neighboring throats is given by

$$K_{\text{local},i}^{\text{eff}} = \frac{\left(\frac{L_{t1,i}}{A_{t1,i}K_{t1,i}} + \frac{L_{p,i}}{A_{p,i}K_{p,i}} + \frac{L_{t2,i}}{A_{t2,i}K_{t2,i}}\right)^{-1}}{\frac{A_{cv,i}}{L_{t1,i} + L_{p,i} + L_{t2,i}}} \quad (4.9)$$

where $L_{j,i}$, $A_{j,i}$, and $K_{j,i}$ are the length, cross-sectional area, and permeability of pore/throat j in i-direction. As shown before in Eq. (4.7), this result is similar to that found for hydraulic conductors in series with a conductivity $g^h = KA/L$.

Among other applications, this modeling approach can be used to examine the effect of GDL heterogeneities on durability. The reduced computational cost of hybrid modeling is useful to simultaneously analyze the effect of short- and long-range heterogeneities, spatial inhomogeneities due to operating conditions and water flooding within a multiphysics CFD framework.

4.3 PEMFC macroscopic modeling

As commented before, volume-averaged macroscopic modeling is the most extended approach for modeling multiphysics and multiphase transport in PEMFCs, and examine performance and durability. In this section, the formulation of a steady-state, two-phase, non-isothermal macroscopic model of a PEMFC is presented. The formulation is divided into four parts: (1) assumptions, (2) governing equations, (3) source terms, and (4) boundary conditions. Typical physico-chemical and effective transport properties of the cell components (PEM, CL, MPL, GDL and BPP) and the channel are listed in Table 4.2. In addition, typical kinetic parameters and geometrical dimensions of a PEMFC are listed in Tables 4.3 and 4.4, respectively. Below, bold symbols denote 3D vectors, for example, the superficial gas-phase velocity, $\boldsymbol{u_g} = \left(u_{g,x}, u_{g,y}, u_{g,z}\right)$, and overlined bold symbols denote second-order tensors, for example, the absolute permeability tensor

$$\overline{\overline{\boldsymbol{K}}} = \begin{pmatrix} K_{xx} & 0 & 0 \\ 0 & K_{yy} & 0 \\ 0 & 0 & K_{zz} \end{pmatrix} = \begin{pmatrix} K_{ip} & 0 & 0 \\ 0 & K_{ip} & 0 \\ 0 & 0 & K_{tp} \end{pmatrix} \quad (4.10)$$

where the subscripts ip and tp are the in- and through-plane orthotropic tensor components, respectively.

Table 4.2 Typical effective transport properties and physico-chemical parameters of a PEMFC.

Parameter	Symbol	Value	Reference
Density of liquid water	ρ_l	$10^3 - 1.78 \times 10^{-2}(T-277.15)^{1.7}$ kg m^{-3}	[133]
Density of electrolyte	ρ_e	1.98×10^5 kg m^{-3}	[122]
Bulk diffusivity of H$_2$ in H$_2$O (equal to H$_2$O in H$_2$)	D_{H_2,H_2O}^{bulk}	$1.05 \times 10^{-4} \left(\frac{T}{333}\right)^{1.5} \left(\frac{10^5}{p_e}\right)$ m^2 s^{-1}	[122]
Bulk diffusivity of O$_2$ in air	$D_{O_2,air}^{bulk}$	$2.65 \times 10^{-5} \left(\frac{T}{333}\right)^{1.5} \left(\frac{10^5}{p_e}\right)$ m^2 s^{-1}	[122]
Bulk diffusivity of H$_2$O in air	$D_{H_2O,air}^{bulk}$	$2.98 \times 10^{-5} \left(\frac{T}{333}\right)^{1.5} \left(\frac{10^5}{p_e}\right)$ m^2 s^{-1}	[122]
Dynamic viscosity of H$_2$	μ_{H_2}	$3.2 \times 10^{-3} \left(\frac{T}{293.85}\right)^{1.5} (T+72)^{-1}$ kg m^{-1} s^{-1}	[122]
Dynamic viscosity of O$_2$	μ_{O_2}	$8.46 \times 10^{-3} \left(\frac{T}{292.25}\right)^{1.5} (T+127)^{-1}$ kg m^{-1} s^{-1}	[122]
Dynamic viscosity of H$_2$O	μ_{H_2O}	$7.51 \times 10^{-3} \left(\frac{T}{291.15}\right)^{1.5} (T+127)^{-1}$ kg m^{-1} s^{-1}	[122]
Dynamic viscosity of N$_2$	μ_{N_2}	$7.33 \times 10^{-3} \left(\frac{T}{300.55}\right)^{1.5} (T+111)^{-1}$ kg m^{-1} s^{-1}	[122]
Dynamic viscosity of liquid water	μ_l	$2.41 \times 10^{-5} 10^{247.8/(T-140)}$ kg m^{-1} s^{-1}	[122]
In- and through-plane normalized gas-phase dry effective diffusivity of MPL/CL	$f_{ip/tp}$	0.12/0.2	[122]
In- and through-plane normalized gas-phase dry effective diffusivity of GDL (Toray)	$f_{gdl,ip/tp}$	0.45/0.21	[42,97,98,122]
Specific heat at constant pressure of H$_2$/O$_2$/H$_2$O/N$_2$	$c_{p,H_2}/c_{p,O_2}/c_{p,H_2O}/c_{p,N_2}$	$1.44 \times 10^4 / 9.28 \times 10^2 / 1.88 \times 10^3 / 1.04 \times 10^3$ J kg^{-1} K^{-1}	[122]
Thermal conductivity of H$_2$/O$_2$/H$_2$O/N$_2$	$k_{H_2}/k_{O_2}/k_{H_2O}/k_{N_2}$	$1.67 \times 10^{-1} / 2.46 \times 10^{-2} / 2.61 \times 10^{-2} / 2.4 \times 10^{-2}$ W m^{-1} K^{-1}	[122]
Thermal conductivity of liquid water	k_l	6.6×10^{-1} W m^{-1} K^{-1}	[122]
In- and through-plane effective thermal conductivity of BPP/MPL/CL/PEM	$k_{ip/tp}^{eff}$	$1.2 \times 10^2 / 10^{-1} / 2.7 \times 10^{-1} / 1.3 \times 10^{-1}$ W m^{-1} K^{-1}	[122]
In- and through-plane effective thermal conductivity of GDL (Toray)	$k_{gdl,ip/tp}^{eff}$	15/ 1 W m^{-1} K^{-1}	[97,122]
Evaporation/condensation rate constant of water	$k_{evp/con}$	10^{-4} Pa^{-1} s^{-1}/10^4 s^{-1}	[122]
Equivalent weight of electrolyte (Nafion)	EW_e	1.1 kg mol^{-1}	[122]
Ionomer volume fraction in CL	ω_e	0.3	[122]
In- and through-plane effective electrical conductivity of BPP/MPL/CL	$\sigma_{e,ip/tp}^{eff}$	$2 \times 10^3 / 3 \times 10^2 / 3 \times 10^2$ S m^{-1}	[122]

(*Continued*)

Table 4.2 (Continued)

Parameter	Symbol	Value	Reference
In- and through-plane effective electrical conductivity of GDL (Toray)	$\sigma_{e-\text{gdl},ip/tp}^{\text{eff}}$	$1.7 \times 10^4 / 10^3$ S m^{-1}	[97,122]
Porosity of GDL/MPL/CL	ε	0.7/0.5/0.4	[37,97,122]
In- and through-plane absolute permeability of MPL/CL	$K_{ip/tp}$	$10^{-13}/10^{-15} - 10^{-14}$ m^2	[122]
In- and through-plane absolute permeability of GDL (Toray)	$K_{\text{gdl},ip/tp}$	$1.7 \times 10^{-11}/5 \times 10^{-12}$ m^2	[60,97,122]
Latent heat of condensation/evaporation (same as for sorption/desorption)	h_{gl}	$3.17 \times 10^6 - 244 \times 10^3 T$ J kg^{-1}	[122]
Entropy change (liquid water production)	ΔS	-1.63×10^2 J mol^{-1} K^{-1}	[122]
Contact angle of GDL/MPL/CL	θ_c	110°/120°/95°	[122]
Surface tension coefficient	σ	$1.22 \times 10^{-1} - 1.67 \times 10^{-4} T$ N m^{-1}	[122]
Effective diffusion coefficient of liquid water in the channel	$D_{th,l}^{\text{eff}}$	$10^{-3} - 10^{-5}$ m^2 s^{-1}	[134,135]

Table 4.3 Typical kinetic parameters of a PEMFC.

Parameter	Symbol	Value	Reference
Exchange current density of HOR	$i_{0,a}$	$5 \times 10^8 \exp\left[-\frac{1.7 \times 10^4}{RT}\left(\frac{1}{T} - \frac{1}{353.15}\right)\right]$ A m^{-3}	[96,122]
Transfer coefficient of HOR	α_a	1	[96,122]
Reference hydrogen concentration	$C_{H_2}^{ref}$	40 mol m^{-3}	[96,122]
Transfer coefficient of ORR	α_c	0.5	[96,122]
Reference prefactor of ORR	j^*	10^{14} A m^{-3}	[96,122]
Activation free energy of RD step	ΔG_{RD}^0	0.278 eV	[96,122]
Activation free energy of RT step	ΔG_{RT}^0	0.5904 eV	[96,122]
Activation free energy of RA step	ΔG_{RA}^0	0.6094 eV	[96,122]
Activation free energy of DA step	ΔG_{DA}^0	0.3907 eV	[96,122]
Adsorption free energy of OH	ΔG_{OH}^0	−0.376 eV	[96,122]
Adsorption free energy of O	ΔG_O^0	−0.3431 eV	[96,122]

Table 4.4 Typical geometrical dimensions of a PEMFC.

Parameter	Symbol	Value
Channel width	w_{ch}	1 mm
Channel height	H_{ch}	1 mm
Rib width	w_{rib}	1 mm
GDL thickness	δ_{gdl}	200 μm
MPL thickness	δ_{mpl}	40 μm
CL thickness	δ_{cl}	10 μm
PEM thickness	δ_{pem}	25 μm
Cell active area	A_{cell}	$\sim$25 cm^2

4.3.1 Assumptions

The model is based on several simplifying hypotheses. The most relevant assumptions are as follows:
1. The flow is laminar and steady.
2. The components of the MEA are macroscopically homogeneous.

3. Interfacial ohmic and thermal resistances are negligible.
4. The membrane is perfectly impermeable to gases.
5. Liquid water produced in the cathode CL is dissolved in the electrolyte.
6. Soret and Dufour effects and viscous heat dissipation are negligible.

Other assumptions are commonly used in PEMFC models and can be found elsewhere [136–138].

4.3.2 Conservations equations

The conservation equations of the model are presented below, indicating the region where each equation is solved (see Fig. 4.12):

- *Gas-phase mass and momentum (x, y and z)*: CLs, MPLs, GDLs and channels. The mass and momentum conservation equations governing the motion of the gas phase are:

$$\nabla \cdot \left(\rho_g \boldsymbol{u}_g\right) = S_{mg}, \quad \nabla \cdot \left(\frac{\rho_g}{[\varepsilon(1-s)]^2} \boldsymbol{u}_g \boldsymbol{u}_g\right) = -\nabla p_g + \nabla \cdot \boldsymbol{\tau} + S_{ug} \quad (4.11)$$

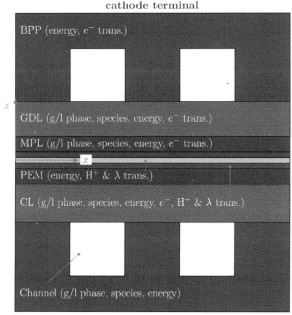

Figure 4.12 Schematic of a cross-section of a PEMFC, showing the anode and the cathode terminals, the MEA components (PEM, CLs, MPLs and GDLs), the BPPs and the channels. The equations that are solved in each region are indicated in brackets. Drawing dimensions are not to scale. *MEA*, membrane electrode assembly; *PEM*, proton exchange membrane; *CLs*, catalyst layers; *MPLs*, microporous layers; *BPPs*, bipolar plates.

where $\rho_g = p_g M_g/RT$ is the density of the ideal gas mixture [with $M_g = \left(\sum_i Y_i/M_i\right)^{-1}$ the molecular weight of the mixture], $\boldsymbol{u_g}$ is the superficial gas-phase velocity, and ε is the porosity ($\varepsilon = 1$ in the channels). The viscous stress tensor is given by

$$\boldsymbol{\tau} = \mu_g \left[\nabla\left(\frac{\boldsymbol{u_g}}{\varepsilon(1-s)}\right) + \left[\nabla\left(\frac{\boldsymbol{u_g}}{\varepsilon(1-s)}\right)\right]^T - \frac{2}{3} \nabla \cdot \left(\frac{\boldsymbol{u_g}}{\varepsilon(1-s)}\boldsymbol{I}\right)\right] \quad (4.12)$$

where $\overline{\overline{\boldsymbol{I}}}$ denotes the second-order unit tensor. According to the kinetic theory of gases, the dynamic viscosity of the gas mixture is equal to

$$\mu_g = \sum_i \frac{X_i \mu_i}{\sum_j X_j \beta_{ij}}; \quad \beta_{ij} = \frac{\left[1 + \left(\frac{\mu_i}{\mu_j}\right)^{1/2}\left(\frac{M_j}{M_i}\right)^{1/4}\right]^2}{\left[8\left(1+\frac{M_i}{M_j}\right)\right]^{1/2}} \quad (4.13)$$

where X_i, μ_i and M_i are the mole fraction, dynamic viscosity and molecular mass of species i, respectively.

- *Species* $\left(Y_{H_2}, Y_{O_2} \text{ and } Y_{H_2O}\right)$: CLs, MPLs, GDLs and channels. Species mass fractions are governed by the convection-diffusion equation:

$$\nabla \cdot \left(\rho_g \boldsymbol{u_g} Y_i\right) - \nabla \cdot \left(\rho_g \overline{\overline{\boldsymbol{D}}}_{i,j}^{\text{eff}} \nabla Y_i\right) = S_i \quad (4.14)$$

where the effective diffusivity tensor of species i in species j is given by

$$\boldsymbol{D}_{ij}^{\text{eff}} = D_{ij}^{\text{bulk}} \begin{pmatrix} f_{ip}(\varepsilon)g_{ip}(s) & 0 & 0 \\ 0 & f_{ip}(\varepsilon)g_{ip}(s) & 0 \\ 0 & 0 & f_{tp}(\varepsilon)g_{tp}(s) \end{pmatrix} \quad (4.15)$$

Here, $f_{ip/tp}(\varepsilon) = D_{ij,ip/tp}^{\text{eff,dry}}/D_{i,j}^{\text{bulk}}$ is the normalized gas-phase dry effective diffusivity, which accounts for the blockage of unsaturated porous media on gas diffusion, and $g_{ip/tp} = D_{ij,ip/tp}^{\text{eff}}/D_{ij,ip/tp}^{\text{eff,dry}}$ is the gas-phase relative effective diffusivity, which accounts for the relative blockage caused by liquid water on gas diffusion. By definition, $f_{ip/tp}$ is lower than one $\left(f_{ip/tp} < 1\right)$ in the porous layers of the MEA, while it is equal to one in the bulk channels $\left(f_{ip} = f_{tp} = 1\right)$. On the other hand, $g_{ip/tp}$ is typically modeled using a power law of the form $g_{ip/tp}(s) = (1-s)^n$, where the power-law exponent is around $n \approx 2.5 - 4$, as reported by García-Salaberri et al. [36,37].

Assuming a perfectly impermeable PEM, at the anode it is only necessary to solve the conservation equation of hydrogen, Y_{H_2}, while the mass fraction of water vapor can be determined as $Y_{H_2O} = 1 - Y_{H_2}$. Similarly, at the cathode, nitrogen is the most abundant species, so that we solve for Y_{O_2} and Y_{H_2O}, and Y_{N_2} is calculated as $Y_{N_2} = 1 - Y_{O_2} - Y_{H_2O}$.

- *Energy*: PEM, CLs, MPLs, GDLs, channels and BPPs. The energy conservation equation determining the temperature field, T, is:

$$\nabla \cdot \left[\boldsymbol{u_g} \left(\rho_g E_g + p_g \right) \right] - \nabla \cdot \left(\boldsymbol{\overline{\overline{k}}^{eff}} \nabla T \right) = S_T \qquad (4.16)$$

where $\boldsymbol{\overline{\overline{k}}^{eff}}$ is the effective thermal conductivity tensor. The total gas-phase energy is equal to

$$E_g = h_g - \frac{p_g}{\rho_g} + \frac{|\boldsymbol{u_g}|^2}{2} \qquad (4.17)$$

where the sensible enthalpy, h_g, is determined as the mass-weighted average of the enthalpy of each species, $h_g = \sum_i Y_i h_i$, $h_i = \int_{T^{ref}}^T c_{p,i} \, dT$, with $c_{p,i}$ the specific heat at constant pressure of species i and $T^{ref} = 298$ K the reference temperature.

In the channels, $k^{eff}_{ch,ip/tp}$ varies with water saturation

$$k^{eff}_{ch,ip/tp} = (1-s) \underbrace{\sum_i \frac{X_i k_i}{\sum_j X_j \beta_{ij}}}_{k^{bulk}_g} + s k_l \qquad (4.18)$$

where k^{bulk}_g is the bulk thermal conductivity of the gas mixture, k_i and k_l are the thermal conductivities of gas species i and of liquid water, respectively, and β_{ij} is given by Eq. (4.13).

In the porous layers of the MEA (GDL, MPL and CL), $k^{eff}_{ip/tp}$ can be written as

$$k^{eff}_{ip/tp} = \frac{k^{eff,dry}_{ip/tp}}{k^{bulk}_{ip/tp}} \frac{k^{eff}_{ip/tp}}{k^{eff,dry}_{ip/tp}} \qquad (4.19)$$

where $k^{eff,dry}_{ip/tp}/k^{bulk}_{ip/tp}$ accounts for the effect of the microstructure of the dry porous medium and $k^{eff}_{ip/tp}/k^{eff,dry}_{ip/tp}$ for the relative effect of water saturation. The dependence of the thermal conductivity on liquid water has a significant effect on water management.

- *Electronic transport*: CLs, MPLs, GDLs and BPPs. Electron transport is governed by the charge conservation equation (Poisson's equation):

$$\nabla \cdot \left(\boldsymbol{\overline{\overline{\sigma}}^{eff}_{e^-}} \nabla \phi_{e^-} \right) = S_{e^-} \qquad (4.20)$$

where ϕ_{e^-} is the electronic potential in the solid phase and $\boldsymbol{\overline{\overline{\sigma}}^{eff}_{e^-}}$ is the effective electrical conductivity tensor. According to Ohm's law, $\boldsymbol{j_{e^-}} = \boldsymbol{\overline{\overline{\sigma}}^{eff}_{e^-}} \nabla \phi_{e^-}$ is the electronic flux directed from low to high electronic potentials.

- *Ionic or protonic transport*: PEM and CLs. As in the case of electron transport, the governing equation for proton transport is given by:

$$-\nabla \cdot \left(\overline{\overline{\sigma}}_{H^+}^{eff} \nabla \phi_{H^+} \right) = S_{H^+} \quad (4.21)$$

where ϕ_{H^+} is the ionic potential in the electrolyte phase and $\overline{\overline{\sigma}}_{H^+}^{eff}$ is the effective ionic conductivity tensor, which depends on the water content, λ, and temperature, T. As an example, the effective ionic conductivity of Nafion NR-211 can be expressed as [139]

$$\sigma_{H^+, ip/tp}^{eff} = (10^{-2}\lambda^2 + 1.05\lambda - 2.06)\exp\left[751.5\left(\frac{1}{303} - \frac{1}{T}\right)\right] \; S \, m^{-1} \quad (4.22)$$

An isotropic Bruggeman correlation, $\omega_e^{1.5}$, is usually considered to correct the ionic conductivity of CLs, i.e., $\sigma_{H^+,cl}^{eff} = \omega_e^{1.5} \sigma_{H^+,pem}^{eff}$ [140,141]. The protonic flux, directed from high to low ionic potentials, is equal to $\boldsymbol{j}_{H^+} = \overline{\overline{\sigma}}_{H^+}^{eff} \nabla \phi_{H^+}$.

- *Dissolved water in electrolyte*: PEM and CLs. The conservation equation of the water content (i.e., water molecules per sulfonic acid group) in the electrolyte, λ, is given by [142]

$$\nabla \cdot \left(\frac{n_d^w}{F}\boldsymbol{j}_{H^+}\right) - \nabla \cdot \left[\left(\frac{\rho_e}{EW_e}\right) \boldsymbol{D}_{\lambda e}^{eff} \nabla \lambda\right] = S_\lambda \quad (4.23)$$

where $V_{m,e} = EW_e/\rho_e$ is the molar volume of the electrolyte.

The first term on the left-hand side of Eq. (4.23) accounts for electroosmotic drag of water due to proton transport from anode to cathode, where the electroosmotic drag coefficient, n_d^w, can be modeled by a continuous piece-wise function [142,143]

$$n_d^w = \begin{cases} \lambda, & \lambda < 1 \\ 1, & 1 \leq \lambda \leq \lambda_{eq}|_{a_{H_2O}=1} \\ 1 + \frac{1}{4}\left(\lambda - \lambda_{eq}|_{a_{H_2O}=1}\right), & \lambda \geq \lambda_{eq}|_{a_{H_2O}=1} \end{cases} \quad (4.24)$$

where λ_{eq} is the equilibrium water content in the electrolyte and $a_{H_2O} = p_{H_2O}/p_{H_2O}^{sat}$ is the water activity. The linear increase of n_d^w for $\lambda \geq \lambda_{eq}|_{a_{H_2O}=1}$ accounts for the raise of n_d^w measured in PEMs equilibrated with liquid water [143].

The second term of Eq. (4.23) accounts for diffusion of dissolved water, where the effective diffusion coefficient of dissolved water in the PEM, $\overline{\overline{D}}_{\lambda e}^{eff}$, increases nonlinearly with λ and T [22,122]

$$D_{pem,\lambda e}^{eff} = 8 \times 10^{-10}\left(\frac{\lambda}{25}\right)^{0.15}\left[1 + \tanh\left(\frac{\lambda - 2.5}{1.4}\right)\right]\exp\left[2416\left(\frac{1}{303} - \frac{1}{T}\right)\right] \; m^2 \, s^{-1} \quad (4.25)$$

In CLs, $\overline{\overline{\mathbf{D}}}_{\lambda e}^{\text{eff}}$ is corrected depending on the ionomer volume fraction, usually considering a Bruggeman correlation, $D_{cl,\lambda e}^{\text{eff}} = \omega_e^{1.5} D_{\text{pem},\lambda e}^{\text{eff}}$.

In addition, hydraulic permeation and thermoosmotic transport can be important when there is a significant pressure and temperature difference between both cell compartments, respectively. The water flux due to convection by hydraulic pressure gradient is given by

$$\mathbf{N}_{\text{conv},\lambda} = -c_w^{\text{avg}} \frac{K_e}{\mu_l} \nabla p_l \qquad (4.26)$$

where $c_w^{\text{avg}} = \lambda/(\lambda V_{m,l} + V_{m,e})$ and $K_e = K_e|_{a_{H_2O}=1}(\lambda/\lambda|_{a_{H_2O}=1})$ are the average concentration and the hydraulic permeability of water in the electrolyte, respectively. Here, $K_e|_{a_{H_2O}=1}$ and $\lambda|_{a_{H_2O}=1}$ are the hydraulic permeability and water content when the electrolyte is in thermodynamic equilibrium with saturated water vapor, while the molar volume of water is equal to $V_{m,l} = M_{H_2O}/\rho_l$. The linear increment of K_e accounts for the increase of the electrolyte permeability by the growth of ionic channels with λ. For Nafion, $K_e|_{a_{H_2O}=1} \sim 10^{-18}$ m². On the other hand, the water flux due to thermoosmotic transport is given by

$$\mathbf{N}_{T,\lambda} = -\frac{D_{\lambda,T}^{\text{eff}}}{M_{H_2O}} \nabla T \qquad (4.27)$$

where $D_{\lambda,T}^{\text{eff}}$ is the thermoosmotic diffusion coefficient in the electrolyte. For Nafion 112, Kim and Mench [144] reported the following correlation:

$$D_{T,\lambda}^{\text{eff}} = -1.04 \times 10^{-5} \exp\left(-\frac{2362}{T}\right) \text{ kg m}^{-1}\text{ s}^{-1}\text{ K}^{-1} \qquad (4.28)$$

where the minus sign considers that water transport occurs from low to high temperatures to increase entropy according to the second principle of thermodynamics [144,145].

Considering these two additional modes of transport, Eq. (4.23) takes the form

$$\nabla \cdot \left(\frac{n_d^w}{F} \mathbf{j}_{\mathbf{H}^+} - \frac{c_w^{\text{avg}} K_e}{\mu_l} \nabla p_l\right) - \nabla \cdot \left[\left(\frac{\rho_e}{EW_e}\right) \overline{\overline{\mathbf{D}}}_{\lambda e}^{\text{eff}} \nabla \lambda + \frac{1}{M_{H_2O}} \overline{\overline{\mathbf{D}}}_{\lambda,T}^{\text{eff}} \nabla T\right] = S_\lambda \qquad (4.29)$$

- *Liquid-phase flow.* CLs, MPLs, GDLs and channels. Transport of liquid water in the porous layers of the MEA is described by the mass conservation equation

$$\nabla \cdot (\rho_l \mathbf{u}_l) = S_{ml} \qquad (4.30)$$

where ρ_l is the density of liquid water and $\mathbf{u}_l$ is the superficial liquid-phase velocity.

Neglecting any effect of gas-phase convection (and gravity), according to Darcy's law, $\boldsymbol{u}_l$ is given by

$$\boldsymbol{u}_l = -\frac{1}{\mu_l}\bar{\bar{\boldsymbol{K}}} \circ \bar{\bar{\boldsymbol{k}}}_{rl} \nabla p_l \tag{4.31}$$

where μ_l is the dynamic viscosity of liquid water, p_l is the liquid-phase pressure, and $\bar{\bar{\boldsymbol{K}}}$ and $\bar{\bar{\boldsymbol{k}}}_{rl}$ are the absolute and liquid-phase relative permeability tensors, respectively. An isotropic power-law correlation is usually considered for k_{rl}, $k_{rl,ip/tp} = s^n$, with $n = 3 - 4$ [122].

Capillary pressure, p_c, is typically expressed as a function of water saturation using the semiempirical correlation proposed by Leverett [136]

$$p_c = p_g - p_l = \sigma \cos\theta_c \left(\frac{\varepsilon}{\left|\bar{\bar{\boldsymbol{K}}}\right|}\right)^{1/2} J(s) \tag{4.32}$$

where σ is the surface tension coefficient of the immiscible liquid-gas pair, θ_c is the contact angle, $\left|\bar{\bar{\boldsymbol{K}}}\right| = \left(\bar{\bar{\boldsymbol{K}}}:\bar{\bar{\boldsymbol{K}}}^T\right)^{1/2}$ is the magnitude of the absolute permeability tensor, and $J(s)$ is the is the Leverett J-function

$$J(s) = \begin{cases} 1.42(1-s) - 2.12(1-s)^2 + 1.26(1-s)^3, & \theta_c < 90^\circ \\ 1.42s - 2.12s^2 + 1.26s^3, & \theta_c \geq 90^\circ \end{cases} \tag{4.33}$$

Although the applicability of this correlation, originally developed for soil and rock porous media, to thin porous media found in PEMFCs has been debated many times (see, e.g., [97], and references therein), it provides a reasonable first approximation to describe capillary transport of water.

Combining Eq. (4.30)-Eq. (4.31), we yield an equation for p_l, so that s can then be determined through Eq. (4.32) using a numerical root-finding method, for example, the Newton–Raphson method. This modeling approach is more physically sound, providing a direct implementation of the saturation jump condition between the GDL, MPL and CL, given by the continuity of capillary pressure through these layers.

Modeling transport of liquid water in the channels poses a complicated two-phase problem due to the formation of liquid droplets, films and/or slugs. This scenario is usually oversimplified assuming a fine mist flow, so that the velocity of the liquid and gas phases are the same, and water transport is modeled with a convection–diffusion equation

$$\nabla \cdot (\rho_l \boldsymbol{u}_g s) - \nabla \cdot \left(\rho_l D_{ch,l}^{eff} \nabla s\right) = S_{ml} \tag{4.34}$$

where $D_{ch,l}^{eff}$ is the effective diffusivity of liquid water in the channel.

4.3.3 Source terms

The source terms include: (1) the electrochemical reaction rates, (2) the mass transfer rates between the different phases of water, (3) the Darcy's viscous resistance of the gas phase, and (4) the release and absorption rates of heat in the various cell components. Table 4.5 summarizes the regions where the different source terms are applied.

- *Reaction rate.* The half-cell reactions corresponding to the fast HOR in the anode CL and the sluggish ORR in the cathode CL are $H_2 \rightarrow 2H^+ + 2e^-$ and $(1/2)O_2 + 2H^+ + 2e^- \rightarrow H_2O$, respectively. The volumetric current density of the ORR can be modeled with a double-trap kinetic model [96]:

$$j_c = j^*(1-s)\left[\exp\left(\frac{-\Delta G^*_{RD}}{k_b T}\right)\theta_{OH} - \exp\left(\frac{-\Delta G^*_{-RD}}{k_b T}\right)(1 - \theta_O - \theta_{OH})\right] \quad (4.35)$$

where j^* is a reference prefactor, $k_b = 8.617 \times 10^{-5}$ eVK^{-1} is the Boltzmann's constant, and ΔG^*_{RD} and ΔG^*_{-RD} are the potential-dependent activation free energies of the forward and backward reactions, respectively. The term $1-s$ accounts for the reduction of the active area available for reaction due to blockage of liquid water. The coverage factors of the intermediate species are given by

$$\theta_{OH} = \frac{C g_{DA}(C g_{RA} + g_{-RD} - g_{RT}) - (C g_{RA} + g_{-RD})(C g_{DA} + g_{-DA} + g_{RT})}{(C g_{DA} - g_{-RT})(C g_{RA} + g_{-RD} - g_{RT}) - (C g_{RA} + g_{-RA} + g_{-RT} + g_{RD} + g_{-RD})(C g_{DA} + g_{-DA} + g_{RT})}$$

$$\theta_O = \frac{C g_{DA}(C g_{RA} + g_{-RA} + g_{-RT} + g_{RD} + g_{-RD}) - (C g_{RA} + g_{-RD})(C g_{DA} - g_{-RT})}{(C g_{DA} + g_{-DA} + g_{RT})(C g_{RA} + g_{-RA} + g_{-RT} + g_{RD} + g_{-RD}) - (C g_{RA} + g_{-RD} - g_{RT})(C g_{DA} - g_{-RT})}$$

(4.36)

Here, C is the oxygen concentration ratio, $C = (C_{O_2}/C^{ref}_{O_2})$, and $g_i = \exp[-\Delta G^*_i/(k_b T)]$, where the reference oxygen concentration is taken equal to the inlet concentration corresponding to the specified working temperature, relative humidity and pressure, i.e., $C^{ref}_{O_2} = C^{in}_{O_2}$. The potential-dependent activation free energies of the ith step are equal to

$$\begin{aligned}
\Delta G^*_{DA} &= \Delta G^0_{DA} \\
\Delta G^*_{RA} &= \Delta G^0_{RA} + \alpha_c e\eta \\
\Delta G^*_{RT} &= \Delta G^0_{RT} + \alpha_c e\eta \\
\Delta G^*_{RD} &= \Delta G^0_{RD} + \alpha_c e\eta \\
\Delta G^*_{-DA} &= \Delta G^0_{DA} - \Delta G^0_O \\
\Delta G^*_{-RA} &= \Delta G^0_{RA} - \Delta G^0_{OH} - (1-\alpha_c)e\eta \\
\Delta G^*_{-RT} &= \Delta G^*_{RT} - \Delta G^0_{OH} + \Delta G^0_O - (1-\alpha_c)e\eta \\
\Delta G^*_{-RD} &= \Delta G^0_{RD} + \Delta G^0_{OH} - (1-\alpha_c)e\eta
\end{aligned} \quad (4.37)$$

Table 4.5 Source terms in a PEMFC.

Source	PEM	CLa	CLc	MPL	GDL	Ch	BPP
S_{mg}	—	$M_{H2}S_{H2} - M_{H2O}(S_{nl} + S_{v\lambda})$	$M_{O_2}S_{O_2} - M_{H2O}(S_{nl} + S_{v\lambda})$	0	0	0	—
S_{ug}	—	Darcy's resistance	Darcy's resistance	Darcy's resistance	Darcy's resistance	0	—
S_{O_2}	—	—	$M_{O_2}S_{O_2}$	0	0	0	—
S_{H2}	—	$M_{H2}S_{H2}$	—	0	0	0	—
S_{H2O}	—	$-M_{H2O}(S_{nl} + S_{v\lambda})$	$-M_{H2O}(S_{nl} + S_{v\lambda})$	$-M_{H2O}S_{nl}$	$-M_{H2O}S_{nl}$	$-M_{H2O}S_{nl}$	—
S_T	$S_{T,H+}$	$S_{T,act} + S_{T,ohm} + M_{H2O}(S_{T,nl} + S_{T,v\lambda})$	$S_{T,rev} + S_{T,act} + S_{T,ohm} + M_{H2O}(S_{T,nl} + S_{T,v\lambda})$	$S_{T,e-} + M_{H2O}S_{T,nl}$	$S_{T,e-} + M_{H2O}S_{T,nl}$	$M_{H2O}S_{T,nl}$	$S_{T,e-}$
S_{e-}	—	j_a	$-j_c$	0	0	—	0
S_{H+}	0	j_a	$-j_c$	—	—	—	—
S_λ	0	$S_{v\lambda}$	$S_{v\lambda}$	—	—	—	—
S_{nl}	—	$M_{H2O}S_{nl}$	$M_{H2O}S_{nl}$	$M_{H2O}S_{nl}$	$M_{H2O}S_{nl}$	$M_{H2O}S_{nl}$	—

BPP, bipolar plate; *Ch*, channel; *CLa*, anode catalyst layer; *CLc*, cathode catalyst layer; *GDL*, gas diffusion layer; *MPL*, microporous layer; *PEM*, polymer electrolyte membrane.

where η is the overpotential, defined as $\eta = \phi_{e^-} - \phi_{H^+}$ in both electrodes (see [137] for further details), $e = 1$ eVV^{-1} is the elementary charge and α_c is the ORR transfer coefficient.

To capture the nanostructure of the CL, an agglomerate model can be used (see, e.g., [146]). Assuming that agglomerates are of constant size and are covered by a thin continuous ionomer film, the agglomerate current density is equal to

$$j_{agg} = 4FV_{agg}^{avg} \frac{p'_{O_2}}{H_{O_2,e}} \left[\frac{1}{E_r k_c} + \frac{\delta_{agg} r_{agg}^2}{3(r_{agg} + \delta_{agg}) D_{O_2,e}} \right]^{-1} \quad (4.38)$$

where r_{agg} and δ_{agg} are the agglomerate radius and thin ionomer film thickness, respectively, $D_{O_2,e}$ is the oxygen diffusivity in the ionomer, $p'_{O_2}/H_{O_2,e}$ is the ratio of oxygen partial pressure and Henry's constant at the thin film-pore interface, and $V_{agg}^{avg} = r_{agg}^3/(r_{agg} + \delta_{agg})^3$ is the ratio of the volume of agglomerate core to the volume of entire agglomerate. Considering first-order oxygen kinetics, the expression for the effectiveness factor, E_r, is

$$E_r = \frac{1}{\phi_L} \left[\frac{1}{\tanh(3\phi_L)} - \frac{1}{3\phi_L} \right] \quad (4.39)$$

where ϕ_L is the Thiele's modulus

$$\phi_L = \frac{r_{agg}}{3} \sqrt{\frac{k_c}{D_{O_2,air}^{eff}}} \quad (4.40)$$

The reaction rate at the surface of agglomerates is given by

$$k_c = \frac{j_c}{4F(1 - \varepsilon_d) V_{agg}^{avg} C_{O_2}^{ref}} \quad (4.41)$$

where $(1 - \varepsilon_d) V_{agg}^{avg}$ is the active area scaling factor.

The macroscopic volumetric current density, j_c, is related with the agglomerate volumetric current density, j_{agg}, as $j_c = j_{agg}/(1 - \varepsilon_d)$. Therefore, combining Eq. (4.35) and Eq. (4.38)-Eq. (4.41) with $C_{O_2} = p'_{O_2}/RT$, p'_{O_2} can be obtained using a numerical root-finding method. Then, j_c can be evaluated using Eq. (4.35), as well as the remaining derived quantities.

The volumetric current density of the HOR is usually modeled with a standard Butler–Volmer kinetics model [96]:

$$j_a = i_{0,a}(1 - s)\left(\frac{C_{H_2}}{C_{H_2}^{ref}}\right)\left[\exp\left(\frac{\alpha_a F}{RT}\eta_a\right) - \exp\left(-\frac{(1 - \alpha_a)F}{RT}\eta_a\right)\right] \quad (4.42)$$

where $i_{0,a}$ is the volumetric exchange current density and α_a is the HOR transfer coefficient. The use of agglomerate models at the anode side is less common, since mass transport losses in this compartment are typically small due to the high diffusivity of hydrogen and fast kinetics of the HOR.

The source terms for the electronic and ionic potentials due to volumetric generation and consumption of electrons and protons are given by $S_{e^-} = S_{H^+} = j_a$ and $S_{e^-} = S_{H^+} = -j_c$ in the anode and the cathode CLs, respectively. According to Faraday's law and the reaction stoichiometry, the production/consumption rates of hydrogen, oxygen and dissolved water in the electrolyte are

$$S_{H_2} = -\frac{j_a}{2F}$$

$$S_{O_2} = -\frac{j_c}{4F} \quad (4.43)$$

$$S_{\lambda,\text{prod}} = \frac{j_c}{2F}$$

where the minus sign denotes consumption.

- *Water evaporation and condensation.* The evaporation/condensation rate of water is typically modeled by the following expression [122]

$$S_{vl} = \begin{cases} \dfrac{k_{evp}\varepsilon s}{V_{m,l}}\left(p_{H_2O} - p_{H_2O}^{sat}\right), & p_{H_2O} < p_{H_2O}^{sat} \\ \dfrac{k_{con}\varepsilon(1-s)X_{H_2O}}{RT}\left(p_{H_2O} - p_{H_2O}^{sat}\right), & p_{H_2O} \geq p_{H_2O}^{sat} \end{cases} \quad (4.44)$$

where k_{evp} and k_{con} are the evaporation and condensation rate constants, $V_{m,l}$ is the molar volume of liquid water, $X_{H_2O} = C_{H_2O}RT/p_g$ is the mole fraction of water vapor, and the driving force is the difference between the partial and saturation pressures of water vapor, p_{H_2O} and $p_{H_2O}^{sat}(T)$, respectively. The saturation pressure of water depends on temperature through the expression

$$\log_{10}\left(p_{H_2O}^{sat}\right) = -2.18 + 2.95 \times 10^{-2}\Delta T - 9.18 \times 10^{-5}\Delta T^2 + 1.44 \times 10^{-7}\Delta T^3 \text{ atm}$$

$$(4.45)$$

where $\Delta T = T - 273.15$.

- *Water sorption and desorption.* The sorption/desorption rate between water vapor and dissolved water in the CLs has the following form

$$S_{v\lambda} = \begin{cases} \dfrac{k_{sor}\rho_e}{EW_e}\left(\lambda_{eq} - \lambda\right), & \lambda < \lambda_{eq} \\ \dfrac{k_{des}\rho_e}{EW_e}\left(\lambda_{eq} - \lambda\right), & \lambda \geq \lambda_{eq} \end{cases} \quad (4.46)$$

where k_{sor} and k_{des} are the sorption and desorption rate constants, and the driving force is the difference between the equilibrium, λ_{eq}, and actual, λ, water contents in the electrolyte. The sorption and desorption rate constants can be modeled as [96,122]

$$k_{des} = \frac{4.59 \times 10^{-5} f_v}{\delta_{cl}} \exp\left[2416\left(\frac{1}{303} - \frac{1}{T}\right)\right] \text{ s}^{-1}$$
$$k_{sor} = \frac{1.14 \times 10^{-5} f_v}{\delta_{cl}} \exp\left[2416\left(\frac{1}{303} - \frac{1}{T}\right)\right] \text{ s}^{-1} \quad (4.47)$$

where

$$f_v = \frac{\lambda V_{m,l}}{V_{m,e} + \lambda V_{m,l}} \quad (4.48)$$

is the water volume fraction in the electrolyte, with $V_{m,e}$ the molar volume of the electrolyte.

The equilibrium water content is given as a function of water activity and temperature by the family of sorption isotherms [122,142]

$$\lambda_{eq} = \begin{cases} \left[1 + 0.2 a_{H_2O}^2 \left(\frac{T - 303.1}{30}\right)\right]\left(14.2 a_{H_2O}^3 - 18.9 a_{H_2O}^2 + 13.4 a_{H_2O}\right), & a_{H_2O} < 1 \\ \lambda_{eq}|_{a_{H_2O}=1} + 6(a_{H_2O} + 2s - 1), & a_{H_2O} \geq 1 \end{cases}$$
$$(4.49)$$

where the term $6(a_{H_2O} + 2s - 1)$ accounts for the raise of the equilibrium water content when the electrolyte is in contact with liquid water rather than with water vapor [142,143].

It should be noted that when the electrolyte is in contact with liquid water ($s > 0$), mass transfer by sorption/desorption between water vapor and dissolved water directly affects the liquid phase, since water is assumed to be close to thermodynamic equilibrium [147]. Consequently, the relative humidity remains close to RH $\approx$ 1, so that when sorption/desorption takes place under partially-saturated conditions, water evaporates/condensates.
- *Darcy's viscous resistance.* The momentum sink term of the gas flow due to Darcy's viscous resistance is equal to

$$S_{u_g} = -\mu_g \left(\overline{1/K} \circ \overline{1/k_{rg}}\right) u_g \quad (4.50)$$

where $\overline{1/K}$ and $\overline{1/k_{rg}}$ are the element-wise inverse gas-phase absolute and relative permeability tensors, the latter accounting for the relative resistance of water saturation on gas convection

$$\overline{1/K} \circ \overline{1/k_{\mathbf{rg}}} = \begin{pmatrix} \dfrac{1}{K_{ip}k_{rg}} & 0 & 0 \\ 0 & \dfrac{1}{K_{ip}k_{rg}} & 0 \\ 0 & 0 & \dfrac{1}{K_{tp}k_{rg}} \end{pmatrix} \qquad (4.51)$$

As in the case of the liquid phase, a power-law function of the form $k_{rg,ip/tp} = (1-s)^n$, with $n = 3 - 4$ is typically used for the porous media of the MEA (GDL, MPL and CL) [122].

- *Heat release and absorption.* Five main heat sources can be found in PEMFCs: (1) heat release by the electrochemical reactions at the reversible cell voltage, $S_{T,\text{rev}}$; (2) activation or irreversible electrochemical heat generation due to cell operation at a voltage different from equilibrium, $S_{T,\text{act}}$; (3) Joule heating due to electronic and ionic transport, $S_{T,\text{ohm}}$; (4) latent heat due to water evaporation/condensation, $S_{T,vl}$; and (5) latent heat due to water sorption/desorption into/from the ionomer, $S_{T,v\lambda}$. Their expressions are as follows:

$$\begin{aligned}
S_{T,\text{rev}} &= \dfrac{j_c}{2F}(-T\Delta S) \\
S_{T,\text{act}} &= |\eta_i j_i| \\
S_{T,\text{ohm}} &= \dfrac{|j_{e^-}|^2}{|\sigma_{e^-}^{\text{eff}}|} + \dfrac{|j_{\mathbf{H}^+}|^2}{|\sigma_{\mathbf{H}^+}^{\text{eff}}|} = S_{T,e^-} + S_{T,\mathbf{H}^+} \\
S_{T,vl} &= h_{gl} S_{vl} \\
S_{T,v\lambda} &= h_{gl} S_{v\lambda}
\end{aligned} \qquad (4.52)$$

where ΔS is the entropy change of the overall cell reaction per mole of hydrogen for liquid water production (assumed here as water dissolved in the ionomer), $|j_{e^-}|^2$ and $|j_{\mathbf{H}^+}|^2$ are the squared magnitudes of the electronic and ionic fluxes, $|\sigma_{e^-}^{\text{eff}}| = \left(\sigma_{e^-}^{\text{eff}} : \sigma_{e^-}^{\text{eff}}\right)^{1/2}$ and $|\sigma_{\mathbf{H}^+}^{\text{eff}}| = \left(\sigma_{\mathbf{H}^+}^{\text{eff}} : \sigma_{\mathbf{H}^+}^{\text{eff}}\right)^{1/2}$ are the magnitudes of the effective electrical and ionic conductivity tensors, h_{lg} is the latent heat of water evaporation/condensation (equal to sorption/desorption). The absolute value in the expression of $S_{T,\text{act}}$ takes into account the negative overpotentials present in the cathode CL ($i = c$).

4.3.4 Boundary conditions

The boundary conditions prescribed at the internal and external surfaces of the computational domain are presented below. The set of internal boundaries includes the fluid–fluid and fluid–solid interfaces between components, whereas the set of external boundaries includes the sidewalls of the cell, the anode and cathode channel inlets/outlets and the bipolar plate terminals (see Fig. 4.12).

- *Internal interfaces.* A continuity boundary condition is prescribed at all internal interfaces between fluid regions (CLs, MPLs, GDLs and channels) for species, Y_i, liquid water, p_l, gas flow, $\boldsymbol{u}_g$, and temperature, T. In contrast, at the walls between fluid and solid regions (i.e., channel/BPP, GDL/BPP, and CL/PEM interfaces) different boundary conditions are prescribed depending on the variable of interest. For species and liquid saturation, a no-flux boundary condition $(\nabla Y_i \cdot \mathbf{n} = 0, \ \nabla p_l \cdot \mathbf{n} = 0)$ is set, where $\mathbf{n}$ is the outward unit normal vector. For temperature, a continuity boundary condition is used since the energy equation holds for all cell components. For gas flow, an impermeable no-slip boundary condition $(\boldsymbol{u}_g = 0)$ is set at all fluid/solid interfaces. In the case of the electronic and ionic potentials, ϕ_{e^-} and ϕ_{H^+}, and dissolved water content, λ, internal boundary conditions vary between variables since they are only solved in certain fluid regions of the computational domain. For the electronic potential, a no-flux boundary condition $(\nabla \phi_{e^-} \cdot \mathbf{n} = 0)$ is set at the CL/PEM and GDL/channel interfaces. For the ionic potential and dissolved water, a no-flux boundary condition $(\nabla \phi_{H^+} \cdot \mathbf{n} = 0, \ \nabla \lambda \cdot \mathbf{n} = 0)$ is set at the CL/GDL interfaces of both compartments.

- *Sidewalls.* The sidewalls of the fluid regions are considered isolated impermeable non-conducting walls, so that impermeable no-slip $(\boldsymbol{u}_g = 0)$ and no-flux $(\nabla \Phi \cdot \mathbf{n} = 0)$ boundary conditions are imposed for the gas flow and the remaining variables $(\Phi = Y_{H_2}, \ Y_{O_2}, \ Y_{H_2O}, \ p_l, \ T, \ \phi_{e^-}, \ \phi_{H^+} \text{ and } \lambda)$, respectively. At the sidewalls of the solid PEM and BPPs where the gas and liquid phases are not defined, a no-flux boundary condition is set for $\Phi = T, \ \phi_{e^-}, \ \phi_{H^+}$ and λ.

- *Anode and cathode channel inlets/outlets.* At the anode channel inlet, the gas mass flux, hydrogen and water vapor mass fractions, liquid saturation and gas temperature are specified

$$\begin{aligned}
(\rho_g \boldsymbol{u}_g) \cdot \mathbf{n} &= -\frac{\dot{m}^{in}_{H_2}}{A_{ch}\left(1 - Y^{in}_{H_2O,a}\right)} \\
Y_{H_2} &= Y^{in}_{H_2} \\
Y_{H_2O} &= Y^{in}_{H_2O,a} \\
s &= 0 \\
T &= T^{in}_a
\end{aligned} \quad (4.53)$$

where $\dot{m}^{in}_{H_2} = \rho^{std}_{H_2} Q_{H_2}$ is the dry hydrogen mass flow rate (i.e., before humidification), Q_{H_2} is the dry hydrogen feed flow rate referenced to standard conditions

($T = 0°C$, $p_g = 1$ bar), A_{ch} is the channel cross-sectional area, and T_a^{in} is the inlet gas temperature. The anode stoichiometric factor (at a reference current density, I^{ref}) is given by

$$\xi_a = \frac{\rho_{H_2}^{std} Q_{H_2} 2F}{A_{cell} I^{ref}} \tag{4.54}$$

where the hydrogen density at standard conditions is $\rho_{H_2}^{std} = 8.99 \times 10^{-2}$ kgm^{-3}. According to the ideal gas law, the inlet mass fractions of hydrogen and water vapor are

$$\begin{aligned} Y_{H_2}^{in} &= \frac{C_{H_2}^{in} M_{H_2}}{C_{H_2O,a}^{in} M_{H_2O} + C_{H_2}^{in} M_{H_2}} \\ Y_{H_2O,a}^{in} &= 1 - Y_{H_2}^{in} \end{aligned} \tag{4.55}$$

with

$$\begin{aligned} C_{H_2}^{in} &= \frac{p_{g,a}^{in} - RH_a^{in} p_{H_2O}^{sat}}{RT_a^{in}} \\ C_{H_2O,a}^{in} &= \frac{RH_a^{in} p_{H_2O}^{sat}}{RT_a^{in}} \end{aligned} \tag{4.56}$$

Here, RH_a^{in} is the anode inlet relative humidity and $p_{g,a}^{in}$ is the inlet gas-phase pressure, which is assumed equal to the back-pressure, $p_{g,a}^{in} \simeq p_{g,a}^{out}$, for the calculation of the inlet gas composition.

Similarly, the gas mass flux, humid air mass fractions, liquid saturation and inlet temperature are prescribed at the cathode channel inlet

$$\begin{aligned} (\rho_g \mathbf{u_g}) \cdot \mathbf{n} &= -\frac{\dot{m}_{air}^{in}}{A_{ch}(1 - Y_{H_2O,c}^{in})} \\ Y_{O_2} &= Y_{O_2}^{in} \\ Y_{H_2O} &= Y_{H_2O,c}^{in} \\ s &= 0 \\ T &= T_c^{in} \end{aligned} \tag{4.57}$$

where $\dot{m}_{air}^{in} = \rho_{air}^{std} Q_{air}$ is the dry air mass flow rate (i.e., before humidification) and Q_{air} is the dry air feed flow rate under standard conditions. The cathode stoichiometric factor is given by

$$\xi_c = \frac{\rho_{O_2}^{std} Q_{air} X_{O_2,air} 4F}{A_{cell} I^{ref}} \tag{4.58}$$

where the oxygen density at standard conditions is $\rho_{O_2}^{std} = 1.43 \text{ kgm}^{-3}$ and the volume fraction of oxygen in air is $X_{O_2,air} = 0.21$. The inlet mass fractions of oxygen and water vapor are

$$Y_{O_2}^{in} = \frac{C_{O_2}^{in} M_{O_2}}{C_{H_2O,c}^{in} M_{H_2O} + C_{N_2}^{in} M_{N_2} + C_{O_2}^{in} M_{O_2}}$$

$$Y_{H_2O,c}^{in} = \frac{C_{H_2O,c}^{in} M_{H_2O}}{C_{H_2O,c}^{in} M_{H_2O} + C_{N_2}^{in} M_{N_2} + C_{O_2}^{in} M_{O_2}} \quad (4.59)$$

with

$$C_{O_2}^{in} = \frac{X_{O_2,air}\left(p_{g,c}^{in} - \text{RH}_c^{in} p_{H_2O}^{sat}\right)}{RT_c^{in}}$$

$$C_{H_2O,c}^{in} = \frac{\text{RH}_c^{in} p_{H_2O}^{sat}}{RT_c^{in}} \quad (4.60)$$

$$C_{N_2}^{in} = C_{O_2}^{in} \frac{1 - X_{O_2,air}}{X_{O_2,air}}$$

At the channel outlets, the back-pressure is imposed on the anode ($i = a$) and the cathode ($i = c$) gas streams, $p_g = p_{g,i}^{out}$, while a zero-gradient boundary condition is set for other variables $\left(\Phi = Y_{H_2}, Y_{O_2}, Y_{H_2O}, p_l \text{ and } T\right)$, $\nabla\Phi \cdot \mathbf{n} = 0$.

- *Anode and cathode terminals.* At the anode and cathode terminals (the outer surfaces of the anode and cathode bipolar plates), the temperature is fixed to the cell operating temperature, $T = T_{cell}$. Dirichlet boundary conditions are also prescribed for the electronic potential. At the anode terminal, the electronic potential is set equal to the electronic potential loss across the cell, i.e., $\phi_{e^-} = E_{rev}^0 - V_{cell}$, where E_{rev}^0 is the reversible voltage, given by

$$E_{rev}^0 \approx 1.23 - 9 \times 10^{-4}(T_{cell} - 298) + \frac{RT_{cell}}{2F}\left(\ln p_{H_2}^{in} + \frac{1}{2}\ln p_{O_2}^{in}\right) \text{ V} \quad (4.61)$$

where T_{cell} is the operating temperature (i.e., volume-average temperature) and $p_{H_2}^{in}$ and $p_{O_2}^{in}$ are the inlet partial pressures of hydrogen and oxygen (in atm). At the cathode terminal, the electronic potential is set equal to zero, $\phi_{e^-} = 0$ (reference value). As discussed in [137], this way of prescribing the boundary conditions for the electronic potential, along with the definition of the overpotentials as $\eta = \phi_{e^-} - \phi_{H^+}$ in both the anode and cathode electrodes, is more computationally efficient and stable.

Acknowledgments

This work was supported by the Spanish Agencia Estatal de Investigación (PID2019-106740RB-I00 and EIN2020-112247) and the project PEM4ENERGY-CM-UC3M funded by the call "Programa de apoyo a la realización de proyectos interdisciplinares de I + D para jóvenes investigadores de la Universidad Carlos III de Madrid 2019–2020" under the frame of the "Convenio Plurianual Comunidad de Madrid-Universidad Carlos III de Madrid".

References

[1] A. Kirubakaran, S. Jain, R.K. Nema, A review on fuel cell technologies and power electronic interface, Renewable and Sustainable Energy Reviews 13 (9) (2009) 2430–2440. Available from: https://doi.org/10.1016/j.rser.2009.04.004.

[2] S. Mekhilef, R. Saidur, A. Safari, Comparative study of different fuel cell technologies, Renewable and Sustainable Energy Reviews 16 (1) (2012) 981–989. Available from: https://doi.org/10.1016/j.rser.2011.09.020.

[3] Energy and Environmental Solutions, Science Applications International Corporation, Fuel Cell Handbook, fifth ed., US Department of Energy, 2000.

[4] O.Z. Sharaf, M.F. Orhan, An overview of fuel cell technology: fundamentals and applications, Renewable and Sustainable Energy Reviews 32 (2014) 810–853. Available from: https://doi.org/10.1016/j.rser.2014.01.012.

[5] M. Winter, R.J. Brodd, What are batteries, fuel cells, and supercapacitors? Chemical Reviews 104 (10) (2004) 4245–4269. Available from: https://doi.org/10.1021/cr020730k.

[6] F. Barbir, PEM Fuel Cells: Theory and Practice, second ed., Elsevier, 2012. Available from: https://www.elsevier.com/books/pem-fuel-cells/barbir/978-0-12-387710-9.

[7] M.W. Ellis, M.R. Von Spakovsky, D.J. Nelson, Fuel cell systems: efficient, flexible energy conversion for the 21st century, Proceedings of the IEEE 89 (12) (2001) 1808–1818. Available from: https://doi.org/10.1109/5.975914.

[8] R. Rashidi, I. Dincer, G.F. Naterer, P. Berg, Performance evaluation of direct methanol fuel cells for portable applications, Journal of Power Sources 187 (2) (2009) 509–516. Available from: https://doi.org/10.1016/j.jpowsour.2008.11.044.

[9] I. Nitta, Inhomogeneous Compression of PEMFC Gas Diffusion Layers, Doctoral Dissertation, Aalto University, 2008. Available from: https://aaltodoc.aalto.fi/handle/123456789/2998.

[10] J.M. Andújar, F. Segura, Fuel cells: history and updating. A walk along two centuries, Renewable and Sustainable Energy Reviews 13 (9) (2009) 2309–2322. Available from: https://doi.org/10.1016/j.rser.2009.03.015.

[11] M.L. Perry, T.F. Fuller, A historical perspective of fuel cell technology in the 20th century, Journal of the Electrochemical Society 149 (7) (2002) S59. Available from: https://iopscience.iop.org/article/10.1149/1.1488651.

[12] K. Joon, Fuel cells—a 21st century power system, Journal of Power Sources 71 (1–2) (1998) 12–18. Available from: https://doi.org/10.1016/s0378-7753(97)02765-1.

[13] P. Costamagna, S. Srinivasan, Quantum jumps in the PEMFC science and technology from the 1960s to the year 2000: Part I. Fundamental scientific aspects, Journal of Power Sources 102 (1–2) (2001) 242–252. Available from: https://doi.org/10.1016/S0378-7753(01)00807-2.

[14] A.B. Stambouli, E. Traversa, Solid oxide fuel cells (SOFCs): a review of an environmentally clean and efficient source of energy, Renewable and Sustainable Energy Reviews 6 (5) (2002) 433–455. Available from: https://doi.org/10.1016/S1364-0321(02)00014-X.

[15] S.K. Kamarudin, F. Achmad, W.R.W. Daud, Overview on the application of direct methanol fuel cell (DMFC) for portable electronic devices, International Journal of Hydrogen Energy 34 (16) (2009) 6902–6916. Available from: https://doi.org/10.1016/j.ijhydene.2009.06.013.

[16] K. Scott, W.M. Taama, P. Argyropoulos, Engineering aspects of the direct methanol fuel cell system, Journal of Power Sources 79 (1) (1999) 43–59. Available from: https://doi.org/10.1016/S0378-7753(98)00198-0.

[17] E.D. Wachsman, K.T. Lee, Lowering the temperature of solid oxide fuel cells, Science 334 (6058) (2011) 935–939. Available from: https://doi.org/10.1126/science.1204090.

[18] J.C. Ho, E.C. Saw, L.Y.Y. Lu, J.S. Liu, Technological barriers and research trends in fuel cell technologies: a citation network analysis, Technological Forecasting and Social Change 82 (1) (2014) 66–79. Available from: https://doi.org/10.1016/j.techfore.2013.06.004.

[19] P. Mock, S.A. Schmid, Fuel cells for automotive powertrains—a techno-economic assessment, Journal of Power Sources 190 (1) (2009) 133–140. Available from: https://doi.org/10.1016/j.jpowsour.2008.10.123.

[20] Y. Wang, K.S. Chen, J. Mishler, S.C. Cho, X.C. Adroher, A review of polymer electrolyte membrane fuel cells: technology, applications, and needs on fundamental research, Applied Energy 88 (4) (2011) 981–1007. Available from: https://doi.org/10.1016/j.apenergy.2010.09.030.

[21] D.L. Greene, G. Duleep, Status and prospects of the global automotive fuel cell industry and plans for deployment of fuel cell vehicles and hydrogen refueling infrastructure. U.S. Department of Energy, 2013.

[22] Q. Zhao, P. Majsztrik, J. Benziger, Diffusion and interfacial transport of water in Nafion, Journal of Physical Chemistry B 115 (12) (2011) 2717–2727. Available from: https://doi.org/10.1021/jp1112125.

[23] M. Momirlan, T.N. Veziroglu, The properties of hydrogen as fuel tomorrow in sustainable energy system for a cleaner planet, International Journal of Hydrogen Energy 30 (7) (2005) 795–802. Available from: https://doi.org/10.1016/j.ijhydene.2004.10.011.

[24] United States DRIVE Partnership. Fuel Cell Technical Team Roadmap, 2013.

[25] J. Huang, D.G. Baird, J.E. McGrath, Development of fuel cell bipolar plates from graphite filled wet-lay thermoplastic composite materials, Journal of Power Sources 150 (1–2) (2005) 110–119. Available from: https://doi.org/10.1016/j.jpowsour.2005.02.074.

[26] S. Sharma, B.G. Pollet, Support materials for PEMFC and DMFC electrocatalysts—a review, Journal of Power Sources 208 (2012) 96–119. Available from: https://doi.org/10.1016/j.jpowsour.2012.02.011.

[27] L. Wu, Z. Zhang, J. Ran, D. Zhou, C. Li, T. Xu, Advances in proton-exchange membranes for fuel cells: an overview on proton conductive channels (PCCs), Physical Chemistry Chemical Physics 15 (14) (2013) 4870–4887. Available from: https://doi.org/10.1039/c3cp50296a.

[28] T. Yamaguchi, H. Zhou, S. Nakazawa, N. Hara, An extremely low methanol crossover and highly durable aromatic pore-filling electrolyte membrane for direct methanol fuel cells, Advanced Materials 19 (4) (2007) 592–596. Available from: https://doi.org/10.1002/adma.200601086.

[29] Z. Yang, H. Nie, X. Chen, X. Chen, S. Huang, Recent progress in doped carbon nanomaterials as effective cathode catalysts for fuel cell oxygen reduction reaction, Journal of Power Sources 236 (2013) 238–249. Available from: https://doi.org/10.1016/j.jpowsour.2013.02.057.

[30] S.G. Kandlikar, M.L. Garofalo, Z. Lu, Water management in a PEMFC: water transport mechanism and material degradation in gas diffusion layers, Fuel Cells 11 (6) (2011) 814–823. Available from: https://doi.org/10.1002/fuce.201000172.

[31] S.G. Kandlikar, Z. Lu, Thermal management issues in a PEMFC stack—a brief review of current status, Applied Thermal Engineering 29 (7) (2009) 1276–1280. Available from: https://doi.org/10.1016/j.applthermaleng.2008.05.009.

[32] T.S. Zhao, C. Xu, R. Chen, W.W. Yang, Mass transport phenomena in direct methanol fuel cells, Progress in Energy and Combustion Science 35 (3) (2009) 275–292. Available from: https://doi.org/10.1016/j.pecs.2009.01.001.

[33] Z. Wan, H. Chang, S. Shu, Y. Wang, H. Tang, A review on cold start of proton exchange membrane fuel cells, Energies 7 (5) (2014) 3179–3203. Available from: https://doi.org/10.3390/en7053179.

[34] G.Q. Lu, C.Y. Wang, Electrochemical and flow characterization of a direct methanol fuel cell, Journal of Power Sources 134 (1) (2004) 33–40. Available from: https://doi.org/10.1016/j.jpowsour.2004.01.055.

[35] A. Arvay, E. Yli-Rantala, C.H. Liu, X.H. Peng, P. Koski, L. Cindrella, et al., Characterization techniques for gas diffusion layers for proton exchange membrane fuel cells—a review, Journal of Power Sources 213 (2012) 317–337. Available from: https://doi.org/10.1016/j.jpowsour.2012.04.026.

[36] P.A. García-Salaberri, J.T. Gostick, G. Hwang, A.Z. Weber, M. Vera, Effective diffusivity in partially-saturated carbon-fiber gas diffusion layers: effect of local saturation and application to macroscopic continuum models, Journal of Power Sources 296 (2015) 440−453. Available from: https://doi.org/10.1016/j.jpowsour.2015.07.034.
[37] P.A. García-Salaberri, G. Hwang, M. Vera, A.Z. Weber, J.T. Gostick, Effective diffusivity in partially-saturated carbon-fiber gas diffusion layers: effect of through-plane saturation distribution, International Journal of Heat and Mass Transfer 86 (2015) 319−333. Available from: https://doi.org/10.1016/j.ijheatmasstransfer.2015.02.073.
[38] Q. He, A. Kusoglu, I.T. Lucas, K. Clark, A.Z. Weber, R. Kostecki, Correlating humidity-dependent ionically conductive surface area with transport phenomena in proton-exchange membranes, Journal of Physical Chemistry B 115 (40) (2011) 11650−11657. Available from: https://doi.org/10.1021/jp206154y.
[39] J. Kleemann, F. Finsterwalder, W. Tillmetz, Characterisation of mechanical behaviour and coupled electrical properties of polymer electrolyte membrane fuel cell gas diffusion layers, Journal of Power Sources 190 (1) (2009) 92−102. Available from: https://doi.org/10.1016/j.jpowsour.2008.09.026.
[40] M. Koz, S.G. Kandlikar, Numerical investigation of interfacial transport resistance due to water droplets in proton exchange membrane fuel cell air channels, Journal of Power Sources 243 (2013) 946−957. Available from: https://doi.org/10.1016/j.jpowsour.2013.06.075.
[41] A. Kusoglu, A.Z. Weber, New insights into perfluorinated sulfonic-acid ionomers, Chemical Reviews 117 (3) (2017) 987−1104. Available from: https://doi.org/10.1021/acs.chemrev.6b00159.
[42] T. Rosén, J. Eller, J. Kang, N.I. Prasianakis, J. Mantzaras, F.N. Büchi, Saturation dependent effective transport properties of PEFC gas diffusion layers, Journal of the Electrochemical Society 159 (9) (2012) F536−F544. Available from: https://doi.org/10.1149/2.005209jes.
[43] A.D. Santamaria, P.K. Das, J.C. MacDonald, A.Z. Weber, Liquid-water interactions with gas-diffusion-layer surfaces, Journal of the Electrochemical Society 161 (12) (2014) F1184−F1193. Available from: https://doi.org/10.1149/2.0321412jes.
[44] N. Zamel, X. Li, Effective transport properties for polymer electrolyte membrane fuel cells—with a focus on the gas diffusion layer, Progress in Energy and Combustion Science 39 (1) (2013) 111−146. Available from: https://doi.org/10.1016/j.pecs.2012.07.002.
[45] X. Li, I. Sabir, Review of bipolar plates in PEM fuel cells: flow-field designs, International Journal of Hydrogen Energy 30 (4) (2005) 359−371. Available from: https://doi.org/10.1016/j.ijhydene.2004.09.019.
[46] S. Litster, G. McLean, PEM fuel cell electrodes, Journal of Power Sources 130 (1−2) (2004) 61−76. Available from: https://doi.org/10.1016/j.jpowsour.2003.12.055.
[47] R. O'Hayre, et al., Fuel Cell Fundamentals, John Wiley and Sons, New York, 2016. Available from: https://onlinelibrary.wiley.com/doi/book/10.1002/9781119191766.
[48] M. Ji, Z. Wei, A review of water management in polymer electrolyte membrane fuel cells, Energies 2 (4) (2009) 1057−1106. Available from: https://doi.org/10.3390/en20401057.
[49] W. Dai, H. Wang, X.Z. Yuan, J.J. Martin, D. Yang, J. Qiao, et al., A review on water balance in the membrane electrode assembly of proton exchange membrane fuel cells, International Journal of Hydrogen Energy 34 (23) (2009) 9461−9478. Available from: https://doi.org/10.1016/j.ijhydene.2009.09.017.
[50] E.C. Kumbur, M.M. Mench, Fuel cells—proton-exchange membrane fuel cells | water management, Encyclopedia of Electrochemical Power Sources, Elsevier, 2009, pp. 828−847. Available from: https://doi.org/10.1016/B978-044452745-5.00862-5.
[51] P.W. Majsztrik, M.B. Satterfield, A.B. Bocarsly, J.B. Benziger, Water sorption, desorption and transport in Nafion membranes, Journal of Membrane Science 301 (1−2) (2007) 93−106. Available from: https://doi.org/10.1016/j.memsci.2007.06.022.
[52] S. Slade, S.A. Campbell, T.R. Ralph, F.C. Walsh, Ionic conductivity of an extruded Nafion 1100 EW series of membranes, Journal of the Electrochemical Society 149 (2002) A1556−A1564. Available from: https://iopscience.iop.org/article/10.1149/1.1517281/meta.
[53] A.A. Kulikovsky, Quasi-3D modeling of water transport in polymer electrolyte fuel cells, Journal of the Electrochemical Society 150 (11) (2003) A1432. Available from: https://iopscience.iop.org/article/10.1149/1.1611489/meta.

[54] M.F. Mathias, J. Roth, J. Fleming, W. Lehnert, Diffusion media materials and characterization, in: W. Vielstich, A. Lamm, H.A. Gastegier (Eds.), Handbook of Fuel Cells: Fundamentals, Technology and Applications, vol. 3, John Wiley & Sons, Ltd., New York, 2003, pp. 517–537. Available from: https://onlinelibrary.wiley.com/doi/abs/10.1002/9780470974001.f303046.

[55] A.S. Aricò, S. Srinivasan, V. Antonucci, DMFCs: from fundamental aspects to technology development, Fuel Cells 1 (2) (2001) 133–161. Available from: https://doi.org/10.1002/1615-6854(200107)1:2<133::AID-FUCE133>3.3.CO;2-X.

[56] L. Cindrella, A.M. Kannan, J.F. Lin, K. Saminathan, Y. Ho, C.W. Lind, et al., Gas diffusion layer for proton exchange membrane fuel cells—a review, Journal of Power Sources 149 (2012) 146–160. Available from: https://www.sciencedirect.com/science/article/pii/S0378775309006399?casa_token = eE-UrT0xchQAAAAA:yFV5QJQXO2x0WscdEBTNHoGLnv7dVc39DB3IFXxVPvgCOJ3EgapH0i3q1fjn-KuGOI1cMqD0xA.

[57] S. Park, J.-W. Lee, B.N. Popov, A review of gas diffusion layer in PEM fuel cells: materials and designs, International Journal of Hydrogen Energy 37 (2012) 5850–5865. Available from: https://www.sciencedirect.com/science/article/pii/S0360319911028825?casa_token = MAIzMSk49OYAAAAA:RBZmZ3m22aAIqGT0x51HeXIXJH-2nqCvS8CciRdhLG1xIW9q2-PB8yBU1toBXVUrfFf8F-Zpmg.

[58] R.R. Rashapov, Unno, J.T. Gostick, Characterization of PEMFC gas diffusion layer porosity, Journal of Power Sources 194 (2009) 146–160. Available from: https://iopscience.iop.org/article/10.1149/2.0921506jes/meta.

[59] R. Flückiger, S.A. Freunberger, D. Kramer, A. Wokaun, G.G. Scherer, F.N. Büchi, Anisotropic, effective diffusivity of porous gas diffusion layer materials for PEFC, Electrochimica Acta 54 (2) (2008) 551–559. Available from: https://doi.org/10.1016/j.electacta.2008.07.034.

[60] J.T. Gostick, M.W. Fowler, M.D. Pritzker, M.A. Ioannidis, L.M. Behra, In-plane and through-plane gas permeability of carbon fiber electrode backing layers, Journal of Power Sources 162 (1) (2006) 228–238. Available from: https://doi.org/10.1016/j.jpowsour.2006.06.096.

[61] J.M. Lamanna, S.G. Kandlikar, Determination of effective water vapor diffusion coefficient in PEMFC gas diffusion layers, International Journal of Hydrogen Energy 36 (8) (2011) 5021–5029. Available from: https://doi.org/10.1016/j.ijhydene.2011.01.036.

[62] R. Omrani, B. Shabani, Gas diffusion layer modifications and treatments for improving the performance of proton exchange membrane fuel cells and electrolysers: a review, International Journal of Hydrogen Energy 42 (47) (2017) 28515–28536. Available from: https://doi.org/10.1016/j.ijhydene.2017.09.132.

[63] A. Ozden, S. Shahgaldi, X. Li, F. Hamdullahpur, A review of gas diffusion layers for proton exchange membrane fuel cells—with a focus on characteristics, characterization techniques, materials and designs, Progress in Energy and Combustion Science 74 (2019) 50–102. Available from: https://doi.org/10.1016/j.pecs.2019.05.002.

[64] J.G. Pharoah, K. Karan, W. Sun, On effective transport coefficients in PEM fuel cell electrodes: anisotropy of the porous transport layers, Journal of Power Sources 161 (1) (2006) 214–224. Available from: https://doi.org/10.1016/j.jpowsour.2006.03.093.

[65] B.P. Solis, J.C.C. Argüello, L.G. Barba, M.P. Gurrola, Z. Zarhri, D.L. TrejoArroyo, Bibliometric analysis of the mass transport in a gas diffusion layer in PEM fuel cells, Sustainability (Switzerland) 11 (23) (2019) 6682. Available from: https://doi.org/10.3390/su11236682.

[66] A. Hermann, T. Chaudhuri, P. Spagnol, Bipolar plates for PEM fuel cells: a review, International Journal of Hydrogen Energy 30 (12) (2005) 1297–1302. Available from: https://doi.org/10.1016/j.ijhydene.2005.04.016.

[67] R. Edinger, S. Kaul, Sustainable Mobility: Renewable Energies for Powering Fuel Cell Vehicles, Praeger, 2003.

[68] A. Iranzo, J.M. Gregorio, P. Boillat, F. Rosa, Bipolar plate research using computational fluid dynamics and neutron radiography for proton exchange membrane fuel cells, International Journal of Hydrogen Energy 45 (22) (2020) 12432–12442. Available from: https://doi.org/10.1016/j.ijhydene.2020.02.183.

[69] T.V. Nguyen, W. He, Interdigitated flow field design, Handbook of Fuel Cells: Fundamentals, Technology and Applications, vol. 3, John Wiley & Sons, 2003. Available from: https://onlinelibrary.wiley.com/doi/abs/10.1002/9780470974001.f303031.

[70] A.M. Chaparro, P. Ferreira-Aparicio, M.A. Folgado, R. Hübscher, C. Lange, N. Weber, Thermal neutron radiography of a passive proton exchange membrane fuel cell for portable hydrogen energy systems, Journal of Power Sources 480 (2020) 228668. Available from: https://doi.org/10.1016/j.jpowsour.2020.228668.

[71] T. Fabian, R. O'Hayre, S. Litster, F.B. Prinz, J.G. Santiago, Passive water management at the cathode of a planar air-breathing proton exchange membrane fuel cell, Journal of Power Sources 195 (10) (2010) 3201−3206. Available from: https://doi.org/10.1016/j.jpowsour.2009.12.030.

[72] J.C. Kurnia, B.A. Chaedir, A.P. Sasmito, T. Shamim, Progress on open cathode proton exchange membrane fuel cell: performance, designs, challenges and future directions, Applied Energy 283 (2020) 116359. Available from: https://doi.org/10.1016/j.apenergy.2020.116359.

[73] H. Yang, T.S. Zhao, Q. Ye, In situ visualization study of CO_2 gas bubble behavior in DMFC anode flow fields, Journal of Power Sources 139 (1−2) (2005) 79−90. Available from: https://doi.org/10.1016/j.jpowsour.2004.05.033.

[74] J.T. Gostick, M.A. Ioannidis, M.W. Fowler, M.D. Pritzker, On the role of the microporous layer in PEMFC operation, Electrochemistry Communications 11 (3) (2009) 576−579. Available from: https://www.sciencedirect.com/science/article/pii/S1388248108006498?casa_token=lorF3fma_OMAAAAA:FHqQTWRsv6Fm0UeJVktAxzF1mkznbJo7eAGkBsJOFKAEyP2cJr8sXny8KNbaB1Gd5ewKBLmlvA.

[75] J.H. Nam, K.J. Lee, G.S. Hwang, C.J. Kim, M. Kaviany, Microporous layer for water morphology control in PEMFC, International Journal of Heat and Mass Transfer 52 (11−12) (2009) 2779−2791. Available from: https://doi.org/10.1016/j.ijheatmasstransfer.2009.01.002.

[76] J.P. Owejan, J.E. Owejan, W. Gu, T.A. Trabold, T.W. Tighe, M.F. Mathias, Water transport mechanisms in PEMFC gas diffusion layers, Journal of the Electrochemical Society 157 (10) (2010) B1456−B1464. Available from: https://doi.org/10.1149/1.3468615.

[77] S. Park, J.W. Lee, B.N. Popov, Effect of carbon loading in microporous layer on PEM fuel cell performance, Journal of Power Sources 163 (1) (2006) 357−363. Available from: https://doi.org/10.1016/j.jpowsour.2006.09.020.

[78] A.Z. Weber, J. Newman, Effects of microporous layers in polymer electrolyte fuel cells, Journal of the Electrochemical Society 152 (4) (2005) A677−A688. Available from: https://doi.org/10.1149/1.1861194.

[79] P.A. García-Salaberri, M. Vera, R. Zaera, Nonlinear orthotropic model of the inhomogeneous assembly compression of PEM fuel cell gas diffusion layers, International Journal of Hydrogen Energy 36 (18) (2011) 11856−11870. Available from: https://doi.org/10.1016/j.ijhydene.2011.05.152.

[80] J.T. Gostick, M.A. Ioannidis, M.D. Pritzker, M.W. Fowler, Impact of liquid water on reactant mass transfer in PEM fuel cell electrodes, Journal of the Electrochemical Society 157 (4) (2010) B563−B571. Available from: https://doi.org/10.1149/1.3291977.

[81] R. Alink, D. Gerteisen, Coupling of a continuum fuel cell model with a discrete liquid water percolation model, International Journal of Hydrogen Energy 39 (16) (2014) 8457−8473. Available from: https://doi.org/10.1016/j.ijhydene.2014.03.192.

[82] P.A. García-Salaberri, M. Vera, On the effects of assembly compression on the performance of liquid-feed DMFCs under methanol-limiting conditions: a 2D numerical study, Journal of Power Sources 285 (2015) 543−558. Available from: https://doi.org/10.1016/j.jpowsour.2015.02.112.

[83] P.A. García-Salaberri, M. Vera, On the effect of operating conditions in liquid-feed direct methanol fuel cells: a multiphysics modeling approach, Energy 113 (2016) 1265−1287. Available from: https://doi.org/10.1016/j.energy.2016.07.074.

[84] P.A. García-Salaberri, M. Vera, I. Iglesias, Modeling of the anode of a liquid-feed DMFC: inhomogeneous compression effects and two-phase transport phenomena, Journal of Power Sources 246 (2014) 239−252. Available from: https://doi.org/10.1016/j.jpowsour.2013.06.166.

[85] M. Secanell, B. Carnes, A. Suleman, N. Djilali, Numerical optimization of proton exchange membrane fuel cell cathodes, Electrochimica Acta 52 (7) (2007) 2668−2682. Available from: https://doi.org/10.1016/j.electacta.2006.09.049.

[86] M. Secanell, K. Karan, A. Suleman, N. Djilali, Multi-variable optimization of PEMFC cathodes using an agglomerate model, Electrochimica Acta 52 (22) (2007) 6318−6337. Available from: https://doi.org/10.1016/j.electacta.2007.04.028.

[87] Z.H. Wang, C.Y. Wang, Mathematical modeling of liquid-feed direct methanol fuel cells, Journal of the Electrochemical Society 150 (4) (2003) A508−A519. Available from: https://doi.org/10.1149/1.1559061.
[88] A.Z. Weber, R.L. Borup, R.M. Darling, P.K. Das, T.J. Dursch, W. Gu, et al., A critical review of modeling transport phenomena in polymer-electrolyte fuel cells, Journal of the Electrochemical Society 161 (12) (2014) F1254. Available from: https://iopscience.iop.org/article/10.1149/2.0751412jes/meta.
[89] J. Wishart, Z. Dong, M. Secanell, Optimization of a PEM fuel cell system based on empirical data and a generalized electrochemical semi-empirical model, Journal of Power Sources 161 (2) (2006) 1041−1055. Available from: https://doi.org/10.1016/j.jpowsour.2006.05.056.
[90] H. Bahrami, A. Faghri, Review and advances of direct methanol fuel cells: part II: modeling and numerical simulation, Journal of Power Sources 230 (2013) 303−320. Available from: https://doi.org/10.1016/j.jpowsour.2012.12.009.
[91] C.Y. Wang, Fundamental models for fuel cell engineering, Chemical Reviews 104 (10) (2004) 4727−4765. Available from: https://doi.org/10.1021/cr020718s.
[92] P.A. García-Salaberri, Modeling diffusion and convection in thin porous transport layers using a composite continuum-network model: application to gas diffusion layers in polymer electrolyte fuel cells, International Journal of Heat and Mass Transfer 167 (2021) 120824. Available from: https://www.sciencedirect.com/science/article/pii/S0017931020337625?casa_token = nXLOON93OskAAAAA:Wh1f3SEHMRa-Bw2_lM07xZelHpo0gY1Tx0tqjyCK3FJvcj50fjVVhtFRNNnvvojzMst7H6Ahwg.
[93] P.P. Mukherjee, Q. Kang, C.Y. Wang, Pore-scale modeling of two-phase transport in polymer electrolyte fuel cells—progress and perspective, Energy and Environmental Science 4 (2) (2011) 346−369. Available from: https://doi.org/10.1039/b926077c.
[94] P.A. García-Salaberri, J.T. Gostick, I.V. Zenyuk, G. Hwang, M. Vera, A.Z. Weber, On the limitations of volume-averaged descriptions of gas diffusion layers in the modeling of polymer electrolyte fuel cells, ECS Transactions 80 (8) (2017) 133. Available from: https://iopscience.iop.org/article/10.1149/08008.0133ecst/meta.
[95] P.A. García-Salaberri, I.V. Zenyuk, J.T. Gostick, A.Z. Weber, Modeling gas diffusion layers in polymer electrolyte fuel cells using a continuum-based pore-network formulation, ECS Transactions 97 (7) (2020) 615−626. Available from: https://doi.org/10.1149/09707.0615ecst.
[96] P.A. García-Salaberri, I.V. Zenyuk, G. Hwang, M. Vera, A.Z. Weber, J.T. Gostick, Implications of inherent inhomogeneities in thin carbon fiber-based gas diffusion layers: a comparative modeling study, Electrochimica Acta 295 (2019) 861−874. Available from: https://doi.org/10.1016/j.electacta.2018.09.089.
[97] P.A. García-Salaberri, I.V. Zenyuk, A.D. Shum, G. Hwang, M. Vera, A.Z. Weber, et al., Analysis of representative elementary volume and through-plane regional characteristics of carbon-fiber papers: diffusivity, permeability and electrical/thermal conductivity, International Journal of Heat and Mass Transfer 127 (2018) 687−703. Available from: https://doi.org/10.1016/j.ijheatmasstransfer.2018.07.030.
[98] J. Hack, P.A. García-Salaberri, M.D.R. Kok, R. Jervis, P.R. Shearing, N. Brandon, et al., X-ray micro-computed tomography of polymer electrolyte fuel cells: what is the representative elementary area? Journal of the Electrochemical Society 167 (1) (2020) 013545. Available from: https://doi.org/10.1149/1945-7111/ab6983.
[99] F.J. Higuera, J. Jiménez, Boltzmann approach to lattice gas simulations, Europhysics Letters (EPL) 9 (1989) 663−668. Available from: https://doi.org/10.1209/0295-5075/9/7/009.
[100] S. Succi, The Lattice Boltzmann Equation: For Fluid Dynamics and Beyond, Princeton University Press, New York, 2001.
[101] Z. Guo, B. Shi, C. Zheng, A coupled lattice BGK model for the Boussinesq equations, International Journal for Numerical Methods in Fluids 39 (4) (2002) 325−342. Available from: https://doi.org/10.1002/fld.337.
[102] D.A. Wolf-Gladrow, Lattice-Gas Cellular Automata and Lattice Boltzmann Models: An Introduction, Springer-Verlag, Berlin, Heidelberg, 2004. Available from: https://www.springer.com/gp/book/9783540669739.

[103] D. Froning, J. Brinkmann, U. Reimer, V. Schmidt, W. Lehnert, D. Stolten, 3D analysis, modeling and simulation of transport processes in compressed fibrous microstructures, using the lattice Boltzmann method, Electrochimica Acta 110 (2013) 325−334. Available from: https://doi.org/10.1016/j.electacta.2013.04.071.

[104] Y. Gao, X. Zhang, P. Rama, R. Chen, H.O.K. Jiang, An improved MRT lattice Boltzmann model for calculating anisotropic permeability of compressed and uncompressed carbon cloth gas diffusion layers based on X-ray computed micro-tomography, Journal of Fuel Cell Science and Technology 9 (4) (2012) 041010. Available from: https://doi.org/10.1115/1.4006796.

[105] Y. Gao, X.X. Zhang, P. Rama, Y. Liu, R. Chen, H. Ostadi, et al., Modeling fluid flow in the gas diffusion layers in PEMFC using the multiple relaxation-time lattice Boltzmann method, Fuel Cells 12 (3) (2012) 365−381. Available from: https://doi.org/10.1002/fuce.201000074.

[106] A. Nabovati, J. Hinebaugh, A. Bazylak, C.H. Amon, Effect of porosity heterogeneity on the permeability and tortuosity of gas diffusion layers in polymer electrolyte membrane fuel cells, Journal of Power Sources 248 (2014) 83−90. Available from: https://doi.org/10.1016/j.jpowsour.2013.09.061.

[107] A. Nabovati, E.W. Llewellin, A.C.M. Sousa, A general model for the permeability of fibrous porous media based on fluid flow simulations using the lattice Boltzmann method, Composites Part A: Applied Science and Manufacturing 40 (6−7) (2009) 860−869. Available from: https://doi.org/10.1016/j.compositesa.2009.04.009.

[108] P. Rama, Y. Liu, R. Chen, H. Ostadi, K. Jiang, Y. Gao, et al., A numerical study of structural change and anisotropic permeability in compressed carbon cloth polymer electrolyte fuel cell gas diffusion layers, Fuel Cells 11 (2) (2011) 274−285. Available from: https://doi.org/10.1002/fuce.201000037.

[109] P. Rama, Y. Liu, R. Chen, H. Ostadi, K. Jiang, X. Zhang, et al., An X-ray tomography based lattice Boltzmann simulation study on gas diffusion layers of polymer electrolyte fuel cells, Journal of Fuel Cell Science and Technology 7 (3) (2010) 0310151−03101512. Available from: https://doi.org/10.1115/1.3211096.

[110] P. Rama, Y. Liu, R. Chen, H. Ostadi, K. Jiang, X. Zhang, et al., Determination of the anisotropic permeability of a carbon cloth gas diffusion layer through X-ray computer micro-tomography and single-phase lattice Boltzmann simulation, International Journal for Numerical Methods in Fluids 67 (4) (2011) 518−530. Available from: https://doi.org/10.1002/fld.2378.

[111] J. Yablecki, A. Nabovati, A. Bazylak, Modeling the effective thermal conductivity of an anisotropic gas diffusion layer in a polymer electrolyte membrane fuel cell, Journal of the Electrochemical Society 159 (6) (2012) B647−B653. Available from: https://doi.org/10.1149/2.013206jes.

[112] B. Han, J. Yu, H. Meng, Lattice Boltzmann simulations of liquid droplets development and interaction in a gas channel of a proton exchange membrane fuel cell, Journal of Power Sources 202 (2012) 175−183. Available from: https://doi.org/10.1016/j.jpowsour.2011.11.071.

[113] L. Hao, P. Cheng, Lattice Boltzmann simulations of water transport in gas diffusion layer of a polymer electrolyte membrane fuel cell, Journal of Power Sources 195 (12) (2010) 3870−3881. Available from: https://doi.org/10.1016/j.jpowsour.2009.11.125.

[114] D.H. Jeon, H. Kim, Effect of compression on water transport in gas diffusion layer of polymer electrolyte membrane fuel cell using lattice Boltzmann method, Journal of Power Sources 294 (2015) 393−405. Available from: https://doi.org/10.1016/j.jpowsour.2015.06.080.

[115] K.N. Kim, J.H. Kang, S.G. Lee, J.H. Nam, C.J. Kim, Lattice Boltzmann simulation of liquid water transport in microporous and gas diffusion layers of polymer electrolyte membrane fuel cells, Journal of Power Sources 278 (2015) 703−717. Available from: https://doi.org/10.1016/j.jpowsour.2014.12.044.

[116] M. Sepe, P. Satjaritanun, S. Hirano, I.V. Zenyuk, N. Tippayawong, S. Shimpalee, Investigating liquid water transport in different pore structure of gas diffusion layers for PEMFC using lattice Boltzmann method, Journal of the Electrochemical Society 167 (10) (2020) 104516. Available from: https://doi.org/10.1149/1945-7111/ab9d13.

[117] P. Satjaritanun, J.W. Weidner, S. Hirano, Z. Lu, Y. Khunatorn, S. Ogawa, et al., Micro-scale analysis of liquid water breakthrough inside gas diffusion layer for PEMFC using X-ray computed

tomography and lattice Boltzmann method, Journal of the Electrochemical Society 164 (11) (2017) E3359—E3371. Available from: https://doi.org/10.1149/2.0391711jes.
[118] Y. Tabe, Y. Lee, T. Chikahisa, M. Kozakai, Numerical simulation of liquid water and gas flow in a channel and a simplified gas diffusion layer model of polymer electrolyte membrane fuel cells using the lattice Boltzmann method, Journal of Power Sources 193 (1) (2009) 24—31. Available from: https://doi.org/10.1016/j.jpowsour.2009.01.068.
[119] T. Arlt, I. Manke, K. Wippermann, H. Riesemeier, J. Mergel, J. Banhart, Investigation of the local catalyst distribution in an aged direct methanol fuel cell MEA by means of differential synchrotron X-ray absorption edge imaging with high energy resolution, Journal of Power Sources 221 (2013) 210—216. Available from: https://doi.org/10.1016/j.jpowsour.2012.08.038.
[120] F. Bresciani, C. Rabissi, A. Casalegno, M. Zago, R. Marchesi, Experimental investigation on DMFC temporary degradation, International Journal of Hydrogen Energy 39 (36) (2014) 21647—21656. Available from: https://doi.org/10.1016/j.ijhydene.2014.09.072.
[121] J. Eller, T. Rosén, F. Marone, M. Stampanoni, A. Wokaun, F.N. Bchi, Progress in in situ X-ray tomographic microscopy of liquid water in gas diffusion layers of PEFC, Journal of the Electrochemical Society 158 (8) (2011) B963—B970. Available from: https://doi.org/10.1149/1.3596556.
[122] P.A. García-Salaberri, D.G. Sánchez, P. Boillat, M. Vera, K.A. Friedrich, Hydration and dehydration cycles in polymer electrolyte fuel cells operated with wet anode and dry cathode feed: a neutron imaging and modeling study, Journal of Power Sources 359 (2017) 634—655. Available from: https://doi.org/10.1016/j.jpowsour.2017.03.155.
[123] C. Hartnig, I. Manke, J. Schloesser, P. Krüger, R. Kuhn, H. Riesemeier, et al., High resolution synchrotron X-ray investigation of carbon dioxide evolution in operating direct methanol fuel cells, Electrochemistry Communications 11 (8) (2009) 1559—1562. Available from: https://doi.org/10.1016/j.elecom.2009.05.047.
[124] L.C. Jean-Marc, R.M. Abouatallah, D.A. Harrington, Detection of membrane drying, fuel cell flooding, and anode catalyst poisoning on PEMFC stacks by electrochemical impedance spectroscopy, Journal of the Electrochemical Society 153 (2006) A857. Available from: https://doi.org/10.1149/1.2179200.
[125] A.A. Kulikovsky, H. Schmitz, K. Wippermann, J. Mergel, B. Fricke, T. Sanders, et al., DMFC: galvanic or electrolytic cell? Electrochemistry Communications 8 (5) (2006) 754—760. Available from: https://doi.org/10.1016/j.elecom.2006.03.011.
[126] M. Schulze, E. Gülzow, S. Schönbauer, T. Knöri, R. Reissner, Segmented cells as tool for development of fuel cells and error prevention/prediagnostic in fuel cell stacks, Journal of Power Sources 173 (1) (2007) 19—27. Available from: https://doi.org/10.1016/j.jpowsour.2007.03.055.
[127] J.M. Sergi, S.G. Kandlikar, Quantification and characterization of water coverage in PEMFC gas channels using simultaneous anode and cathode visualization and image processing, International Journal of Hydrogen Energy 36 (19) (2011) 12381—12392. Available from: https://doi.org/10.1016/j.ijhydene.2011.06.092.
[128] M. Zago, A. Casalegno, A physical model of direct methanol fuel cell anode impedance, Journal of Power Sources 248 (2014) 1181—1190. Available from: https://doi.org/10.1016/j.jpowsour.2013.10.047.
[129] M. Zago, A. Casalegno, C. Santoro, R. Marchesi, Water transport and flooding in DMFC: experimental and modeling analyses, Journal of Power Sources 217 (2012) 381—391. Available from: https://doi.org/10.1016/j.jpowsour.2012.06.022.
[130] I.V. Zenyuk, D.Y. Parkinson, G. Hwang, A.Z. Weber, Probing water distribution in compressed fuel-cell gas-diffusion layers using X-ray computed tomography, Electrochemistry Communications 53 (2015) 24—28. Available from: https://doi.org/10.1016/j.elecom.2015.02.005.
[131] M. Sabharwal, L.M. Pant, A. Putz, D. Susac, J. Jankovic, M. Secanell, Analysis of catalyst layer microstructures: from imaging to performance, Fuel Cells 16 (6) (2016) 734—753. Available from: https://doi.org/10.1002/fuce.201600008.
[132] L. Holden, B.F. Nielsen, Global upscaling of permeability in heterogeneous reservoirs; the output least squares (OLS) method, Transport in Porous Media 40 (2) (2000) 115—143. Available from: https://doi.org/10.1023/A:1006657515753.
[133] F.M. White, Fluid Mechanics, McGraw-Hill, New York, 1994.

[134] T. Berning, N. Djilali, A 3D, multiphase, multicomponent model of the cathode and anode of a PEM fuel cell, Journal of the Electrochemical Society 150 (12) (2003) A1589. Available from: https://iopscience.iop.org/article/10.1149/1.1621412/meta.

[135] Q. Ye, T.V. Nguyen, Three-dimensional simulation of liquid water distribution in a PEMFC with experimentally measured capillary functions, Journal of the Electrochemical Society 154 (12) (2007) B1242. Available from: https://iopscience.iop.org/article/10.1149/1.2783775/meta.

[136] K. Jiao, X. Li, Water transport in polymer electrolyte membrane fuel cells, Progress in Energy and Combustion Science 37 (3) (2011) 221−291. Available from: https://doi.org/10.1016/j.pecs.2010.06.002.

[137] H. Wu, X. Li, P. Berg, On the modeling of water transport in polymer electrolyte membrane fuel cells, Electrochimica Acta 54 (27) (2009) 6913−6927. Available from: https://doi.org/10.1016/j.electacta.2009.06.070.

[138] H. Wu, P. Berg, X. Li, Steady and unsteady 3D non-isothermal modeling of PEM fuel cells with the effect of non-equilibrium phase transfer, Applied Energy 87 (9) (2010) 2778−2784. Available from: https://doi.org/10.1016/j.apenergy.2009.06.024.

[139] P. Dobson, C. Lei, T. Navessin, M. Secanell, Characterization of the PEM fuel cell catalyst layer microstructure by nonlinear least-squares parameter estimation, Journal of the Electrochemical Society 159 (5) (2012) B514−B523. Available from: https://doi.org/10.1149/2.041205jes.

[140] B. Tjaden, S.J. Cooper, D.J. Brett, D. Kramer, P.R. Shearing, On the origin and application of the Bruggeman correlation for analysing transport phenomena in electrochemical systems, Current Opinion in Chemical Engineering 12 (2016) 44−51. Available from: https://doi.org/10.1016/j.coche.2016.02.006.

[141] D.A.G. Bruggeman, Dielectric constant and conductivity of mixtures of isotropic materials, Annals of Physics (Leipzig) 24 (1935) 636−679. Available from: https://onlinelibrary.wiley.com/doi/abs/10.1002/andp.19354160705.

[142] T.E. Springer, T.A. Zawodzinski, S. Gottesfeld, Polymer electrolyte fuel cell model, Journal of the Electrochemical Society 138 (1991) 2334−2342. Available from: https://doi.org/10.1149/1.2085971.

[143] A.Z. Weber, J. Newman, Transport in polymer-electrolyte membranes: I. Physical model, Journal of the Electrochemical Society 150 (7) (2003) A1008. Available from: https://iopscience.iop.org/article/10.1149/1.1580822/meta.

[144] S. Kim, M.M. Mench, Investigation of temperature-driven water transport in polymer electrolyte fuel cell: phase-change-induced flow, Journal of The Electrochemical Society 156 (3) (2009) B353. Available from: https://iopscience.iop.org/article/10.1149/1.3046136/meta.

[145] M. Bhaiya, A. Putz, M. Secanell, Analysis of non-isothermal effects on polymer electrolyte fuel cell electrode assemblies, Electrochimica Acta 147 (2014) 294−309. Available from: https://www.sciencedirect.com/science/article/pii/S0013468614018696?casa_token = jhkoVtbosFkAAAAA:JVYlVBCCwTgD1ph1CAUd2ogwtnAFgx_gImrXFofMP1PHU0yeeGW34uJQql8EpF2BytkaGnEXw.

[146] I.V. Zenyuk, P.K. Das, A.Z. Weber, Understanding impacts of catalyst-layer thickness on fuel-cell performance via mathematical modeling, Journal of the Electrochemical Society 163 (7) (2016) F691. Available from: https://iopscience.iop.org/article/10.1149/2.1161607jes/meta.

[147] I.V. Zenyuk, et al., Investigating evaporation in gas diffusion layers for fuel cells with X-ray computed tomography, The Journal of Physical Chemistry C 120 (50) (2016) 28701−28711. Available from: https://pubs.acs.org/doi/abs/10.1021/acs.jpcc.6b10658.

CHAPTER 5

Mathematical modeling for fuel cells

Abdalla M. Abdalla[1], Mohamed K. Dawood[1], Shahzad Hossain[2], Mohamed El-Sabahy[1], Basem E. Elnaghi[3], Shabana P.S. Shaikh[4] and Abul K. Azad[5]

[1]Mechanical Engineering Department, Faculty of Engineering, Suez Canal University, Ismailia, Egypt
[2]Bangladesh Atomic Energy Commission, Dakha, Bangladesh
[3]Electrical Engineering Department, Faculty of Engineering, Suez Canal University, Ismailia, Egypt
[4]Department of Physics, SP Pune University, Pune, India
[5]Faculty of Integrated Technologies, Universiti Brunei Darussalam, Bandar Seri Begawen, Brunei

5.1 Introduction

Fuel/green energy cells devices are electrochemical ones, converting the chemical energy to the electrical energy. FCs types have merits over other types used in various applications to obtain energy. Where, the electrochemical reactions occurred through feeding of hydrogen and oxygen source (usually air). Surely, hydrogen source is the fuel and chemical reduction of hydrogen electrochemically results in a very high performance. The most, important thing about fuel cells is the used material specifically a catalyst and its function to speed up the process of reactions at the electrodes. Fuel cell types are mainly categorized into several kinds based on the nature of electrolyte used. These types include proton exchange membrane fuel cells (PEMFC), direct methanol fuel cell (DMFC), phosphoric acid fuel cell (PAFC), alkaline fuel cell (AFC), molten carbonate fuel cell (MCFC), and solid oxide fuel cell (SOFC). In the previous decades scientists paid great attention and concentration of fuel cells for their unexpected performance. However the wide and various varieties of applications, fuel cells have many advantages like the nature of their theoretical efficiency, their possibility of using either natural and biogas or methane as a fuel in addition to the higher performance [1]. Many researchers and scientists cited the significance of fuel cell devices because of their modular and distributed nature, and zero noise pollution [1–4].

Many experiments have done and are still ongoing today by developments of used materials including electrolytes and electrodes. The main target of such developments is to reach a high performance with lower cost [4]. The need of higher performance and efficiency requires a tremendous effort in order to manage the economic changes in the whole world. This can only be adjusted by doing an optimization to the fuel cell types through simulation or modeling techniques [5]. Theoretical work and mathematical modeling analysis of fuel cells are very limited compared to experimental ones. It needs more focus especially with new material synthesis at different scales (macro, micro, and nano). According to different applications of fuel cells and based on the most common classification of these devices, their main features can be noticed [6].

5.2 Fuel cells simulation and modeling

Extreme progress and attention has been achieved in the recent decades on fuel cells devices through experimental and theoretical modeling approaches. Hence, fuel cell devices are considered the most important sources of clean and sustainable energy. Mathematical and numerical analysis studies were proposed [7−12] to expect a high performance of the fuel cell. When improvement of the new design or the synthesized material is utilized, the optimization of this process is required. However, these efforts that have been done theoretically need more development, and there is still no complete simulation using a finite element technique of the whole fuel cell systems that can be suitable for different scales. Some trials were achieved [8−10] to simulate the different fuel cell types with different mathematical geometry models. Trials tried to show how the cells could be affected by multiple parameters like cell durability, chemical stability, used fuel, used materials, temperature ranges, output power, scales and nature of the application.

5.3 Process design and mathematical formulation

Fuel cell devices' main target is to get higher performance at different ranges of temperatures, low, intermediate and high with a reasonable cost. The process design of any fuel cell types mainly depends on the following:
- The electrolyte material which basically define the kind of used fuel cell, it can be made from a list of substances as; potassium hydroxide, salt carbonates, and phosphoric acid.
- Utilized fuel in the device, the most fuel applied in such devices is hydrogen.
- Anode/anode catalyst, mainly it can be a fine platinum powder, breaks down the fuel into electrons and ions.
- Cathode/ cathode catalyst, like nickel, converts ions into waste chemicals, with water being the most applied type of waste.
- Gas diffusion layers that are designed to resist oxidization.
- Normally, a typical fuel cell produces a voltage from 0.6 to 0.7 V at full rated load. Voltage decreases as current increases, due to several factors like,
- Activation loss
- Ohmic loss
- Mass transport loss.

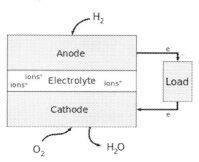

Figure 5.1 Schematic representation of fuel cell device.

Based on what have mentioned above and for the various types of fuel cells devices, it is very essential to be aware of the electrochemical reactions occurred inside each type of these fuel cell devices as shown in Fig. 5.1. Thereby, mathematical modeling and simulation as a software tool that can save time, energy and cost is essentially to be considered with any experimental investigation. The simulation techniques will give more view about each.

Mathematical formulation in fuel cells depends on fuel cell type, utilized fuel, temperature range, and application to be applied in. Some researchers have proposed it in another term through model characteristics [13] which mainly includes model, cell numbers, nominal power, hydrogen pressure in the cell inlet, hydrogen purity, and max temperature. Therefore, the following subsections will present the common equations and mathematical formulation applied in different types of fuel cell devices.

It is very important to create a mathematical modeling of a fuel cell system that is comparable to the phases used for the scientific method. The software tool used in the modeling and simulation must include observing, researching, building, and predicting the results. Once starting the creation of the model by using the most two common ways (empirical and non-imperical models), therefore, the whole boundary conditions based on the investigated case must be available. Also, the whole number of experiments would be too costly or time-consuming, then using a simulation will save the time and the cost as well and the validation of both mathematical and experimental is guaranteed through the agreement between the observations results and the model predictions.

5.3.1 Solid oxide fuel cells

This type uses ceramic materials in electrolyte and electrodes as well. Because SOFCs are made entirely of solid materials, they are not limited to the flat plane configuration of other types of fuel cells and often designed as rolled tubes. SOFCs requires a range of operating temperatures (500°C–1000°C). Also it can operate with a variety of fuels including natural gas [14]. Table 5.1 [15] shows the common governing equations used in SOFC modeling.

Table 5.1 The governing equations used for modelling the solid oxide fuel cell.

Subsystem 1 (SS1)	
Mass balance	$L\frac{d\rho_{air}^{cat}}{dt} = u_{air_{in}}^{cat}\rho_{air_{in}}^{cat} - u_{air}^{cat}\rho_{air}^{cat} - N_{O_2}M_{O_2}\left(\frac{2r_{ct_i}L}{r_{ct_i}^2 - r_{ct_o}^2}\right)$
	$L\frac{d\rho_{air}^{cat}\xi_{O_2}^{cat}}{dt} = u_{air_{in}}^{cat}\rho_{air_{in}}^{cat}\xi_{O_{2_{in}}}^{cat} - u_{air}^{cat}\rho_{air}^{cat}\xi_{O_2}^{cat} - N_{O_2}M_{O_2}\left(\frac{2r_{ct_i}L}{r_{ct_i}^2 - r_{ct_o}^2}\right)$
Momentum balance	$L\frac{d(u_{air}^{cat}\rho_{air}^{cat})}{dt} = \left(u_{air_{in}}^{cat}\right)^2\rho_{air_{in}}^{cat} - \left(u_{air}^{cat}\right)^2\rho_{air}^{cat} + \frac{\rho_{air_{in}}^{cat}R*T_{air}}{M_{air}} - \frac{\rho_{air}^{cat}R*T_{air}}{M_{air}}$
Subsystem 2 (SS2)	
Model equations for cathodic-side diffusion layer	$\frac{\Delta_{cat}}{T_{ct}R}\frac{d}{dt}\left[P_{O_2}^{TPB}\right] = N_{O_2} - \left(\frac{1}{2\pi r_{ct_i}L}\right)R_{O_2}$
Subsystem 3 (SS3)	
The cell output voltage is governed by:	$\frac{dV_{tl}}{dt} = \left(\frac{1}{R_{tct}C_{ct}}\right)E - \frac{1}{C_{ct}}\left(\frac{1}{R_{tct}} + \frac{1}{R_{tct} + R_{load}}\right)V_{tl}$
	$V_{out} = \left(\frac{R_{load}}{R_{to} + R_{load}}\right)V_{tl}$
	$I = \left(\frac{1}{R_{to} + R_{load}}\right)V_{tl}$
Subsystem 4 (SS4)	
Model equations for the anodic-side diffusion layer	$\frac{\Delta_{ano}}{T_{ct}R}\frac{d}{dt}\left[P_{H_2}^{TPB}\right] = N_{H_2} - \left(\frac{1}{2\pi r_{ct_o}L}\right)R_{H_2}$
	$\frac{\Delta_{ano}}{T_{ct}R}\frac{d}{dt}\left[P_{H_2O}^{TPB}\right] = -N_{H_2O} + \left(\frac{1}{2\pi r_{ct_o}L}\right)R_{H_2O}$
Subsystem 5 (SS5)	
Total mass balance on the fuel	$L\frac{d\rho_{fuel}^{ano}}{dt} = u_{fuel_{in}}^{ano}\rho_{fuel_{in}}^{ano} - u_{fuel}^{ano}\rho_{fuel}^{ano} - \left(\frac{2\pi r_{ct_o}L}{A_{ano}}\right) \times [M_{H_2O}N_{H_2O} - M_{H_2}N_{H_2}]$
Mass balance on the fuel components	$L\frac{d\left(\rho_{fuel}^{ano}\xi_{H_2}^{ano}\right)}{dt} = u_{fuel_{in}}^{ano}\rho_{fuel_{in}}^{ano}\xi_{H_{2_{in}}}^{ano} - u_{fuel}^{ano}\rho_{fuel}^{ano}\xi_{H_2}^{ano} + \left(\frac{2\pi r_{ct_o}L}{A_{ano}}\right)M_{H_2}[-N_{H_2}]$
	$L\frac{d\left(\rho_{fuel}^{ano}\xi_{H_2O}^{ano}\right)}{dt} = u_{fuel_{in}}^{ano}\rho_{fuel_{in}}^{ano}\xi_{H_2O_{in}}^{ano} - u_{fuel}^{ano}\rho_{fuel}^{ano}\xi_{H_2O}^{ano} + \left(\frac{2\pi r_{ct_o}L}{A_{ano}}\right)M_{H_2O}[N_{H_2O}]$
Momentum balance	$L\frac{d(u_{fuel}^{ano}\rho_{fuel}^{ano})}{dt} = \left(u_{fuel_{in}}^{ano}\right)^2\rho_{fuel_{in}}^{ano} - \left(u_{fuel}^{ano}\right)^2\rho_{fuel}^{ano} + \frac{\rho_{fuel_{in}}^{ano}R*T_{fuel}^{ano}}{M_{fuel_{in}}^{ano}} - \frac{\rho_{fuel}^{ano}R*T_{fuel}^{ano}}{M_{fuel}^{ano}}$

5.3.2 Molten carbonate fuel cells

The MCFCs [16] is very similar to SOFC. It is run in a moderate and high temperature range of 600°C and above. MCFCs are currently being suited for natural gas, biogas, and coal-based power plants for electrical utility, industrial, and military applications.

MCFCs are high-temperature fuel cells that use an electrolyte composed of a molten carbonate salt mixture suspended in a porous, chemically inert ceramic matrix. This type is composed of many unit fuel cells which are connected in a series.

In the MCFC, the main electrochemical reactions are as follows:

$$\text{(anode)} \quad H_2 + CO_3^{2-} \rightarrow CO_2 + H_2O + 2e^- \tag{5.1}$$

$$\text{(cathode)} \quad CO_2 + \frac{1}{2}O_2 + 2e^- \rightarrow CO_3^{2-} \tag{5.2}$$

$$CO + H_2O \leftrightarrow H_2 + CO_2 \tag{5.3}$$

Mass balance:

$$\frac{1}{\Delta y}\frac{\partial(n_a^k X_a^k)}{\partial X} + \gamma_{E,\beta}\frac{j^k}{2F} + VS, a\frac{n_s^k}{\Delta X \Delta Y} = 0 \tag{5.4}$$

$$\frac{1}{\Delta y}\frac{\partial(n_c^k X_{a\beta}^k)}{\partial X} + \gamma_{E,\beta}\frac{j^k}{F} = 0 \tag{5.5}$$

$$K_{S,k,y}^k = \frac{(P_{H_2}^k)(P_{CO_2}^k)}{(P_{H_2O}^k)(P_{CO}^k)} = \frac{(X_{H_{2xy}}^k)(P_{CO_2}^k)}{(P_{H_2Ox,y}^k)(P_{CO}^k)} (\approx 2 \text{ at } 650°C) \tag{5.6}$$

$$= 157.02 - 0.4447 K_{x,y}^k + 4.2777 \times 10^{-4} \quad K_{x,y}^k - 1.3871 \times 10^{-7} \quad K_{x,y}^{k3} \tag{5.7}$$

Here, $Txyk\,[k]$ is the local anode gas temperature at x and y in the k-th cell

Energy balance:

$$\frac{\partial^2 T_3^k}{\partial X^2} + \frac{\partial^2 T_3^k}{\partial y^2} = \frac{h_{reS}}{b_s k_s}(T_s^k - T_c^k) + \frac{h_{res}^k}{b_s k_s}(T_c^{k+1} - T_s^k) \\ + \frac{h_{s,ga}}{b_s k_s}(T_c^{k+1} - T_{ga}^k) + \frac{h_{s,ga}^{k+1}}{b_s k_s}(T_{gc}^{k+1} - T_s^k) \tag{5.8}$$

$$\frac{\partial}{\partial x}(m_a C_{pga} T_{ga}^k) = h_{cga}(T_c^k - T_{ga}^k) + h_{cga}(T_s^k - T_{ga}^k) + \sum_a C_{ax} C_{p,ga,a} b_{ga} T_g^k + \frac{Q_s^k}{\Delta x \Delta y} \tag{5.9}$$

$$Q_{S_{xy}}^k = \Delta H_{S_{xy}}^k \cdot \Delta n_{CO_{xy}}^k \qquad (5.10)$$

$$\Delta H_{S_{xy}}^k = -9932.5 - 0.515 T_{xy}^k + (3.117 \times 10^{-3}) T_{xy}^{k2} - (1.05 \times 10^{-6}) T_{xy}^{k3} \qquad (5.11)$$

$$\frac{\partial}{\partial x} + \frac{\partial}{\partial x} = \frac{h_{ega}}{b_e k_e} (T_e^k - T_s^k) + \frac{h_{ega}^{k-1}}{b_e k_e} (T_e^k - T_{ga}^{k-1}) + \frac{h_{egc}}{b_e k_e} (T_e^k - T_{gc}^k) + \frac{Q_E^k}{b_e k_e} \qquad (5.12)$$

$$Q_{E_{xy}}^k = i_{xy}^k \cdot \left(-\frac{\Delta H_{E_{xy}}^k}{2F} + V^k \right) \qquad (5.13)$$

$$\Delta H_{S_{xy}}^k = -57.018 - 2.738 T_{xy}^k + (0.474 \times 10^{-3}) T_{xy}^{k2} + (2.637 \times 10^{-6}) T_{xy}^{k3} \qquad (5.14)$$

Current density and cell voltage

$$V_{x,y}^k = \left(V_{cN_{xy}}^k - V_{aN_{xy}}^k \right) - j_{x,y}^k Z_{x,y}^k \qquad (5.15)$$

$$V_{aN_{xy}}^k = V_a^{0k} + \frac{RT_{xy}^k}{2F} \ln \left(\frac{X_{aCO_{2xy}}^k X_{H_2O_{xy}}^k}{X_{H_{2xy}}^k} \right) \qquad (5.16)$$

$$V_{cN_{xy}}^k = V_c^{0k} + \frac{RT_{xy}^k}{2F} \ln \left(X_{cCO_{2xy}}^k X_{O_{2xy}}^{k 1/2} \right) \qquad (5.17)$$

$$Q_{xy}^k = R_{ohm_{xy}}^k + Z_{a_{xy}}^k + Z_{c_{xy}}^k \qquad (5.18)$$

5.3.3 Proton exchange membrane fuel cells

This type of fuel cells device (see Fig. 5.2), has another name called a polymer electrolyte membrane (PEM) fuel cells [17,18]. It is being developed basically for transport applications. Moreover, it is well suited for both stationary and portable fuel cell applications. The most essential features of such type, is the lower temperature/pressure ranges (50°C−110°C) and a special proton-conducting PEM.

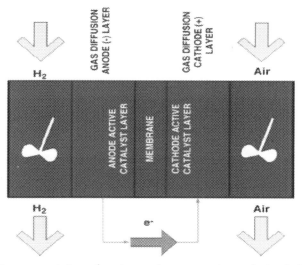

Figure 5.2 Schematic representation of proton exchange membrane fuel cells [17].

PEMFCs generate electricity and operate on the opposite principle to PEM electrolysis, which consumes electricity. They are a leading candidate to replace the aging alkaline fuel cell technology, which was used in the aerospace applications like in Space Shuttle.

Based on the assumptions, the following fuel cell optimization formulation is derived:

$$v = \min_{x \in \Omega} f(x) = \min \frac{C_1 A}{t} + C_2 y_{H_2}^i F^i - C_3 y_{H_2}^0 F^0 \quad (5.19)$$

$$S.t. W = E \quad A_i \quad (5.20)$$

$$i = ZFN_{H^+} \quad (5.21)$$

$$N_{H^+} = 2N_{H2} \quad (5.22)$$

$$i = i_a^o \left(e^{\bar{\alpha}_a F \eta_a / RT} - e^{-\bar{\alpha}_a F \eta_a / RT} \right) \quad (5.23)$$

$$i = -i_c^o \left(e^{\bar{\alpha}_c F \eta_c / RT} - e^{-\bar{\alpha}_c F \eta_c / RT} \right) \quad (5.24)$$

$$E_{Nernst} = E^o - \frac{RT}{nF} \ln \frac{(P_{H_2O})^2}{(P_{H_2}^*)^2 (P_{O_2})} \tag{5.25}$$

$$E = E_{Nernst} + \eta_c - \eta_a - i\rho\delta \tag{5.26}$$

$$N_{H_2} = \frac{k_g}{RT} \left(\gamma_{H_2}^o P - P_{H_2}^* \right) \tag{5.27}$$

$$F^i = N_{H_2} A + F^o \tag{5.28}$$

$$\gamma_{H_2}^i F^i = N_{H_2} A + \gamma_{H_2}^o F^o \tag{5.29}$$

$$O \leq P_{H_2}^* \leq \gamma_{H_2}^o P \leq \gamma_{H_2}^i P \tag{5.30}$$

$$-E^o \eta_c \leq 0 \leq \eta_a \leq E^o \tag{5.31}$$

$$0 \leq E \leq E^o \tag{5.32}$$

$$0 \leq F^o \leq F^i \tag{5.33}$$

where

$$x^T = \left[P_{H_2}^*, \gamma_{H_2}^o, A, i, \eta_a, \eta_c, E, E_{Nerst}, F^i, F^o, N_{H_2}, N_{H^+} \right]$$

$$\begin{aligned} i &= mFk_a e^{-[\Delta G_a(0) - m(1-\alpha)FE_c]/RT}[\text{Red}] \\ &- mFk_c e^{-[\Delta G_c(0) + m\alpha FE_c]/RT}[\text{O}_x] \end{aligned} \tag{5.34}$$

$$\begin{aligned} i_c^o &\triangleq mFk_a e^{-[\Delta G_a(0) - m(1-\alpha)FE_c^o]/RT}[\text{Red}] \\ &- mFk_c e^{-[\Delta G_c(0) + m\alpha FE_c^o]/RT}[\text{O}_x] \end{aligned} \tag{5.35}$$

where E_c^o is the potential at zero net current. Defining the electrode's overpotential as $\eta_e = E_e - E_e^o$, then yields

$$i_c^o e^{\frac{m(1-\alpha)F\eta_c}{RT}} = mFk_a e^{-[\Delta G_a(0) - m(1-\alpha)FE_c]/RT}[\text{Red}]$$

And $\quad i_c^o e^{-m\alpha F\eta_c/RT} = mFk_c e^{-[\Delta G_c(0) + m\alpha FE_c]/RT}[\text{O}_x]$

$$i = i_c^o e^{m(1-\alpha)F\eta_c/RT} - i_c^o e^{-\frac{m\alpha F\eta_c}{RT}} \qquad (5.36)$$

$$\Delta G^o = -nFE^o \qquad (5.37)$$

5.3.4 Direct alkaline fuel cell

The AFC [19] started to be developed since more than 40 years and it was the first fuel cell technology applied in various applications, especially the aerospace ones like Apollo space missions and automobiles. Unfortunately, this type faced some technical, commercial, and safety problems. Recently, in the last few decades the interest in AFC increased due to more favorable oxygen reduction and fuel oxidation reactions in alkaline condition. In addition to the use of many software tools, it makes it very easy to expect any problem that can occur, and enables it to increase performance of these types of cells. The main governing equations in AFCs can be noticed from the following:

$$E_{cell} = E - (\eta_{ac} + \eta_{oh} + \eta_{conc}) \qquad (5.38)$$

$$\text{Alcohol} \rightarrow \text{Reaction intermediate} \rightarrow \text{Acid(in anionic form)}$$
$$\downarrow$$
$$\text{Poisoning} \qquad \text{species} \rightarrow CO_2 \qquad (5.39)$$
$$(CO_2)$$

$$CH_3OH \Leftrightarrow HCO_{ad} + 3H^+ + 3e^-$$
$$\xrightarrow{+3OH^-} 3H_2O \qquad (5.40)$$

$$OH^- \Leftrightarrow OH_{ad} + e^- \qquad (5.41)$$

$$HCO_{ad} + OH_{ad} \rightarrow HCOOH \qquad (5.42)$$

$$CH_3OH \xrightarrow{-H(H^+ + e^-)} CH_3O_{ad} \xrightarrow{-H(H^+ + e^-)} CH_2O_{ad} \xrightarrow{-H(H^+ + e^-)} CHO_{ad} \qquad (5.43)$$

$$C_2H_5OH \Leftrightarrow CH_3CO_{ad} + 3H^+ + 3e^-$$
$$\xrightarrow{+3OH^-} 3H_2O \qquad (5.44)$$

$$OH^- \Leftrightarrow OH_{ad} + e^- \qquad (5.45)$$

$$CH_3CO_{ad} + OH_{ad} \rightarrow CH_3COOH \tag{5.46}$$

$$BH_4^- \Leftrightarrow HBO_{2ad} + 5H_2O + 8e^- \tag{5.47}$$

$$OH^- \Leftrightarrow OH_{ad} + e^- \tag{5.48}$$

$$HBO_{2ad} + OH_{ad} + e^- \rightarrow BO^-_2 + H_2O \tag{5.49}$$

$$O_2 + H_2O + 2e^- \rightarrow HO^-_2 + OH^- \tag{5.50}$$

$$HO^-_2 + H_2O + 2e^- \rightarrow 3OH^- \tag{5.51}$$

Anode and cathode reactions in direct methanol are given by:

$$CH_3OH + O_2 \rightarrow HCOOH + H_2O \tag{5.52}$$

$$E_{cell} = E - \left(\frac{RT}{\alpha nF}\right) \ln\left(\frac{jC_M^{-1} C_{OH}^{-0.5}}{K}\right) - jR_{oh} - \frac{RT}{\alpha nF} \ln\left(\frac{j}{j_o(1-jM)}\right) \tag{5.53}$$

Anode and cathode reactions in direct ethanol given by:

$$C_2H_5OH + O_2 \rightarrow CH_3COOH + H_2O \tag{5.54}$$

$$BH_4^- + 2O \rightarrow BO_2^- + 2H_2O \tag{5.55}$$

$$E_{cell} = E - \left(\frac{RT}{\alpha nF}\right) \ln\left(\frac{jC_E^{-1} C_{OH}^{-0.5}}{K'}\right) - jR_{oh} - \frac{RT}{\alpha nF} \ln\left(\frac{j}{j_o(1-jM)}\right) \tag{5.56}$$

5.3.5 Microbial fuel cells

Among the different types of fuel cell devices, there is a microbial fuel cell (MFC) [20] which is mainly known as a bioelectrochemical system generates electricity with bacteria in nature and a high-energy oxidant (see Fig. 5.3). MFCs are categorized into mediated and unmediated systems. The first type depends on the use of a mediator where electrons transfers from the bacteria in the cell to the anode. While, the second type the bacteria typically have electrochemically active redox

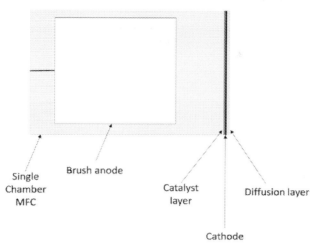

Figure 5.3 Schematic representation of microbial fuel cells cell based on cathode support [20].

proteins that can transfer electrons to the anode directly. Many efforts were done, to widespread this type of fuel cells to generate hydrogen and electricity as well. In addition, simulation and modeling were studied to make more validation of these types of devices.

The governing equations applied in MFCs mathematical modeling could be shown as in the table below:

$I = \sigma \nabla \phi$
$i = \nabla I$
$i_x = \nabla I_x = \nabla(\sigma_{x,\text{eff}} \nabla \phi_x)$
$i_l = \nabla I_l = \nabla(\sigma_{l,\text{eff}} \nabla \phi_l)$
$\sigma_{x,\text{eff}} = \varepsilon_x^{1.5} \sigma_x$
$\sigma_{l,\text{eff}} = \varepsilon_l^{1.5} \sigma_l$
$i_a = k_{o,a} \gamma_l f_e^o F\left(\frac{C_x}{C_x + K_{x,a}}\right) x_a \frac{A_s}{V_a} \exp\left(\frac{\alpha_a F \eta}{RT}\right)$
$i_c = i_{o,c}\left(-\frac{C_{O_2}}{C_{O_2,\text{ref}}} \exp\left[\frac{-\alpha_c F \eta}{RT}\right]\right)$
$\eta = \phi_x - \phi_l - E_{eq}$
$\varepsilon_{p\text{eff}} \frac{\partial C_x}{\partial t} + \nabla\left(-D_{\text{eff},a} \nabla C_x\right) = \frac{-\alpha_a i_a}{nF} - q_m x_m$
$\frac{dx_a}{dt} = Y_a \frac{\alpha_a i_a}{nF} x_a - K_{d_a} x_a$
$\frac{dx_m}{dt} = Y_m q_m x_m - K_{d_m} x_m$
$q_m = q_{\max_m} \frac{C_x}{C_x + K_{s_m}}$
$\varepsilon_p \frac{\partial}{\partial t}(\rho \omega_i) = \nabla \cdot \left(\rho D_i^m \nabla \omega_i + \rho \omega_i D_i^m \frac{\nabla M}{M}\right) + R_i$

where,

$$D_i^m = \frac{1-\omega_i}{\sum_{k\neq i}^M \frac{x_g}{D_{xg}}}$$

$$D_{e,k} = \frac{\varepsilon_{\mu,c}}{\tau_F} D_{i,k}$$

$$x_k = \frac{\omega_k}{M_k} M$$

$$\frac{1}{M} = \sum_{i=1}^q \frac{\omega_i}{M_i}$$

$$R_{O_2} = -\frac{\alpha_c i_c}{4F}$$

$$R_{H_2O} = \frac{2\alpha_c i_c}{4F} = \frac{\alpha_c i_c}{2F}$$

$$R_{N_2} = 0$$

$$J = \sum_l \left(V_{\exp(l)} - V_{\sim(l)}\right)^2$$

In the case of nanowire applied in MFCs, there was, a modling [21] based on the Nernst–Monod relationship using combinations of the Monod relationship through the calculation of an electron acceptor utilization rate. The used equations in such case as follow:

Anode: $CH_3COO^- + 4H_2O \rightarrow 8e^- + 8H^+ + 2HCO_3^-$
Cathode: $Fe(CN)_6^{3-} + e^- \rightarrow Fe(CN)_6^{4-}$

$$J = j_{max}\left(\frac{1}{1+\exp\left[-\frac{F}{RT}\eta\right]}\right)$$

$$j_{max} = \gamma_s q_{max} X_f L_{fa}$$

$$J = -\frac{K_{blo}(E_{om} - E_{interface})}{\Delta x}$$

$$J = j_o\left\{\exp\left[\frac{\alpha nF}{RT}(E - E_n^o)\right] - \exp\left[\frac{(1-\alpha)nF}{RT}(E - E_n^o)\right]\right\}$$

$$J = -j_o\left[\frac{nF(1-\alpha)(E_{anode} - E_{interface}^o)}{RT}\right]$$

$$E_{cell} = E_{cell} - \eta_{act} - \eta_{anc} - \eta_{ohm}$$

$$\eta_{conc} = \frac{RT}{F} \ln\left(\frac{b_{lna} + d_{det}}{(Y q_{max} - b_{rex}) - (b_{lna} + d_{det})}\right)$$

$$\eta_{act} = \frac{i_{cell}}{A_{surf} i_o} \frac{RT}{nF}\left(\frac{C_{red}}{C_{ax}}\right)$$

$$R_{cell} = \frac{\delta^M}{\kappa}$$

$$i = j_{max}\left(\frac{1}{1+\exp\left(-\frac{F}{RT}\eta\right)}\right) - \frac{K_{blo}(E_{om} - E_{interface})}{\Delta x} + i_o\exp\left[\frac{\alpha nF}{RT}(E-E_n^o)\right] - i_o\exp\left[\frac{-(1-\alpha)nF}{RT}(E-E_n^o)\right]$$

5.4 Key challenges of mathematical modeling

The energy efficiency of fuel cells systems is measured by the ratio of the amount of useful energy produced by the system to the total amount of input energy or it may calculated by the percentage of both. The main goal of mathematical modeling applied in fuel cells is to get the theoretical maximum efficiency of any type of power generation system that cannot be reached experimentally. However, the numerical/mathematical calculation allows the comparison of different types of power generation obtained through the electrochemical reactions in the fuel cell devices.

The mathematical modeling in fuel cells that are extremely important to be aware of and the important challenges that appear and are faced in the fuel cell system include (1) the amount of heat that a fuel cell generates, (2) the amount of fuel needed for a fuel cell, (3) the amount of electricity generated by a fuel cell, (4) the amount of voltage and current required to produce a certain amount of power through the electrochemical reactions, (5) the flow rate through a fuel cell flow channel, and finally (6) the whole stack design. Hence, finding a reliable model that is able to simulate the actual process is challenging and requires more research and investigations.

5.5 Conclusion

With the rapid increase of technology today in various applications, mathematical modeling has become a very useful tool in saving time and cost. In this chapter, we have checked a good number of literature equations and parameters used in multiple software. The model description and identification is too important to understand the behavior of the main parameters and variables in the modeled system. It is also required to give an effective engineering solution saving, efforts, time, cost, and give high validation with actual results.

References

[1] N.H. Menzler, F. Tietz, S. Uhlenbruck, H.P. Buchkremer, D. Stöver, Materials and manufacturing technologies for solid oxide fuel cells, Journal of Materials Science 45 (2010) 3109–3135.
[2] S.M. Haile, Fuel cell materials and components, Acta Materialia 51 (2003) 5981–6000.
[3] V. Ramani, Fuel cells, The Electrochemical Society (2006) 41–44. Available from: https://www.electrochem.org/dl/interface/spr/spr06/spr06_p41-44.pdf.
[4] S.A.N.P. Jiang, S.H.W.A. Chan, A review of anode materials development in solid oxide fuel cells, Journal of Materials Science 9 (2004) 4405–4439.
[5] J.J. Baschuk, X. Li, A general formulation for a mathematical PEM fuel cell model, Journal of Power Sources 142 (2004) 134–153.
[6] S. Edition, W. Virginia, Fuel cell handbook (2004).
[7] J. Zhang, L. Lei, D. Liu, F. Zhao, M. Ni, F. Chen, Mathematical modeling of a proton-conducting solid oxide fuel cell with current leakage, Journal of Power Sources. 400 (2018) 333–340.
[8] S. Ou, H. Kashima, D.S. Aaron, J.M. Regan, M.M. Mench, Multi-variable mathematical models for the air-cathode microbial fuel cell system, Journal of Power Sources 314 (2016) 49–57.
[9] S.A. Hajimolana, M.A. Hussain, W.M.A. Wan, M. Soroush, A. Shamiri, Mathematical modeling of solid oxide fuel cells : a review, Renewable Sustainable Energy Reviews 15 (2011) 1893–1917.
[10] M.L. Aksu, B. Zu, Advanced mathematical model for the passive direct borohydride/peroxide fuel cell. 6 (2011) 2–9.

[11] F. Gonzatti, F.A. Farret, Mathematical and experimental basis to model energy storage systems composed of electrolyzer, metal hydrides and fuel cells, Energy Conversion Management 132 (2017) 241−250.
[12] Z. Pan, Y. Bi, L. An, Mathematical modeling of direct ethylene glycol fuel cells incorporating the effect of the competitive adsorption, Appllied Thermal Engineering 147 (2019) 1115−1124.
[13] R.S. Gomes, A.L. De Bortoli, A three-dimensional mathematical model for the anode of a direct ethanol fuel cell, Appllied Energy 183 (2016) 1292−1301.
[14] M.S. Ismail, D.B. Ingham, K.J. Hughes, L. Ma, M. Pourkashanian, An efficient mathematical model for air-breathing PEM fuel cells. 135 (2014) 490−503.
[15] S. Tonekabonimoghadam, R.K. Akikur, M.A. Hussain, S. Hajimolana, R. Saidur, M.A. Hashim, Mathematical modelling and experimental validation of an anode-supported tubular solid oxide fuel cell for heat and power generation, Energy (2015).
[16] S. Lee, D. Kim, H. Lim, G. Chung, Mathematical modeling of a molten carbonate fuel cell, International Journal of Hydrogen Energy 35 (2010) 13096−13103.
[17] K.I. Chen, J. Winnick, V.I. Manousiouthakis, Global optimization of a simple mathematical model for a proton exchange membrane fuel cell. 30 (2006) 1226−1234.
[18] J.J. Baschuk, X. Li, A comprehensive, consistent and systematic mathematical model of PEM fuel cells. 86 (2009) 181−193.
[19] A. Verma, S. Basu, Experimental evaluation and mathematical modeling of a direct alkaline fuel cell. 168 (2007) 200−210.
[20] S. Gadkari, S. Gu, J. Sadhukhan, Two-dimensional mathematical model of an air-cathode microbial fuel cell with graphite fiber brush anode, Journal of Power Sources. 441 (2019) 227145.
[21] T. Lan, C. Wang, T. Sangeetha, Y. Yang, A. Garg, Constructed mathematical model for nanowire electron transfer in microbial fuel cells, Journal of Power Sources. (2018) 483−488.

SECTION 2:

NANOSTRUCTURES AND NANOMATERIALS FOR FUEL CELLS

CHAPTER 6

Nanostructures and nanomaterials in microbial fuel cells

Saranya Narayanasamy and Jayapriya Jayaprakash
Department of Applied Science and Technology, Alagappa College of Technology, Anna University, Chennai, India

6.1 Introduction

Bioelectrochemical systems (BES) is an emerging renewable technology utilized to generate green energy through bioelectrocatalytic processes. In recent times, BESs have gained increased interest owing to low carbon energy production, and application in waste valorization and the sustainable bioremediation of environmental pollutants. Microbial fuel cells (MFCs) are sustainable BES that makes use of biofilms as catalysts to convert the chemical energy stored in biodegradable matters directly into electrical energy [1]. The concept of MFCs was discovered by Potter [2], who showed that bacteria can transfer electrons externally to electrodes [2]. In the last decade, this technology has been developed practically to generate electricity while simultaneously treating wastewater. The incredible development of MFC systems in the past 30 years has nurtured a wide variety of its applications, such as biosensors, robotics, desalination, water treatment, and electricity generation. MFCs are primarily made up of two components, the two conductive electrodes, which are placed in the anodic and cathodic chambers, respectively. The electrons and protons are produced at the anode by the oxidation of biodegradable matter using the biocatalyst, that is, bacteria. In the cathode chamber, similar to conventional fuel cells, reduction of the electron acceptor takes place when electrons move from the anode through an external circuit and when protons move through the membrane [3]. Despite the innovative involvement of biological energy sources, the power output from MFCs is still not enough for practical applications. This is mainly due to (1) difficulty in achieving electron transfer between the extracellular electrode and the bacterial cells at the anode and (2) the slow kinetic rate and high overpotential of oxygen reduction on the surface of the cathode, especially in neutral pH, as well as (3) poor adsorption of O_2 at the cathode electrode [4].

The performance of MFCs primarily depends on their key components: (1) anode material; (2) cathode material; (3) the substrate used as a catalyst; and (4) polymer electrolyte membrane (PEM) [5]. Several factors also determine the MFC power density:

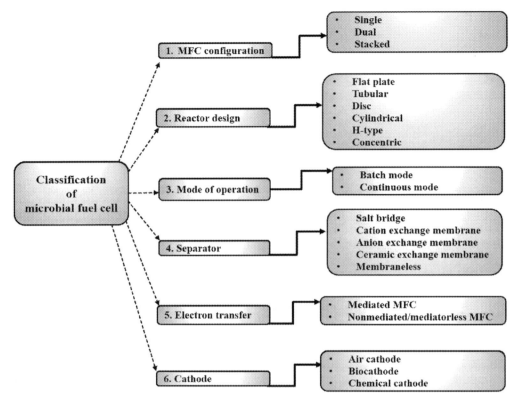

Figure 6.1 Schematic of MFC classification based on performance-determining factors (*MFC*, Microbial fuel cell).

the extracellular electron-transfer rate from the bacteria to the electrode, substrate type and degradation rate, reactor design, operating conditions, and external resistance. The broad classification of MFCs based on various factors is presented in Fig. 6.1.

In the last decade, research in this field has been focused on the electrode material because it directly influences the performance and thus the commercialization potential of MFCs [6]. Various electrode materials have been used to enhance the performance of MFCs ranging from single-metal electrodes to composite nanomaterials [1]. In this chapter, the important features of nanostructures are explained using carbon nanostructures as examples in Section 6.2. Anodes, cathodes and membranes made of their composites with other materials (metal-based, carbon-based, and polymer-based) are discussed in Section 6.3.

6.2 Nanostructured materials in microbial fuel cells

Carbon-structured electrode materials [e.g., carbon paper (CP), carbon cloth (CC), activated carbon (AC), reticulated vitrified carbon, carbon fiber, carbon brush, carbon felt (CF), carbon foam, graphite felt, and graphite rod] [7] and metal-based electrodes [e.g., molybdenum, platinum (plate and mesh), stainless steel (plate, foam, mesh, wool), silver, titanium, copper, and nickel] have been widely investigated and extensively used as MFC anodes [8,9]. These electrode materials have their own merits, yet their demerits, such as low electricity generation due to low level of microbial inoculum and electron transfer efficiency and poor stability should be overcome. One of the promising solutions to resolve this problem is to modify carbon-based anodes to enhance microbial immobilization, reduce the electron transfer barrier, and thereby promote electricity generation [10]. Thus electrode modification with nanoparticles/nanostructures can be a state-of-the-art approach to decrease the electron transfer resistance [11].

6.2.1 Nanostructures as electrode materials

Nanomaterials are characterized by an ultrafine grain size (<50 nm) or dimensionality limited to 50 nm and can be created with various modulation dimensionalities as defined by Richard W. Siegel [12]: zero [zero-dimensional (0D)—atomic clusters, filaments, and cluster assemblies], one [one-dimensional (1D)—multilayers], two [two-dimensional (2D)—ultrafine-grained overlayers or buried layers], and three [three-dimensional (3D)—nanophase materials consisting of equiaxed nanometer-sized grains].

Two primary methodologies are used for the fabrication of nanostructures. One is the bottom-up approach, which involves assembling atoms and molecules to synthesize nanoparticles of diverse shape and size, such as those used in sol—gel processing, chemical vapor deposition, laser pyrolysis, and inert gas condensation. The other is the top-down approach, which involves working from bigger to smaller objects and is extensively used in mechanical milling, laser ablation, electro-explosion, electron beam treatment, and plasma etching. The three ways of nanoparticle synthesis, that is, physical, chemical, and biological, work on either of these two methodologies. Nanoparticle synthesis by physical techniques employs the use of high energy radiation, thermal or electrical energy, mechanical pressure for material etching, melting, or evaporation. In the following section, different carbon-based nanomaterials of various dimensions are discussed in order to discern the importance of nanostructures in MFCs.

6.2.1.1 Carbon-based structures

Carbon-based nanomaterials are one of the most widely studied materials in the field of nanotechnology due to low-cost production, low intrinsic toxicity, and multifunctional surface functionalization. They can be 0D [quantum dots (QDs), AC, and mesoporous carbon], 1D [carbon nanotubes (CNT) and carbon nanofibers (CNF)], 2D (graphene-based materials), and 3D (macroporous carbon) materials.

6.2.1.1.1 Zero-dimensional (0D) carbon nanostructures (nanoparticles/nanospheres)

Unlike the bulk high-dimensional nanomaterials, 0D nanomaterials are mostly spherical or quasispherical nanoparticles with an ultrasmall diameter of less than 100 nm and high surface-to-volume ratios; they have more active edge sites per unit mass. The 0D nanostructures used in MFCs are nanoparticles or nanospheres, such as AC [13–15], mesoporous carbon [16–18], and QDs.

Compared with other nanomaterials, carbon dots (CDs) have drawn significant scientific focus because of the multitude synthesis methods. CDs are nanometer-scale (2–10 nm) semiconductor materials with special optical properties and quantum mechanical properties, such as size-tunable emission spectra, broad-range UV excitation, bright fluorescence, and high photostability [19]. Generally, the approaches for CDs preparation can be mainly divided into top-down and bottom-up routes. Carbon QDs are a group of benign, abundant, and inexpensive carbon nanomaterials. CDs rich in hydroxyl and carboxyl groups which have reducing property and thus facilitate electron transfer by forming strong hydrogen bonds with bacterial cytochrome [20]. The addition of CDs as a suspension in the anode chamber of an MFC resulted in a 22.5% enhancement in the maximum power density. The inherent ability of carbon QDs to donate and accept electrons in microbial electron transfer renders it crucial as an electron shuttle in the anode chamber of MFCs [8]. The application of QDs has attracted great attention in the photocatalytic field especially as photosensitizers in solar cells and photo-MFCs [21,22].

Guan et al. supported biocompatible 0D N-doped CDs on CP to make an MFC anode. The developed electrode exhibited a highly hydrophilic surface ($\sim 10°$) compared with the raw CP electrode ($\sim 112°$) and benefited microbial immobilization. In addition, this electrode displayed a smaller charge transfer barrier, suggesting enhanced interaction between the microbes and the anodes. The MFC with the N-doped CDs@carbon paper anode displayed double the power density of the CP anode (0.32 vs 0.15 mW). Microbial analysis showed that the relative abundance of Pseudomonas was significantly enhanced on the as-prepared electrode (40% higher than that on the unmodified electrode) (Fig. 6.2) [23]. Similarly, when polypyrrole nanowires (PPy-NWs) modified by carbon dots (CDs/PPy-NW) with a unique dot-line structure was used as the anode, the electron transfer rate increased sharply by 26% over pure PPy-NWs, and the resistance of the CDs/PPy-NW electrode was only one-third that of the pure PPy-NWs electrode. The mini-MFC equipped with the CDs/PPy-NW composite anode exhibited a high open-circuit voltage (630 mV) and twice the maximum power density of that equipped with pure PPy-NWs as the anode [24].

6.2.1.1.2 One-dimensional carbon nanostructures (Nanowires/nanorods/nanofibers/nanotubes)

One-dimensional (1D) carbon materials are usually available as nanowires, nanorods, nanofibers, or nanotubes. CNFs and CNTs are capable of interacting with various inorganic species at the surface that enhance separation, adsorption and catalysis and thus

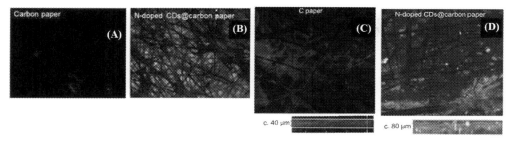

Figure 6.2 Fluorescence microscopy images of (A) the carbon paper electrode and (B) the N-doped CDs@carbon paper electrode under UV light. The confocal laser scanning microscopy (CLSM) images of the biofilms inoculated on (C) the carbon paper electrode and (D) the N-doped CDs@carbon paper electrode showcase increased biofilm thickness with modification [20]. From Y.F. Guan, F. Zhang, B.C. Huang, H.Q. Yu, Enhancing electricity generation of microbial fuel cell for wastewater treatment using nitrogen-doped carbon dots-supported carbon paper anode, Journal of Cleaner Production 229 (2019) 412–419. https://doi.org/10.1016/j.jclepro.2019.05.0402

Figure 6.3 FE-SEM images of (A) CNT-coated bacterial cellulose and (B) after bacterial colonization. (C) SEM image of the surface of plain CNT-textile (*CNT*, Carbon nanotube; *FE-SEM*, field-emission scanning electron microscopy; *SEM*, scanning electron microscopy) [25,26]. *Reprinted with permission from M. Mashkour, M. Rahimnejad, M. Mashkour, F. Soavi, Electro-polymerized polyaniline modified conductive bacterial cellulose anode for supercapacitive microbial fuel cells and studying the role of anodic biofilm in the capacitive behavior, Journal of Power Sources 478(August) (2020) 228822. https://doi.org/10.1016/j.jpowsour.2020.2288223; X. Xie, L. Hu, M. Pasta, G.F. Wells, D. Kong, C.S. Criddle, et al., Three-dimensional carbon nanotube-textile anode for high-performance microbial fuel cells, Nano Letters 11(1) (2011) 291–296. Copyright 2020, Elsevier & 2011, Royal Society of Chemistry.*

have been considered as highly promising nanomaterials. CNTs have received increasing attention in research since their discovery in 1991 due to their unique mechanical, electrical and chemical properties. According to the characteristics of the tubular structure, they can be divided into single-walled CNTs and multiwalled carbon nanotubes (MWCNTs). They have high elastic modulus, tensile strength and thermal/electrical conductivity, surface-area-to-volume ratio, and excellent mechanical and chemical stability. The excellent electrical and biocompatible properties of CNTs promote electron-transfer reactions, microbial attachment, substrate diffusion, and substrate oxidation, making CNTs extremely attractive as an electrode material in MFCs (Fig. 6.3).

On the other hand, CNFs have high porosity, large specific surface area and interconnected filaments, which can satisfy many anode material requirements after stabilization and carbonization. Electrospinning is a straight-forward, simple and low-cost technology to obtain CNFs. Electrospun nanofibers have good application prospects in energy storage and conversion, filtration/separation and sensor. In particular, they are binder-free and do not need secondary processing using a conductivity material. Composite material with excellent conductivity, surface area, mechanical properties and low costs can be formed by hybridizing CNFs with functional materials. Moreover, they also facilitate substrate transport, intercellular communication and charge transfer. For instance, Cai et al. studied a carboxylated multiwalled CNTs/CNFs composite electrode fabricated by electrospinning. The heat pressing process was applied to improve the interconnection of the fiber aggregates and mechanical stability and to reduce contact resistance. The optimal dose of CNTs was selected to fabricate the anode for MFCs after comparison with the plain CNF electrospun anode and a commercial CF anode. The optimal anode delivered a maximum power density that was more than twofold higher than those of the CNF and CF anodes, respectively. The as-prepared anode also exhibited better conductivity, excellent biocompatibility, good hydrophilicity and superior electrocatalytic activity, which were not only beneficial for the attachment and reproduction of microorganisms but also promoted extracellular electron transfer between the bacteria and the anode [27].

The activated carbon nanofiber nonwoven (ACNFN) is an ultrathin interconnected structure fabricated by the pyrolysis of electrospun polyacrylonitrile (PAN) and subsequent steam activation. It possesses a large bioaccessible surface area and has an open porous structure that supports biofilm growth greatly. The power density obtained using ACNFN (758 W m^{-3}) was dramatically higher than those that obtained with CC and GAC (161 and 3.4 W m^{-3}, respectively) [28].

Electrodes composed of carbon nanofibers along with carbon nanotubes (CNTs/CNFs) combine both oxygen reduction reaction (ORR) catalytic activity and electrochemical capacitance behavior. The interconnectivity of the fiber aggregates and thorn-like structure of the as-prepared electrodes exhibited low internal resistance (0.18 Ω cm^{-2}) and high exchange current density (13.68 A m^{-2}). When MFCs were equipped with the carbon nanotubes (CNTs/CNFs) cathode, a maximum power density of 306 ± 14 mW m^{-2} was obtained, which was 140% higher than that using Pt/C [29].

6.2.1.1.3 Two-dimensional carbon nanostructures (Nanosheets, nanobelts, cyclic and spiral nanostructures)

2D nanostructures are sheet-like materials with a high aspect ratio. They can also exist as nanobelts and cyclic or spiral nanostructures. Graphene, graphene oxide (GO), and reduced graphene oxide (rGO) are the best representatives of 2-D carbon nanosheets. Graphene (G), a single-atom-thick sheet of hexagonally arranged sp^2 hybridized

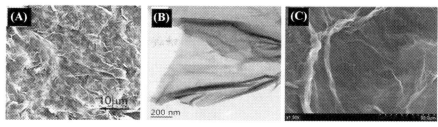

Figure 6.4 (A) SEM image of the GO electrode; (B) TEM image of the graphene layers on the GO electrode surface; and (C) SEM image of the GO electrode after discharge (*GO*, Graphene oxide; *SEM*, scanning electron microscopy; *TEM*, transmission electron microscopy). From Q.-Q. Wang, X.-Y. Wu, Y.-Y. Yu, D.-Z. Sun, H.-H. Jia, Y.-C. Yong, Facile in-situ fabrication of graphene/riboflavin electrode for microbial fuel cells, Electrochimica Acta 232 (2017) 439–444. https://doi.org/10.1016/j.electacta.2017.03.0084

carbon atoms, can be freely suspended or adhered on a conducting substrate [30]. Various forms of graphene [32] and related graphene-based materials, such as GO [33], rGO [34], graphite oxide [31], exfoliated graphite, graphene nanosheets, graphite nanoplatelets, graphite nanosheets, and graphite nanoflakes, have been explored in MFCs. Graphene and GO exhibit excellent electronic conductivity, high surface area, good biocompatibility, high electrical conductivity, robustness, and great mechanical stability [35]. Graphene can be synthesized by both top-down processes (e.g., electrochemical/ mechanical/ chemical exfoliation of graphite) and bottom-up methods (e.g., chemical vapor deposition and chemical synthesis) [11]. Hummers and Offeman defined a method in 1958 to produce GO from graphene that is still used widely [36].

GO is an important derivative of graphene with abundant epoxy and hydroxyl functional groups on the basal planes, and carbonyl/carboxyl groups at the sheet edges (Fig. 6.4). Therefore the groups attached to the plane influence its conductivity, while those attached to the edges may not. Although these functional groups increase its hydrophilic character, the conjugated sp^2 network of the individual graphene basal planes is disrupted, which impairs the electrical properties of GO. However, these can be restored by varying the synthesis and processing procedures as they are nonstoichiometric in nature. A commonly used approach is to attach polymers or polymerization initiators onto the GO platelets and reduce the oxidized surface. Eliminating only the epoxy and hydroxyl groups on the plane while retaining the carboxyl, carbonyl, and ester groups at the edges, can enhance the electrical conductivity of graphene-based materials. rGO offers advantages over graphene and graphite, which is more conductive and stronger compared with GO. It can be stored for longer duration without agglomeration and is more stable in organic solvents. However, the single component graphene materials have certain limitations, such as weak electrochemical activity, easy agglomeration, and difficult processing, which greatly limit their application. Therefore functional modification of graphene and GO is crucial to expanding their applications.

Pareek, Shanthi, and Venkata Mohan [31] reported that a chemically reduced graphene oxide (CGO) anode provided higher power density (6 mW m^{-2}) from the same amount of substrate mass compared with CC (0.8 mW m^{-2}) and GO (1.6 mW m^{-2}). This could be attributed to the larger surface area for better bacterial colonization and the higher conductivity of the CGO electrode [31]. Electrochemical exfoliation of a graphite plate (GP) leads to the in situ formation of graphene layers on its surface (GL/GP), and when used as the anode in an MFC, it could achieve a 1.72-, 1.56-, and 1.26-fold higher maximum power density than those of the original GP, exfoliated-graphene-modified GP (EG/GP) and chemically reduced-graphene-modified GP (rGO/GP) anodes, respectively. The high performance of the GL/GP anode was due to its macroporous structure, improved electron transfer and high electrochemical capacitance [37]. Similarly, a thin-layer reduced graphene oxide-based carbon brush (RGO-CB) anode fabricated using an environmentally friendly technique greatly enhanced the maximum power and current densities by 10 and 6 times, respectively, compared with those of the plain carbon brush (PCB). While the rGO layer did not affect the biofilm formation, and thus the COD removal, it increased the columbic efficiency increased by 12 times. These results demonstrate that modifying the electrode surface with rGO is useful for improving MFC performance through enhancing electron transfer and increasing the electrochemically active surface area [38]. Chen et al. studied the antibacterial activity of graphene on Shewanella Mr biofilms and proved that biofilm viability and electrochemical activity in the graphene-modified anodes were distinctly improved and exceeded that of the bare graphite anodes after mature biofilms were established. However, the antibacterial activity of graphene at the initial stages had no negative effect on the electricity generation of the MFC with graphene-modified anodes which also increases bacterial attachments and improvement in electron transfer between the bacteria and the anode. The maximum power densities of graphene-modified MFC reached 102 mW m^{-2}, which was 3.52 times larger than those of the MFCs with bare graphite [39].

Reduced graphene oxide (rGO)-based cathodes exhibit better structural characteristics and electron transfer than GO-based cathodes, leading to improved catalysis compared with graphene [40]. When rGO mixed with carbon black was employed as an air cathode for oxygen reduction in MFCs [41], the cathode performance improved due to the comparatively flatter and wider shape of rGO and $\sim$200% higher electrical conductivity than carbon black (CB). However, the complete replacement of CB with rGO decreased the cathodic performance (rGO) due to increased thickness and crack formation. Thus the structure and morphology of the surface play a vital role in the performance of graphene-based electrodes in MFCs.

Graphene-modified electrodes are usually fabricated from GO sheets using different reduction methods [42], such as hydrothermal reduction, chemical vapor deposition, chemical reduction [31,34], electrochemical reduction [43], and microbial reduction [44,45]. Nanostructured graphene materials have been used in MFCs to facilitate anodic electron transfer by taking advantage of the unique properties of graphene, including high surface-area-to-volume ratio, faster electron mobility, and excellent electron transfer characteristics. Its biocompatibility, large surface area, improved bioadhesion, and improved extracellular electron transfer (EET) favor anodic efficiency. Moreover, the increased number of active sites, high conductivity, oxidative stability, and improved ORR catalytic activity favor cathodic efficiency [33].

6.2.1.1.4 Three-dimensional carbon nanostructures: (Hierarchical/macroporous structure)

Nanoporous carbon is an emerging candidate for MFC applications. Based on pore-size distribution, porous carbons can be classified as microporous (<2 nm), mesoporous ($2-50$ nm), and macroporous (>50 nm). Compared with 2D electrodes, the porosity of the 3D architectures enhances the bioelectrochemical properties owing to the possibility for internal colonization with efficient transport of substrates, large surface-area-to-volume ratio and increased bacterial cell attachment [46]. The 3D macropores also facilitate efficient mass transfer without the concern of bioblocking and biofouling [47,48]. Recently, a number of new porous materials with nanocomponents and 3D structures, including CNTs, metal nanoparticles, pyrolyzed polymeric materials, graphene-coated sponges, and rGO-covered nickel foam, have been developed to improve the performance of MFCs [49]. Open and macroporous structures offer high specific catalyst-electrolyte interfacial surface area, improved catalyst loading and enhanced cathode catalyst activity for the ORR, yielding a superior performance [47].

N-doped carbon foam (NCF) formed from resorcinol-formaldehyde (RF) sol-gels on melamine supports through the freeze-drying and pyrolysis approach [50] possessed a rigid nanoporous 3D structure comprised of interconnected carbon wires and flat carbon particles, along with rough texture (Fig. 6.5). The NCF prepared with 10% RF exhibited unique porous characteristics with sufficient open macropores (easily accessible to microorganisms) and a maximum external surface area, making it a suitable anode for MFCs. In contrast, 15% RF produced an NCF with a compact carbon texture and less intrinsic resistance, making it a more preferable cathode with high ORR activity and enhanced maximum power density (35.74 W m^{-3}), which was 1.56 and 1.67 times higher than those of the NCF-5 (22.84 W m^{-3}) and NCF-10 (21.37 W m^{-3}) cathodes, respectively. In particular, the overall performance of the MFC (based on the anode area; NCF-10) operated with NCF-15 as the cathode was 1.15 times higher than that of the MFC operated with Pt-CP as a cathode.

A comparison of the MFC performance of 3D carbon-coated Berl saddles (specific surface area 0.5 m^2 g^{-1}) and a 2D commercial CF material (specific surface area 1.5 m^2 g^{-1}) by Hidalgo et al. confirmed enhanced charge-transfer in the 3D electrode

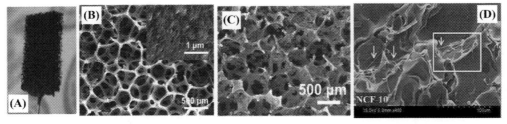

Figure 6.5 (A) CNT-sponge electrode; SEM image of the electrode (B) before and (C) after microbial colonization; (D) NCF-10, the interconnected carbon surface with carbon wires indicated by the arrows and highlighted (*CNT*, Carbon nanotube; *NCF*, N-doped carbon fiber; *SEM*, scanning electron microscopy) [26,51]. *Reprinted with permission from X. Xie, M. Ye, L. Hu, N. Liu, J.R. McDonough, W. Chen, H.N. Alshareef, C. S. Criddle, Y. Cui, Carbon nanotube-coated macroporous sponge for microbial fuel cell electrodes, Energy & Environmental Science 5(1) (2012) 5265. https://doi.org/10.1039/c1ee02122b5; S.Y. Sawant, T.H. Han, S.A. Ansari, J.H. Shim, A.T.N. Nguyen, J.J. Shim, et al., A metal-free and non-precious multifunctional 3D carbon foam for high-energy density supercapacitors and enhanced power generation in microbial fuel cells, Journal of Industrial and Engineering Chemistry 60 (2018b) 431–440. Copyright 2018, Elsevier & 2012, Royal Society of Chemistry.*

due to its high contact surface area, which supports an increased bacteria growth, thereby producing a higher power density. The main contribution to internal resistance was from the mass-transport limitations at the biofilm/electrolyte interface. The maximum power density of the felt-based cell reduced at 5 months (from 7 to 4 µW cm^{-2}), whereas the saddle-based MFC exceeded 9 µW cm^{-2} (after 2 months). On the other hand, [32] fabricated a 3D graphene structure by the chemical reduction of GO and the formation of hydrogels. These 3D graphene electrodes showed remarkably high capacitive current, charge storage, and lower charge transfer resistance, as estimated through electrochemical analysis, which exhibited a maximum power density of 0.49 mW m^{-2} in potassium ferricyanide [32]. Santoro et al. developed 3D graphene nanosheets (3D-GNS) as cathode catalysts for MFCs and reported high performance toward oxygen electroreduction in neutral media with high current densities. The addition of 3D-GNS decreased the ohmic losses by 14%–25% and allowed the SC-MFC with 3D-GNS (loading 10 mg cm^{-2}) to generate the maximum power (P_{max}) of 5.746 ± 0.186 W m^{-2} [52]. Thus 3D electrodes are better than 2D electrodes in terms of electrochemical, stability, and biofilm development.

6.3 Nanocomposite materials

Materials such as CNTs, graphene, conducting polymers, metal oxides, and biochar have been employed as composite electrode modifiers with varying degree of success. The upcoming section highlights the latest research advances made in MFC composite electrodes using the aforementioned nanomaterials in combination with metal, polymer or bulk carbon materials. Therefore this section is organized according to the role of the carbon nanostructures in the composite electrodes (anode and cathode) and composite membranes.

6.3.1 Anode materials

The characteristics of the anode material play an important role and are the main cause for the low efficiency of several MFC prototypes. A good anode material should perform reasonably in terms of specific characteristics including bacterial adhesion/biofilm formation ability, electrical conductivity, surface area and porosity, besides being low-cost, stable, and available. Anode characteristics affect bacterial growth, colonization, the density of bacterial adhesion, and the extent of biofilm formation, and consequently affect extracellular electron transfer. Materials, such as conducting polymers, metal, and metal oxides, have all been used as anode modifiers/composites with varying degrees of success. To a certain extent, carbon materials meet all these requirements. Thus their possible use in MFC anodes is widely studied. However, their inherent hydrophobic nature makes them unsuitable for robust microbial adhesion, resulting in poor electron transfer capability when an unmodified electrode is used. Beneficial electrode modification is required to improve the efficiency of electricity generation in MFCs. The following section deals with the various anode modification methodologies and discusses their development. In Table 6.1 the various nanostructured composites as anode material used in MFCs were summarized.

6.3.1.1 Polymer carbon nanocomposites (P-C anode)

Anode surface modifications play an essential role in enhancing the overall performance of MFCs and also ensure optimal biocatalytic activity. The anode surface can be modified to provide a beneficial environment for the microbial population, which results in improved electron transfer from the bacterial biofilm to the anodic surface. Conducting polymers with nanostructured carbon materials have achieved these goals in MFC anodes.

For instance, polypyrrole (PPy) is known to possess excellent electrochemical properties, good biocompatibility and environmental stability for application in MFCs [63]. A hierarchically structured textile polypyrrole/poly(vinyl alcohol-co-polyethylene) nanofibers/poly(ethylene terephthalate) (referred to PPy/NFs/PET) has been shown to be an excellent anode for MFCs as it affords an open, porous, and 3D interconnecting conductive scaffold with high surface roughness, which facilitates microbial colonization and electron transfer from the exoelectrogens to the anode. The mediator-less MFC with the PPy/NFs/PET anode and *Escherichia coli* (*E. coli*) as the microbial catalyst achieved a remarkable maximum power density of 2420 mW m^{-2}, which is approximately 17 times higher than that of the reference anode PPy/PET (144 mW m^{-2}) [64]. The pyrrole monomers are effectively adsorbed over the

Table 6.1 Summary and comparison of various nanostructured composites used as anode materials in MFCs.

Nanostructures	Anode modifications Coating/support	Method	Cathode Coating/support	Microorganism	Substrate	Reactor type	Cell voltage	Max PD (mW m^{-2})/ (Improved)	Ref.
Polymer									
0D: PPy and PTh Nps	PPy-NPs and PTh-NPs/GF	Chemical polymerization	0.5 mg cm^{-2} Pt/C on CC	*Shewanella putrefaciens*	Glucose	SC	0.842	1220 (52.5%)	[103]
1D: PANI Nw arrays	(BL-PANI) NW/CC	Electropolymerization	0.5 mg cm^{-2} Pt/C on CP	Enriched effluent	Acetate	DC	0.570	567.2 (58.1%)	[54]
2D: PANI/GO	Polypyrrole/GO/CF	Electropolymerization	CF	*Shewanella oneidensis*	Lactate	DC	—	1326 (698.7%)	[55]
3D: Macroporous PANI	N-doped Carbon loofah sponge	Chemical polymerization	0.5 mg cm^{-2} Pt/C on CC	Anaerobic sludge	Acetate	SC	0.580	1090 ± 72 (184.5% w.r.t. GP)	[56]
Metal oxide									
0D: FeO Nps	FeO nanoparticles-coated carbon/CP	Biosynthesis	CP	Bacteria from distillery WW	Distillery WW	DC	0.765 ± 0.05	140.5 (31%)	[57]
1D: Nts	MWCNTs-SnO$_2$/GCE	Ultrasonication	Pt rod	*Escherichia coli*	Glucose	DC	0.8	1000 (2.8 times)	[58]
2D: TiO$_2$-Ns	PANI/TiO$_2$-Ns/CP	Electropolymerization	CP	*Shewanella loihica*	Defined media	DC	0.732	813 (64%)	[59]
3D: Fe$_3$O$_4$-Nt composite arrays	N-doped C/Fe$_3$O$_4$-Nt composite (CC@N-C/Fe$_3$O$_4$)/CC	Hydrothermal method; itching	0.5 mg cm^{-2} Pt/C on CC	Preacclimated bacteria	Sodium acetate	SC	0.671	1210 ± 40	[60]

Carbon

0D: Quantum dots	(FeS$_2$-S, N codoped GQDs)/CC	Hydrothermal method	CC	Activated sludge	Sodium acetate	DC	0.600	1128 (76% w.r.t. CC)	[8]
1D: Carbon nanofibers	(N-CNFs)/CF	Electros pinning and pyrolysis	CF	Mixed culture	Sodium acetate	SC	—	2153 ± 11.6 (1680%)	[61]
2D: Graphene layers on Gr	GP	Electrochemical exfoliation	0.5 mg cm^{-2} Pt/C on CC	Activated sludge	Sodium acetate	SC	—	677 ± 34 (1.72 times)	[33]
3D: 3D rGO	—	Electrochemical exfoliation of CP	40% Pt/C on CP	*Escherichia coli*	LB medium	SC	0.623	897.1	[62]

Abbreviations: C, Carbon; CB, carbon brush; CC, carbon cloth; CF, carbon felt; DC, double chamber; G, graphene; GCE, glassy carbon electrode; GF, graphite felt; GO, graphene oxide; GP, graphite plate; Gr, graphite; Nps, nanoparticles; Ns, nanosheets; NW, nanowires; PPy, polypyrrole; Pt, platinum; PTh, polythiophene; QD, quantum dots; rGO, reduced graphene oxide; SC, single chamber; WW, wastewater.

negatively charged GO sheets through electrostatic and π-π interactions and polymerized into PPy in the adsorbed state, forming a stable composite electrode [65]. For example, conductive PPy/rGO composite synthesized through simple, in situ polymerization and bioreduction techniques achieved almost twofold higher maximum MFC power density of (1068 mW m^{-2}) than that of bare PPy. The strong active carbon support could prohibit the swelling and shrinkage of the conductive polymer PPy and provide strong physical and electrochemical robustness to the rGO/PPy composite, which increased the durability of MFC performance upto 300 h.

Likewise, Roh [66] fabricated PAN nanofibers, with and without embedded CNTs by electrospinning. PPy was coated on the activated PAN/CNT nanofiber by in situ chemical polymerization in order to improve the electrochemical performance. In comparison with unmodified CC anodes, PPy-PAN/CNT nanofiber composite showed 40% improvement in the maximum power density, thereby exhibiting good prospects for application in MFCs [66]. A similar outcome was obtained with PPy polymerized on vertically grown CNTs (VCNTs/PPy) to avoid direct contact between the CNTs and electricigen [67]. The maximum power density of the MFC using this anode was approximately 2.63-fold higher than the unmodified carbon fiber brush anode.

Liao et al. evaluated tartaric acid-doped polyaniline (PANI) nanowire-modified CC (PANI-TA) as the anode and achieved 1.6 times higher voltage output (435 ± 15 mV) and 4.1 times greater power output (490 ± 12 mW m^{-2}) than that of the MFC with plain CC anode. Strikingly, the performance of PANI-TA MFC was superior to that of the MFCs with inorganic acid-doped PANI-modified anodes [68]. Similarly, a bacterial cellulose (BC)/PANI anode synthesized with BC as the continuous phase and PANI as the dispersed phase produced a maximum power density of 117.76 mW m^{-2}. Such efficiency was favored by the hydrogel nature of BC and PANI conductivity, which are beneficial to achieving active microbial biofilm on the anode surface [69].

Moreover, the PANI/rGO/CNTs/S capacitive bioanode displayed a very good ability to synchronously produce electricity, store energy and release the stored charge in short bursts with the maximum power density of 571.5 mW m^{-2}, which is expected to meet the demand of electrical equipment. This can be attributed to several factors, including great specific surface area, the distinct macroporous architecture of the sponge substrate, the effect of two carbon nanomaterials on decreasing the bioanode resistance, the energy storage characteristic of PANI, and the good biocompatibility of the coating materials [70].

Graphene, like other carbon nanomaterials is highly hydrophobic and unfavorable for bacteria adhesion. Thus Liu et al. [71] decorated the surface of graphene with the hydrophilic conducting polymer PANI through in situ polymerization in order to promote bacterial adhesion and biofilm formation. Graphene enables insulator-to-conductor transition at significantly lower loading by providing percolated pathways for electron transfer and making the polymers composite electrically conductive [71]. A similar case was observed when PANI together with graphene was chosen to modify oxidized CC in situ using secondary bond forces (such as pep stacking, hydrogen bonds and electrostatic forces) [72]. PANI and MWCNTs can be modified in a controllable, stable and binder-free manner on the surface of macroporous GF by electrochemical techniques. With the modification of a rough, nanocilia containing PANI film, the hydrophobic surface of GF becomes hydrophilic [73].

6.3.1.2 Metal-carbon nanocomposites (M-C anode)

The conductivity and electrocatalytic activity of carbon-based materials can be improved when composited with metal-based materials, such as nonnoble metals. Currently, nanostructured metal materials with high catalytic activity, conductivity and biocompatibility are popularly utilized for anode decoration in MFCs as they increase the number of electroactive sites and act as electron transfer bridge between the bacteria and the anode, thereby enhancing power generation. Various anode types made of nano-based metal-composite materials are discussed in the following section.

6.3.1.2.1 Metal-based carbon nanocomposite anodes

The poor wettability, low charge transfer capacity, and high cost of carbonaceous electrodes proscribe their real-life applications in large-scale MFCs. The transition metals most commonly used in nanoparticle layers are Fe, Zn, Ni, Cu, and Co, because they enhance both the electrochemical activity and bioactivity of carbonaceous anode materials. Among these metals, Fe and Co are considered the most promising elements because of their ability to stimulate microbial growth and accelerate their adhesion on the anode surface. A thin layer of metal/metal oxide coating on carbon materials may significantly enhance microbial adhesion and the wettability of the anode surface and decrease electron transfer resistance.

For instance, Mohamed et al. [74] proposed the electro-deposition of a novel metal nanoparticle sheathed with metal oxide (cobalt oxide) on CP, which generated almost 140% power density (138.4 mW m^{-2}) and 210% current density (690 mA m^{-2}) as those of pristine CP (57.6 mW m^{-2} and 222 mA m^{-2}) [74]. Similarly, when the surface of a stainless-steel mesh (SSM) was modified using an acrylic-based graphite conductive paint (GP) and applied as a low-cost anode, the membrane-less SCMFC was able to generate a maximum power density of 463.88 mW m^{-3} at 199 mA m^{-3} using this low-cost high-performance electrodes [75]. Qiao et al. [76] reported a hierarchical porous graphene/nickel (G/Ni) composite anode with pore sizes 20 nm to 50 mm developed by freeze-drying assisted self-assembly process using a low weight percentage of graphene (5 wt.%) to nickel [76]. Its hierarchical porous structure could allow for high biocatalyst loading and offer a large porous surface for electron transfer/transport, thus simultaneously boosting the biocatalysis and electrocatalysis for high MFC performance. This binder-free hierarchical porous G/Ni anode delivered a maximum power density of 3903 mW m^{-2} with Shewanella putrefaciens, which is approximately 13-fold higher than that of the conventional CC MFC anode. Similar results were obtained by Gupta et al. with an alumina (AA)/nickel (Ni) nanoparticle (NPs)-dispersed carbon nanoparticles (NPs)-dispersed CNF-based anode and E. coli [77].

Very recently, Saadi et al. [78] coated polydimethylsiloxane (PDMS)-doped commercial CNFs over SS to improve the interfacial electron transfer, biocompatibility and corrosion resistance of stainless steel (SS) anodes. Coating the SS anodes with CNF-PDMS improved the stability and increased the start-up period of MFCs due to the hydrophobic properties of PDMS. The comb-like conducting structures on the CNF-PDM layers facilitated direct EET of EAB to the electrode and consequently accelerated biofilm formation [78].

6.3.1.2.2 Metal oxide-based carbon nanocomposite anode

Titanium dioxide (TiO$_2$) is one of the most attractive metal oxides owing to its unique physical, chemical, and photocatalytic properties. It also has various advantages, such as abundance, low cost, and structural stability, and is applied in many fields, including photocatalysis, fuel cells, and sensors. For instance, the anatase TiO$_2$@CNTs-MFC hybrid nanostructures combine the advantages of CNTs and TiO$_2$: excellent electrical conductivity, high surface area, and good biocompatibility. This anode could produce 1.5 and 1.7 times the maximum power density obtained with the CNTs and TiO$_2$ anodes (1120 vs 730 and 670 mW m^{-2}), respectively [79].

Titania nanowire structures TiO$_2$ (TiO$_2$-NWs) deposited on a CP substrate through a hydrothermal reaction demonstrated the role of semiconductive nano-TiO$_2$ and promoted long-distance EET in the bacterial network. The TiO$_2$-NWs/CP anode greatly outperformed the carbon electrode owing to the open porous interface and electrochemically active surface area, resulting in 49.5% higher power density output than that of raw CP (198 mW m^{-2}) [80]. Moreover, lysine-modified TiO$_2$ nanotube arrays (TNT/HL) with high conductivity and corrosion resistance were synthesized by the anodization and thermal treatment of titanium. The MFC with this metal-based anode showed a 1.4 times higher maximum power density (0.88 W m^{-2}) than that of the CC anode (about 0.61 W m^{-2}) [81].

Shan et al. [82] decorated MoS$_2$ nanosheets with TiO$_2$ via liquid-phase exfoliation and in situ hydrolysis to form a nanocomposite photocathode, in which the MoS$_2$ nanosheets acted as a cocatalyst in the photocathode and expanded the absorption range of visible light, thereby leading to increased electronic participation in Cr (VI) reduction. A maximum power density (147.4 mW m^{-2}) achieved under illumination was approximately 1.9, 2.4, and 3.7 times higher than those of the TiO$_2$, MoS$_2$, and graphite cathodes, respectively [82].

A three-dimensional ordered porous carbon (3DOM-C) modified with TiO$_2$ on a CC anode (TiO$_2$/3DOM-C) provided an easily accessible surface area for bacterial adherence, which significantly improved the power output (2.3-fold higher power density; 973 mW m^{-2}) compared with MFCs using CC. This was due to the higher EET efficiency of TiO$_2$/3DOM-C than CC due to the unique structure of 3DOM-C and excellent properties of hydrophilic TiO$_2$ [83].

A MFC equipped with an rGO/MnO$_2$/CF anode fabricated by a simple dip-coating and electro-deposition process significantly increased the anodic reaction rates and facilitated electron transfer from the bacteria to the anode. The rGO/MnO$_2$ composite delivered a maximum power density (2065 mW m^{-2}) 154% higher than that with a bare CF anode [84].

6.3.1.3 Metal-polymer nanocomposite (M-P anode)

Conducting polymer coatings, such as PANI and PPy, have been shown to be effective against corrosion in metal- and composite-based MFC anodes. These polymers possess high electron mobility, stability, biocompatibility, anticorrosive nature, excellent electrokinetics, and can be produced by facile chemical or electrochemical processes. In particular, PANI-modified anodes enhance microbial adhesion at the surface and consequently improve the bioelectrocatalytic performance. It is anticipated that coating the metal with conducting polymers will minimize corrosion and improve the anodic performance through enhanced microbial adhesion. Stainless steel (SS) is recognized to have outstanding mechanical properties and electrical conductivity; however, its poor biocompatibility, biocorrosion, and low current output limit its extensive application as a bioanode.

Surface modifications of SS electrodes by flame-oxidization, flame deposition, and nanocarbon blended polymer coating was recently explored to produce a more biocompatible electrode surface with enhanced microbial bioelectrocatalytic activity. Of these, the coating of nanocarbon using polymer binder resulted in high internal resistance and hindered electron transfer from the bacteria to the SS surface; however, the coating of nanocarbon onto SS without a polymer binder resulted in unstable current generation due to less interaction. The microbial bioelectrocatalysis of the stainless steel mesh (SSM) was greatly enhanced by a surface modification process that included acid etching, binder-free carbon black (CB) coating and low-temperature heat treatment [85]. The modified SSM, could generate a stable and nearly three orders of magnitude higher current density, than the untreated SSM electrode. Similarly, Liang et al. [86] attempted different types of modifications on the SS anode and found that functionalization with hydrophilic groups resulted in 1.5-fold to 2-fold higher current output through enhanced bioadhesion and electron transport rate, respectively. Conducting polymers like PANI and PPy were the promising choices for SS-modification, that offers higher compatibility, charge transferability and low activation [86,87]. For instance, when stainless steel wool (SS-W) was coated with PANI and PPy, both SS/PANI-W and SS/PPy-W achieved superior maximum power densities that obtained with pristine SS-W. Further, in comparison with SS-P-based anodes, all the SS-W based anodes gave improved power densities (at least by 70%). The observed enhancement in the bioelectrocatalytic performance was due to the tunable shape, size, design, high surface area, and mechanical properties of the composite anode [88].

6.3.2 Cathode materials

Although precious metal catalysts, such as platinum, palladium, and gold, present high catalytic activity for oxygen reduction, their high cost, limited availability, and high sensitivity toward biological and chemical fouling hinder their large-scale applications in MFCs. Thus various nonnoble metal-based and polymer-based catalysts have been investigated as cathode catalysts for MFCs to lower the cost of cathodes and simultaneously improve ORR kinetics. The upcoming section discusses such nanocomposite cathode materials, including polymer-carbon nanocomposite (P-C cathode), metal-carbon nanocomposite (M-C cathode), and metal-polymer anocomposite (M-P cathode).

6.3.2.1 Polymer-carbon nanocomposites (P-C cathode)

Optimization of the cathode catalyst is critical for MFCs performance. Various composites have been developed as alternatives for the precious Pt cathode catalyst. In this regard, conducting polymers play an important role as ORR catalysts when combined with carbon materials. For instance, chemically synthesized polypyrrole nanoparticles (PPy-NP) were coated on carbon cloth (CC) and used as the cathode for bioenergy production and iron removal. The CC cathode coated with PPy showed lower charge transfer resistance and higher conductivity achieving a lower but comparable maximum power density than the conventional Pt/C cathode [89].

Similarly, the 75% PANI/MWCNT composite cathode generated a higher maximum power density (476 mW m^{-2}) than that of pure MWCNT (367 mW m^{-2}) but lower than that of Pt/C cathode (541 mW m^{-2}). Thus the PANI/MWCNT composites may be suitable low-cost alternatives to the Pt/C catalyst in MFCs [90].

Furthermore, Conductive polyaniline nanofibers (PANInf) synthesized by interfacial polymerization and composited with carbon black exhibited higher electrocatalytic activity than pristine PANInf, when used as the cathode for oxygen reduction, resulting in a large enhancement in power density from 185 to 496 mW m^{-2}. Although the power density is still lower than that of the conventional Pt catalyst (604.3 mW m^{-2}) the facile bulk synthetic process, low cost, and porous nature would make PANInf/C an effective alternative to Pt in large-scale applications [91]. Additionally, the highly porous nanostructure and thickness of the PANInf film may increase the diffusion resistance of oxygen to the bulk, consequently reducing electron loss at the anode.

6.3.2.2 Metal-carbon nanocomposites (M-C cathode)

Recently, low-cost nonnoble metals and carbon-based materials with relatively high catalytic activity, are emerging as promising Pt alternatives. Especially, transitional metal-based catalysts have attracted growing attention for their outstanding catalytic activity, ease of fabrication, and high stability.

6.3.2.2.1 Metal-based carbon nanocomposites cathode

Even though Pt is well known for its high catalytic activity, its poor cycle stability caused by catalyst poisoning and biofouling dwindles its potential. Research has shown that the incorporation of Co, Cu, Ni, and Fe can significantly facilitate the ORR activity of carbon-based catalysts [92]. Furthermore, bimetallic catalysts show even better catalytic activity than the monometallic cathode catalysts in MFCs. However, simpler and more facile preparation methods for bimetallic nanocatalysts could benefit their practical applications. To achieve this, Guo et al. [93] hydrothermally synthesized novel Fe-modified α-MnO_2 (Fe-Mn) bimetallic nanocatalysts, with Fe:Mn (atom%) ratios of = 1:4, 1:2, and 1:1. The $FeMn_2$ (Fe:Mn = 1:2) achieved a maximum power density of 1940 ± 31 mW m^{-2} in MFCs with Pt/C. The high catalytic activity of $FeMn_2$ is likely due to the synergistic effect of the weak Mn-O bonds, a large number of defects, large specific surface area, and high Mn(III):Mn(IV) ratio. This work not only puts forward an easy-to-accomplish method to design and prepare bimetallic nanocatalysts but also provides a potential alternative to Pt/C for sustainable energy generation in MFCs [94]. Likewise, Liu and Vipulanandan (2016) improved MFC performance by using rod-shaped metallic nanoparticles (Fe, Ni, and Fe/Ni) produced by the precipitation/coprecipitation method. Fe nanoparticles had the biggest influence on bacterial growth, while the influence decreased with time probably because of the deterioration of the Fe catalyst over time. Ni nanoparticles increased the bacterial growth the most in the later growth stage compared with the Fe and Fe/Ni nanoparticles, which could be due to the relatively better stability of the Ni nanoparticles than the others. This study found that rod-shaped nanoparticles have larger specific surface areas compared with spherical and cubic nanoparticles and hence show increased catalytic properties. The addition of 1.5 mg cm^{-2} Fe nanoparticles to the cathode surface-enhanced power production by over 500% to 66.4 mW m^{-3}, and the order of bacterial growth promotion was as follows: Fe > Ni > Fe/Ni [93].

In addition, graphene or carbon black-supported expensive metal nanoparticle catalysts enhance electrocatalytic activity toward ORR in the alkaline medium at a low cost. For instance, the expensive silver nanoparticles electrodeposited on AC by Pu et al. [95] and used as an efficient low-cost cathodic catalyst alternative to Pt in ORR generated 69% higher maximum power density than the bare AC cathode at the electro-deposition time of 50 s (Ag-50) [63]. Li et al. [14] employed uniformly distributed silver nanoparticles on activated carbon powder (Ag/AC) as the cathode catalyst in MFCs. The higher mass transport of oxygen and hydroxide favored better catalytic activity toward oxygen reduction. Ag/AC enhanced the maximum power density by 14.6% (1100 ± 10 mW m^{-2}) compared with that of a bare AC cathode due to the additional silver catalysis [14]. A maximum power density of 1791 mW m^{-2} was obtained with the Ag/Fe/N/C composite prepared from Ag and Fe-chelated melamine at 630°C, which is far higher than that of Pt/C (1192 mW m^{-2}). The Fe-bonded N and the cooperation of pyridinic N and pyrrolic N in Ag/Fe/N/C contributed equally to the high ORR catalytic activity. The OH or O_2 species originating from the catalysis of O_2 can suppress the biofilm growth on Ag/Fe/N/C cathodes. However, the introduction of Ag eliminated the biofilm-inhibiting blockage of the proton/OH$^-$ transport channels, which correspondingly contributes to a lower overpotential in SC-MFCs [96].

Similarly, heteroatom-doped carbon nanomaterials possess greater cathodic efficiency. For example, Modi et al. [97] experimented with in situ nitrogen (N)-doped nickel (Ni) nanoparticle (NP)-dispersed CNFs coated using catalytic chemical vapor deposition on an activated carbon fiber (ACF) substrate. The performance of the MFCs using the N-doped material improved approximately by twofolds compared with that using the undoped and Ni-CNF/ACF cathode. N-doping contributed to facile electron transfer by lowering the solution resistance (Rs) favoring bacterial growth and improving electrical conductivity of the electrode and thereby the ORR catalytic at the cathode [97]. Nitrogen-doped carbon nanotube-embedded cobalt nanoparticles nanopolyhedra (Co-NCNTNP) was used as an electrocatalyst on AC to synthesize bimetallic metal-organic framework-based air-cathode for MFCs. The as-prepared electrocatalyst retained high specific surface area, complex textural properties, transition metal element, and nitrogen chemical state, which resulted in superior ORR activity and remarkable power generation (154% vs control) [98].

6.3.2.2.2 Metal oxide-based carbon nanocomposite cathode

Traditional platinum (Pt) has always been the best ORR catalyst as it reduces oxygen via a four-electron reduction pathway and exhibits chemical inertness. However, the high cost of platinum (Pt) catalysts and other precious metals used as catalysts for the ORR further reduce their practical applications. Alternatively, different transition metal oxides with unique properties have been investigated as low-cost and environment-friendly ORR catalysts [99]. Composite materials containing metal oxides, such as lead oxide [100], cobalt oxide [101], titanium dioxide [102], vanadium oxide [103,104], and manganese oxide [105], are widely studied as potential for MFC cathode materials. They improve the MFC performance by increasing the surface area and decreasing the potential losses. Manganese oxides (MnO_x) have attracted wide attention as cathode catalysts for MFCs, and the sequence of their catalyst performance is in the decreasing order Mn_5O_8, Mn_3O_4, Mn_2O_3, MnOOH, and among the MnO_2

phases, the performance sequence is β-, λ-, γ-, and α-MnO$_2$ [106]. Various types of carbon-based conductive supporting materials have been employed to enhance the electrochemical ORR performance [107]. For instance, the composite of α-MnO$_2$/GO/AC with 10% GO achieved a maximum power density of 148.4 mW m^{-2} due to its great electrocatalytic properties and low resistance and potential losses, which lead to significantly improved electrochemical activity and ORR [108].

6.3.2.3 Metal polymer nanocomposite (M-P cathode)

A ternary polyaniline-titanium dioxide-graphene (PANI-TiO$_2$-GN) nanocomposite with a nanoporous structure, when used as a bifunctional catalyst, facilitated better electrode/electrolyte contact and substrate diffusion to the electrode. Furthermore, it possessed a high specific area, excellent biocompatibility, good conductivity, and chemical stability, rendering excellent ORR and EET catalytic property. The PANI-TiO$_2$-GN Shewanella-immobilized anode MFC generated a high-power density 1.3 and 2.7 times higher than that of the PANI-TiO$_2$-GN nonimmobilized and the plain CP immobilized anode, respectively. PANI/V$_2$O$_5$ also produced as much power as Pt (80%) and can be a good alternative to Pt in MFCs [109] owing to the smaller size of the nanocomposite and the significant electronic interactions between V$_2$O$_5$ and PANI in the nanocomposite. The comparison of various nanostructured composites used as cathode materials used in MFCs were presented in the table (Table 6.2).

6.3.3 Membranes

The anode and cathode chambers in an MFC are separated by a crucial component, namely the membrane (except in membrane-less MFCs). The membrane materials have a great impact on the mass transport and ohmic losses and, thus limit the voltage and performance of the cell. In addition, the most common designs make use of expensive polymeric cation exchange membranes (CEMs), such as nylon, because of their high proton conductivity. MFC scale-up will require the application of cost-effective separators because the subsequent need for hydration and the price of CEMs make them inappropriate for large-scale air-cathode-based systems [120]. A suitable membrane must have good ionic conduction, ion selectivity (e.g., proton conductivity), insulation, durability, biocompatibility, and chemical stability, besides being impervious to fouling and clogging and involving a low-cost manufacturing strategy. To achieve an efficient anaerobic treatment system that is economically viable and generate renewable energy (such as bioelectricity), the most important requirement is a low-cost, durable, and fouling-resistant MFC membrane [121].

An efficient membrane should have low internal membrane resistance, minimal or no oxygen diffusion, antifouling property, and the ability to evade substrate crossover and mitigate pH splitting. Perfluorosulfonic acid polymers consist of a polytetrafluoroethylene backbone and side chains made of vinyl ethers (e.g., -O-CF$_2$-CF-O-CF$_2$-CF$_2$-) which

Table 6.2 Comparison of the various nanostructured composites used as cathode materials in MFCs.

Anode	Nanostructures	Cathode Coating/support	Method	Electron acceptor	MFC type	OCV (V)	Max PD (mW m^{-2}) (Improved)	Ref.
Polymer								
PPy/GF	0D: PPy Nps	CC	Brush coating	FeCl$_3$	SC	0.599	190 ± 4 (29.9% less than Pt/C)	[48]
GF	1D: PANI-derived Fe-N-C-Nr	(Fe-N-C/G)/GF	Brush coating	O$_2$	SC	0.552 ± 0.012	1601 ± 59 (9.06% w.r.t. Pt/C)	[110]
Gr	2D: PANI-G Ns	PANI-G-Ns	Chemical polymerization	O$_2$	SC	0.640	99 (1547%)	[111]
GF	3D: Gr	3D: Gr-PANI/ tourmaline	Electrochemical polymerization	O$_2$	DC	—	266 (393%)	[112]
Metal oxide								
CF	0D: CoFe$_2$O$_4$ Nps	N-AC/SSM	Rolling press	O$_2$	SC	—	1770.8 (2.39 times)	[113]
CC	1D: α-MnO$_2$/C Nw	α-MnO$_2$/C/CC	Brush coating	O$_2$	SC	0.46 ± 0.01	180 ± 4.65 (62.16%)	[114]
CF	2D: (MnO$_2$/GNS)	(MnO$_2$/GNS)/CP	Brush coating	O$_2$	SC	0.771	2084 (21.58%)	[115]
3D CNF	3D: CNF-skinned 3D Ni/C MPs	Ni-CNF-skinned 3D C-MPs	Laser ablation; chemical vapor deposition	K$_3$Fe(CN)$_6$	DC	0.39	2496 (399% w.r.t. Ni)	[116]
Carbon								
SS	0D: Carbon Nps	SSM	Combustion	K$_3$Fe(CN)$_6$	DC	0.68	1650 ± 50	[117]
CP	1D: Activated CNF (AECNF)	CP	Brush coating	O$_2$	DC	0.629	61.3 (3.17 times)	[118]
CFB	2D: N-doped CNT/rGO Ns hybrids	(N-CNT-rGO)/ CFB	Drop casting	O$_2$	SC	0.600	1329 (1.37 times)	[119]
CB	3D: 3D-G Ns	SSM	Rolling press	O$_2$	SC	—	2059 ± 0.3	[52]

Abbreviations: C, Carbon; CB, carbon brush; CC, carbon cloth; CF, carbon felt; CNF, carbon nanofiber; CNT, carbon nanotube; DC, double chamber; G, graphene; GF, graphite felt; GO, graphene oxide; GP, graphite plate; Gr, graphite; MPs, micropillars; Nps, nanoparticles; Nr, nanorod; Ns, nanosheets; Nt, nanotubes; Nw, nanowires; PPy, polypyrrole; PTh, polythiophene; QD, quantum dots; rGO, reduced graphene oxide; SC, single chamber; WW, wastewater.

terminate in sulfonic acid groups in a cluster region. The most used membrane materials (trade names) are Nafion (DuPont), 3M, Flemion (Asahi Glass), Hyflon (Solvay-Solexis), and Aquivion (Solvay). Nafion membranes are the most commercialized membranes for MFCs because the sulfonic group in Nafion is hydrolyzed to deliver H_3O^+ at the end of the membrane. Subsequently, H^+ is released to the cathodic chamber and reacts with O_2 to form H_2O. This is possible only due to the selective permeability of cations (H^+) to the cathode chamber from the anode chamber. Although these commercial polymeric nanocomposites have been found to be efficient, advanced, and economical membranes, the main problems of using Nafion and other CEMs as separators in MFCs are pH splitting between the two chambers caused by proton accumulation in the anode chamber, oxygen diffusion from the cathode to the anode chamber, substrate loss and biofouling [122]. Nanocomposite membranes are economical and have proton transfer capability and higher fouling resistance. However, adding a polymer or nanoparticles into polymeric membranes can alter the original membrane structure, for example, by increasing the surface roughness and chemical stability. Therefore we present an overview of various composite membranes used in MFCs. They are classified into three categories: polymer-metal nanocomposites (P-M membrane), polymer-carbon nanocomposites (P-C membrane), and polymer-polymer nanocomposites (P-P membrane).

6.3.3.1 Polymer-carbon nanocomposites (P-C membrane)

Carbon nanocomposites change membrane roughness, pore size and porosity and increase power generation in MFCs. The decreased pore size and roughness of these nanocomposite membranes result in the blockage of oxygen transfer from the cathode to the anode and hinder the migration of bacteria and other elements from the anode to the cathode. On the other hand, the reduced roughness decreases fouling problems in the membrane while augmenting its conductivity (a result of the inherent nature of CNFs). For instance, Ghasemi et al. [123] studied an activated CNF/Nafion composite membrane, showing higher MFC performance (maximum power density 57.64 mW m^{-2}) compared with the nonactivated CNF/Nafion composite, Nafion 112, and Nafion 117 membranes owing to smaller pore size, higher porosity, and lower surface roughness [123].

6.3.3.2 Polymer-polymer nanocomposites (P-P membrane)

The low-conducting polymer membrane PES, when blended with high-conducting SPEEK to form a low-cost PES/SPEEK 5% composite membrane, exhibited better MFC performance in terms of maximum power density (170 mW m^{-2}) and CE (68%−76%) than Nafion membranes because of the presence of sulfonate groups [124]. Studies have also shown that Nafion 112/PANI composite membrane could produce approximately nine times (124.03 mW m^{-2}) to the maximum power density of the Nafion 112 membrane (13.98 mW m^{-2}), which is comparable to that of the Nafion 117 membrane (133.3 mW m^{-2}) [125].

6.3.3.3 Polymer-metal nanocomposites (P-M membrane)

Below are some key advantages of polymer-metal nanocomposites as membranes are.

1. It is possible to develop membranes with suitable structure and properties (e.g., to avoid the permeation of certain unwanted species and achieve specific separating ability) by manipulating the interactions between the nanoparticles surface and polymer chains.
2. Significant improvement in membrane hydrophilicity and antifouling capability can be achieved by introducing nanoparticles into the polymer matrix to modify the chemical and physical properties.

Among the metal oxide composite materials, titania (TiO_2), and magnetite (Fe_3O_4) play a crucial role in modifying sulfonated poly(ether ether ketone) (SPEEK), the alternative polymer for Nafion. SPEEK is well known for its excellent efficiency, particularly in MFCs, since its oxygen mass transfer coefficient value is an order lesser than that of Nafion.

Titanium dioxide (TiO_2) is a special semiconductor with photocatalytic and desirable hydrophilic properties. It has been studied widely as a nanofiller for it has intrinsic hydrogen diffusion ability and can impart hydrophilic characteristics, which reduce hydrophobic interactions between the membrane surface and microorganisms. Thus compared with other surface-bound nanoparticles, titania at the surface mitigates fouling better. A nanocomposite SPEEK membrane incorporated with sulfonated TiO_2 particles (7.5 wt.% TiO_2-SO_3H) displayed higher peak power density in a single-chamber microbial fuel cell (SCMFC) compared with those with SPEEK-TiO_2 and SPEEK membranes. The oxygen mass transfer coefficient and the internal resistance of the SPEEK membrane decreased following TiO_2-SO_3H incorporation [126]. TiO_2-SPEEK composite membrane prepared by the solvent-casting method showed improved oxygen mass transfer coefficient values and transport of cations besides protons along with antifouling property. The improved ion exchange and water absorption capacity of the composite membrane resulted in enhanced proton conductivity and electrochemical properties, achieving higher maximum power density and current density than those of commercially available Nafion 117 [127]. On the other end, the Fe nanoparticles on the synthesized membrane were able to increase the maximum power of the MFC by over 29% higher than when Nafion 117 was used as PEM [128,129].

6.4 Conclusion

MFCs are novel BESs, which represent a pioneering technology, owing to their potential in treating wastewater and simultaneously generating bioelectricity. This chapter accentuates the importance of various dimensional carbon-based nanostructures that are used in MFCs. Various composite anodes, cathodes, and membranes made in combination with other materials (metal-based, carbon-based, and polymer-based) and the enumeration of their important characteristics that influence the MFC performance are also discussed. Even though individual nanostructures elevate the efficiency of MFCs, their composites provide be remarkable advantages. Being biocompatible and stable, carbon-based nanostructures can be inexorable materials for MFCs. Among these, graphene- and CNT-based composite materials form promising MFC electrodes.

Acknowledgment

N. Saranya sincerely thanks CSIR for the Senior Research Fellowship (SRF) and Anna University for the Anna Centenary Research Fellowship (ACRF).

References

[1] S. Narayanasamy, J. Jayaprakash, Application of carbon-polymer based composite electrodes for microbial fuel cells, Reviews in Environmental Science and Biotechnology 19 (3) (2020) 595–620. Available from: https://doi.org/10.1007/s11157-020-09545-x.
[2] M.C. Potter, Electrical effects accompanying the decomposition of organic compounds, Proceedings of the Royal Society of London. Series B 84 (571) (1911) 260–276. Available from: https://doi.org/10.1098/rspb.1911.0073.
[3] J. Jayapriya, V. Ramamurthy, Use of non-native phenazines to improve the performance of *Pseudomonas aeruginosa* MTCC 2474 catalysed fuel cells, Bioresource Technology 124 (2012) 23–28. Available from: https://doi.org/10.1016/j.biortech.2012.08.034.
[4] C. Zhao, P. Gai, R. Song, Y. Chen, J. Zhang, J.-J. Zhu, Nanostructured material-based biofuel cells: recent advances and future prospects, Chemical Society Review 46 (5) (2017) 1545–1564. Available from: https://doi.org/10.1039/C6CS00044D.
[5] N. Saranya, J. Jayapriya, V. Ramamurthy, Unsaturated polyesters in microbial fuel cells and biosensors, Unsaturated Polyester Resins: Fundamentals, Design, Fabrication, and Applications, Elsevier, 2019, pp. 557–578. Available from: https://doi.org/10.1016/B978-0-12-816129-6.00021-1.
[6] J. Jayaprakash, A. Parthasarathy, R. Viraraghavan, Decolorization and degradation of Monoazo and Diazo dyes in Pseudomonas catalyzed microbial fuel cell, Environmental Progress & Sustainable Energy 35 (6) (2016) 1623–1628. Available from: https://doi.org/10.1002/ep.
[7] J. Jayapriya, J. Gopal, V. Ramamurthy, U. Kamachi Mudali, B. Raj, Preparation and characterization of biocompatible carbon electrodes, Composites Part B: Engineering 43 (3) (2012) 1329–1335. Available from: https://doi.org/10.1016/j.compositesb.2011.10.019.
[8] M. Farahmand Habibi, M. Arvand, U. Schröder, S. Sohrabnezhad, Self-assembled cauliflower-like pyrite-S, N co-doped graphene quantum dots as free-standing anode with high conductivity and biocompatibility for bioelectricity production, Fuel 286 (Part 1) (2021) 119291. Available from: https://doi.org/10.1016/j.fuel.2020.119291.
[9] J. Jayapriya, V. Ramamurthy, The role of electrode material in capturing power generated in Pseudomonas catalysed fuel cells, The Canadian Journal of Chemical Engineering 92 (4) (2014) 610–614. Available from: https://doi.org/10.1002/cjce.21895.
[10] S. Narayanasamy, J. Jayaprakash, Improved performance of *Pseudomonas aeruginosa* catalyzed MFCs with graphite/polyester composite electrodes doped with metal ions for azo dye degradation, Chemical Engineering Journal 343 (2018) 258–269. Available from: https://doi.org/10.1016/j.cej.2018.02.123.
[11] Y. Zhang, L. Liu, B. Van der Bruggen, F. Yang, Nanocarbon based composite electrodes and their application in microbial fuel cells, Journal of Materials Chemistry A 5 (25) (2017) 12673–12698. Available from: https://doi.org/10.1039/C7TA01511A.

[12] W. Richard, Siegel, Synthesis and properties of nanophase materials, Materials Science and Engineering: B 168 (2) (1993) 189−197. Available from: https://doi.org/10.1016/0921-5093(93)90726-U.

[13] I. Gajda, J. You, C. Santoro, J. Greenman, I.A. Ieropoulos, A new method for urine electrofiltration and long term power enhancement using surface modified anodes with activated carbon in ceramic microbial fuel cells, Electrochimica Acta 353 (2020) 136388. Available from: https://doi.org/10.1016/j.electacta.2020.136388.

[14] D. Li, Y. Qu, J. Liu, G. Liu, J. Zhang, Y. Feng, Enhanced oxygen and hydroxide transport in a cathode interface by efficient antibacterial property of a silver nanoparticle-modified, activated carbon cathode in microbial fuel cells, ACS Applied Materials and Interfaces 8 (32) (2016) 20814−20821. Available from: https://doi.org/10.1021/acsami.6b06419.

[15] V.J. Watson, C. Nieto Delgado, B.E. Logan, Influence of chemical and physical properties of activated carbon powders on oxygen reduction and microbial fuel cell performance, Environmental Science and Technology 47 (12) (2013) 6704−6710. Available from: https://doi.org/10.1021/es401722j.

[16] Y. Ahn, I. Ivanov, T.C. Nagaiah, A. Bordoloi, B.E. Logan, Mesoporous nitrogen-rich carbon materials as cathode catalysts in microbial fuel cells, Journal of Power Sources 269 (2014) 212−215. Available from: https://doi.org/10.1016/j.jpowsour.2014.06.115.

[17] Y. Zhang, J. Sun, B. Hou, Y. Hu, Performance improvement of air-cathode single-chamber microbial fuel cell using a mesoporous carbon modified anode, Journal of Power Sources 196 (18) (2011) 7458−7464. Available from: https://doi.org/10.1016/j.jpowsour.2011.05.004.

[18] M. Liu, M. Zhou, L. Ma, H. Yang, Y. Zhao, Architectural design of hierarchically meso-macroporous carbon for microbial fuel cell anodes, RSC Advances 6 (33) (2016) 27993−27998. Available from: https://doi.org/10.1039/c5ra26420k.

[19] L. Zeng, X. Li, S. Fan, J. Li, J. Mu, M. Qin, et al., Bioelectrochemical synthesis of high-quality carbon dots with strengthened electricity output and excellent catalytic performance, Nanoscale 11 (10) (2019) 4428−4437. Available from: https://doi.org/10.1039/c8nr10510c.

[20] A.S. Vishwanathan, K.S. Aiyer, L.A.A. Chunduri, K. Venkataramaniah, S. Siva Sankara Sai, G. Rao, Carbon quantum dots shuttle electrons to the anode of a microbial fuel cell, 3 Biotech 6 (2) (2016) 1−6. Available from: https://doi.org/10.1007/s13205-016-0552-1.

[21] E.F. Talooki, M. Ghorbani, M. Rahimnejad, M.S. Lashkenari, Evaluation of a visible light-responsive polyaniline nanofiber-cadmium sulfide quantum dots photocathode for simultaneous hexavalent chromium reduction and electricity generation in photo-microbial fuel cell, Journal of Electroanalytical Chemistry 873 (2020) 114469. Available from: https://doi.org/10.1016/j.jelechem.2020.114469.

[22] B. Wang, H. Zhang, X.Y. Lu, J. Xuan, M.K.H. Leung, Solar photocatalytic fuel cell using CdS-TiO$_2$ photoanode and air-breathing cathode for wastewater treatment and simultaneous electricity production, Chemical Engineering Journal 253 (2014) 174−182. Available from: https://doi.org/10.1016/j.cej.2014.05.041.

[23] Y.F. Guan, F. Zhang, B.C. Huang, H.Q. Yu, Enhancing electricity generation of microbial fuel cell for wastewater treatment using nitrogen-doped carbon dots-supported carbon paper anode, Journal of Cleaner Production 229 (2019) 412−419. Available from: https://doi.org/10.1016/j.jclepro.2019.05.040.

[24] B. Li, Z. Zhao, Z. Weng, Y. Fang, W. Lei, D. Zhu, et al., Modification of PPy-NW anode by carbon dots for high-performance mini-microbial fuel cells, Fuel Cells 20 (2) (2020) 203−211. Available from: https://doi.org/10.1002/fuce.201900210.

[25] M. Mashkour, M. Rahimnejad, M. Mashkour, F. Soavi, Electro-polymerized polyaniline modified conductive bacterial cellulose anode for supercapacitive microbial fuel cells and studying the role of anodic biofilm in the capacitive behavior, Journal of Power Sources 478 (August) (2020) 228822. Available from: https://doi.org/10.1016/j.jpowsour.2020.228822.

[26] X. Xie, L. Hu, M. Pasta, G.F. Wells, D. Kong, C.S. Criddle, et al., Three-dimensional carbon nanotube-textile anode for high-performance microbial fuel cells, Nano Letters 11 (1) (2011) 291−296. Available from: https://doi.org/10.1021/nl103905t.

[27] T. Cai, M. Huang, Y. Huang, W. Zheng, Enhanced performance of microbial fuel cells by electrospinning carbon nanofibers hybrid carbon nanotubes composite anode, International Journal of

Hydrogen Energy 44 (5) (2019) 3088−3098. Available from: https://doi.org/10.1016/j.ijhydene.2018.11.205.
[28] S.S. Manickam, U. Karra, L. Huang, N.N. Bui, B. Li, J.R. McCutcheon, Activated carbon nanofiber anodes for microbial fuel cells, Carbon 53 (2013) 19−28. Available from: https://doi.org/10.1016/j.carbon.2012.10.009.
[29] T. Cai, Y. Huang, M. Huang, Y. Xi, D. Pang, W. Zhang, Enhancing oxygen reduction reaction of supercapacitor microbial fuel cells with electrospun carbon nanofibers composite cathode, Chemical Engineering Journal 371 (April) (2019) 544−553. Available from: https://doi.org/10.1016/j.cej.2019.04.025.
[30] A. Bianco, H.M. Cheng, T. Enoki, Y. Gogotsi, R.H. Hurt, N. Koratkar, et al., All in the graphene family—A recommended nomenclature for two-dimensional carbon materials, Carbon 65 (2013) 1−6. Available from: https://doi.org/10.1016/j.carbon.2013.08.038.
[31] A. Pareek, J. Shanthi Sravan, S. Venkata Mohan, Exploring chemically reduced graphene oxide electrode for power generation in microbial fuel cell, Materials Science for Energy Technologies 2 (3) (2019) 600−606. Available from: https://doi.org/10.1016/j.mset.2019.06.006.
[32] A. Pareek, J. Shanthi Sravan, S. Venkata Mohan, Fabrication of three-dimensional graphene anode for augmenting performance in microbial fuel cells, Carbon Resources Conversion 2 (2) (2019) 134−140. Available from: https://doi.org/10.1016/j.crcon.2019.06.003.
[33] N. Yang, Y. Ren, X. Li, X. Wang, Effect of graphene-graphene oxide modified anode on the performance of microbial fuel cell, Nanomaterials 6 (9) (2016) 174. Available from: https://doi.org/10.3390/nano6090174.
[34] J. Islam, G. Chilkoor, K. Jawaharraj, S.S. Dhiman, R. Sani, V. Gadhamshetty, Vitamin-C-enabled reduced graphene oxide chemistry for tuning biofilm phenotypes of methylotrophs on nickel electrodes in microbial fuel cells, Bioresource Technology 300 (December 2019) (2020) 122642. Available from: https://doi.org/10.1016/j.biortech.2019.122642.
[35] S. Ci, P. Cai, Z. Wen, J. Li, Graphene-based electrode materials for microbial fuel cells, Science China Materials 58 (6) (2015) 496−509. Available from: https://doi.org/10.1007/s40843-015-0061-2.
[36] V.B. Mohan, K.T. Lau, D. Hui, D. Bhattacharyya, Graphene-based materials and their composites: a review on production, applications and product limitations, Composites Part B: Engineering 142 (December 2017) (2018) 200−220. Available from: https://doi.org/10.1016/j.compositesb.2018.01.013.
[37] J. Tang, S. Chen, Y. Yuan, X. Cai, S. Zhou, In situ formation of graphene layers on graphite surfaces for efficient anodes of microbial fuel cells, Biosensors and Bioelectronics 71 (2015) 387−395. Available from: https://doi.org/10.1016/j.bios.2015.04.074.
[38] E.T. Sayed, H. Alawadhi, A.G. Olabi, A. Jamal, M.S. Almahdi, J. Khalid, et al., Electrophoretic deposition of graphene oxide on carbon brush as bioanode for microbial fuel cell operated with real wastewater, International Journal of Hydrogen Energy 46 (8) (2021) 5975−5983. Available from: https://doi.org/10.1016/j.ijhydene.2020.10.043.
[39] J. Chen, F. Deng, Y. Hu, J. Sun, Y. Yang, Antibacterial activity of graphene-modified anode on *Shewanella oneidensis* MR-1 biofilm in microbial fuel cell, Journal of Power Sources 290 (2015) 80−86. Available from: https://doi.org/10.1016/j.jpowsour.2015.03.033.
[40] Y. Wu, L. Wang, M. Jin, F. Kong, H. Qi, J. Nan, Reduced graphene oxide and biofilms as cathode catalysts to enhance energy and metal recovery in microbial fuel cell, Bioresource Technology 283 (March) (2019) 129−137. Available from: https://doi.org/10.1016/j.biortech.2019.03.080.
[41] B. Koo, S.-M. Lee, S.-E. Oh, E.J. Kim, Y. Hwang, D. Seo, et al., Addition of reduced graphene oxide to an activated-carbon cathode increases electrical power generation of a microbial fuel cell by enhancing cathodic performance, Electrochimica Acta 297 (2019) 613−622. Available from: https://doi.org/10.1016/j.electacta.2018.12.024.
[42] T.S. Song, Y. Jin, J. Bao, D. Kang, J. Xie, Graphene/biofilm composites for enhancement of hexavalent chromium reduction and electricity production in a biocathode microbial fuel cell, Journal of Hazardous Materials 317 (2016) 73−80. Available from: https://doi.org/10.1016/j.jhazmat.2016.05.055.
[43] A.T. Najafabadi, N. Ng, E. Gyenge, Electrochemically exfoliated graphene anodes with enhanced biocurrent production in single-chamber air-breathing microbial fuel cells, Biosensors and Bioelectronics 81 (2016) 103−110. Available from: https://doi.org/10.1016/j.bios.2016.02.054.

[44] Y. Cheng, M. Mallavarapu, R. Naidu, Z. Chen, In situ fabrication of green reduced graphene-based biocompatible anode for efficient energy recycle, Chemosphere 193 (November) (2018) 618–624. Available from: https://doi.org/10.1016/j.chemosphere.2017.11.057.

[45] Y. Yuan, S. Zhou, B. Zhao, L. Zhuang, Y. Wang, Microbially-reduced graphene scaffolds to facilitate extracellular electron transfer in microbial fuel cells, Bioresource Technology 116 (2012) 453–458. Available from: https://doi.org/10.1016/j.biortech.2012.03.118.

[46] Y.Y. Yu, D.D. Zhai, R.W. Si, J.Z. Sun, X. Liu, Y.C. Yong, Three-dimensional electrodes for high-performance bioelectrochemical systems, International Journal of Molecular Sciences 18 (1) (2017) 90. Available from: https://doi.org/10.3390/ijms18010090.

[47] X. Xie, M. Pasta, L. Hu, Y. Yang, J. McDonough, J. Cha, et al., Nano-structured textiles as high-performance aqueous cathodes for microbial fuel cells, Energy & Environmental Science 4 (4) (2011) 1293. Available from: https://doi.org/10.1039/c0ee00793e.

[48] X. Xie, M. Ye, L. Hu, N. Liu, J.R. McDonough, W. Chen, Carbon nanotube-coated macroporous sponge for microbial fuel cell electrodes, Energy & Environmental Science 5 (1) (2012) 5265–5270. Available from: https://doi.org/10.1039/c1ee02122b.

[49] S. You, M. Ma, W. Wang, D. Qi, X. Chen, J. Qu, et al., 3D Macroporous nitrogen-enriched graphitic carbon scaffold for efficient bioelectricity generation in microbial fuel cells, Advanced Energy Materials 7 (4) (2017) 1601364. Available from: https://doi.org/10.1002/aenm.201601364.

[50] S.Y. Sawant, T.H. Han, S.A. Ansari, J.H. Shim, A.T.N. Nguyen, J.J. Shim, et al., A metal-free and non-precious multifunctional 3D carbon foam for high-energy density supercapacitors and enhanced power generation in microbial fuel cells, Journal of Industrial and Engineering Chemistry 60 (2018) 431–440. Available from: https://doi.org/10.1016/j.jiec.2017.11.030.

[51] S.Y. Sawant, T.H. Han, S.A. Ansari, J.H. Shim, A.T.N. Nguyen, J.J. Shim, et al., A metal-free and non-precious multifunctional 3D carbon foam for high-energy density supercapacitors and enhanced power generation in microbial fuel cells, Journal of Industrial and Engineering Chemistry 60 (2018) 431–440. Available from: https://doi.org/10.1016/j.jiec.2017.11.030.

[52] C. Santoro, M. Kodali, S. Kabir, F. Soavi, A. Serov, P. Atanassov, Three-dimensional graphene nanosheets as cathode catalysts in standard and supercapacitive microbial fuel cell, J. Power Sources 356 (2017) 371–380. Available from: https://doi.org/10.1016/j.jpowsour.2017.03.135.

[53] A. Sumisha, K. Haribabu, Modification of graphite felt using nano polypyrrole and polythiophene for microbial fuel cell applications-a comparative study, Int. J. Hydrogen Energy 43 (2018) 3308–3316. Available from: https://doi.org/10.1016/j.ijhydene.2017.12.175.

[54] W. Zhang, B. Xie, L. Yang, D. Liang, Y. Zhu, H. Liu, Brush-like polyaniline nanoarray modified anode for improvement of power output in microbial fuel cell, Bioresour. Technol. 233 (2017) 291–295. Available from: https://doi.org/10.1016/j.biortech.2017.02.124.

[55] Z. Lv, Y. Chen, H. Wei, F. Li, Y. Hu, C. Wei, C. Feng, One-step electrosynthesis of polypyrrole/graphene oxide composites for microbial fuel cell application, Electrochim. Acta 111 (2013) 366–373. Available from: https://doi.org/10.1016/j.electacta.2013.08.022.

[56] Y. Yuan, S. Zhou, Y. Liu, J. Tang, Nanostructured macroporous bioanode based on polyaniline-modified natural loofah sponge for high-performance microbial fuel cells, Environ. Sci. Technol. 47 (2013) 14525–14532. Available from: https://doi.org/10.1021/es404163g.

[57] M. Harshiny, N. Samsudeen, R.J. Kameswara, M. Matheswaran, Biosynthesized FeO nanoparticles coated carbon anode for improving the performance of microbial fuel cell, Int. J. Hydrogen Energy 42 (2017) 26488–26495. Available from: https://doi.org/10.1016/j.ijhydene.2017.07.084.

[58] A. Mehdinia, E. Ziaei, A. Jabbari, Multi-walled carbon nanotube/SnO2 nanocomposite: A novel anode material for microbial fuel cells, Electrochim. Acta 130 (2014) 512–518. Available from: https://doi.org/10.1016/j.electacta.2014.03.011.

[59] T. Yin, H. Zhang, G. Yang, L. Wang, Polyaniline composite TiO2 nanosheets modified carbon paper electrode as a high performance bioanode for microbial fuel cells, Synth. Met. 252 (2019) 8–14. Available from: https://doi.org/10.1016/j.synthmet.2019.03.027.

[60] Y. Wang, C. Liu, S. Zhou, R. Hou, L. Zhou, F. Guan, R. Chen, Y. Yuan, Hierarchical N-doped C/Fe3O4 nanotube composite arrays grown on the carbon fiber cloth as a bioanode for

high-performance bioelectrochemical system, Chem. Eng. J. 406 (2021) 126832. Available from: https://doi.org/10.1016/j.cej.2020.126832.
[61] G. Massaglia, V. Margaria, M.R. Fiorentin, K. Pasha, A. Sacco, M. Castellino, A. Chiodoni, S. Bianco, F.C. Pirri, M. Quaglio, Nonwoven mats of N-doped carbon nanofibers as high-performing anodes in microbial fuel cells, Mater. Today Energy 16 (2020) 100385. Available from: https://doi.org/10.1016/j.mtener.2020.100385.
[62] M. Chen, Y. Zeng, Y. Zhao, M. Yu, F. Cheng, X. Lu, Y. Tong, Monolithic three-dimensional graphene frameworks derived from inexpensive graphite paper as advanced anodes for microbial fuel cells, J. Mater. Chem. A 4 (2016) 6342–6349. Available from: https://doi.org/10.1039/C6TA00992A.
[63] K.B. Pu, C.X. Lu, K. Zhang, H. Zhang, Q.Y. Chen, Y.H. Wang, In situ synthesis of polypyrrole on graphite felt as bio-anode to enhance the start-up performance of microbial fuel cells, Bioprocess and Biosystems Engineering 43 (3) (2020) 429–437. Available from: https://doi.org/10.1007/s00449-019-02238-y.
[64] Y. Tao, Q. Liu, J. Chen, B. Wang, Y. Wang, K. Liu, et al., Hierarchically three-dimensional nanofiber based textile with high conductivity and biocompatibility as a microbial fuel cell anode, Environmental Science and Technology 50 (14) (2016) 7889–7895. Available from: https://doi.org/10.1021/acs.est.6b00648.
[65] G. Gnana Kumar, C.J. Kirubaharan, S. Udhayakumar, K. Ramachandran, C. Karthikeyan, R. Renganathan, et al., Synthesis, structural, and morphological characterizations of reduced graphene oxide-supported polypyrrole anode catalysts for improved microbial fuel cell performances, ACS Sustainable Chemistry and Engineering 2 (10) (2014) 2283–2290. Available from: https://doi.org/10.1021/sc500244f.
[66] S.H. Roh, Electricity generation from microbial fuel cell with polypyrrole-coated carbon nanofiber composite, Journal of Nanoscience and Nanotechnology 15 (2) (2015) 1700–1703. Available from: https://doi.org/10.1166/jnn.2015.9317.
[67] N. Zhao, Z. Ma, H. Song, Y. Xie, M. Zhang, Enhancement of bioelectricity generation by synergistic modification of vertical carbon nanotubes/polypyrrole for the carbon fibers anode in microbial fuel cell, Electrochimica Acta (2019) 69–74. Available from: https://doi.org/10.1016/j.electacta.2018.11.039.
[68] Z.H. Liao, J.Z. Sun, D.Z. Sun, R.W. Si, Y.C. Yong, Enhancement of power production with tartaric acid doped polyaniline nanowire network modified anode in microbial fuel cells, Bioresource Technology 192 (2015) 831–834. Available from: https://doi.org/10.1016/j.biortech.2015.05.105.
[69] M. Mashkour, M. Rahimnejad, M. Mashkour, Bacterial cellulose-polyaniline nano-biocomposite: a porous media hydrogel bioanode enhancing the performance of microbial fuel cell, Journal of Power Sources 325 (2016) 322–328. Available from: https://doi.org/10.1016/j.jpowsour.2016.06.063.
[70] H. Xu, J. Wu, L. Qi, Y. Chen, Q. Wen, T. Duan, et al., Preparation and microbial fuel cell application of sponge-structured hierarchical polyaniline-texture bioanode with an integration of electricity generation and energy storage, Journal of Applied Electrochemistry 48 (11) (2018) 1285–1295. Available from: https://doi.org/10.1007/s10800-018-1252-9.
[71] X. Liu, X. Zhao, Y.Y. Yu, Y.Z. Wang, Y.T. Shi, Q.W. Cheng, et al., Facile fabrication of conductive polyaniline nanoflower modified electrode and its application for microbial energy harvesting, Electrochimica Acta 255 (2017) 41–47. Available from: https://doi.org/10.1016/j.electacta.2017.09.153.
[72] L. Huang, X. Li, Y. Ren, X. Wang, In-situ modified carbon cloth with polyaniline/graphene as anode to enhance performance of microbial fuel cell, International Journal of Hydrogen Energy 41 (26) (2016) 11369–11379. Available from: https://doi.org/10.1016/j.ijhydene.2016.05.048.
[73] H.-F. Cui, L. Du, P.-B. Guo, B. Zhu, J.H.T. Luong, Controlled modification of carbon nanotubes and polyaniline on macroporous graphite felt for high-performance microbial fuel cell anode, Journal of Power Sources 283 (2015) 46–53. Available from: https://doi.org/10.1016/j.jpowsour.2015.02.088.
[74] H.O. Mohamed, E.T. Sayed, M. Obaid, Y.J. Choi, S.G. Park, S. Al-Qaradawi, et al., Transition metal nanoparticles doped carbon paper as a cost-effective anode in a microbial fuel cell powered by pure and mixed biocatalyst cultures, International Journal of Hydrogen Energy 43 (46) (2018) 21560–21571. Available from: https://doi.org/10.1016/j.ijhydene.2018.09.199.

[75] M. Masoudi, M. Rahimnejad, M. Mashkour, Fabrication of anode electrode by a novel acrylic based graphite paint on stainless steel mesh and investigating biofilm effect on electrochemical behavior of anode in a single chamber microbial fuel cell, Electrochimica Acta 344 (2020) 136168. Available from: https://doi.org/10.1016/j.electacta.2020.136168.

[76] Y. Qiao, X.-S. Wu, C.-X. Ma, H. He, C.M. Li, A hierarchical porous graphene/nickel anode that simultaneously boosts the bio- and electro-catalysis for high-performance microbial fuel cells, RSC Advances 4 (42) (2014) 21788. Available from: https://doi.org/10.1039/c4ra03082f.

[77] S. Gupta, A. Yadav, N. Verma, Simultaneous Cr(VI) reduction and bioelectricity generation using microbial fuel cell based on alumina-nickel nanoparticles-dispersed carbon nanofiber electrode, Chemical Engineering Journal 307 (2017) 729—738. Available from: https://doi.org/10.1016/j.cej.2016.08.130.

[78] M. Saadi, J. Pézard, N. Haddour, M. Erouel, T.M. Vogel, K. Khirouni, Stainless steel coated with carbon nanofiber/PDMS composite as anodes in microbial fuel cells, Materials Research Express 7 (2) (2020) 0—8. Available from: https://doi.org/10.1088/2053-1591/ab6c99.

[79] Z. Wen, S. Ci, S. Mao, S. Cui, G. Lu, K. Yu, et al., TiO_2 nanoparticles-decorated carbon nanotubes for significantly improved bioelectricity generation in microbial fuel cells, Journal of Power Sources 234 (2013) 100—106. Available from: https://doi.org/10.1016/j.jpowsour.2013.01.146.

[80] X. Jia, Z. He, X. Zhang, X. Tian, Carbon paper electrode modified with TiO_2 nanowires enhancement bioelectricity generation in microbial fuel cell, Synthetic Metals 215 (2016) 170—175. Available from: https://doi.org/10.1016/j.synthmet.2016.02.015.

[81] L. Deng, G. Dong, Y. Zhang, D. Li, T. Lu, Y. Chen, et al., Lysine-modified TiO_2 nanotube array for optimizing bioelectricity generation in microbial fuel cells, Electrochimica Acta 300 (2019) 163—170. Available from: https://doi.org/10.1016/j.electacta.2019.01.105.

[82] Y. Shan, J. Cui, Y. Liu, W. Zhao, TiO_2 anchored on MoS_2 nanosheets based on molybdenite exfoliation as an efficient cathode for enhanced Cr (VI) reduction in microbial fuel cell, Environmental Research Vi (2020) 110010. Available from: https://doi.org/10.1016/j.envres.2020.110010.

[83] M. Liu, M. Zhou, H. Yang, G. Ren, Y. Zhao, Titanium dioxide nanoparticles modified three dimensional ordered macroporous carbon for improved energy output in microbial fuel cells, Electrochimica Acta 190 (2016) 463—470. Available from: https://doi.org/10.1016/j.electacta.2015.12.131.

[84] C. Zhang, P. Liang, X. Yang, Y. Jiang, Y. Bian, C. Chen, Binder-free graphene and manganese oxide coated carbon felt anode for high-performance microbial fuel cells, Biosensors and Bioelectronics (2016) 32—38. Available from: https://doi.org/10.1016/j.bios.2016.02.051.

[85] X. Peng, S. Chen, L. Liu, S. Zheng, M. Li, Modified stainless steel for high performance and stable anode in microbial fuel cells, Electrochimica Acta 194 (2016) 246—252. Available from: https://doi.org/10.1016/j.electacta.2016.02.127.

[86] Y. Liang, H. Feng, D. Shen, N. Li, K. Guo, Y. Zhou, et al., Enhancement of anodic biofilm formation and current output in microbial fuel cells by composite modification of stainless steel electrodes, Journal of Power Sources 342 (2017) 98—104. Available from: https://doi.org/10.1016/j.jpowsour.2016.12.020.

[87] J.M. Sonawane, S.A. Patil, P.C. Ghosh, S.B. Adeloju, Low-cost stainless-steel wool anodes modified with polyaniline and polypyrrole for high-performance microbial fuel cells, Journal of Power Sources 379 (December 2017) (2018) 103—114. Available from: https://doi.org/10.1016/j.jpowsour.2018.01.001.

[88] J.M. Sonawane, P.C. Ghosh, S.B. Adeloju, Electrokinetic behaviour of conducting polymer modified stainless steel anodes during the enrichment phase in microbial fuel cells, Electrochimica Acta 287 (2018) 96—105. Available from: https://doi.org/10.1016/j.electacta.2018.07.077.

[89] A. Sumisha, K. Haribabu, Nanostructured polypyrrole as cathode catalyst for Fe (III) removal in single chamber microbial fuel cell, Biotechnology and Bioprocess Engineering 25 (1) (2020) 78—85. Available from: https://doi.org/10.1007/s12257-019-0288-y.

[90] Y. Jiang, Y. Xu, Q. Yang, Y. Chen, S. Zhu, S. Shen, Power generation using polyaniline/multi-walled carbon nanotubes as an alternative cathode catalyst in microbial fuel cells, International Journal of Energy Research 38 (11) (2014) 1416—1423. Available from: https://doi.org/10.1002/er.3155.

[91] J. Ahmed, H.J. Kim, S. Kim, Polyaniline nanofiber/carbon black composite as oxygen reduction catalyst for air cathode microbial fuel cells, Journal of the Electrochemical Society 159 (5) (2012) B497−B501. Available from: https://doi.org/10.1149/2.049205jes.

[92] L. Liu, X. Sun, W. Li, Y. An, H. Li, Electrochemical hydrodechlorination of perchloroethylene in groundwater on a Ni-doped graphene composite cathode driven by a microbial fuel cell, RSC Advances 8 (63) (2018) 36142−36149. Available from: https://doi.org/10.1039/C8RA06951D.

[93] J. Liu, C. Vipulanandan, Effects of Fe, Ni, and Fe/Ni metallic nanoparticles on power production and biosurfactant production from used vegetable oil in the anode chamber of a microbial fuel cell, Waste Management 66 (2017) 169−177. Available from: https://doi.org/10.1016/j.wasman.2017.04.004.

[94] X. Guo, J. Jia, H. Dong, Q. Wang, T. Xu, B. Fu, R. Ran, P. Liang, X. Huang, X. Zhang, Hydrothermal synthesis of Fe−Mn bimetallic nanocatalysts as high-efficiency cathode catalysts for microbial fuel cells, J. Power Sources. 414 (2019) 444−452. Available from: https://doi.org/10.1016/j.jpowsour.2019.01.024.

[95] L. Pu, K. Li, Z. Chen, P. Zhang, X. Zhang, Z. Fu, Silver electrodeposition on the activated carbon air cathode for performance improvement in microbial fuel cells, Journal of Power Sources 268 (2014) 476−481. Available from: https://doi.org/10.1016/j.jpowsour.2014.06.071.

[96] Y. Dai, Y. Chan, B. Jiang, L. Wang, J. Zou, K. Pan, et al., Bifunctional Ag/Fe/N/C catalysts for enhancing oxygen reduction via cathodic biofilm inhibition in microbial fuel cells, ACS Applied Materials & Interfaces 8 (11) (2016) 6992−7002. Available from: https://doi.org/10.1021/acsami.5b11561.

[97] A. Modi, S. Singh, N. Verma, In situ nitrogen-doping of nickel nanoparticle-dispersed carbon nanofiber-based electrodes: its positive effects on the performance of a microbial fuel cell, Electrochimica Acta 190 (2015) 620−627. Available from: https://doi.org/10.1016/j.electacta.2015.12.191.

[98] S. Zhang, W. Su, X. Wang, K. Li, Y. Li, Bimetallic metal-organic frameworks derived cobalt nanoparticles embedded in nitrogen-doped carbon nanotube nanopolyhedra as advanced electrocatalyst for high-performance of activated carbon air-cathode microbial fuel cell, Biosensors and Bioelectronics 127 (2019) 181−187. Available from: https://doi.org/10.1016/j.bios.2018.12.028.

[99] S. Narayanasamy, J. Jayaprakash, Carbon cloth/nickel cobaltite (NiCo2O4)/polyaniline (PANI) composite electrodes: Preparation, characterization, and application in microbial fuel cells, Fuel 301 (2021) 121016. Available from: https://doi.org/10.1016/j.fuel.2021.121016.

[100] J.M. Morris, S. Jin, J. Wang, C. Zhu, M.A. Urynowicz, Lead dioxide as an alternative catalyst to platinum in microbial fuel cells, Electrochemistry Communications 9 (7) (2007) 1730−1734. Available from: https://doi.org/10.1016/j.elecom.2007.03.028.

[101] M. Masalovich, A. Ivanova, O. Zagrebelnyy, A. Nikolaev, A. Galushko, A. Baranchikov, et al., Fabrication of composite electrodes based on cobalt (II) hydroxide for microbiological fuel cells, Journal of Sol-Gel Science and Technology 92 (2) (2019) 506−514. Available from: https://doi.org/10.1007/s10971-019-04977-6.

[102] P.N. Venkatesan, S. Dharmalingam, Synthesis and characterization of Pt, Pt−Fe/TiO$_2$ cathode catalysts and its evaluation in microbial fuel cell, Materials for Renewable and Sustainable Energy 5 (3) (2016) 1−9. Available from: https://doi.org/10.1007/s40243-016-0074-0.

[103] S. Mahalingam, S. Ayyaru, Y.-H. Ahn, Enhanced cathode performance of a rGO−V$_2$O$_5$ nanocomposite catalyst for microbial fuel cell applications, Dalton Transactions 47 (46) (2018) 16777−16788. Available from: https://doi.org/10.1039/C8DT02445F.

[104] Md. T. Noori, M.M. Ghangrekar, C.K. Mukherjee, V$_2$O$_5$ microflower decorated cathode for enhancing power generation in air-cathode microbial fuel cell treating fish market wastewater, International Journal of Hydrogen Energy 41 (5) (2016) 3638−3645. Available from: https://doi.org/10.1016/j.ijhydene.2015.12.163.

[105] M. Lu, S. Kharkwal, H.Y. Ng, S.F.Y. Li, Carbon nanotube supported MnO$_2$ catalyst for oxygen reduction reaction and their applications in microbial fuel cells, Biosensors and Bioelectronics 26 (12) (2011) 4728−4732. Available from: https://doi.org/10.1016/j.bios.2011.05.036.

[106] L. Zhang, C. Liu, L. Zhuang, W. Li, S. Zhou, J. Zhang, Manganese dioxide as an alternative cathodic catalyst to platinum in microbial fuel cells, Biosensors and Bioelectronics 24 (9) (2009) 2825−2829. Available from: https://doi.org/10.1016/j.bios.2009.02.010.

[107] Y. Zhang, Y. Hu, S. Li, J. Sun, B. Hou, Manganese dioxide-coated carbon nanotubes as an improved cathodic catalyst for oxygen reduction in a microbial fuel cell, Journal of Power Sources 196 (22) (2011) 9284−9289. Available from: https://doi.org/10.1016/j.jpowsour.2011.07.069.

[108] A. Tofighi, M. Rahimnejad, M. Ghorbani, Ternary nanotube α-MnO$_2$/GO/AC as an excellent alternative composite modifier for cathode electrode of microbial fuel cell, Journal of Thermal Analysis and Calorimetry 135 (3) (2019) 1667−1675. Available from: https://doi.org/10.1007/s10973-018-7198-7.

[109] K.B. Ghoreishi, M. Ghasemi, M. Rahimnejad, M.A. Yarmo, W.R.W. Daud, N. Asim, et al., Development and application of vanadium oxide/polyaniline composite as a novel cathode catalyst in microbial fuel cell, International Journal of Energy Research 38 (1) (2014) 70−77. Available from: https://doi.org/10.1002/er.3082.

[110] C. Cao, L. Wei, G. Wang, J. Liu, Q. Zhai, J. Shen, A polyaniline-derived iron-nitrogen-carbon nanorod network anchored on graphene as a cost-effective air-cathode electrocatalyst for microbial fuel cells, Inorg. Chem. Front. 4 (2017) 1930−1938. Available from: https://doi.org/10.1039/c7qi00452d.

[111] Y. Ren, D. Pan, X. Li, F. Fu, Y. Zhao, X. Wang, Effect of polyaniline-graphene nanosheets modified cathode on the performance of sediment microbial fuel cell, J. Chem. Technol. Biotechnol. 88 (2013) 1946−1950. Available from: https://doi.org/10.1002/jctb.4146.

[112] H. Zhang, R. Zhang, G. Zhang, F. Yang, F. Gao, Modified graphite electrode by polyaniline/tourmaline improves the performance of bio-cathode microbial fuel cell, Int. J. Hydrogen Energy 39 (2014) 11250−11257. Available from: https://doi.org/10.1016/j.ijhydene.2014.05.057.

[113] Q. Huang, P. Zhou, H. Yang, L. Zhu, H. Wu, In situ generation of inverse spinel CoFe2O4 nanoparticles onto nitrogen-doped activated carbon for an effective cathode electrocatalyst of microbial fuel cells, Chem. Eng. J. 325 (2017) 466−473. Available from: https://doi.org/10.1016/j.cej.2017.05.079.

[114] M.R. Majidi, F. Shahbazi Farahani, M. Hosseini, I. Ahadzadeh, Low-cost nanowired α-MnO2/C as an ORR catalyst in air-cathode microbial fuel cell, Bioelectrochemistry 125 (2019) 38−45. Available from: https://doi.org/10.1016/j.bioelechem.2018.09.004.

[115] Q. Wen, S. Wang, J. Yan, L. Cong, Z. Pan, Y. Ren, Z. Fan, MnO2-graphene hybrid as an alternative cathodic catalyst to platinum in microbial fuel cells, J. Power Sources 216 (2012) 187−191. Available from: https://doi.org/10.1016/j.jpowsour.2012.05.023.

[116] P. Khare, J. Ramkumar, N. Verma, Carbon nanofiber-skinned three dimensional Ni/carbon micropillars: high performance electrodes of a microbial fuel cell, Electrochim. Acta 219 (2016) 88−98. Available from: https://doi.org/10.1016/j.electacta.2016.09.140.

[117] S. Singh, P.K. Bairagi, N. Verma, Candle soot-derived carbon nanoparticles: An inexpensive and efficient electrode for microbial fuel cells, Electrochim. Acta 264 (2018) 119−127. Available from: https://doi.org/10.1016/j.electacta.2018.01.110.

[118] M. Ghasemi, S. Shahgaldi, M. Ismail, B.H. Kim, Z. Yaakob, W.R. Wan Daud, Activated carbon nanofibers as an alternative cathode catalyst to platinum in a two-chamber microbial fuel cell, Int. J. Hydrogen Energy 36 (2011) 13746−13752. Available from: https://doi.org/10.1016/j.ijhydene.2011.07.118.

[119] Y. Du, F.X. Ma, C.Y. Xu, J. Yu, D. Li, Y. Feng, L. Zhen, Nitrogen-doped carbon nanotubes/reduced graphene oxide nanosheet hybrids towards enhanced cathodic oxygen reduction and power generation of microbial fuel cells, Nano Energy 61 (2019) 533−539. Available from: https://doi.org/10.1016/j.nanoen.2019.05.001.

[120] E. Antolini, Composite materials for polymer electrolyte membrane microbial fuel cells, Biosensors & Bioelectronics 69C (2015) 54−70. Available from: https://doi.org/10.1016/j.bios.2015.02.013.

[121] M. Shabani, H. Younesi, M. Pontié, A. Rahimpour, M. Rahimnejad, A.A. Zinatizadeh, A critical review on recent proton exchange membranes applied in microbial fuel cells for renewable energy recovery, Journal of Cleaner Production 264 (2020) 121446. Available from: https://doi.org/10.1016/j.jclepro.2020.121446.

[122] M. Oliot, S. Galier, H. Roux de Balmann, A. Bergel, Ion transport in microbial fuel cells: key roles, theory and critical review, Applied Energy, 183, 2016, pp. 1682−1704. Available from: https://doi.org/10.1016/j.apenergy.2016.09.043.

[123] M. Ghasemi, S. Shahgaldi, M. Ismail, Z. Yaakob, W.R.W. Daud, New generation of carbon nanocomposite proton exchange membranes in microbial fuel cell systems, Chemical Engineering Journal 184 (2012) 82–89. Available from: https://doi.org/10.1016/j.cej.2012.01.001.

[124] S.S. Lim, W.R.W. Daud, J. Md Jahim, M. Ghasemi, P.S. Chong, M. Ismail, Sulfonated poly(ether ether ketone)/poly(ether sulfone) composite membranes as an alternative proton exchange membrane in microbial fuel cells, International Journal of Hydrogen Energy 37 (15) (2012) 11409–11424. Available from: https://doi.org/10.1016/j.ijhydene.2012.04.155.

[125] N. Mokhtarian, M. Ghasemi, W.R. Wan Daud, M. Ismail, G. Najafpour, J. Alam, Improvement of microbial fuel cell performance by using nafion polyaniline composite membranes as a separator, Journal of Fuel Cell Science and Technology 10 (4) (2013) 1–6. Available from: https://doi.org/10.1115/1.4024866.

[126] S. Ayyaru, S. Dharmalingam, Improved performance of microbial fuel cells using sulfonated polyether ether ketone (SPEEK) TiO_2-SO_3H nanocomposite membrane, RSC Advances 3 (47) (2013) 25243–25251. Available from: https://doi.org/10.1039/c3ra44212h.

[127] P.N. Venkatesan, S. Dharmalingam, Effect of cation transport of SPEEK—rutile TiO_2 electrolyte on microbial fuel cell performance, Journal of Membrane Science 492 (2015) 518–527. Available from: https://doi.org/10.1016/j.memsci.2015.06.025.

[128] N.V. Prabhu, D. Sangeetha, Characterization and performance study of sulfonated poly ether ether ketone/Fe_3O_4 nano composite membrane as electrolyte for microbial fuel cell, Chemical Engineering Journal 243 (2014) 564–571. Available from: https://doi.org/10.1016/j.cej.2013.12.103.

[129] M. Rahimnejad, M. Ghasemi, G.D. Najafpour, M. Ismail, A.W. Mohammad, A.A. Ghoreyshi, et al., Synthesis, characterization and application studies of self-made Fe_3O_4/PES nanocomposite membranes in microbial fuel cell, Electrochimica Acta 85 (2012) 700–706. Available from: https://doi.org/10.1016/j.electacta.2011.08.036.

CHAPTER 7

Metal-organic frameworks for fuel cell technologies

Muhammad Rizwan Sulaiman and Ram K. Gupta
Department of Chemistry, Kansas Polymer Research Center, Pittsburg State University, Pittsburg, KS, United States

7.1 Introduction

Today, innovative discoveries and advanced technologies have bestowed us the leverage of the modern world. Cool summer and warm winter days, sparkling nights, swift transportation, and everything has become possible with technological advancements. But to ensure the availability of these perks, we need reliable and sustainable energy sources. Unlike human history, where our ancestors used to rely on the fundamental form of energy such as human muscles and the burning of woods and biomasses, the modern world has completely revolutionized the concept of energy. Presently, fossil fuel is considered the focal point of energy, contributing around 84% of the world's total energy needs, and about 64% of the total electricity is produced from fossil fuels. However, the exhausted reservoirs and severe environmental hazards associated with the use of fossil fuels, such as CO_2 emission and global warming, hinder the extensive reliability of fossil fuels. Global energy consumption has also been intensified in the last decade and is expected to boost by around 56% from 2010 to 2040. Specifically, there will a need for 840 quadrillions British thermal units (BTUs) of energy each year by 2040, producing about 43 gigatons of CO_2 each year by 2035, which will lead to an intense energy crisis and severe global warming [1,2]. These concerns encouraged scientists to figure out alternative, reliable, and clean energy sources for everyday activities.

Clean energy has various advantages over conventional fossil fuels, such as plenty of sustainable energy resources, lowered carbon emission, and reduced global warming. This clean energy can be obtained by different renewable sources, including wind, solar, hydropower, or fuel cells, etc. Due to the zero CO_2 emission and excellent sustainability, renewable energy techniques have been thriving with an annual growth of 2.5% and is projected to contribute around 15% of the world's energy usage by 2040. Among these renewable energy sources, hydrogen-powered fuel cells are considered a promising candidate to replace fossil fuels. They provide energy by transforming chemical energy into electrical energy without releasing any gas. Additionally, their efficiency is much better than fossil fuels. Specifically, the fuel cell efficiency is 60%, which is twice as much as fossil fuel 34% [3].

There are various energy storage and conversion systems that work on different phenomena. Among these techniques, fuel cells have garnered enormous consideration because of plenty of energy sources and higher efficiency [4,5]. A fuel cell is a device that converts chemical energy into electrical energy. The electrochemical energy conversion system is considered a promising alternative to meet future energy needs and is capable of tackling growing environmental concerns simultaneously [6]. The consideration of any energy conversion system or fuel cells is generally based on two critical criteria: (1) the availability of primary energy sources and (2) environment-related problems associated with it. The catalytic fuel cells and other similar systems effectively fulfill these criteria with huge resources, excellent energy density, high productivity, and nonhazardous emission [7]. The electrochemical energy conversion systems can efficiently outperform other conventional heat engines. The heat engines employ pistons that work based on Carnot cycles, which require the combustion of fuels and produce multiple greenhouse gases. Whereas, the catalytic conversion systems generate clean electrical energy that is ready to use by direct conversion of chemical energy without undergoing combustion reactions. Additionally, a typical engine wastes tons of energy due to friction between piston parts, making it less efficient than a fuel cell [8].

Scientists have fabricated several types of the fuel cell such as proton–exchange membrane (PEM), alkaline, direct methanol, solid oxide, molten carbonate, and phosphoric acid fuel cells [9]. These devices have a definite mechanism different from each other, but the basic working principle for each fuel cell remains similar. Basically, each device converts energy from chemical to electrical. Hydrogen is used as a fuel in many fuel cells; still, some fuel cells, such as direct methanol fuel cells, use different fuels to generate electricity. The main advantage of using H_2 as a fuel is that it produces water as the only by-product. The generated water can be utilized in a water splitter to generate H_2 again. So, the cycle will go on without any environmental concerns. Therefore more attention is paid to the fuel cells that have closed H_2-H_2O process, such as proton–exchange membrane fuel cells (PEMFCs).

PEMFCs use H_2 as a fuel, which is converted to generate protons and electrons. These electrons are pushed to move through the external wire to the cathodic side and reduce oxygen, completing the circuit. The protons are transmitted to the cathodic side through a PEM and combined with electrons and oxygen to form water and heat as the by-products, as shown in Fig. 7.1. Hence the restoration of water is done that can be used to be split again, making the process more sustainable. But, sustainability can only be achieved if renewable techniques are being used for the breaking of water. The methods include photocatalytic or electrocatalytic reactions, wind energy, wave energy, etc. Renewable sources prevent the involvement of fossil fuels and provide sustainability to the process. Moreover, the existing water-splitting processes demand more energy compared to its integration. Currently, the water-splitting system works at 80% productivity but only manages to produce about

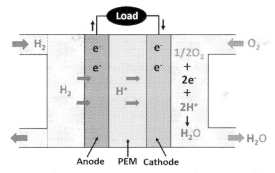

Figure 7.1 Schematic diagram of PEMFC (*PEMFC*, Proton-exchange membrane fuel cell).

60% of the electricity. However, it is noteworthy that this technique is about 34% more proficient than conventional fossil fuel-based systems implying that it is still an appealing alternative [10].

PEMFCs provide many benefits, including low working temperature, sustainable process, superior current density, heat and water management, compactness, quick start, and noncontinuous operability. Regardless of these great advantages, PEMFCs have some constraints that restrict their performance. The main hurdle lies in the high activation of electrodes, making the anodic and cathodic reaction sluggish. As a result, the employment of precious and expensive metals becomes inevitable. Currently, platinum and platinum-based materials have been employed to expedite the reaction. But, the scarcity and enormous price of the noble metal electrocatalysts obstruct their commercialization. The PEM holds a critical significance in PEMFCs. The PEM is the membrane that passes the proton through it and prevents electron transfers. For commercializing the fuel cells, the PEM must be inexpensive, anticorrosive, and have high proton conductivity.

Carbon monoxide (CO) poisoning is also an important issue to be addressed. The production of H_2 from fossil fuel leaves traces of CO, which can deteriorate platinum-based electrocatalyst during the operation. Therefore the fabrication of nonprecious metal electrocatalysts has become a center of interest. Additionally, the production of fuels such as H_2 and O_2 from fossil fuels affects the fuel cell's carbon-free approach, which originated the fuel cell's necessity in the first place. Consequently obtaining H_2 from fossil fuel would break the H_2-H_2O cycle, which affects the sustainability and zero-carbon strategy. So the best way to get H_2 is by using renewable resources, which can ensure the sustainability and cleanliness of the PEMFCs. It seems obvious that the fabrication of low-cost, productive, and safe fuel cell systems arise critical material challenges. Currently, the focal point of the research is to fabricate appropriate electrocatalysts that can effectively replace expansive noble metal-based electrocatalysts having improved catalytic performance. The replacement of costly electrocatalysts with inexpensive and better-performing electrocatalysts would greatly assist in commercializing the PEMFCs. For this reason, Different types of materials have been studied over the last few decades. In recent times, metal-organic frameworks (MOFs) have become top researched material and observed tremendous publications in top-notch journals.

MOFs are a kind of crystalline material having high porosity and extremely organized pore morphology that has garnered immense interest. MOF was first introduced by Yaghi and Li during the 1990s [11,12]. Since then, more than 20,000 varieties of MOFs have been prepared for different technical operations. However, a minimal number of MOFs have been carefully characterized by porosity, and only a few of them have been investigated for fuel cell operations. As indicated by the term, MOFs can be prepared by joining metal-based inorganic units, called secondary building units (SBUs), with polydentate, which form 1D-, 2D-, and 3D-networked structures. The unique design of MOF contributes to its superior surface area, which is far greater than other counterparts like active carbons and zeolite [13]. Owing to their special morphology, MOFs hold exceptional characteristics, including well-specified pore size distributions, highly available pore volumes, and textural adjustability. Furthermore, the synergistic effects of central metal units and organic linkers could produce various unique characteristics, making them a prime candidate for different applications, such as catalysis, gas storage and separation, sensors, luminescence, ion transmission, drug delivery, magnetism, electrochemistry, and fuel cells. Post calcination of MOF is another way to obtain highly functional large surface area, as well as doped/undoped carbon nanomaterials and composites. Consequently various synthetic techniques can be employed to obtain MOFs on different substrates with great conductivity, such as carbon nanotubes (CNTs) and graphene. In this chapter, we will discuss the morphology and structure of various MOFs as a promising candidate used for the application of PEM fuel cells. The designing, structural characteristics, and electrocatalytic performances of MOF-derived material will be reviewed in detail.

7.2 Structure of metal-organic frameworks

MOFs are the kind of porous materials that were first developed about twenty years ago. Since their introduction, scientists have found new hopes in them to solve the future's energy problems. They are among the fastest-growing research fields in chemistry and material science for more than 10 years with tremendous applications. MOFs are also known as porous coordination polymers. They are crystalline materials composed of inorganic/metal nodes, also called SBUs, and one or more organic linkers attached by coordination bonding [14]. Several MOFs morphologies are displayed in Fig. 7.2. MOFs have a crystalline morphology and possess a long-range arrangement. The uniqueness of MOFs lies in their remarkably ordered pores that substantially

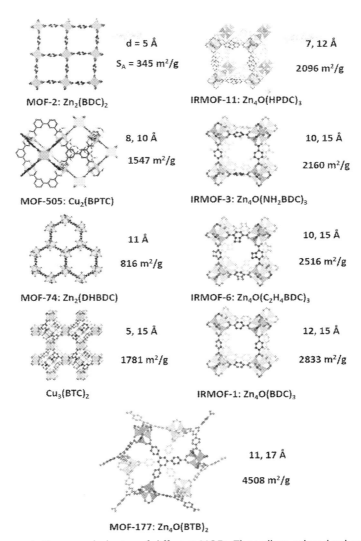

Figure 7.2 The crystalline morphologies of different MOFs. The yellow-colored sphere indicates the void space in MOFs. Also, the structural formula, specific pore area, and pore size are specified (*MOFs*, Metal-organic frameworks). *Adapted with permission from A.R. Millward, O.M. Yaghi [15], Metal-organic frameworks with exceptionally high capacity for storage of carbon dioxide at room temperature, Journal of the American Chemical Society 127 (2005) 17998−17999. Copyright (2005) American Chemical Society.*

present across the structure. These pores or channels contain guest molecules (from the solvent added at the time of synthesis) or opposite ions that equalize the charge of metal nodes and provide charge equilibrium to the structure. MOFs can be tailor-made to achieve desirable properties by precisely choosing metal nodes and organic linkers. This capability makes MOFs superior and highly efficient porous materials compared to other carbon and zeolite-based materials [16]. This significant characteristic enables MOFs to be easily modified and optimized their pores for a particular application, which does not reflect in the case of other porous materials, such as zeolites. In zeolite-based materials, pores are restrained by the hard tetrahedral oxide structure that resists any modification. Their chemical framework directly impacts the properties of MOFs. Since MOFs are formed through the powerful coordination bonded structure of organic linkers and central metal atoms, they can withstand high temperatures around 500°C. Although it is hard to produce chemical resistant MOFs because they undergo the ligand exchange process in most of the chemicals, but various excellent chemical stable MOFs have also been synthesized. So, some of the significant elements of MOFs are necessary to be discussed.

7.2.1 Secondary building units

There are different types of MOFs, whose topologies are directed by the SBUs. The shape as well as chemical properties of SBUs along with organic linkers significantly contribute to formation and design of MOFs. It is noted that polydentate ligands can accumulate and hold the metal ions at a specific spot under some precise conditions, which resulted in SBUs (Fig. 7.3). Afterward, these SBUs are connected by organic linkers and produce an extensive network with exceptional morphological stability. For the continuous growth and polymerization of SBU, three factors are considered crucial: (1) the central structures of SBUs are directed by metal nodes, which also work as a point of propagation for SBUs; (2) the way the organic linkers coordinating each other gives essential information for the anticipated topology; and (3) solvent molecules act as the fillers that fill out space within the structure. When these solvent

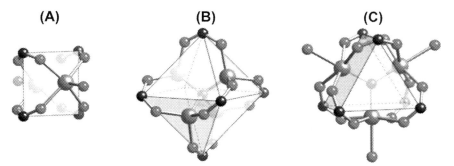

Figure 7.3 SBUs are generally present in metal carboxylates in the shape of (A) square, (B) octahedral, and (C) trigonal (color indications: blue = metal ions, black = C, red = O) (*SBUs*, Secondary building units). *Adapted with permission from J.L.C. Rowsell, O.M. Yaghi, Metal-organic frameworks: a new class of porous materials, Microporous Mesoporous Mater 73 (2004) 3–14. Copyright (2004) Elsevier.*

molecules withdraw from the structure, it leaves the highly ordered metal nodes. Consequently the shape of SBUs also relies on the ratio of metal and ligand, solvent, and origin of anions to equalize the metal ions charge [18].

7.2.2 Open metal sites

Metal sites are crucial in determining the properties of MOFs. Mostly, the MOFs are synthesized by using the transition metals, such as Ni, Cu, Co, Fe, Mn, and Cr. These metals usually behave as a Lewis acid inside the MOFs and act as hosts for solvent molecules and opposite ions throughout the preparation process. After synthesizing, MOFs are activated by the removal of loosely bonded atoms or molecules from the metal sites and leaving behind the open metal sites arranged in an orderly manner onto the pores' surface. These Lewis acid characteristics of metal spots trigger the catalytic activity of the interconnected organic part of the MOFs. Generally, Ni-, Fe-, or Co-containing MOFs have displayed excellent ability to be used as noble metal-free electrocatalysts. Various studies suggest that open metal spot-containing MOFs showed excellent heterogeneous catalytic activity [19]. The density functional theory estimations have suggested that various metals may possess varying electrochemical activity because of their specific electronic conformation and central D-band. Furthermore, MOFs containing more than one metal can be produced for improved controllability over the structural arrangement. Ten different metals have been successfully incorporated into the same MOF simultaneously without any change in its topology [20]. Nanostructures MOFs with increased complexities can also be produced for a higher activity or unique characteristics, such as regioselectivity. Moreover, the electron insufficiency of the metal spots provides excellent oxidation activities toward alcohol and thioethers. While the adsorption characteristics of the MOFs also improved significantly due to partially positive-charged metal spots. Due to the enhanced adsorption characteristics, MOFs are considered a promising candidate for hydrogen production.

7.2.3 Pores

Pores are the free spaces created inside the MOFs when the molecules from the solvent and the counterions are removed. Pores are categorized based on their opening diameter. In MOFs, mesopores and micropores are common and having different functions. Mesopores, having a diameter of 20–500 Å, provide enhanced sites for the heterogeneous catalytic reaction. Whereas micropores have diameters less than 20 Å and have the advantage to add more specific surface area to the material. Micropores trap gases and vapors and help in enhanced interaction between them. Due to the large pore size of mesopores, the structure is more likely to collapse during further activation and the long run, while the micropores remain stable throughout the process and result in durable porosity. Furthermore, MOFs can overlap and infiltrate each other, providing increased packing capacity and reducing the pore volume. Therefore, by careful design, MOFs can be used as an efficient hydrogen storage material.

7.2.4 Functional groups

The structure of MOFs can be modified by carefully selecting the organic linkers and metal-containing SBUs. But, the selection of these components significantly impacts the functionality of the material. MOFs can possess a significant number of functional groups with different diverse functionalities in a mixed culture. These arrangements make the MOFs special among different materials because the functional dispersity provides unique properties that cannot be found in separated functional group materials. However, MOFs can undergo significant and unwanted modifications by small changes. For example, slightly altering copolymers' side chains can affect the system's entropy along with some unwanted changes in the morphology [21]. But, careful alteration in functional groups can enable MOFs to be employed in various applications.

The functional groups in MOFs can also be modified by simply functionalizing the organic linkers before the synthesis of MOFs. By this method, various functional groups can be added to the MOFs structure, and properties of the resulting MOFs can be tuned accordingly. Various MOFs with a large number of specific functional groups, like methyl, halogens, amine, and different substituents covered on the pore surface, have been studied [22–26]. Although the method of modifying organic linker before MOF preparation becomes effective in producing functionalized MOFs, the extent of functionalities inside the pores remains low. The main problem is the incompatibility of the different functionalities with designed structures and their stability during the solvothermal environment. One important consideration that needs attention is that the MOFs' functional groups should not affect the MOFs' framework, which usually takes place during the interaction of functional groups and the metal ions in a certain manner that results in altered MOFs topology.

Other methodologies for the synthesis of functional group-bearing MOFs involves the modification after the synthesis of MOFs. These methodologies include functionalizing the synthesized MOFs structure by the action of chemicals. This method has several advantages over presynthesized functionalizing methods. One is the MOFs' structural stability since functionalization can be done with no harm to the MOFs framework. Second, the functional groups that can coordinate with the metal ions and hold them within the structure can be easily added to the MOFs after synthesis, providing extra strength to the framework. Furthermore, the postfunctionalization method provides a facility to produce different species and quantity of functional groups within MOFs. For example, MOFs having the same functionalities but different frameworks and MOFs with the same framework, but different functionalities can also be prepared [27,28]. Thereby, structural to the functional correlation of MOFs can be carefully investigated.

Postfunctionalization of MOFs is carried out in three ways, such as covalently, coordinatively, and by both coordinate and covalent. In the covalent method, the organic linkers get altered, while in the coordinative method, only the coordinating condition of SBUs inside the MOF's framework gets altered with no alteration in topology [29,30]. Whereas in the combined covalent-coordinative method, more ligands and metal ions can be attached in the MOFs [31]. For instance, by transforming loose anime groups to Schiff base ligands, vanadyl-acetylacetonate was inserted into $Zn_4O(bdc-NH_2)_3$, bdc-NH$_2$ = 2-amino 1,4, benzene dicarboxylate, and the synthesized compound was employed as a catalyst in a chemical reaction [32].

7.2.5 Development of porous structure

MOFs are the porous materials developed by joining organic linkers with SBUs in a net-shaped structure. Unlike other materials, MOFs can be easily identified because of their unique characteristics of powerful bonding, robust structure, containing connecting units, and possession of precisely ordered crystalline morphology [17]. Thus MOFs are prepared by the traditional methods in which metal nodes and organic linkers are self-organized to produce highly porous and structurally crystalline material. Mostly, the MOFs' preparation procedures involve solvothermal treatments, which provide ease and promote the creation of metal-organic building units. The network structure of MOFs can be determined by the properties they possess. However, these net-shaped material's chemistry appears very simple but contains the toughest and complicated segment of improving the experimental conditions for its production. A considerable change can occur in MOFs' development just by doing little alteration in the experimental conditions. Furthermore, a simple metal coordinated structure can grow into multiple varieties of structures, which cause complications in acknowledging the morphology as a whole [33]. Additionally, the metal ions are so unstable that they cannot prescribe any specific structure, resulting in the unpredictability of the structure. Similarly, the flexible organic linkers may lead to several potential shapes, which restricts the prediction of the structure. The stated problem can be solved by utilizing organic linkers having inflexible and rigid units in their structure, which eventually lessen the loose positioning of ligand's lone pairs. Exploiting the repeated coordinative design can also assist in determining the network structure, which can be done by available polymerized structure or by using individual model complexes. Additional factors can also affect the morphological foreseeability of MOFs. These factors include the ring-opening polymerizability and charge equalizing anions. Various new methodologies have been introduced that assist in MOF's design, such as the PLUXter technique [34]. In this technique, the hierarchized cluster study and primary constituent study are applied to produce classification identification, which allows additional morphology-characteristics investigations of typical morphologies present in the cluster.

The MOFs are porous materials that possess net-like morphologies, which is uncommon in other conventional zeolitic and carbonaceous pores-bearing materials. They hold a special feature of producing remarkable passages and cavities with vast diameter. But, developing mesopores in MOFs from the extensive linkers is also a challenge, and significant caution is necessary due to the possibility of the interpenetration of MOF structures, which leads to a decrease in active pore size. To overcome this problem, the MOFs can be designed by producing the void in a specific manner. For this reason, the boron network in CaB_6, graphene network, and the network in diamonds are commonly employed in producing MOFs with high porosity [35]. This strategy is known as the node and spacer technique and turned out as an effective technique in producing a wide range of different structures. It also helped in categorizing MOFs having similar connectivity but stable pore dimensionality. Also, bulkier counterions inside the MOFs' cavity play an important role in controlling interpenetration [36].

The MOFs framework and guest molecules interaction are crucial for the voids within the MOFs. If they interact strongly and extensively, then the voids or cavities may be occupied by the guest molecules instead of extra polymer strings. Yaghi et al. produced the MOF by using Ni metal [37]. The x-ray imaging of the MOF's crystals showed squared shaped pores that were sufficiently big and can undergo interpenetration. But, the cavities were occupied by the guest molecules connected by hydrogen bonding. These guest molecules include H_2O, perchlorate anions, and uncoordinated ligands that were free from structural interpenetration. The interpenetration can also be overcome by employing bulkier organic linkers together with the mild reaction environment. The other difficulty experienced in developing MOFs is the removal of solvent molecules, which are diffused within the pores and templates. Removing these molecules from their diffused state resulted in instability of the MOFs structure. During the guest removal. The framework can be preserved by exploiting the chelate effect.

Many MOFs have been investigated that maintain their framework during the removal of the guest solvent from the structure. In comparison, some of the solvent-free materials have been witnessed to carry out reabsorption of tiny molecules with a great selective approach. In particular, Kitaura and colleagues prepared a Cu-containing MOF using 1,2 dipyridylglycol as a ligand [38]. The vacant MOF displayed good structural maintenance with minor variations in the interlayer gap. X-ray diffraction characterizations of the synthesized MOF revealed about the importance of structural rearrangement in regulating the ability to absorb and desorb. The above mentioned intrinsic characteristics of MOFs, such as porosity, crystallinity, functionality, and high specific surface area, enable them to be utilized in various applications such as sensors [39], gas adsorption and storage [40], catalysis [41], and luminescence [42]. The MOFs' structural predictability provides a systematized technique for structural modification by logically changing the particular element to get the required result. For example, the pore size can be controlled by choosing short and long linkers. If short linkers are employed, pore size will be narrow, and if long linkers are used, pore size will be broad. Similarly, hydrophilicity can be managed in MOF by employing polar and nonpolar linkers. These unique characteristics of MOFs enable MOF to be used in fuel cell applications because of its ability to be tailor-made as per the requirement of the fuel cell. In this chapter, we will discuss the application of MOF in fuel cells.

7.2.6 Design of metal-organic framework derivative

MOFs have been extensively employed as a precursor for the controlled development of additional complex nanomorphologies. Besides the exploitation of MOF's intrinsic characteristics, the free surface can be packed with metals having exceptional activity, such as Pt, Pd, and Rh. The additional complex nanomorphologies can also be achieved from MOFs through the pyrolysis process under N_2, O_2, or N_2/H_2 environment. MOFs are also utilized as self-sacrificing templates, which produce a variety of MOF derivatives, such as metal oxides [43], nanostructured carbon [44], metal composites [45], and different metal-based materials, including metal nitride and metal carbide as illustrated in Fig. 7.4. Furthermore, along with acquiring MOF nanostructures, it can also be doped with heteroatoms, such as N, P, and B to further increase catalytic activity. These MOFs derivative have shown better catalytic activity than the original MOFs and hence find significantly increased applications as a catalyst. Furthermore, these MOF derivatives possess a critical characteristic of withstanding harsh environments, such as highly acidic and highly basic, which are normal electrochemical applications.

7.3 Metal-organic frameworks for fuel cells applications

Fuels cells are electrochemical devices and are holds great significance among energy conversion techniques. It produces electricity by the conversion of fuels, like H_2, natural gas, and methanol. These devices can be utilized to provide power to automobiles, any immobile premises, or any portable devices. However, the existing fuel cells are yet to meet the requirement of the United States Department of Energy (United States DOE) put forward in 2016 to produce fuel cells. According to DOE, the exciting fuel cells do not meet the planned requirement in pricing, performance, and stability. To overcome these problems, the efficiency of the fuel cell's electrode and membrane need to be improved. For this reason, MOFs and MOFs-based materials are considered promising materials for electrode and or polymer electrolyte membranes.

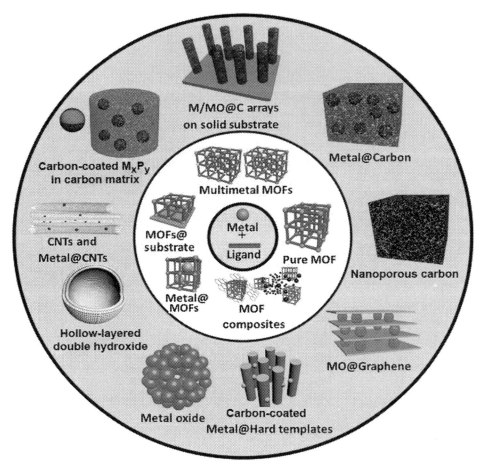

Figure 7.4 Illustrative analysis of different MOFs and MOFs-based materials (*MOFs*, Metal-organic frameworks) [46]. *Adapted with permission from A. Mahmood, W. Guo, H. Tabassum, R. Zou, Metal-organic framework-based nanomaterials for electrocatalysis, Advanced Energy Materials 6 (2016) 1600423. Copyright (2016) John Wiley and Sons.*

7.3.1 Proton-conducting metal-organic frameworks

The PEMFCs have garnered considerable importance because of their wide range of practical applications. The well-known PEM is known as Nafion, which is a sulfurized fluoropolymer. Nafion has exceptional protons conductivity and is also thermally and chemically stable. But, fuel cells require a working temperature of less than 80°C and a humid atmosphere, which limits the efficiency and usability of Nafion PEM [47]. Therefore developing a new membrane with high proton conductivity became crucial. MOFs are considered a good alternative for the Nafion membrane because of crystallinity, high proton-conductive nature, and low fabrication price. Two categories for proton transfer in MOFs are the low temperature (less than 100°C and in the moist state) and elevated temperature (above 100°C and dry state).

7.3.1.1 Proton transfer under low temperature

The first low-temperature proton-transferring MOFs was introduced in 2009 by Masaaki et al. [48]. The chemical formulation of proton-transferring MOFs was [(NH$_4$)$_2$(ada){Zn$_2$(oxa)$_3$0.3H$_2$O]$_n$ (oxa = oxalic acid and ada = adipic acid). The synthesized MOFs have an efficiency similar to that of Nafion. The H$_2$O molecules, ammonium ions, and -COOH groups present in the adipic acid worked as a conduction media, as shown in Fig. 7.5A. The synthesized MOFs achieved an elevated proton conductivity of 8×10^{-1} S cm^{-1} at 25°C under a relative humidity of 98%. The increased activation energy (E$_{ac}$) of MOF (0.63 eV) as compared to Nafion (0.22 eV) suggested that the proton transfer was initiated from the vehicle and Grotthus techniques. The employment of 2D oxalate-containing porous structures as the host allowed the insertion of various proton carriers within the MOF's porous structures, which assisted in investigating efficient proton conductors. Various oxalate-containing MOFs have been introduced, but [{(Me$_2$NH$_2$)$_3$(SO$_4$)}$_2${Zn$_2$(ox)$_3$}]$_n$ is an exceptional specimen that demonstrated excellent conductivity under both conditions, such as low and high temperatures [50]. Additionally, the insertion of various components to manage the pores surrounding offers a good technique to formulate and produce proton-transferring MOF (Fig. 7.5B) [49]. By this technique, the type, concentration, and movement of proton carriers are controlled proficiently. The proton conductivity depends upon different factors, such as

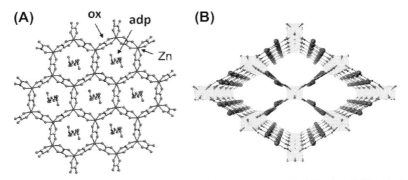

Figure 7.5 MOF-containing proton conductors (A) structure of {(NH$_4$)$_2$(adp)[Zn$_2$(ox)$_3$]0.3H$_2$O}$_n$ excluding ammonium ions, electrolytes, or H$_2$ and (B) schematic of MIL-53 (*MOF*, Metal-organic framework). *Adapted with permission from M. Sadakiyo, T. Yamada, H. Kitagawa, Rational designs for highly proton-conductive metal-organic frameworks, Journal of the American Chemical Society 131 (2009) 9906–9907 [48]. Copyright (2009) American Chemical Society; adapted with permission from A. Shigematsu, T. Yamada, H. Kitagawa, Wide control of proton conductivity in porous coordination polymers, Journal of the American Chemical Society 133 (2011) 2034–2036 [49]. Copyright (2011) American Chemical Society.*

coordinating water molecules, guest molecules, and ligands' functionality. Recently, a technique for the preparation of proton-transferring MOF has been developed, which utilizes organic phosphonic acid and sulfonic acid as a ligand [47,51]. The large quantities of O_2 atoms in phosphonates and sulfonates are connected to H_2O molecule in the electrolyte through H-H bonding, which offered a simple proton-transferring route. With the motivations from these advancements, various MOFs with proficient proton-transferring routes have been studied.

The proton conductivity can also be influenced by the defects and the extremely ordered film of MOFs. The impact of defects on the proton conductivity can be demonstrated using a zirconium-sulphoterephthalate (ZST) MOFs. The defects in ZST MOFs can be produced in two ways; by utilizing a surplus amount of acetic/sulphoacetic acid during the fabrication or dipping the as-synthesized sample in 0.1M sulfuric acid [52]. The defective samples showed enhanced proton conduction as compared to as-fabricated ZST MOF. The conduction for defective samples ranged from 2.4×10^{-3} to 5.6×10^{-3} S cm^{-1} under a relative humidity of 95% and at 65°C. The effect of defects is an important parameter for the development and application of proton-transferring MOFs. The extremely oriented MOF nanofilms are comprised of $[Cu(TCPP)]_n$, where TCPP = tetrakis(4-carboxyphenyl)porphyrin. The film was fabricated via deposition of material on SiO_2 (300 nm thick)@Si disk by using Cr and Au electrodes, as shown in Fig. 7.6 [53]. Superior proton conductivity (3.9×10^{-3} S cm^{-1}) was witnessed under the relative humidity of 98%. The increased proton conduction may be associated with a large number of pendant groups, which includes the H_2O and carboxylic groups oriented on the nanosheet, and the immensely ordered crystalline structure. However, in pristine and film structure, MOFs displayed high protonic conduction at reduced temperatures. But, little studies have been done on employing MOF-based proton conductors in the fuel cells.

7.3.1.2 Proton transfer at high temperature

The water-facilitated MOF-based proton conductors that work at low temperatures have limited application for the fuel cells operating at high temperatures ($>100°C$). Therefore various researches have been done to produce water-free MOF-based proton conductors. Such MOFs contain water-free guest molecules that assist in providing proton delocalization passages for efficient proton conduction above the water evaporation temperature. For this purpose, different guest compounds, including imidazole, histamine, ionic liquids, and triazole, have been efficiently inserted in MOFs to study their proton conduction behavior. Compared to water-facilitated MOF-based proton conductors, the relative humidity does not affect the proton conductivity of the water-free MOF-based proton conductors.

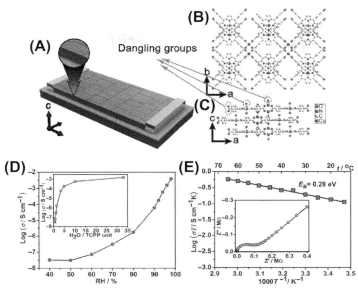

Figure 7.6 MOF-derived nanofilms for the proton conduction: (A) diagrammatic representation of nanosheets, (B and C) structural depiction, (D) proton conductivity characterization of the synthesized nanofilm in different relative humidity (RH), and (E) least-squares fitting is shown as a solid line. Inset: typical Nyquist plot of the MOF nanofilm studied under 98% RH at ambient temperature (MOF, Metal-organic framework). *Adapted with permission from G. Xu, K. Otsubo, T. Yamada, S. Sakaida, H. Kitagawa, Superprotonic conductivity in a highly oriented crystalline metal-organic framework nanofilm, Journal of the American Chemical Society 135 (2013) 7438–7441. Copyright (2013) American Chemical Society.*

Bureekaew et al. studied two imidazole-encapsulated Al-containing MOFs Imi@Al(μ2-OH)(ndc) (ndc = 1,4-naphthalenedicarboxylate) and Imi@Al(μ2-OH)(bdc) (bdc = 1,4-benzene-dicarboxylate) [54]. Imidazole is a colorless nonvolatile solid having an elevated boiling point. It contains nitrogen atoms through which it gets its proton conductivity. Each imi@Al-containing MOF displayed the proton conductivity ranging from 10^{-8} to 10^{-10} S cm^{-1} at ambient temperature, and at 120°C, the conduction improved extensively to 2.2×10^{-5} for Al(μ2-OH)(ndc) and 1.0×10^{-7} S cm^{-1} for Al(μ2-OH)(bdc). There is a significant proton conductivity variation between both imidazole-containing MOFs. This variation due to the varying movement of imidazole molecules within the channels. The hydrophobic nature of Al(μ2-OH)(ndc) channels provides poor engagement with the polar imidazoles, which allows the unrestricted movement of imidazole molecules within the channels and transfer protons. In contrast, the Al(μ2-OH)(bdc) channels have amphipathic characteristics, which results in firm engagement of polar imidazole molecules at the channels' hydrophilic parts and hinders its free movement within the structures, which reduce its proton conductivity.

Another study employed histamine instead of imidazole for the proton conduction in Al(μ2-OH)(ndc) [55]. The histamine contains an imidazole ring and an amine group. The imidazole ring offers two proton-hopping spots to histamine, and amine offers one proton-hopping spot. When packed in a MOF, there is an equivalent of one histamine molecule for each Al^{+3} ions. At 150°C and water-free conditions, the histamine@Al (μ2-OH)(ndc) showed exceptional proton conductivity of 1.7×10^{-3} S cm^{-1} that is considerably high compared to pure histamine. Moreover, the conductivity decreased to 2.1×10^{-4} S cm^{-1} by reducing the histamine quantity to half at the same conditions. The histamine-based MOF showed better proton conductance compared to the imidazole-based MOF mentioned above. This variation is due to the increased histamine concentration, which is double of imidazole concentration in Al-containing MOFs. Thus the amount or concentration of proton-carrying materials directly affected the proton conductivity of a material. The histamine-containing MOFs is considered a promising material for developing hybrid conductors. Munehiro et al. investigated the effect of histamine in the proton conductivity of Zn-MOF-74. The displayed inferior proton conductivity of 4.3×10^{-9} S cm^{-1} at a temperature of 146°C. The low conductance was ascribed to the poor histamine movement within the MOF framework, which restricted the motion of protons between the imidazole rings [55].

Jeff et al. investigated three-dimensional proton-conducting MOF (β-PCMOF2) [56]. The sulfonate-containing ligand (2,4,6-trihydroxy-1,3,5-benzene trisulfonate) was employed to synthesize MOF, whose channels contain triazole, which significantly improved the proton-conducting behavior and also increased the working temperature. The triazole is also a nonvolatile compound and possesses a high boiling temperature. The proton conductivity of the synthesized MOF material was ranging from 2×10^{-4} to 5×10^{-4} S cm^{-1} at 150°C under a dehydrated H_2 environment, which was more than that of both individuals, such as pristine (PCMOF2) and triazole. The conductivity can be modified by changing the triazole quantity within the channels. Partially loaded triazole MOF {β-PCMOF2(Tz)$_{0.45}$} was employed as a proton-conducting membrane in an H_2-air fuel cell, having a high operating temperature. The cell produced an open-circuit voltage (V_{oc}) of 1.18 V with the stability of 72 h at 100°C and displayed gas-proof characteristics during operation. These results indicated the compatibility of MOFs to be employed as a membrane in fuel cells.

Another technique to produce water-free proton conductors is to develop a MOF with inherent proton-conducting characteristics. This technique was introduced by Kitagawa and his colleagues and prepare a two-dimensional MOF {Zn(H$_2$PO$_4$)$_2$(TzH)$_2$}$_n$ [57]. The synthesized MOF has multilayer morphologies that are developed by octagonally interconnected Zn^{+2} ions linked to orthophosphates and triazole molecules. The connection between the layers through the hydrogen bonding between triazoles and orthophosphates delivers a proton passage. The material offered a proton conductivity of 1.2×10^{-4} S cm^{-1} when studied through pellets without any

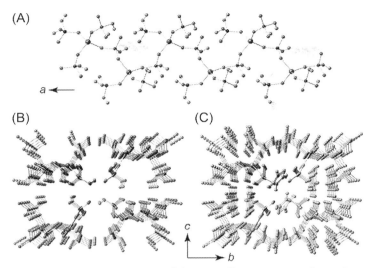

Figure 7.7 (A) Diagrammatic representation of the crystalline structure of one-dimensional coordination chains of [Zn(HPO$_4$)(H$_2$PO$_4$)$_2$]$_2$ in [Zn(HPO$_4$)(H$_2$PO$_4$)$_2$](ImH$_2$)$_2$, (B) schematic of four one-dimensional, and (C) [Zn(HPO$_4$)(H$_2$PO$_4$)$_2$](ImH$_2$)$_2$. (Blue = ImH$_2$$^+$ ions, Gray = Networks.). *Adapted with permission S. Horike, D. Umeyama, M. Inukai, T. Itakura, S. Kitagawa, Coordination-network-based ionic plastic crystal for anhydrous proton conductivity, Journal of the American Chemical Society 134 (2012) 7612–7615. Copyright (2012) American Chemical Society.*

proton-transmitting mediator. Moreover, the proton conductivity characterizations through the signal crystal indicated the anisotropic nature of proton conductance. Another Zn-phosphate-based MOF [{Zn(HPO$_4$)(H$_2$PO$_4$)$_2$}(ImH$_2$)$_2$] has also been produced by Kitagawa and colleagues [58]. The synthesized MOFs are composed of Zn^{+2} ions and two orthophosphates, forming single-dimensional chains. In the middle of two chains, there are two imidazole molecules, providing an ionic crystalline system, as shown in Fig. 7.7. Substantial H-H bonding takes place between the imidazole cations and orthophosphates. The proton conductance in the synthesized MOF significantly relies on the temperature, which improves as the temperature increases attaining the conductivity of 2.6×10^{-4} S cm^{-1} at 130°C and attributed to the local movement of the imidazole cations.

7.4 Metal-organic frameworks as oxygen reduction reaction catalyst

Oxygen reduction reaction (ORR) is a vital process for PEMFCs and takes place at the cathodic side of the fuel cells. During the reaction, the O$_2$ molecules are reduced to produce H$_2$O, as displayed as follows:

$$O_2 + 4H^+ + 4e^- \rightarrow 2H_2O \tag{7.1}$$

Eq. (7.1) demonstrates the complete ORR that takes place on the surface of electrocatalysts. The process is not a simple one-step reaction and involves various intermediate stages. The reaction requires the adsorption of oxygen molecules on the electrocatalyst, which is quite a simple step that produces adsorbed O_2 ($O_{2(ads)}$). However, the reduction of the adsorbed O_2 is a slow reaction and can occur through associated, dissociated, or peroxy-routes. Specifically, the ORR can take place in either of two different ways: the two-electron transfer and four-electron transfer. The proficiency of the fuel cell relies on the way the reaction will take place. The ORR reaction, involving the two-electron transfer leads to the production of stable H_2O_2 molecules, which eventually diminishes the catalyst's activity. Therefore four-electron transferring ORR reaction are favorable for the higher performance of fuel cells.

To further increase the fuel cell performance, the electrocatalysts must have ease in exposing their active spots to the oxygen molecules with high accessibility to the electrocatalyst surface molecules. The MOFs provide superior specific surface area along with the central active metal spot. The superior specific area aims to provide greater accessibility of oxygen molecules while the central metal serves as the ORR's active spot. However, the MOF's applicability as electrocatalysts in the fuel cells is limited because of their inferior stability at the operating voltage. Basolite was the first MOF employed to reveal its potential catalytic activity under alkaline electrolytes [59]. Later, Mao et al. displayed the effective employment of Cu-containing MOFs (Cu(II)@benzene-1,3,5-tricarboxylate) as ORR electrocatalyst [60]. The Cu-containing MOFs displayed superior activity displaying the onset potential (E_{on}) of -0.1 V against the reference electrode of Ag/AgCl. The MOF carried out oxygen reduction through the four-electron transfer route. The different redox states of Cu, such as Cu^{+1} and Cu^{+2}, reflect on the cyclic voltammetry analysis with the highest current witnessed at -0.15 V and also promote oxygen adsorption with no effects on coordination chemistry. Although the Cu-containing MOFs showed excellent catalytic performance, the insufficient electron conductivity of pristine MOFs significantly lowers their catalytic activity. Various techniques have been investigated to improve the MOF's conductance ability and enhance electrocatalyst/electrolyte interaction. These techniques include: (1) employing substantially conductive substrate to produce MOF composites, (2) packing ionized species or nanoparticles within the MOF structure, and (3) utilizing intermediaries on glass electrode together with MOFs at the time of analyzing electrochemical behavior.

The employment of intercalating reduced graphene oxide (rGO) layers within the Fe-containing MOFs has been reported by Maryam et al. [61]. The rGO worked as an exceptional electron conductor, as illustrated in Fig. 7.8. The functionalized moiety of rGO and pyridine was found to offer the foundation for MOF development on the rGO. Furthermore, during characterization, it is found that the amount of rGO has an ample impact on the catalytic performance of the synthesized MOF, as displayed in

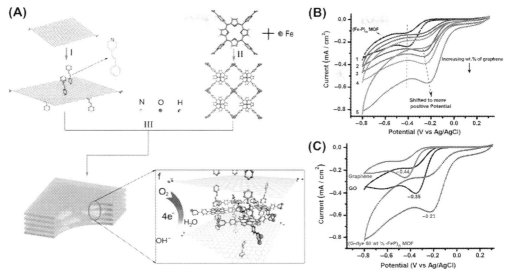

Figure 7.8 (A) Diagrammatic representation of r-GO, G-dye, and (G-dye-FeP)$_n$ structures, (B) Cyclic Voltammetry (CV) curves of ORR for (1) (Fe-P)n MOF, (2) (G-dye 5 wt.%-FeP)n MOF, (3) (G-dye 10 wt.%-FeP)n MOF, (4) (G-dye 25 wt.%-FeP)n MOF, and (5) (G-dye 50 wt.%-FeP)n and (C) CV curves of ORR MOF electrodes (*MOF*, Metal-organic framework; *ORR*, oxygen reduction reaction, *r-GO*, reduced graphene oxide). *Adapted with permission from M. Jahan, Q. Bao, K.P. Loh, Electrocatalytically active graphene-porphyrin MOF composite for oxygen reduction reaction, Journal of the American Chemical Society 134 (2012) 6707–6713. Copyright (2012) American Chemical Society.*

Fig. 7.8B and C. The synthesized electrocatalyst followed the four-electron transfer route and displayed superior E$_{on}$ of −0.23 V versus Ag/AgCl and exceptional durability time in alkaline electrolyte. In similar efforts, the rGO remains fixed as a binder on the electrocatalyst surface, resisting the withdrawal of MOFs material from the electrode foundation and enhancing the total electron conductance.

Lately, nanoporous carbon materials (NPCs) have gained popularity as a nonmetallic electrocatalyst because of their high catalytic performance, abundance, and excellent stability. Due to the superior surface area, intrinsic spongy and highly organized structures, MOFs have been widely applied as a precursor to obtaining NPCs. Generally, Zn has been utilized predominantly for the preparation of metal-free NPCs. The carbon produced by the transition of organic ligand acts as a reducing agent of Zn, which evaporated at high temperatures, such as 900°C. In the organic ligand, the occurrence of different heteroatoms, such as nitrogen, sulfur, boron, and phosphorus, offers an excellent advantage of direct incorporation of heteroatom to the NPCs. Zn-based MOF (ZIF-8) was employed by Zhang et al. to produce extremely graphitized nitrogen-doped polygonal NPC material with a surface area (A$_s$) of 932 m^2 g^{-1} [62]. The material maintained the symmetrical polygonal structures of 80 nm in size with higher pyridinic N abundance. These characteristics enabled the material to display high catalytic activity through a four-electron transfer route, which was similar to Pt-based electrocatalysts. Furthermore, Peng et al. employed secondary carbon to enhance porosity, heteroatom doping, and electrical conductivity [63]. For this reason, ZIF-7 (Zn-benzimidazole) was utilized with glucose as a secondary source of carbon. The fabricated material offered high catalytic performance with an E_{on} of 0.86 and $E_{1/2}$ of 0.70 V versus reversible hydrogen electrode (RHE).

The catalytic performance of MOF-based carbonaceous material can be further increased by applying a layer of high conductivity heteroatom-doped carbon. The exterior side of the resulting material exhibits enhanced catalytic performance, while the interior part of the material offers excellent electron conductivity. For example, the NPC layered with nitrogen-doped carbon was prepared by Pandiaraj et al. [64]. The fabrication procedure involves two stages. Firstly, the MOF-5 is calcined at 1000°C to acquire porosity in the carbon material. Secondly, the coating of the nitrogen-based melamine solution on the synthesized NPC was done, followed by the calcination process, which eventually transformed the top-layered melamine into the g-C_3N_4, as illustrated in Fig. 7.9A. The fabricated electrocatalyst displayed superior ORR performance with an E_{on} of 0.035 V versus Hg/HgO in alkaline electrolyte. Because of higher efficiency and excellent conductivity, MOF-based nitrogen-doped CNTs can be obtained.

Zhang et al. put forward a general technique for preparing MOF-5-derived N-CNTs through three-stage pyrolysis at various temperatures under an N_2 environment [65]. The first-stage pyrolysis was taken place at a low temperature; the second-stage pyrolysis involves Ni incorporation into the structure as a catalyst to produce CNTs; and the third-stage pyrolysis involves nitrogen-doping, as displayed in Fig. 7.9B. The disintegration of urea to produce some intermediate compounds, such as $C_2N_2^+$, $C_3N_2^+$, and $C_3N_3^+$, promotes the shrinking of the carbonaceous structure that eventually resulted in the development of CNTs. The nitrogen-doped CNTs showed superior A_s of 1053 $m^2 g^{-1}$ with enhanced E_{on} of -0.051 V versus Ag/AgCl.

Metal/nitrogen@carbon (MNC) framework is also a major category of electrocatalysts and has extensively been investigated for their superior catalytic performance. Usually, the MNC framework is produced by carrying out the pyrolysis process of either transition metal-containing compounds or by combining C, N, and M as a precursor. In MNC-based electrocatalysts, the MN_4 sites are believed to act as active sites. Moreover, for an efficient MNC catalyst, the M_0 NPs contents should be minimal. The support for C is also a significant factor for achieving effective catalysis. The carbon support influences many factors, such as active spot allocation, mass transport, and electrode conductivity. The carbon support should possess superior graphitization, wide channels, and higher A_s for enhanced three-phase interaction on electrocatalyst. Due to the intrinsic existence of M, N, and C, MOFs precursors are believed to be

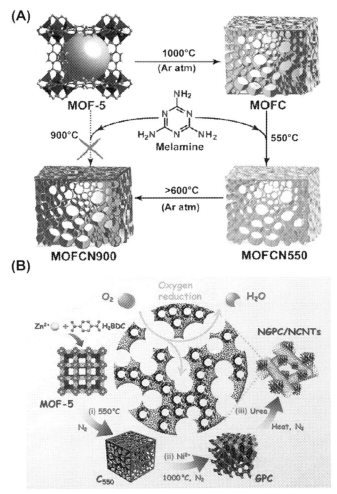

Figure 7.9 (A) Fabrication of nitrogen-doped porous carbonaceous material g-C$_3$N$_4$ and (B) depiction of the progressively evolving structure of MOF-5 into NGPC/NCNTs (*MOF*, Metal-organic framework; *NCNTs*, nitrogen-doped carbon nanotubes; *NGPC*, nitrogen-doped graphitic porous carbon). *Adapted with permission from S. Pandiaraj, H.B. Aiyappa, R. Banerjee, S. Kurungot Post modification of MOF derived carbon via g-C3N4 entrapment for efficient metal-free oxygen reduction reaction, Chemical Communications 50 (2014) 3363. Copyright (2014) Royal Society of Chemistry; adapted with permission from L. Zhang, X. Wang, R. Wang, M. Hong, Structural evolution from metal-organic framework to hybrids of nitrogen-doped porous carbon and carbon nanotubes for enhanced oxygen reduction activity, Chemistry of Materials 27 (2015) 7610–7618. Copyright (2015) American Chemical Society.*

ideal for acquiring MNC electrocatalysts with improved graphite content, high A_s, and wide channels. The pyrolysis transforms organic ligand into pyridinic-N-containing carbonaceous material, which later interacts with metal to produce MN_4 throughout the extremely graphitized material. The Fe-containing NC materials have been extensively studied for their enhanced ORR activity. The employment of Fe during the pyrolysis also helps in controlling the morphology of the carbonaceous nanomaterials. Furthermore, the pyrolysis temperature also influences the structure of the resulting material. For instance, Li et al. prepared Fe/N/C electrocatalysts at varying pyrolysis temperatures, such as 800°C, 900°C, and 1000°C. At 800°C, the material transformed into anion-like carbonaceous nanoparticles [66]. At 900°C, the material changed into CNTs, and at 1000°C, the material converted into graphene@CNTs, as displayed in Fig. 7.10A. The changing morphology of the material displayed the importance of the pyrolysis temperature. The resulted electrocatalyst had bamboo-like CNTs morphology with Fe_3C particles enclosed in the tube-shaped structures. The electrocatalyst presented excellent $E_{1/2}$ of 0.79 V in the acidic condition and 0.88 V in the alkaline condition versus RHE.

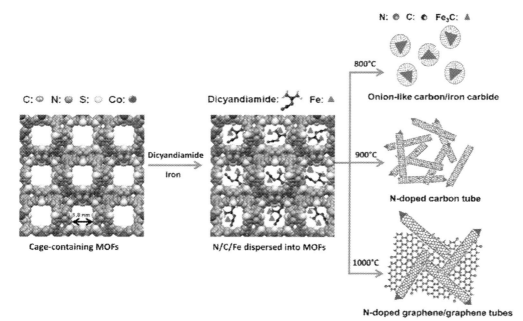

Figure 7.10 Schematic representation for the production of C nanostructures present in N-Fe-containing MOF electrocatalysts produced at various temperatures (*MOF*, Metal-organic framework). *Adapted with permission from Q. Li, P. Xu, W. Gao, S. Ma, G. Zhang, R. Cao, et al., Graphene/graphene-tube nanocomposites templated from cage-containing metal-organic frameworks for oxygen reduction in Li-O2 batteries, Advanced Materials 26 (2014) 1378–1386. Copyright (2013) John Wiley and Sons.*

Metal sulfides (M_aS_b) are also a crucial category of MOF-based electrocatalysts for ORR. These metal sulfide-based electrocatalysts are known for their ability to offer catalysis at a wide range of temperatures. Furthermore, metal sulfides are found to be more stable in the acidic medium than other transition metal-based materials. The metal sulfides are also produced through the pyrolysis process in the presence of sulfur. For example, $CoS_{1.097}/C$ was produced by Bali et al. [67]. The synthetic route for the preparation of the electrocatalyst involves the pyrolysis of ZIF-9 in the presence of pulverized sulfur at 900°C. The produced material showed relatively lower onset potential but offered a current density comparable to Pt/C electrocatalysts and also followed the four-electrons transfer route.

Recently, the carbon-layered core-shell supported on a highly conductive substrate was produced by Xia et al. [68]. CMK-3 was employed as a substrate owing to its superior A_s, rigid framework, higher conduction, and highly arranged morphology. The substrate's exterior was layered with Co-containing ZIF-9 and then passed through a series of pyrolysis processes to transform ZIF-9 into Co/Co_3O_4@C core/bishell morphology. The synthesized material showed high A_s of $616\,m^2\,g^{-1}$, as displayed in Fig. 7.11. The electrocatalyst revealed an excellent E_{on} of 0.85 V and a good $E_{1/2}$ of 0.70 V versus RHE. The investigation indicated a significance of the core-shell morphology for ORR. The material Co_3O_4@C-CMK displayed superior E_{on} but inadequate current density comparing to Pt/C. But, when the material is oxidized externally, a core-shell-like morphology of Co/Co_3O_4@C-CMK was obtained. The synthesized core-shell material displayed high E_{on} as well as improved current density. The results indicated that the significance of core-shell-like morphology for ORR electrocatalysis.

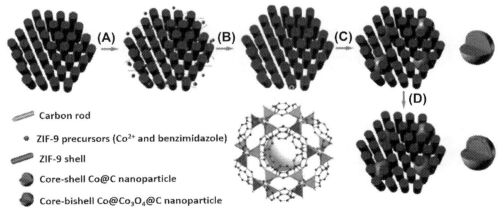

Figure 7.11 Depiction for the creation of the Co@Co3O4@C-CM. Adapted with permission from W. Xia, R. Zou, L. An, D. Xia, S. Guo, Metal-organic framework route to in-situ encapsulation of Co@Co3O4@C Core@Bishell nanoparticles into highly ordered porous carbon matrix for oxygen reduction, Energy & Environmental Science 8 (2015) 568. Copyright (2014) Royal Society of Chemistry.

7.5 Conclusion

The quest for a low-priced electrocatalyst with superior electrocatalytic activity has been seeking the attention of researchers all around the globe. Today, developing highly efficient material for harvesting clean and renewable energy is the focal point of scientists. PEMFCs are a promising route to obtain clean energy and are a vital part hydrogen energy chain. These FCs transform the chemical energy of the H_2 into electrical energy through electrochemical reactions. MOFs contains a combination of organic and inorganic parts and are considered a promising material for FCs application. MOFs exhibit extraordinary properties, such as superior porosity, wide channels, tunable functionality, highly arranged crystalline structure, and high stability. MOFs can either be used as a polymer electrolyte member (PEM) or as an ORR electrocatalyst. Although, MOFs electrocatalysts obtained via the different process and synthetic routes have exhibited properties and electrochemical activities similar to Pt-based catalysts. But, there are very few studies on the direct application of MOFs for the ORR catalysis. Therefore more investigations are necessary for this field to obtain a material that can be applied industrially. In conclusion, substantial advancements have been made in the MOFs, but there are a lot of challenges that need to be addressed and resolved. Throughout previous years, MOFs have transformed into a material that exhibits comparable electrocatalytic performance to the precious metal and becomes an exceptional candidate that has the potential to replace expensive and precious metals for the application of FCs electrocatalysts.

References

[1] D.Y. Chung, J.M. Yoo, Y.E. Sung, Highly durable and active Pt-based nanoscale design for fuel-cell oxygen-reduction electrocatalysts, Advanced Materials 30 (2018) 1704123.
[2] Energy Information Administration (United States), International Energy Outlook 2013, Energy Information Administration, Washington, DC, 2013.
[3] W. Vielstich, Handbook of Fuel Cells, Wiley, 2009.
[4] W.J. Ong, L.L. Tan, S.P. Chai, S.T. Yong, A.R. Mohamed, Surface charge modification via protonation of graphitic carbon nitride (g-C_3N_4) for electrostatic self-assembly construction of 2D/2D reduced graphene oxide (rGO)/g-C_3N_4 nanostructures toward enhanced photocatalytic reduction of carbon dioxide to methane, Nano Energy 13 (2015) 757−770.
[5] W.J. Ong, L.K. Putri, L.L. Tan, S.P. Chai, S.T. Yong, Heterostructured AgX/g-C_3N_4 (X = Cl and Br) nanocomposites via a sonication-assisted deposition-precipitation approach: emerging role of halide ions in the synergistic photocatalytic reduction of carbon dioxide, Applied Catalysis B: Environmental 180 (2016) 530−543.
[6] J. Lee, B. Jeong, J.D. Ocon, Oxygen electrocatalysis in chemical energy conversion and storage technologies, Current Applied Physics 13 (2013) 309−321.
[7] R. Borup, J. Meyers, B. Pivovar, Y.S. Kim, R. Mukundan, N. Garland, et al., Scientific aspects of polymer electrolyte fuel cell durability and degradation, Chemical Reviews 107 (2007) 3904−3951.
[8] J. Kunze, U. Stimming, Electrochemical versus heat-engine energy technology: A tribute to wilhelm ostwald's visionary statements, Angewandte Chemie International Edition 48 (2009) 9230−9237.
[9] V. Dusastre, Materials for sustainable energy, Nature Publishing Group., 2010.

[10] Y. Ren, G.H. Chia, Z. Gao, Metal-organic frameworks in fuel cell technologies, Nano Today 8 (2013) 577−597.

[11] O.M. Yaghi, G. Li, H. Li, Selective binding and removal of guests in a microporous metal-organic framework, Nature 378 (1995) 703−706.

[12] O.M. Yaghi, H. Li, Hydrothermal synthesis of a metal-organic framework containing large rectangular channels, Journal of the American Chemical Society 117 (1995) 10401−10402.

[13] H. Furukawa, N. Ko, Y.B. Go, N. Aratani, S.B. Choi, E. Choi, et al., Ultrahigh porosity in metal-organic frameworks, Science 329 (2010) 424−428.

[14] H. Furukawa, K.E. Cordova, M. O'Keeffe, O.M. Yaghi, The chemistry and applications of metal-organic frameworks, Science 341 (2013) 1230444.

[15] A.R. Millward, O.M. Yaghi, Metal-organic frameworks with exceptionally high capacity for storage of carbon dioxide at room temperature, Journal of the American Chemical Society 127 (2005) 17998−17999.

[16] H.J. Son, S. Jin, S. Patwardhan, S.J. Wezenberg, N.C. Jeong, M. So, et al., Light-harvesting and ultrafast energy migration in porphyrin-based metal-organic frameworks, Journal of the American Chemical Society 135 (2013) 862−869.

[17] J.L.C. Rowsell, O.M. Yaghi, Metal-organic frameworks: a new class of porous materials, Microporous Mesoporous Mater 73 (2004) 3−14.

[18] C.S. Collins, D. Sun, W. Liu, J.L. Zuo, H.C. Zhou, Reaction-condition-controlled formation of secondary-building-units in three cadmium metal-organic frameworks with an orthogonal tetrakis (tetrazolate) ligand, Journal of Molecular Structure 890 (2008) 163−169.

[19] S. Horike, M. Dincă, K. Tamaki, J.R. Long, Size-selective Lewis acid catalysis in a microporous metal-organic framework with exposed Mn^{2+} coordination sites, Journal of the American Chemical Society 130 (2008) 5854−5855.

[20] L.J. Wang, H. Deng, H. Furukawa, F. Gándara, K.E. Cordova, D. Peri, et al., Synthesis and characterization of metal-organic framework-74 containing 2, 4, 6, 8, and 10 different metals, Inorganic Chemistry 53 (2014) 5881−5883.

[21] S.B. Darling, Directing the self-assembly of block copolymers, Progress in Polymer Science 32 (2007) 1152−1204.

[22] R. Banerjee, H. Furukawa, D. Britt, C. Knobler, M. O'Keeffe, O.M. Yaghi, Control of pore size and functionality in isoreticular zeolitic imidazolate frameworks and their carbon dioxide selective capture properties, Journal of the American Chemical Society 131 (2009) 3875−3877.

[23] T. Devic, P. Horcajada, C. Serre, F. Salles, G. Maurin, B. Moulin, et al., Functionalization in flexible porous solids: effects on the pore opening and the host-guest interactions, Journal of the American Chemical Society 132 (2010) 1127−1136.

[24] M. Eddaoudi, J. Kim, N. Rosi, D. Vodak, J. Wachter, M. O'Keeffe, et al., Systematic design of pore size and functionality in isoreticular MOFs and their application in methane storage, Science 295 (2002) 469−472.

[25] Y.K. Hwang, D.Y. Hong, J.S. Chang, S.H. Jhung, Y.K. Seo, J. Kim, et al., Amine grafting on coordinatively unsaturated metal centers of MOFs: consequences for catalysis and metal encapsulation, Angewandte Chemie International Edition, 47, 2008, pp. 4144−4148.

[26] S.S. Kaye, J.R. Long, Matrix isolation chemistry in a porous metal-organic framework: photochemical substitutions of N_2 and H_2 in $Zn_4O[(\eta 6-1,4-benzenedicarboxylate)Cr(CO)_3]_3$, Journal of the American Chemical Society 130 (2008) 806−807.

[27] Z. Wang, K.K. Tanabe, S.M. Cohen, Accessing postsynthetic modification in a series of metal-organic frameworks and the influence of framework topology on reactivity, Inorganic Chemistry 48 (2009) 296−306.

[28] K.K. Tanabe, Z. Wang, S.M. Cohen, Systematic functionalization of a metal-organic framework via a postsynthetic modification approach, Journal of the American Chemical Society 130 (2008) 8508−8517.

[29] J.S. Seo, D. Whang, H. Lee, S.I. Jun, J. Oh, Y.J. Jeon, et al., A homochiral metal-organic porous material for enantioselective separation and catalysis, Nature 404 (2000) 982−986.

[30] O.K. Farha, K.L. Mulfort, J.T. Hupp, An example of node-based postassembly elaboration of a hydrogen-sorbing, metal-organic framework material, Inorganic Chemistry 47 (2008) 10223−10225.

[31] K.K. Tanabe, S.M. Cohen, Engineering a metal-organic framework catalyst by using postsynthetic modification, Angewandte Chemie International Edition, 48, 2009, pp. 7424–7427.

[32] M.J. Ingleson, J. Perez Barrio, J.B. Guilbaud, Y.Z. Khimyak, M.J. Rosseinsky, Framework functionalisation triggers metal complex binding, Chemical Communications (2008) 2680–2682.

[33] S.L. James, Metal-organic frameworks, Chemical Society Reviews 32 (2003) 276–288.

[34] D. Lau, G. Hay D, R. Hill M, W. Muir B, A. Furman S, F. Kennedy D, PLUXter: rapid discovery of metal-organic framework structures using PCA and HCA of high throughput synchrotron powder diffraction data, Combinatorial Chemistry & High Throughput Screening 14 (2012) 28–35.

[35] S. Noro, S. Kitagawa, M. Kondo, K. Seki, A new, methane adsorbent, porous coordination polymer [n], Angewandte Chemie International Edition 39 (2000) 2081–2084.

[36] B.F. Hoskins, R. Robson, Design and construction of a new class of scaffolding-like materials comprising infinite polymeric frameworks of 3D-linked molecular rods. A reappraisal of the $Zn(CN)_2$ and $Cd(CN)_2$ structures and the synthesis and structure of the diamond-related framework, Journal of the American Chemical Society 112 (1990) 1546–1554.

[37] O.M. Yaghi, H. Li, T.L. Groy, A molecular railroad with large pores: synthesis and structure of Ni (4,4′-bpy)2.5(H_2O)2(ClO_4)2–1.5(4,4′-bpy)-$2H_2O$ 1, Inorganic Chemistry 36 (1997) 4292–4293.

[38] R. Kitaura, K. Fujimoto, S.I. Noro, M. Kondo, S. Kitagawa, A pillared-layer coordination polymer network displaying hysteretic sorption: [Cu_2(pzdc)2(dpyg)]n (pzdc = pyrazine-2, 3-dicarboxylate; dpyg = 1, 2-Di(4-pyridyl)glycol), Angewandte Chemie International Edition, 41, 2002, pp. 133–135.

[39] L.E. Kreno, K. Leong, O.K. Farha, M. Allendorf, R.P. Van Duyne, J.T. Hupp, Metal-organic framework materials as chemical sensors, Chemical Reviews 112 (2012) 1105–1125.

[40] X. Lin, N.R. Champness, M. Schröder, Hydrogen, methane and carbon dioxide adsorption in metal-organic framework materials, Topics in Current Chemistry 293 (2009) 35–76.

[41] V.I. Isaeva, L.M. Kustov, The application of metal-organic frameworks in catalysis (review), Petroleum Chemistry 50 (2010) 167–180.

[42] M.D. Allendorf, C.A. Bauer, R.K. Bhakta, R.J.T. Houk, Luminescent metal-organic frameworks, Chemical Society Reviews 38 (2009) 1330–1352.

[43] W. Cho, S. Park, M. Oh, Coordination polymer nanorods of Fe-MIL-88B and their utilization for selective preparation of hematite and magnetite nanorods, Chemical Communications 47 (2011) 4138–4140.

[44] J. Tang, R.R. Salunkhe, J. Liu, N.L. Torad, M. Imura, S. Furukawa, et al., Thermal conversion of core-shell metal-organic frameworks: a new method for selectively functionalized nanoporous hybrid carbon, Journal of the American Chemical Society 137 (2015) 1572–1580.

[45] S.J. Yang, S. Nam, T. Kim, J.H. Im, H. Jung, J.H. Kang, et al., Preparation and exceptional lithium anodic performance of porous carbon-coated ZnO quantum dots derived from a metal-organic framework, Journal of the American Chemical Society 135 (2013) 7394–7397.

[46] A. Mahmood, W. Guo, H. Tabassum, R. Zou, Metal-organic framework-based nanomaterials for electrocatalysis, Advanced Energy Materials 6 (2016) 1600423.

[47] P. Ramaswamy, N.E. Wong, G.K.H. Shimizu, MOFs as proton conductors-challenges and opportunities, Chemical Society Reviews 43 (2014) 5913–5932.

[48] M. Sadakiyo, T. Yamada, H. Kitagawa, Rational designs for highly proton-conductive metal-organic frameworks, Journal of the American Chemical Society 131 (2009) 9906–9907.

[49] A. Shigematsu, T. Yamada, H. Kitagawa, Wide control of proton conductivity in porous coordination polymers, Journal of the American Chemical Society 133 (2011) 2034–2036.

[50] S.S. Nagarkar, S.M. Unni, A. Sharma, S. Kurungot, S.K. Ghosh, Two-in-one: inherent anhydrous and water-assisted high proton conduction in a 3D metal-organic framework, Angewandte Chemie International Edition, 53, 2014, pp. 2638–2642.

[51] J.M. Taylor, R.K. Mah, I.L. Moudrakovski, C.I. Ratcliffe, R. Vaidhyanathan, G.K.H. Shimizu, Facile proton conduction via ordered water molecules in a phosphonate metal-organic framework, Journal of the American Chemical Society 132 (2010) 14055–14057.

[52] J.M. Taylor, T. Komatsu, S. Dekura, K. Otsubo, M. Takata, H. Kitagawa, The role of a three dimensionally ordered defect sublattice on the acidity of a sulfonated metal-organic framework, Journal of the American Chemical Society 137 (2015) 11498–11506.

[53] G. Xu, K. Otsubo, T. Yamada, S. Sakaida, H. Kitagawa, Superprotonic conductivity in a highly oriented crystalline metal-organic framework nanofilm, Journal of the American Chemical Society 135 (2013) 7438−7441.
[54] S. Bureekaew, S. Horike, M. Higuchi, M. Mizuno, T. Kawamura, D. Tanaka, et al., One-dimensional imidazole aggregate in aluminium porous coordination polymers with high proton conductivity, Materials for Sustainable Energy: A Collection of Peer-reviewed Research and Review, Nature Publishing Group, 2010, pp. 232−237.
[55] D. Umeyama, S. Horike, M. Inukai, Y. Hijikata, S. Kitagawa, Confinement of mobile histamine in coordination nanochannels for fast proton transfer, Angewandte Chemie International Edition 123 (2011) 11910−11913.
[56] J.A. Hurd, R. Vaidhyanathan, V. Thangadurai, C.I. Ratcliffe, I.L. Moudrakovski, G.K.H. Shimizu, Anhydrous proton conduction at 150°C in a crystalline metal-organic framework, Nature Chemistry 1 (2009) 705−710.
[57] D. Umeyama, S. Horike, M. Inukai, T. Itakura, S. Kitagawa, Inherent proton conduction in a 2D coordination framework, Journal of the American Chemical Society 134 (2012) 12780−12785.
[58] S. Horike, D. Umeyama, M. Inukai, T. Itakura, S. Kitagawa, Coordination-network-based ionic plastic crystal for anhydrous proton conductivity, Journal of the American Chemical Society 134 (2012) 7612−7615.
[59] K.F. Babu, M.A. Kulandainathan, I. Katsounaros, L. Rassaei, A.D. Burrows, P.R. Raithby, et al., Electrocatalytic activity of BasoliteTM F300 metal-organic-framework structures, Electrochemistry Communications 12 (2010) 632−635.
[60] J. Mao, L. Yang, P. Yu, X. Wei, L. Mao, Electrocatalytic four-electron reduction of oxygen with copper (II)-based metal-organic frameworks, Electrochemistry Communications 19 (2012) 29−31.
[61] M. Jahan, Q. Bao, K.P. Loh, Electrocatalytically active graphene-porphyrin MOF composite for oxygen reduction reaction, Journal of the American Chemical Society 134 (2012) 6707−6713.
[62] L. Zhang, Z. Su, F. Jiang, Y. Lingling, J. Qian, Y. Zhou, et al., Highly graphitized nitrogen-doped porous carbon nanopolyhedra derived from ZIF-8 nanocrystals as efficient electrocatalysts for oxygen reduction reactions, Nanoscale 6 (2014) 6590.
[63] P. Zhang, F. Sun, Z. Xiang, Z. Shen, J. Yun, D. Cao, ZIF-derived in situ nitrogen-doped porous carbons as efficient metal-free electrocatalysts for oxygen reduction reaction, Energy & Environmental Science 7 (2014) 442−450.
[64] S. Pandiaraj, H.B. Aiyappa, R. Banerjee, S. Kurungot, Post modification of MOF derived carbon via g-C_3N_4 entrapment for efficient metal-free oxygen reduction reaction, Chemical Communications 50 (2014) 3363.
[65] L. Zhang, X. Wang, R. Wang, M. Hong, Structural evolution from metal-organic framework to hybrids of nitrogen-doped porous carbon and carbon nanotubes for enhanced oxygen reduction activity, Chemistry of Materials 27 (2015) 7610−7618.
[66] Q. Li, P. Xu, W. Gao, S. Ma, G. Zhang, R. Cao, et al., Graphene/graphene-tube nanocomposites templated from cage-containing metal-organic frameworks for oxygen reduction in Li-O2 batteries, Advanced Materials 26 (2014) 1378−1386.
[67] F. Bai, H. Huang, C. Hou, P. Zhang, Porous carbon-coated cobalt sulfide nanocomposites derived from metal organic frameworks (MOFs) as an advanced oxygen reduction electrocatalyst, New Journal of Chemistry 40 (2016) 1679−1684.
[68] W. Xia, R. Zou, L. An, D. Xia, S. Guo, Metal-organic framework route to in-situ encapsulation of Co@Co_3O_4@C Core@Bishell nanoparticles into highly ordered porous carbon matrix for oxygen reduction, Energy & Environmental Science 8 (2015) 568.

CHAPTER 8

Advanced carbon-based nanostructured materials for fuel cells

Muhammad Rizwan Sulaiman[1,2] and Ram K. Gupta[1]

[1]Department of Chemistry, Kansas Polymer Research Center, Pittsburg State University, Pittsburg, KS, United States
[2]Department of Plastic Engineering, Kansas Technology Center, Pittsburg State University, Pittsburg, KS, United States

8.1 Introduction

Carbon is the sixth element in the periodic table and is one of the most abundant elements in the universe, earth, and all living organisms present on earth, including human beings. Carbon has a unique ability to bind with itself and with other elements or molecules, which attracts scientists to investigate its potential in vast fields. Carbon also served as a great source of energy in human history, starting with the use of wood as a primary source of energy. With industrial advancement and the invention of the steam engine, coal became the primary energy source. Then coal was substituted with electrical energy and fossil fuel dominated as the primary energy source in the 1970s. All these energy components are also the main source of contamination of the environment, releasing tons of CO_2 and other greenhouse gases, which eventually result in undesirable global warming and other greenhouse effects. Additionally, these carbon-based energy sources are not renewable, and their reservoirs are depleting at a faster rate as the world's appetite for energy increases day by day.

To overcome the problems of environmental unfriendliness and continuously depleting reservoirs, scientists have been investigating efficient, renewable, and clean energy sources. Recently, fuel cells (FCs) have garnered tremendous attention owing to their ability to produce electricity directly from chemical fuels through electrochemical reactions. FCs are appealing technology due to various reasons, such as unique operating features, superior efficiency, reliability, and minimized environmental consequences. FCs are categorized based on the operating conditions and the configuration of the FCs. Most commonly, FCs are classified based on electrolytes utilized in the FCs and can be of the following types: (1) alkaline fuel cells having an alkaline solution as an electrolyte, normally KOH, (2) solid oxide fuel cells include solid oxide-based ceramic ion transferring electrolyte, (3) phosphoric acid fuel cells having H_2SO_4 solution as an electrolyte, (4) polymer electrolyte membrane fuel cells (PEMFCs) with proton exchange membrane as an electrolyte, and (5) molten carbonate fuel cells having molten carbonate as an electrolyte. Among these FCs, PEMFCs are very popular among researchers owing to their many benefits such as low operating

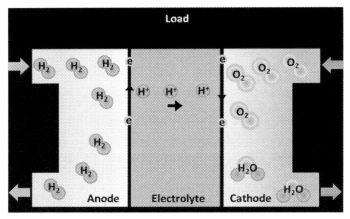

Figure 8.1 Adiagrammatic illustration of PEMFCs (*PEMFCs*, Polymer electrolyte membrane fuel cells).

temperatures, consistent in providing high current density, lower weight, compact framework, quick startups, and suitability for noncontinuous functioning. Hydrogen is used as a fuel for the PEMFCs that can be produced via water splitting through water electrolysis. Direct methanol fuel cell (DMFC) is a type of PEMFCs and is a potential candidate for generating clean energy. In DMFCs, methanol is used as a fuel that can be produced from biomass and utilizes a polymer exchange membrane for protons transfer. The renewable fuel sources of these FCs make them an efficient energy conversion device with the least contamination into the environment.

The basic operating principle for the FC remains the same. In any PEMFCs, the fuel is fed into the anodic compartment, which is transformed into its ions with the generation of electrons. These electrons flow via an external circuit toward the cathodic compartment producing electric energy, and the ions (usual protons) are moved through the polymer electrolyte membrane (PEM) to the cathode and combined to form a by-product. The PEMFC having hydrogen as a fuel in the anode is illustrated in Fig. 8.1, and the reactions can be described in the chemical equations:

$$\text{At anode: } H_2 \rightarrow 2H^+ + e^-$$

$$\text{At cathode: } O_2 + 4H^+ + 4e^- \rightarrow 2H_2O$$

$$\text{Overall PEMFC reaction: } 2H_2O + O_2 \rightarrow 2H_2O$$

Similarly, DMFCs with methanol as the fuel has the same working principle, as shown in Fig. 8.2. But, DMFCs also involve fuel oxidation reaction at the anodic side of the FC. The chemical equation representing the reaction of DMFCs are as following:

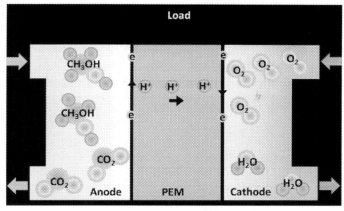

Figure 8.2 Operating mechanism of DMFCs (*DMFCs*, Direct methanol fuel cells).

At anode: $CH_3OH + H_2O \rightarrow CO_2 + 6H^+ + 6e^-$

At cathode: $3/2O_2 + 6H^+ + 6e^- \rightarrow 3H_2O$

Overall DMFC reaction: $CH_3OH + 3/2O_2 \rightarrow CO_2 + 2H_2O$

Regardless of the benefits of these FCs, various obstacles such as increased costs, operation performance, and stability, must be bridged to achieve commercial feasibility. Almost all of these FCs employ expensive metals, such as Pt and Pt-containing materials, as electrocatalysts to speed up the sluggish kinetics of these FCs, such as the oxygen reduction process at the cathodic side of the PEMFCs and the methanol oxidation reaction at the anode of the DMFCs. Furthermore, unsatisfactory operational stability is due to the electrocatalyst oxidation, catalyst relocation, reduction in electrocatalyst's active surface area, and carbon support decay [1,2]. Also, PEMFCs are susceptible to CO poisoning, which can be found in the fuel obtained from fossil fuels or produced during operations, such as in methanol oxidation reaction in DMFCs. Similarly, the slow kinetics of the oxygen reduction reaction (ORR) and the methanol oxidation reaction requires the employment of electrocatalysts. It accounts for the main factor for voltage loss in PEMFCs despite the presence of Pt electrocatalyst [3]. Currently, the Pt or Pt-containing electrocatalysts have been the state of the art catalysts for most FCs. Tremendous researches have been taken place to find cheap, efficient, abundant, and highly stable electrocatalysts. Similarly, scientists are also investigating new PEMs for FCs.

Carbon-based materials have recently garnered enormous attention due to their earth abundance, environmental friendliness, good stability in both acidic and alkaline conditions, and exceptional structural properties. Carbon is mainly present in its allotropic forms, such as amorphous, graphite, diamond. They also have the unique capability to produce various allotropes with different properties, such as three-dimensional (3D) carbon materials arranged tetrahedrally possess diamond-like hardness and transparency. Whereas carbon atoms in black graphite configured in countless layered structures are slippery, as shown in Fig. 8.3. The recent advancements in low-dimensional carbonaceous systems, such as fullerenes [zero-dimensional (0D)], carbon nanotubes (CNTs) [one-dimensional (1D)], and graphene [two-dimensional (2D)], have promoted many studies for the development of highly efficient and active electrocatalyst. In this chapter, we will discuss efficient carbon-based electrocatalyst materials for ORR.

8.2 Important oxygen reduction reaction characterization notions

For the determination and performance comparison of the ORR electrocatalysts, understanding the fundamental characterization concept is essential, such as current density, onset potential, Tafel slope, turnover frequency, voltage gap, and electrochemical surface area.

8.2.1 Onset potential (E_{ons})

The onset potential is usually used to indicate the potential at which the current deflects from the baseline. In an experiment, identifying the E_{ons} can be very random. A common method to find the E_{ons} is by intersecting the tangents between the baseline and the increasing current in the current voltammetry curve. It should be noted that the background must be accurate to prevent any overvaluation of E_{ons}. For instance, the ORR current can be measured by obtaining polarization curves in the nitrogen and oxygen environment through similar experiments, such as scan rate,

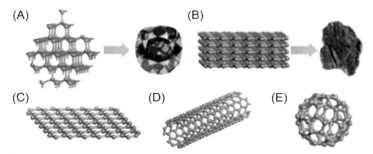

Figure 8.3 Different allotropes of carbon: (A) 3D diamond, (B) 3D graphite, (C) 2D graphene, (D) 1D carbon nanotube, and (E) 0D fullerene [4]. *Adapted with permission from H. Huang, X. Wang, Recent progress on carbon-based support materials for electrocatalysts of direct methanol fuel cells, Journal of Materials Chemistry A 2 (2014) 6266–6291. Copyright (2014) Royal Society of Chemistry.*

direction, and rotation rate, and later remove the curve obtained in nitrogen from the curve obtained in oxygen. There is a kinetic energy barrier that needs to be bridged to achieve the ORR. The E_{ons} for ORR is always greater than the oxygen's standard redox potential, which is 1.23 V at 298K temperature and a pH of 0.0591 against normal hydrogen electrode.

8.2.2 Current density

Current density (j) of any electrode is normally standardized by the surface area of the material or by the mass of the active material present in the electrode. The overall current density has a strong connection with the oxygen generation rate. The current density is a critical parameter that reflects the effectiveness of a catalyst in any FC device.

8.2.3 Tafel slope

The overpotential (η) is typically associated with the logarithm of current density. Its linear part is provided as the Tafel equation: $\eta = a + b \log(j)$, where b represents the Tafel slope. The measurement indicates the reaction path and also the rate-determining step. Furthermore, this equation also gives a current density at $\eta = 0$, known as exchange current density (j_o). The exchanged current density is the representation of inherent catalyst activity at an equilibrium condition. Another way of obtaining the Tafel slope is through plotting the graph of log R_{ct}^{-1} versus η. This method gives a pure Tafel slope value through charge transfer kinetics compared to the Tafel slopes calculated from the cyclic voltammetry data. Because in cyclic voltammetry data, electrocatalyst resistance may also contribute to the Tafel slope, which results in contamination of the Tafel slope measurements.

8.2.4 Electron transfer number and HO_2^- percentage

The ORR can take place either by a two-electron transfer process or by a four-electron transfer process. The two-electron transfer process involves the production of peroxide that can corrode the electrocatalysts and are highly unfavorable for ORR. In contrast, the four-electron transfer processes do not produce peroxide and are highly preferred over the two-electron transfer process. The measurement of electron transfer number (n) can be done using the following equation:

$$n = 4j_d(j_d + j_r/N)^{-1}$$

In this equation, j_d represents the disk current, j_r represents the ring current, and N represents the collection efficiency. The equation for the measurement of HO_2^- is given as follows:

$$HO_2^- = 200(j_r N^{-1})(j_d + j_r N^{-1})^{-1}$$

8.2.5 Turnover frequency

The turnover frequency is measured by the following equation:

$$\text{TOF} = (j \times A)(n \times F \times m)^{-1}.$$

In this equation, j = current density at a provided η, A = electrode's surface area, n = electron transfer number at a provided η, F = Faraday constant, and m = metal content's number of moles. The active material, for example, metal quantity, can be obtained by the induced coupled plasma method or by the cyclic voltammetry method. The TOF shows the inherent catalytic activity of the electrocatalyst.

8.3 Approaches to synthesize nanocarbons

Carbon-derived nanomaterials can be categorized into four structural dimensionalities, such as 0D, 1D, 2D, and 3D, as shown in Fig. 8.4. Carbonaceous 0D materials consist of nanosized 3D structures, for example, fullerene and atomic clusters. The 1D materials comprise 2D nanosized structures, such as nanotubes, nanorods, and nanofibers. Similarly, 2D materials contain one dimension in the range of nanometers, such as graphene. There are granular structures in the 3D materials with all three equal dimensions, such as carbon sponges. The synthetic approaches and the applications of these carbon materials of different nanosized dimensions in the FCs are discussed further.

8.3.1 Zero-dimensional carbonaceous materials

Fullerenes are identified as the first 0D carbon-derived material. In 1985, Curl, Kroto, and Smalley discovered Buckminsterfullerene (C_{60}) during graphite vaporization. The founders also won the Nobel Prize for their contribution in 1996. Inspired by Fuller's round roof structure, the discovered material's molecule was called Fullerene. It has a cage-like morphology with a molecular diameter ranging from 1 nm and dramatically resembles the shape of a soccer ball. The Buckminsterfullerene has the tiniest molecules among other fullerene species and is symbolized as a great 0D nanomaterial. Fullerenes can be produced in several ways, but it is generally produced by the application of a high voltage between the two graphitic electrodes, resulting in a carbon plasma arc, which produces black residue. From this residue, different fullerene can be separated easily. Fullerenes can also be prepared by vaporizing graphite or by catalytic chemical vapor deposition (CVD) from any carbon source. Fullerenes exhibit unique properties and have an excellent ability to be modified. With the modification, the electronic structure, physical properties, and solubility of the fullerene are also affected.

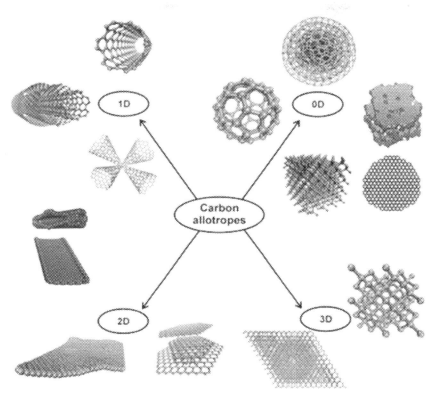

Figure 8.4 Classification of carbon nanomaterials based on dimensionality [5]. *Adapted with permission from V. Georgakilas, J.A. Perman, J. Tucek, R. Zboril, Broad family of carbon nanoallotropes: classification, chemistry, and applications of fullerenes, carbon dots, nanotubes, graphene, nanodiamonds, and combined superstructures, Chemical Reviews 115 (2015) 4744–4822. Copyright (2014) Royal Society of Chemistry.*

8.3.1.1 Fullerene-based electrocatalysts

Buckminsterfullerene is one of the most studied and stable materials in the fullerene family. It has been employed in different applications such as biosensors [6,7], screen printed systems [8], solar cells [9], and adsorption electrodes [10]. Furthermore, C_{60} derivatives mixed with polyanilines (PAnis) have also been utilized in different devices, such as mobile phones and microwaves. Various reports also claimed to use fullerene derivatives in FC applications. Fullerenes have unique characteristics that enable them to be employed in FCs. These characteristics include superior electrochemical stability, nanosized particles, organized structure, specific morphology, superior electrical and thermal conductivity, high specific surface area, and unique mechanical strength. Moreover, the C_{60} offers the possibility to modify the pore structure and the end functional groups, allowing them to be tailor-made with variable characteristics, which is challenging in other material.

The employment of fullerenes for catalyzing ORR in the FC is very seldom. Mostly the fullerenes are used to provide support to the metals in the electrocatalysts. The preparation of fullerene for electrocatalyst was first reported in 2007 by heat treatment of carbon, which produced C_{60} surface. Shioyama et al. synthesized a thermally treated mixture of mesopore-carbon microbeads (MCMBs) or glassy carbon power (GCP) (80%), C_{60} (10%), and acenaphthylene (10%) at 1000°C under the Ar flow for 1 hour [11]. The electrocatalysts were prepared by introducing Pt salt, such as $Pt(NO_2)_2(NH_3)_2$, dissolved in ethanol, to the mixture prepared earlier. At 250°C, the reduction of the Pt was carried out. Different samples with this formulation were prepared, and their activity toward ORR was examined. It was observed that the Pt/MCMB and Pt/GCP electrocatalysts were performed well as compared to Pt/C (carbon black). Furthermore, it is also observed that the inclusion of C_{60} in Pt/MCMB and Pt/GCP was unfavorable and negatively impacted the electrocatalytic activity of the electrocatalyst. The metallofullerenes were utilized for the theoretical studies on the stability of ORR [12]. The metallofullerene, such as PtC_{59}, is a C_{60} molecule with the C atom replaced with the Pt atom. The study was based on the data obtained in the form of enthalpies of the ORR adsorbates, such as H atom, OH, OOH, O_2, and H_2O molecules on PtC_{59}. The results showed that H_2 and the adsorbates contaminated the active spots in the electrocatalysts and can also decelerate the water generation on the PEMFC's cathode.

Nitrogen is considered a good doping agent that helps catalyze the ORR in FCs [13]. With this consideration, Wu et al. produced an N-doped NPM ORR electrocatalyst [14]. The N-doped onion-like carbonaceous materials were fabricated by heat treatment of a hybrid material comprised of hexamethylene diamine complex with Co and Fe species. The graphitized carbon material obtained from the Co offers a strong network to keep NPM active sites. This strong attraction can avoid water flooding and enhance the opposition to oxidative attack in the cathode. Thus resulted in enhanced durability. The hybrid of C_{60}-graphene was produced by a solid-state mechanochemical process. The process involves grinding the pristine graphite, C_{60}, and lithium hydroxide, which acted as a catalyst. As a result, the graphene nanoplatelets and the C_{60} formed direct bonds by two CeC single bonds. The catalytic performance obtained from the hybrid was superior to the pristine graphite [15].

DMFCs are another type of FC that has the potential to be used on commercial products. However, due to their low functionality, DMFCs have not been utilized in any commercial products, such as laptops and mobile phones. The catalytic activities linked with the methanol oxidation in DMFCs can be improved using fullerene and its derivatives. DMFCs are very simple to operate, easy to handle, produce high energy density, and operate at a lower temperature. The efficiency of DMFCs can be increased by enhancing the methanol electrooxidation activity at the anodic side of FCs. The anodic electrocatalysts for DMFCs are mostly fabricated by employing nanoparticles obtained through the electrical deposition of Pt on a high surface area of conducting carbon film.

The challenges faced include reducing precious metal usage, obtaining a large surface area, and downsizing the FC. For this purpose, different approaches are employed along with fullerene to obtain the maximum activity for methanol oxidation in DMFCs. In one study, fullerene, altered with PAni, was employed along with Pt-Ru in an electrocatalyst [16]. PAni is mostly known for its high charge transfer characteristics, which enhance stability and offer superior capacitance results [17]. The electrocatalyst was prepared by a two-step process involving the composition of PAni-C_{60} and Pt-Ru nanoparticle electrocatalyst production, as displayed in Fig. 8.5. In the first step, PAni-C_{60} composites were produced by the oxidative polymerization of aniline monomer on the powdered C_{60} surface with H_2O_2 as an oxidant. Later, the metal ions were inserted to be attached to the PAni-C_{60} surface. Lastly, the in situ reductions of the metal ions was occurred by using KBH_4 to produce Pt-Ru nanoparticles. The prepared material was electrochemically tested and displayed excellent catalytic activity for methanol oxidation.

Recently, the pyridine and fullerene composite have been characterized for the application in methanol oxidation [18]. The fulleropyrrolidine nanosheets (PyC_{60}) were produced on the glassy carbon (GC) via a dip-attach method followed by the electrical deposition of Pt by immersing the GC altered electrode in K_2PtCl_6 and K_2SO_4. The fulleropyrrolidine offers excellent support to the Pt nanoparticles, increases active surface area, and enhances the stability of the electrocatalyst by providing substantial support to the catalyst. Therefore the Pt/PyC_{60} electrocatalyst showed excellent catalytic activity for methanol oxidation reaction, surpassing the activity obtained from C_{60}/nonfunctionalized Pt and the commercially available Pt/C catalyst.

8.3.1.2 Carbon dots-based electrocatalysts

Carbon dots (CDs) are another 0D carbonaceous materials, which was unintentionally discovered in 2004 during the CNT purification process and was named carbon quantum dots. The CDs are quasispherical nanoparticles and contain sp^3 hybridized carbon

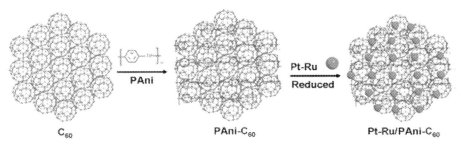

Figure 8.5 Schematic illustration of the synthetic mechanism of Pt-Ru/PAni-C_{60} electrocatalyst (*PAni*, Polyaniline). *Adapted with permission from Z. Bai, M. Shi, L. Niu, Z. Li, L. Jiang, L. Yang, A facile preparation of Pt-Ru nanoparticles supported on polyaniline modified fullerene [60] for methanol oxidation, Journal of Nanoparticle Research 15 (2013). Copyright (2013) Springer Nature.*

and oxygen/nitrogen groups. Recently, CDs have been considered promising materials to be applied in energy conversion and storage devices. To obtain an effectively controlled size, structure, and dimensionality of CDs, fabricating methods have a crucial role. Various research articles are published related to the employment of CDs in photocatalysis. However, a limited amount of work has been done on CDs for FCs. Some works will be discussed in this chapter with their working mechanism. The preparation of CDs can be done either by top-down approach or bottom-up approach. The top-down approach is breaking down a bulkier material into smaller particles. The bottom-up technique is nothing but the opposite of top-down. Furthermore, the CDs can also be doped with heteroatoms to increase their catalytic activity. The doped CDs, such as nitrogen-doped CDs (N-CDs), contain numerous defects on the surfaces and edges that eventually helps in having a large number of active sites.

Different types of metal-free electrocatalysts have been synthesized for the application of FCs and successfully obtained higher catalytic activity than Pt/C. The ORR activity has been further improved by using various dimensional carbon materials, such as graphene. Along with their advantages, some drawbacks mainly arise from their synthetic approaches, such as aggregation, inhomogeneous distribution, and lower doped-atom ratio, which significantly affect the electrocatalytic efficiency. For addressing these issues, researchers are working on CDs to produce green and improved electrocatalysts. CDs possess high electron transfer and high surface area that reflects in their higher catalytic activity. The —COOH, —NH_3, and —OH functional groups on the surface of CDs promotes improved multicomponent electrocatalytic behavior [19]. For this reason, Niu et al. synthesized N-CDs garnished on graphene oxide (GO) by a facile single-step process [20]. The synthetic route involves the in situ preparation of intermediates similar to polymer followed by the carbonization process to decorate the N-CDs on GO for the application as ORR electrocatalyst. The synthetic process is displayed in Fig. 8.6. The synthesized N-CDs/GO electrocatalyst displayed superior electrical conductivity and current density, which was about 1.4-folds than the Pt/C electrocatalyst [20]. The excellent catalytic activity may be due to the improved N-CDs and GO interrelation that promotes fast electron transfer, which ultimately enhanced the catalytic activity.

In general, monodoping of N disturbs the electroneutrality of the material and helps in the absorption of oxygen molecules, and the transformation of reactive species takes place. In addition to this, heteroatoms codoping of sulfur, phosphorus, and nitrogen in CDs are also considered to improve catalytic performance. It can tune the chemical and physical characteristics of the material and facilitates in providing additional active spots to enhance the electrochemical system [21]. For this reason, Zhou et al. came up with a new approach to develop a highly efficient and low-cost electrocatalyst by a two-step process [22]. Briefly, the microbial reduction process was used

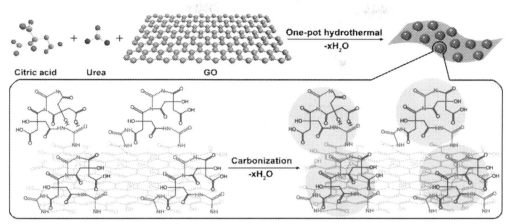

Figure 8.6 Schematic representation of a single-step synthetic process of N-doped CDs garnished on GO (*CDs*, Carbon dots; *GO*, graphene oxide). *Adapted with permission from W.J. Niu, R.H. Zhu, Yan-Hua, H.B. Zeng et al., One-pot synthesis of nitrogen-rich carbon dots decorated graphene oxide as metal-free electrocatalyst for oxygen reduction reaction, Carbon 109 (2016) 402–410. Copyright (2016) Elsevier.*

to produce microorganism-assisted graphene-based carbon dots (CDs-M@rGO) by a facile hydrothermal process. The technique used in this study was to synthesize CDs via Shewanella oneidensis bacterial cell. The cell is comprised of nitrogen, sulfur, and phosphorus sources. Upon hydrothermal treatment, the rod-like morphology of the cell transformed into tiny particles. These particles undergo slight polymerization and carbonization to form CDs having a size ranging from 2 to 6 mm. The as-synthesized CDs were garnished on the surface of microorganism-assisted graphene oxide (M-rGO) itself, as shown in Fig. 8.7A. The diagrammatic representation of the synthetic method is shown in Fig. 8.7B–D. In an alkaline electrolyte, the CD-M@rGO displayed superior catalytic activity for ORR. LSV characterization revealed that E_{ons} and $E_{1/2}$ of CD-M@rGO were significantly superior to carbon dot-based graphene oxide (C-rGO) and M-rGO and similar to Pt/C. The electrocatalyst showed excellent electron transfer, which was about 3.7 and 4.2 at 800–2400 rpm, and exhibited a four-electron transfer route. Furthermore, the accelerated degradation test of CD-M@rGO revealed a loss of only 3% in current density, which is around 15% for Pt/C. These results indicated higher stability of CD-M@rGO. These examples revealed that the 0D carbonaceous materials, such as fullerene and CDs, are promising materials in developing efficient and low-cost electrocatalysts, which can lower Pt metal usage and improve the catalytic activity.

8.3.2 One-dimensional carbonaceous materials

The 1D nanostructures, including nanowires, nanorods, nanotubes, and nanochains, offer a favorable structural model with many advantages over other

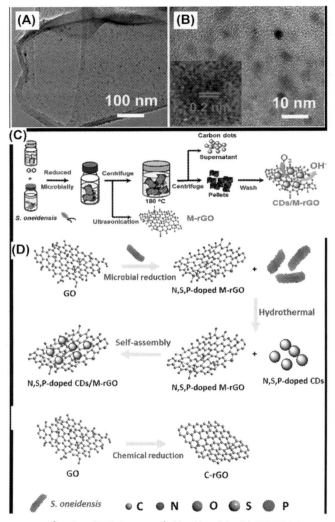

Figure 8.7 (A) Low-magnification TEM image of CDs-M@rGO, (B) HRTEM image of CD-M@rGO at high magnification, (C) schematic illustration for the preparation of CD-M@rGO, and (D) fabrication reaction and activities of CDs-M@rGO and preparation process of C-rGO (*CDs-M@rGO*, microorganism-assisted graphene-based carbon dots; *TEM*, transmission electron microscopy). *Adapted with permission from L. Zhou, P. Fu, Y. Wang, L. Sun, Y. Yuan, Microbe-engaged synthesis of carbon dot-decorated reduced graphene oxide as high-performance oxygen reduction catalysts, Journal of Materials Chemistry A 4 (2016) 7222–7229. Copyright (2016) Royal Society of Chemistry.*

counterparts. 1D designs allow the valuable exposure of low-energy crystal surface and lattice planes with lower lattice borders. The indicating surface would assist decrease the surface energy of the entire system and actively participate in the catalysis. The individual crystalline planes restrict the unwanted low-coordinating defect spots that have low catalytic activity and susceptible to oxidation and degradation. Furthermore, 1D nanomaterials promote the electron transfer via path guiding effect in electrocatalysts and improve the reaction kinetics on the electrocatalyst surface. Moreover, the nanostructures having less than 2 mm diameter are considered to have enhanced catalytic activity because of the surface shrinking effect [23]. The substantial scale length and the withheld electric conduction help reduce the reliance on carbon support for dispersion of nanoparticles and electron conductivity, which eventually tackle the support decay challenges confronted by the Pt/C electrocatalysts [24]. These remarkable benefits of 1D nanostructures allow the reduction of employment of precious metals with no effects on the catalytic activity but have improved stability. By considering all these characteristics, 1D nanostructures are emerging as a new pathway toward efficient electrocatalysts for FC applications and have drawn a significant focus of the researchers worldwide.

8.3.2.1 Carbon nanotubes-based 1D nanostructures

The CNTs are sp^2-hybridized carbon atoms, which restricts the high catalytic activity of the pristine CNTs, regardless of their superior electron conduction, chemical stability, and special 1D morphology. Various methods have been developed to overcome this barrier and improve the activity of the materials, which can be split into two main strategies. (1) The introduction of heteroatom to obtain heteroatom-doped CNTs, which can modify the electronic arrangement of the material and improve the catalytic activity, as displayed in Fig. 8.8A. (2) Utilize the substantial surface area, chemical stability, high electron conduction of CNT and produce a CNT matrix to decorate catalytically active material on it, such as Pt (Fig. 8.8B) and heteroatom–doped materials (Fig. 8.8C).

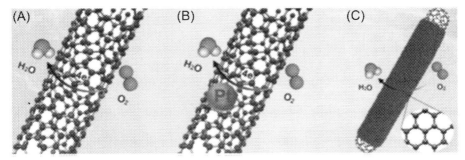

Figure 8.8 Illustration of ORR catalysis on (A) heteroatom-doped CNT, (B) Pt/CNT, and (C) CNT@heteroatom-doped carbon [25] (*CNT*, Carbon nanotube; *ORR*, oxygen reduction reaction). *Adapted with permission from J.C. Li, P.X. Hou, C. Liu, Heteroatom-doped carbon nanotube and graphene-based electrocatalysts for oxygen reduction reaction, Small 13 (2017) 1702002. Copyright (2017) John Wiley and Sons.*

8.3.2.1.1 Heteroatom-doped CNTs

The heteroatom-doped CNTs can be prepared by two synthetic routes, which are in situ process and posttreatment. The in situ process involves the replacement of carbon atoms with the heteroatoms within the carbon structure in the course of CNTs development. Usually, the in situ process requires Fe/Co catalysts to disintegrate C atoms or/and dopants and speed up the CNTs development (Fig. 8.9). Chen et al. developed N-doped CNTs via an injection CVD process [26]. In this method, ethylenediamine was used as a carbon and nitrogen source precursor. The prepared electrocatalyst ethylenediamine-nitrogen-containing carbon nanotube (NCNT) showed excellent results, which are comparable to Pt/C regarding E_{ons} and $E_{1/2}$. N-doping and defects were considered the main reason for the excellent catalytic activity. Similarly, Su et al. prepared N-doped CNTs via direct pyrolysis of Zn-Fe-zeolitic imidazolate framework (ZIF), as illustrated in Fig. 8.10 [27]. Ferric (Fe) was also employed as a metal source, which was also presented in the final product. A large quantity of graphitic N and the presence of Fe species demonstrated superior electrocatalytic activity than the Pt/C in an alkaline environment.

In a recent time, a composite of NCNT and NPs were prepared via a direct pyrolysis process of a mixture containing iron salt and N-based organic compound [28]. In an alkaline environment, the synthesized electrocatalysts displayed excellent catalytic activity compared to or superior to Pt/C. But, in an acidic environment, the catalyst showed a more negative $E_{1/2}$ value than Pt/C (>100 mV). Li and coworkers proposed and synthesized a hierarchical mesoporous/microporous Fe-/N-doped CNT with the help of Fe(NO$_3$) [29]. They employed anodic aluminum oxide (AAO) as a template for CNT. AAO also provided Fe to produce Fe-N$_x$ active spots, as illustrated in Fig. 8.11. The synthesized material exhibits a superior surface area of about 2137 m^2, ample Fe-N$_x$ active sites, excellent

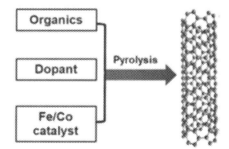

Figure 8.9 Schematic showing in situ growth of heteroatom-doped CNTs (*CNTs*, Carbon nanotubes) [25]. *Adapted with permission from J.C. Li, P.X. Hou, C. Liu, Heteroatom-doped carbon nanotube and graphene-based electrocatalysts for oxygen reduction reaction, Small 13 (2017) 1702002. Copyright (2017) John Wiley and Sons.*

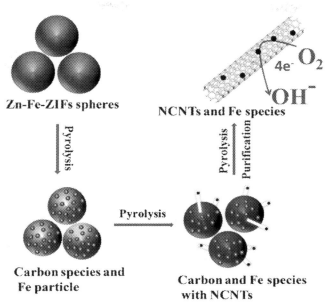

Figure 8.10 Schematic illustration of the synthesis procedure for N-CNTs (all N-CNTs contain Fe species) through the pyrolysis of Zn-Fe-ZIF spheres (*N-CNTs*, nitrogen-doped carbon nanotubes; *ZIF*, zeolitic imidazolate framework) [27]. *Adapted with permission from P. Su, H. Xiao, J. Zhao, Y. Yao, Z. Shao, C. Li, et al., Nitrogen-doped carbon nanotubes derived from Zn-Fe-ZIF nanospheres and their application as efficient oxygen reduction electrocatalysts with in situ generated iron species, Chemical Science 4 (2013) 2941–2946. Copyright (2013) Royal Society of Chemistry.*

conductivity, and meso/micropores, and displayed exceptional ORR activity in an alkaline and acidic environment. Additionally, different template techniques have been investigated to produce Fe- and N-doped CNTs having superior electrocatalytic activity. The samples produced by these techniques have plentiful porous structures, which offer exceptional advantages to FC reactions with ammonia [30]. The sample exhibited an excellent ORR performance in an alkaline environment, similar to Pt/C, and has around 80 mV more negative $E_{1/2}$ than Pt/C in an acidic environment. Moreover, TEM and elemental mapping characterizations were employed to explore the distribution of Fe and N arrangement atomically. The results revealed that the Fe species were in the near vicinity of N atoms, which suggested that the Fe-N_x acts as active spots.

8.3.2.1.2 Carbon nanotube-supported active materials

Highly conductive CNT are extensively utilized as a carrier for active species, and this technique has widely been employed to produce super active electrocatalysts for FCs. CNTs excellently support highly active noble-metal NPs and improve their performance by limiting active material accumulation and improving the electric conductance through their

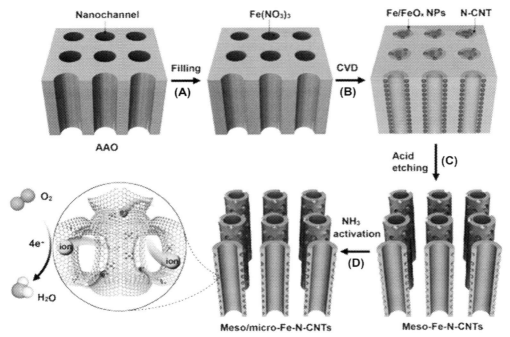

Figure 8.11 Schematic illustration of the synthetic procedure of the meso-micro-Fe-N-CNTs (*CNTs*, Carbon nanotubes) [29]. *Adapted with permission from J.C. Li, P.X. Hou, C. Shi, S.Y. Zhao, D.M. Tang, M. Cheng, et al., Hierarchically porous Fe-N-doped carbon nanotubes as efficient electrocatalyst for oxygen reduction, Carbon 109 (2016) 632–639. Copyright (2016) Elsevier.*

inherently strong conductivity and exceptional surface area. But, the high costs of noble metals turn the direction of researches toward nonprecious metal-based electrocatalysts. For this reason, new materials were developed in past with comparable or even better catalytic performances, as mentioned above. For instance, CNTs are also capable of supporting nonmetallic species. Highly active nonmetallic CNT-based electrocatalysts were fabricated by layering a heteroatom-doped carbon shell on the CNTs core (Fig. 8.8C) [31,32]. The results revealed that the synthesized electrocatalysts performed comparably to Pt/C in an alkaline environment and even outperformed in an acidic environment. Similarly, bioinspired active species are also considered as a potential material for FC catalysis. They are produced by low-temperature reaction rather than high-temperature carbonization. Recently, bioinspired nickel garnered on CNTs have been synthesized and obtained excellent results for H_2-air FCs.

8.3.3 Graphene-based two-dimensional carbonaceous materials

The most common 2D carbonaceous material is graphene. It is a 2D one-atom-thick carbon layer arranged in a hexagonal structure having sp^2 hybridization. All other graphitic forms of carbon are made up of graphene, which also includes CNTs. Exhibiting various similarities with CNTs features, such as superior aspect ratio, superior surface area, plentiful electronic states, and excellent mechanical strengths, graphene is considered a promising material for various applications. Comparative to CNTs, the 2D planar conformation of graphene aids improves the electronic conductivity and is the best candidate to be employed in ORR electrocatalyst.

8.3.3.1 Heteroatom-doped graphene materials

Graphene materials are usually doped with different heteroatoms to obtain improved catalytic activities. Among the heteroatoms, nitrogen doping has proved to be an efficient method to enhance the catalytic activities of graphene. The insertion of nitrogen atoms into the sp^2-hybridized carbonaceous structure modifies the surficial electronic state of graphene. For example, the electronic lone pair of N atoms can develop a delocalized conjugated structure with the carbon structure [33], enhancing the reaction capabilities and catalytic activity of graphene. Different types of nitrogen functionalities, such as pyridinic, pyrrolic, and graphitic nitrogen, are produced within the graphitic structure during the synthesis, as shown in Fig. 8.12. Soon after the employment of vertically aligned nitrogen-containing carbon nanotubes (VA-NCNTs) as an ORR electrocatalyst [33], Qu et al. produced nitrogen-doped graphene films synthesized via CVD with the help of ammonia [35]. The synthesized material showed excellent electrocatalytic activity in an alkaline environment and was

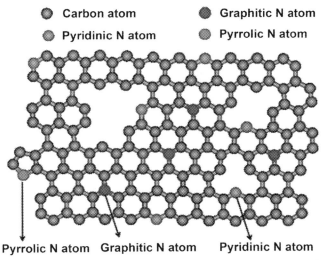

Figure 8.12 Schematic structure of N-doped grapheme [34]. *Adapted with permission from Z. Sheng, L. Shao, J. Chen, W. Bao, F. Wang, X. Xia, Catalyst-free synthesis of nitrogen-doped graphene via thermal annealing graphite oxide with melamine and its excellent electrocatalysis, ACS Nano 5 (2011) 4350–4358. Copyright (2011) American Chemical Society.*

similar to the activity exhibited by VA-NCNTs having similar N-doping content [35]. Similar to N-doped CNTs. It was found out that heteroatom doping can also modify the chemical properties of graphene.

Graphene-based nitrogen-doped material with various sizes, characteristics, and costs for different applications can be obtained via CVD and arc discharge of graphite electrodes and several extensive techniques, including molecular assembling and liquid-phase exploitation, as illustrated in Fig. 8.13 [36]. For example, a unique approach for single-step one-pot direct preparation of nitrogen-doped graphene has been published [37]. The samples were prepared by the reaction between tetrachloromethane and lithium nitride under moderate conditions, which exhibited the nitrogen content ranging from 4.5% to 16.4%. An improved ORR electrocatalytic activity was observed, which was better than pure graphene and carbon black available in the market (XC-72). Furthermore, multilayered N-doped graphene films have been produced, having an N content of 12.5% and a uniform structure [38]. The material was obtained via a reaction between cyanuric chloride and trinitrophenol. Results obtained from the material showed superior catalytic performance and high stability in a phosphate buffer solution with a mixed two- and four-electron transportation route.

Moreover, Sheng et al. described that the pyridine-N in nitrogen-doped graphene controls the ORR catalytic activity of graphene, and the nitrogen content in graphene does not influence the catalytic activity to a large extent [34]. Similarly, Deng et al.

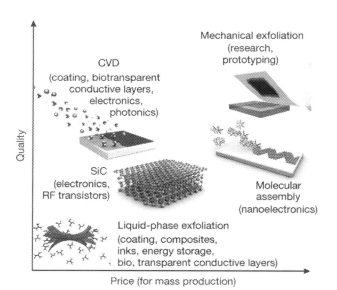

Figure 8.13 Scheme illustration of methods for mass production of graphene. *Adapted with permission from K.S. Novoselov, V.I. Fal'Ko, L. Colombo, P.R. Gellert, M.G. Schwab, K. Kim, A roadmap for graphene, Nature 490 (2012) 192−200. Copyright (2012) Springer Nature.*

demonstrated that the ORR catalytic activity of the graphene relies on both graphitic and pyridinic nitrogen [37]. The catalytic mechanism of nitrogen-doped graphene was also investigated in an acidic condition [39]. The density functional theory method was employed for this study. The results obtained during the electronic transportation revealed that the ORR followed a four-electron transfer route in the presence of graphitic nitrogen. But, in the case of pure graphene, the dissimilar electrocatalytic performance was witnessed. When graphene is doped with nitrogen, irregular spin and atomic charge density are established, which helps N-doped graphene exhibit superior electrocatalytic activity. Despite the indication of different studies for the potential impact of nitrogen on ORR performance, further investigation is essential to fully grasp the actual relationship between electrocatalytic performance and the N-doped graphene.

Similarly, graphene materials have also been doped with other heteroatoms, including sulfur, phosphorus, boron, and halogens. Among these heteroatoms, sulfur atoms possess the same electronegativity as carbon atoms and do not exhibit doping-triggered charge transportation. However, sulfur-doped graphene also displayed superior electrocatalytic performance than the Pt/C in alkaline media [40]. The better catalytic activity is probably induced because of the dope-triggered spin redistributive effect [41]. Liu et al. doped graphene with phosphorus, which has similar valence shell electrons and chemical properties that of nitrogen [42]. The synthetic route involves pyrolyzing toluene and TPP at $1000°C$ under an argon environment. The synthesized material displayed excellent catalytic performance, long-term stability, and superior capability to tolerate methanol crossover in alkaline conditions. The results obtained from these researches are significant in obtaining metal-free doped-carbon material for ORR and also helpful in welcoming new synthetic routes for extensive production of new low-priced catalysts for ORR.

8.3.3.2 Graphene-supported carbonaceous materials

The catalytic activities of the material are firmly associated with their surficial characteristics, particularly for graphene-based materials. Therefore their catalytic performance can be further enhanced by immobilizing active materials on the graphene surface. The catalytic performance will also be affected by the composition, particle size, exposed surface, and interaction between the supporting material and the graphene. Till now, various synthetic approaches have been employed to prepare graphene-supported catalysts. Most of them are Pt-based electrocatalysts supported on graphene, which improves the catalytic performance and reduced corrosion and oxidation of the materials. But, the scarcity and the high cost of precious metals are the main hurdles for large scale production. Instead of precious metals, nonprecious metals, such as Fe, Cu, and Co, have also been supported on graphene and demonstrated comparable electrocatalytic performance. The interaction between the graphene surface and the non-Pt metal induces a charge transport process. This charge transport relies on the space and Fermi-level variance between supporting graphene and the nanocatalysts, which is advantageous for the improved electrocatalytic performance of the graphene-based electrocatalyst.

Tsai and colleagues synthesized nanoparticles (FeCN) containing C and Fe_3N_2 supported on N-doped graphene nanosheets [43]. The graphene nanosheets were produced via encapsulation and heat treatment methods. Comparing to other samples, such as N-doped GN, FeN, FeN@C, and the FeCN@N-GNs displayed superior ORR catalytic performance with a four-electron transfer route and possessed exceptional stability. The Fe-, N-, and C-based active spots in the FeCN@N-GNs are a potential reason for the improved electrocatalytic activity. Furthermore, Co-containing nanomaterials anchored on graphene sheets have been garnered much interest from the researchers due to their highly active electrocatalytic performance. Various techniques have been formulated to obtain higher performance. Kim et al. synthesized Co [tetrakis{oaminophenyl}porphyrin] (CoTAPP) nanomaterials and supported it on the carbon nanostructures, such as graphene, uniwalled CNTs, and multiwalled CNTs by diazonium salt reaction [44]. By comparing their electrocatalytic activity toward ORR, it is observed that the CoTAPP@G outperformed other electrocatalysts with excellent catalytic performance without any use of precious metals.

8.3.4 Three-dimensional carbon materials

Along with ongoing research on creating innovative 0D, 1D, and 2D ORR electrocatalysts, 3D carbonaceous materials are also investigated for developing an efficient performance. Developing precise 3D porous carbon-based structures is a potential strategy to improve the ORR activity by offering superior electrolyte permeability, mass transfer, and electron transportation pathway. 3D porous nanostructures can be categorized into three groups based on pore size, such as microporous (less than 2 nm), mesoporous (greater than 2 nm and less than 50 nm), and macroporous (greater than 50 nm) nanomaterials. 3D nanomaterials have been developed into various types of hierarchically porous morphology by extending the pores throughout the length scale range. The main advantages of 3D nanostructures are their large surface area, plentiful active spots, and superior pore volume. These benefits allow 3D nanomaterials to produce high catalytic activity without the need for precious metals. Furthermore, porous carbon materials with the active species offer an excellent possibility to improve catalytic activity and stability.

8.3.4.1 Heteroatom-doped porous carbon nanomaterials

The carbon nanomaterials are usually doped with heteroatom to increase the catalytic performance of the materials. They can be doped with nitrogen, boron, phosphorus, and sulfur, which act as active spots and facilitate catalytic performance.

For example, Liang et al. synthesized hierarchically porous carbonaceous nanostructures doped with N via template technique and employed them as ORR electrocatalysts, as shown in Fig. 8.14A [45]. Colloidal silica was employed as the hard template, and a polymer of o-phenylenediamine was utilized as a precursor. The as-synthesized mesoporous carbon nanomaterial was activated in ammonia at an elevated temperature to further increase the surface area, which was improved to 1280 m^2 g^{-1}, having mesoporous and microporous morphologies with a large number of micropores. The hierarchy of the pores can be verified by the presence of ultramicro, microporous structure integrated into the mesoporous structures, as displayed in Fig. 8.14B and C. Furthermore, the ammonia activation drastically improves the quarternary N content from 38% to 71% (Fig. 8.14D). The electrochemical characterization of the material demonstrated enhanced E$_{1/2}$, superior stability, and a four-electron transfer pathway (Fig. 8.14E and F). Overall, the improved surface area, enhanced quarternary-N content, and the decreased inherent resistance facilitated ORR performance. Along with the hierarchical porous structure, the catalytic activity of the electrocatalysts can be further improved by

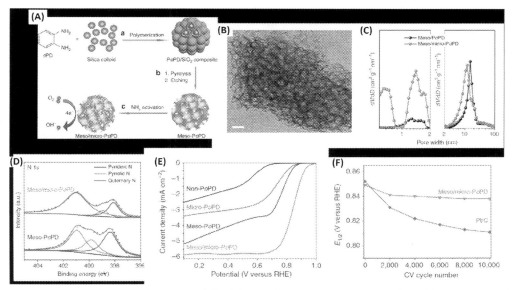

Figure 8.14 (A) Schematic depiction of the fabricated meso-micro-PoPD catalyst, (B) TEM image of meso-micro-PoPD, (C) related pore size distribution plots, (D) high-resolution N 1s spectra of meso-PoPD and meso-micro-PoPD electrocatalysts, (E) ORR polarization curves of meso-micro-PoPD with control samples, and (F) E$_{1/2}$ as a function of the number of potential cycles (*ORR*, Oxygen reduction reaction; *PoPD*, polymer of o-phenylenediamine; *TEM*, transmission electron microscopy). *Adapted with permission from H.W. Liang, X. Zhuang, S. Brüller, X. Feng, K. Müllen, Hierarchically porous carbons with optimized nitrogen doping as highly active electrocatalysts for oxygen reduction, Nature Communications 5 (2014) 4973. Copyright (2014) Springer Nature.*

codoping. For instance, Zhao et al. fabricated sulfur and nitrogen-doped carbon catalysts by employing silica nanospheres template, having perfect trimodal porous morphology [46]. The excellent ORR activity was observed due to their excellent S- and N-doping content and superior electronic and mass transfer characteristics of the synthesized material.

8.3.4.2 Metal-organic framework-derived porous carbon materials

In recent times, metal-organic frameworks (MOFs) have successfully attracted broad interest in different applications. MOFs are the coordination between the organic and inorganic moieties through covalent bonding. They offer varieties of different structures and exceptional benefits of high surface area, superior pore volume, and their availability in different pore sizes. MOFs are considered a substitute precursor to obtaining nanoporous carbonaceous material, allowing them to be employed for various applications. Fu et al. fabricated N-doped hierarchical porous carbon material from nitrogen-based isoreticular MOF-3 via direct pyrolysis [47]. The porosity, surface area, and the content of N-doping were managed by the pyrolysis temperature. The material synthesized at 950°C demonstrated superior ORR catalytic performance, near to Pt/C. The actual interest for these materials emanates from the adjustability of structure, pore size and volume, and controllable doping content. Furthermore, the influence of pyrolysis temperature and different variables can be adjusted to regulate the extent of graphitization, pore size and volume, amount of N-doping, and consequently, the reaction pathway.

8.3.4.3 Three-dimensional porous metal-nitrogen-carbon materials

Currently, lots of attempts have been made to develop efficient nonprecious metal-based electrocatalysts with different active spots. Among these materials, transition metal-nitrogen-carbon (MNC)-based materials have been considered a potential category of nonprecious metal catalysts for ORR catalysis due to their superior performance in acidic and alkaline environments. Two general techniques have been employed to prepare MNC electrocatalysts. One includes the metal-nitrogen moieties, and the second necessitates the N dopants inside the carbon network. It is believed that both metal-nitrogen moiety and N-doping are significant to improve the ORR activity. However, the working principle for the active spots in MNC catalysts is wholly known. It is believed that the surface area and the morphology of the electrocatalysts mainly control the availability of the active spots and improves the catalytic activity accordingly. By now, the preparation of MNC catalysts is performed by precisely selecting the appropriate transition metal/nitrogen precursors and carbon support accompanied by the heating process. Based on widespread research to improve the ORR activity, careful regulation of the metal-nitrogen moieties throughout the 3D porous carbon material is essential for acquiring the required composition and morphology, which is important to attain high performance and stability analogous to Pt/C.

For obtaining better porosity and pore size, mesoporous silica and silica nanoparticles are generally employed as a template to synthesize hierarchically arranged porous MNC electrocatalysts with suitable metal-nitrogen precursors. Kong et al. produced a range of various hierarchically porous Fe-N-C electrocatalysts by pyrolyzing various N-based materials and FeCl$_3$ with the Santa Barbara amorphous-15 (SBA-15) templates [48]. The author carried out a systematic study to determine the influence of N precursor and the pyrolyzing temperature on ORR activity. It is observed that the employment of 2,2 bipyridine and iron chelates at an annealing temperature of 900°C displayed excellent E$_{ons}$ and E$_{1/2}$ in acidic and alkaline conditions. Superior ORR performance was accomplished due to the optimum equilibrium of active spot density and efficient charge and mass transfer. Similarly, Liang and his colleagues employed silica nanoparticles, SBA-15, and montmorillonite (MMT) as a template for fabricating structures with different porosity, as shown in Fig. 8.15A—C [49]. The vitamin B12

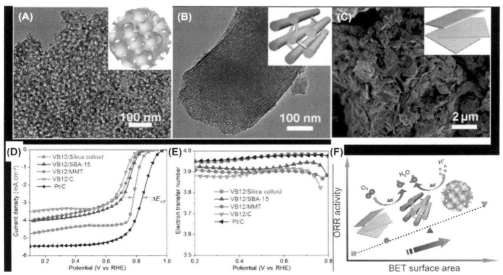

Figure 8.15 [(A) and (B)] TEM image, (C) SEM images of as-synthesized Co-N-C catalysts: (A) VB12-silica colloid, (B) VB12-SBA-15, and (C) VB12-MMT; (D) ORR polarization curve of various Co-N-C electrocatalysts in 0.5M oxygen-saturated sulfuric acid, (E) electron-transfer characterization of various Co-N-C and Pt/C electrocatalysts, and (F) the reliance of ORR catalytic performance on the Brunauer—Emmett—Teller (BET) surface area (*MMT*, Montmorillonite; *ORR*, oxygen reduction reaction; *SBA-15*, Santa Barbara amorphous-15; *SEM*, scanning electron microscopy; *TEM*, transmission electron microscopy; *VB12*, vitamin B12) [49]. *Adapted with permission from H.-W. Liang, W. Wei, Z.-S. Wu, X. Feng, K. Müllen, Mesoporous metal-nitrogen-doped carbon electrocatalysts for highly efficient oxygen reduction reaction, Journal of the American Chemical Society 135 (2013) 16002—16005. Copyright (2013) American Chemical Society.*

(VB12) and colloidal silica exhibited interlinked bubble-like morphology, VB12 and SBA-15 offered well-established rectilinearly arranged mesoporous morphology, and VB12 and MMT composition presented nanosheet-like structures. Furthermore, a high amount of N/Co and plentiful active spots were also identified in the synthesized samples. Among all three samples, VB12/colloidal silica demonstrated better ORR performance in acidic condition, which indicated that the well-established porous morphology and superior surface area played a critical part in improving ORR activity, as demonstrated in Fig. 8.15D−F.

8.4 Conclusion

The significance of synthesizing renewable materials for energy harvesting applications has been obvious since worldwide energy utilization, and greenhouse gas emissions are expediting at an alarming rate. In this scenario, FCs are considered to be an efficient and environment-friendly device for energy conversion that can fulfill the demand for substitutive energy sources. The state-of-the-art platinum metal and platinum oxides displayed unprecedented ORR activity. But, the high cost and less availability restrict the broad-scale application of FCs. The comprehensive research works have been performed to reduce the cost and improve the activity of nonprecious metal-based electrocatalysts. These researches demonstrated that carbon-based materials exhibit essential properties that can be utilized for attaining higher catalytic activity. These properties include high surface area, porous structure, various pore size, large pore volume, superior charge transfer, mass transport, enhanced electrical conductance, anchoring ability, different synthetic techniques, and their availability in all four-dimensional systems, for example, fullerene as 0D, CNTs as 1D, graphene as 2D, and various other 3D carbonaceous materials. These materials showed excellent electrocatalytic performance without any incorporation of precious metals. Heteroatom doping, incorporation of transition metal species, and improving the porosity by different techniques to obtain high surface area and improved accessibility are the critical factors for attaining good catalytic performance. With all these advancements and techniques, carbon-based materials still need to be improved to obtain Pt-like catalytic activity. For this reason, further researches are being done to produce electrocatalysts with higher activity and stability so that it can be employed in a wide range of applications.

References

[1] R. Bashyam, P. Zelenay, A class of non-precious metal composite catalysts for fuel cells, Nature 443 (2006) 63−66.
[2] H. Xu, R. Kunz, J.M. Fenton, Investigation of platinum oxidation in PEM fuel cells at various relative humidities, Electrochemical and Solid-State Letters 10 (2007) 1−5.

[3] H.A. Gasteiger, S.S. Kocha, B. Sompalli, F.T. Wagner, Activity benchmarks and requirements for Pt, Pt-alloy, and non-Pt oxygen reduction catalysts for PEMFCs, Applied Catalysis B: Environmental 56 (2005) 9−35.

[4] H. Huang, X. Wang, Recent progress on carbon-based support materials for electrocatalysts of direct methanol fuel cells, Journal of Materials Chemistry A 2 (2014) 6266−6291.

[5] V. Georgakilas, J.A. Perman, J. Tucek, R. Zboril, Broad family of carbon nanoallotropes: classification, chemistry, and applications of fullerenes, carbon dots, nanotubes, graphene, nanodiamonds, and combined superstructures, Chemical Reviews 115 (2015) 4744−4822.

[6] V.G. Gavalas, N.A. Chaniotakis, [60]Fullerene-mediated amperometric biosensors, Analytica Chimica Acta 409 (2000) 131−135.

[7] S. Sotiropoulou, V. Gavalas, V. Vamvakaki, N.A. Chaniotakis, Novel carbon materials in biosensor systems, Biosensors and Bioelectronics 18 (2002) 211−215.

[8] S. Palanisamy, B. Thirumalraj, S.M. Chen, M.A. Ali, F.M.A. Al-Hemaid, Palladium nanoparticles decorated on activated fullerene modified screen printed carbon electrode for enhanced electrochemical sensing of dopamine, Journal of Colloid and Interface Science 448 (2015) 251−256.

[9] S.E. Shaheen, C.J. Brabec, N.S. Sariftci, F. Padinger, T. Fromherz, J.C. Hummelen, 2.5% efficient organic plastic solar cells, Applied Physics Letters 78 (2001) 841−843.

[10] M. Noked, A. Soffer, D. Aurbach, The electrochemistry of activated carbonaceous materials: past, present, and future, Journal of Solid State Electrochemistry 15 (2011) 1563−1578.

[11] H. Shioyama, A. Ueda, N. Kuriyama, Surface treatment of carbon supports for PEM fuel cell electrocatalyst, Journal of New Materials for Electrochemical Systems 10 (2007) 201−204.

[12] M.A. Gabriel, L. Genovese, G. Krosnicki, O. Lemaire, T. Deutsch, A.A. Franco, Metallofullerenes as fuel cell electrocatalysts: a theoretical investigation of adsorbates on $C_{59}Pt$, Physical Chemistry Chemical Physics 12 (2010) 9406−9412.

[13] S.M. Shamsunnahar, M. Nagai, Nitrogen doping of ash-free coal and effect of ash components on properties and oxygen reduction reaction in fuel cell, Fuel 126 (2014) 134−142.

[14] G. Wu, M. Nelson, S. Ma, H. Meng, G. Cui, P.K. Shen, Synthesis of nitrogen-doped onion-like carbon and its use in carbon-based CoFe binary non-precious-metal catalysts for oxygen-reduction, Carbon 49 (2011) 3972−3982.

[15] J. Guan, X. Chen, T. Wei, F. Liu, S. Wang, Q. Yang, et al., Directly bonded hybrid of graphene nanoplatelets and fullerene, Journal of Materials Chemistry A 3 (2015) 4139−4146.

[16] Z. Bai, M. Shi, L. Niu, Z. Li, L. Jiang, L. Yang, A facile preparation of Pt-Ru nanoparticles supported on polyaniline modified fullerene [60] for methanol oxidation, Journal of Nanoparticle Research 15 (2013).

[17] C. Hu, S. Chen, Y. Wang, X. Peng, W. Zhang, J. Chen, Excellent electrochemical performances of cabbage-like polyaniline fabricated by template synthesis, Journal of Power Sources 321 (2016) 94−101.

[18] X. Zhang, L.X. Ma, Electrochemical fabrication of platinum nanoflakes on fulleropyrrolidine nanosheets and their enhanced electrocatalytic activity and stability for methanol oxidation reaction, Journal of Power Sources 286 (2015) 400−405.

[19] M.F. Gomes, Y.F. Gomes, A. Lopes-Moriyama, E.L. de Barros Neto, C.P. de Souza, Design of carbon quantum dots via hydrothermal carbonization synthesis from renewable precursors, Biomass Conversion and Biorefinery 9 (2019) 689−694.

[20] W.J. Niu, R.H. Zhu, Yan-Hua, H.B. Zeng, et al., One-pot synthesis of nitrogen-rich carbon dots decorated graphene oxide as metal-free electrocatalyst for oxygen reduction reaction, Carbon 109 (2016) 402−410.

[21] V.C. Hoang, K. Dave, V.G. Gomes, Carbon quantum dot-based composites for energy storage and electrocatalysis: mechanism, applications and future prospects, Nano Energy 66 (2019) 104093.

[22] L. Zhou, P. Fu, Y. Wang, L. Sun, Y. Yuan, Microbe-engaged synthesis of carbon dot-decorated reduced graphene oxide as high-performance oxygen reduction catalysts, Journal of Materials Chemistry A 4 (2016) 7222−7229.

[23] C. Koenigsmann, W.P. Zhou, R.R. Adzic, E. Sutter, S.S. Wong, Size-dependent enhancement of electrocatalytic performance in relatively defect-free, processed ultrathin platinum nanowires, Nano Letters 10 (2010) 2806−2811.

[24] B.Y. Xia, W.T. Ng, H. Bin Wu, X. Wang, X.W. Lou, Self-supported interconnected Pt nanoassemblies as highly stable electrocatalysts for low-temperature fuel cells, Angewandte Chemie International Edition, 51, 2012, pp. 7213–7216.
[25] J.C. Li, P.X. Hou, C. Liu, Heteroatom-doped carbon nanotube and graphene-based electrocatalysts for oxygen reduction reaction, Small 13 (2017) 1702002.
[26] Z. Chen, D. Higgins, H. Tao, R.S. Hsu, Z. Chen, Highly active nitrogen-doped carbon nanotubes for oxygen reduction reaction in fuel cell applications, Journal of Physical Chemistry C 113 (2009) 21008–21013.
[27] P. Su, H. Xiao, J. Zhao, Y. Yao, Z. Shao, C. Li, et al., Nitrogen-doped carbon nanotubes derived from Zn-Fe-ZIF nanospheres and their application as efficient oxygen reduction electrocatalysts with in situ generated iron species, Chemical Science 4 (2013) 2941–2946.
[28] W. Yang, X. Liu, X. Yue, J. Jia, S. Guo, Bamboo-like carbon nanotube/Fe$_3$C nanoparticle hybrids and their highly efficient catalysis for oxygen reduction, Journal of the American Chemical Society 137 (2015) 1436–1439.
[29] J.C. Li, P.X. Hou, C. Shi, S.Y. Zhao, D.M. Tang, M. Cheng, et al., Hierarchically porous Fe-N-doped carbon nanotubes as efficient electrocatalyst for oxygen reduction, Carbon 109 (2016) 632–639.
[30] Y. Li, W. Zhou, H. Wang, L. Xie, Y. Liang, F. Wei, et al., An oxygen reduction electrocatalyst based on carbon nanotube-graphene complexes, Nature Nanotechnology 7 (2012) 394–400.
[31] H. An, R. Zhang, Z. Li, L. Zhou, M. Shao, M. Wei, Highly efficient metal-free electrocatalysts toward oxygen reduction derived from carbon nanotubes@polypyrrole core-shell hybrids, Journal of Materials Chemistry A 4 (2016) 18008–18014.
[32] Y.J. Sa, C. Park, H.Y. Jeong, S.H. Park, Z. Lee, K.T. Kim, et al., Carbon nanotubes/heteroatom-doped carbon core-sheath nanostructures as highly active, metal-free oxygen reduction electrocatalysts for alkaline fuel cells, Angewandte Chemie International Edition 53 (2014) 4102–4106.
[33] K. Gong, F. Du, Z. Xia, M. Durstock, L. Dai, Nitrogen-doped carbon nanotube arrays with high electrocatalytic activity for oxygen reduction, Science 323 (2009) 760–764.
[34] Z. Sheng, L. Shao, J. Chen, W. Bao, F. Wang, X. Xia, Catalyst-free synthesis of nitrogen-doped graphene via thermal annealing graphite oxide with melamine and its excellent electrocatalysis, ACS Nano 5 (2011) 4350–4358.
[35] L. Qu, Y. Liu, J.-B. Baek, L. Dai, Nitrogen-doped graphene as efficient metal-free electrocatalyst for oxygen reduction in fuel cells, ACS Nano 4 (2010) 1321–1326.
[36] K.S. Novoselov, V.I. Fal'Ko, L. Colombo, P.R. Gellert, M.G. Schwab, K. Kim, A roadmap for graphene, Nature 490 (2012) 192–200.
[37] D. Deng, X. Pan, L. Yu, Y. Cui, Y. Jiang, J. Qi, et al., Toward N-doped graphene via solvothermal synthesis, Chemistry of Materials 23 (2011) 1188–1193.
[38] L. Feng, Y. Chen, L. Chen, Easy-to-operate and low-temperature synthesis of gram-scale nitrogen-doped graphene and its application as cathode catalyst in microbial fuel cells, ACS Nano 5 (2011) 9611–9618.
[39] L. Zhang, Z. Xia, Mechanisms of oxygen reduction reaction on nitrogen-doped graphene for fuel cells, Journal of Physical Chemistry C 115 (2011) 11170–11176.
[40] Z. Yang, Z. Yao, G. Li, G. Fang, H. Nie, Z. Liu, et al., Sulfur-doped graphene as an efficient metal-free cathode catalyst for oxygen reduction, ACS Nano 6 (2012) 205–211.
[41] I.Y. Jeon, S. Zhang, L. Zhang, H.J. Choi, J.M. Seo, Z. Xia, et al., Edge-selectively sulfurized graphene nanoplatelets as efficient metal-free electrocatalysts for oxygen reduction reaction: the electron spin effect, Advanced Materials 25 (2013) 6138–6145.
[42] Z.W. Liu, F. Peng, H.J. Wang, H. Yu, W.X. Zheng, J. Yang, Phosphorus-doped graphite layers with high electrocatalytic activity for the O_2 reduction in an alkaline medium, Angewandte Chemie International Edition 50 (2011) 3257–3261.
[43] C.W. Tsai, M.H. Tu, C.J. Chen, T.F. Hung, R.S. Liu, W.R. Liu, et al., Nitrogen-doped graphene nanosheet-supported non-precious iron nitride nanoparticles as an efficient electrocatalyst for oxygen reduction, RSC Advances 1 (2011) 1349–1357.
[44] S.K. Kim, S. Jeon, Improved electrocatalytic effect of carbon nanomaterials by covalently anchoring with CoTAPP via diazonium salt reactions, Electrochemistry Communications 22 (2012) 141–144.

[45] H.W. Liang, X. Zhuang, S. Brüller, X. Feng, K. Müllen, Hierarchically porous carbons with optimized nitrogen doping as highly active electrocatalysts for oxygen reduction, Nature Communications 5 (2014) 4973.
[46] Y. Li, H. Zhang, Y. Wang, P. Liu, H. Yang, X. Yao, et al., A self-sponsored doping approach for controllable synthesis of S and N co-doped trimodal-porous structured graphitic carbon electrocatalysts, Energy & Environmental Science 7 (2014) 3720–3726.
[47] S. Fu, C. Zhu, Y. Zhou, G. Yang, J.W. Jeon, J. Lemmon, et al., Metal-organic framework derived hierarchically porous nitrogen-doped carbon nanostructures as novel electrocatalyst for oxygen reduction reaction, Electrochimica Acta 178 (2015) 287–293.
[48] A. Kong, X. Zhu, Z. Han, Y. Yu, Y. Zhang, B. Dong, et al., Ordered hierarchically micro- and mesoporous Fe-Nx-embedded graphitic architectures as efficient electrocatalysts for oxygen reduction reaction, ACS Catalysis 4 (2014) 1793–1800.
[49] H.-W. Liang, W. Wei, Z.-S. Wu, X. Feng, K. Müllen, Mesoporous metal-nitrogen-doped carbon electrocatalysts for highly efficient oxygen reduction reaction, Journal of the American Chemical Society 135 (2013) 16002–16005.

CHAPTER 9

Covalent organic framework-based materials as electrocatalysts for fuel cells

Anuj Kumar[1], Shashank Sundriyal[2], Tribani Boruah[3], Charu Goyal[1], Sonali Gautam[1], Dipak Kumar Das[1] and Tuan Anh Nguyen[4]

[1]Department of Chemistry, GLA University, Mathura, India
[2]Advanced Carbon Products Department, CSIR-National Physical Laboratory, New Delhi, India
[3]Northeast Hill University (NEHU), Umshing Mawkynroh, Shillong, India
[4]Institute for Tropical Technology, Vietnam Academy of Science and Technology, Hanoi, Vietnam

9.1 Introduction

9.1.1 Motivations and scope

During the last few decades, considerable effort has been devoted to achieve clean, safe, and renewable energy sources to meet the incessantly rising energy demands of the globe due to the fast depletion of existing fossil fuels [1]. In this perspective, hydrogen fuel cells (involve H_2 oxidation and O_2 reduction) combined with water electrolyzers (involve H_2-evolution and O_2-evolution reaction) could be regarded as the sustainable energy conversion devices, offering a fully close loop of clean energy system [2]. However, hydrogen fuel cells in particular possess extremely sluggish oxygen reduction reaction (ORR) at cathode, which involves the large overpotential, thereby displaying major efficiency loss, and thus limits further practical applications of fuel cells [3]. Noble Pt-based materials are quite effective for ORR electrocatalysis. Nonetheless, shortage of this valuable metal and higher price raise several constraints to their common usage and wide-scale implementations. Hence searching cost effective, efficient, and reliable electrocatalysts for ORR process is highly desired [4].

Recently, extensive attempts were made to establish effective and durable precursors of inexpensive and abundant metallic elements worthy of catalyzing such processes at substantial levels and low overpotentials. Similar efforts have contributed to the recognition of several metal-based materials like single atoms catalysts (SACs) [5], peroversuskite oxides [6], metal phosphorus [7], and metal sulfides [8] as electrocatalysts with low cost and high performance. However, for such electrocatalysts, in-depth knowledge of electrocatalytic mechanism is yet to be explored.

Herein, the efforts have been made to provide a discussion on covalent organic frameworks (COFs)-based materials as ORR electrocatalysts [9]. COFs are versatile class of crystalline porous molecular-based materials [10]. Due to their flexible, adjustable, and highly conjugated molecular structure along with significant porosity, specific surface area, and facile synthetic strategies, COFs have established as the promising alternatives not only in fuel cells but also in water electrolyzers [11], metal-ion batteries [12], supercapacitors [13], and photolytic cells [14].

In this chapter, we proposed a brief overview on chemistry of COFs and the relevant critical factors that are essential for the development of COF-based materials for electrocatalysis, which has a foundation for further discussed sections (in Section 10.1.2). In Section 10.2, we reviewed the recent advancements on COF-based materials for ORR for the development of fuel cell technology. Furthermore, in Section 10.3, we addressed the several goals of molecular and material engineering strategies to design the COF-based materials. Moreover, some unresolved issues related to COF-based materials with specific challenges and directions for further investigations are also highlighted.

9.1.2 Fuel cell and its chemistry

Fuel cells are the eco-friendly devices that convert chemical and thermal energies into electrical energy and work, respectively, under the thermodynamics laws. The chemical processes by which electric current is generated are electrochemical, rather than thermochemical. Without moving components inside the fuel cells, there is no noise during fuel cells operation. In 1839 William Grove demonstrated H_2-O_2 fuel cell [15] while working on water electrolysis. This produces H_2 and O_2 by a power supply and when he used an ammeter instead of power supplier, a small current was observed, due to the flow of electrons from anode to cathode. Furthermore, he postulated that if H_2 and O_2 was supplied to Pt electrodes (anode and cathode-dipped in electrolyte solution), electricity is generated. He coined the word fuel cell (gaseous voltaic battery).

Further, in 1939, Francis T. Becon used Pt catalyst in a H_2-O_2 fuel cell [16], which was later developed and patented by Parsons in 1954. In 1965 [17] fuel cell was utilized by NASA in Apollo space mission. In the 1990s tremendous interest arises in potential use of fuel cells due to their high-energy conversion efficiency [18] and to date, this technology is entering more and more into daily life. Especially the microfuel cells are the hopeful successor of today's battery; although there are still some obstacles in the path of fuel cell commercialization.

The H_2-O_2 fuel cells are composed of a sandwiched electrolyte resides between both anode and cathode electrodes. Basically, two electrochemical reactions, hydrogen oxidation reaction (HOR) and ORR occurs to produce water as product and electricity, respectively. A systematic structure of fuel cell and its chemistry is shown in Fig. 9.1. Usually, in aqueous solutions, ORR takes place through two key pathways: (1) $2e^-$ transfer pathway to give H_2O_2 [Eqs. (9.1) and (9.2)] and (2) $4e^-$ pathway to

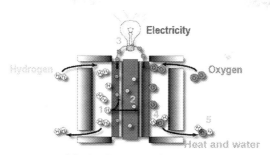

Figure 9.1 A systematic structure of fuel cell and its chemistry. *Image was taken from https://www.greenspec.co.uk/building-design/fuel-cells/.*

give H_2O [Eq. (9.3)]. The maximum free energy or maximum oxidizing power of O_2 is accessed when O_2 interacts at the cathode of fuel cell through $4e^-$ transfer. The catalyst should, therefore, facilitate the reduction of O_2 to water via the $4e^-$ process for a complete ORR. Two electron ORR results in the low power conversion efficiency and generates a reactive intermediate, which can be transformed into harmful radical species and consequently results in stability loss of the electrocatalysts. The four-electron ORR enables the rupture of O—O bond, involving the association of O_2 with one active site/two active sites at the same time at the electrode surface [19].

$$2e^- \text{ Pathway: } O_2 + 2H^+ + 2e^- \rightarrow H_2O_2 (E^0 = 0.695 \text{V vs SHE}) \quad (9.1)$$

$$H_2O_2 + 2H^+ + 2e^- \rightarrow H_2O (E^0 = 1.77 \text{ V vs SHE}) \quad (9.2)$$

$$4e^- \text{ Pathway: } O_2 + 4H^+ + 4e^- \rightarrow H_2O \ (E^0 = 1.229 \text{ V vs SHE}) \quad (9.3)$$

Where, E^0 and SHE, denotes the respective normal electromotive force and saturated hydrogen electrode.

On the other hand, HOR is also sluggish and basically one electron reaction producing protons and electron. The Pt/C catalyst is shown to have significant HOR activity and the detail chemistry of the HOR electrocatalysis at Pt electrode is represented by Eqs. (9.4)–(9.6).

$$O_2 + Pt \rightarrow PtH_2 \quad (9.4)$$

$$PtH_2 \rightarrow PtH_{ads} \quad (9.5)$$

$$PtH_{ads} \rightarrow Pt + H^+ + e^- \quad (9.6)$$

9.1.3 Chemistry of covalent organic frameworks

The chemistry behind connecting the molecular building blocks through the covalent bonding has generated novel classes of porous material and COFs is perhaps the latest among them [20]. Covalent bond at atomic level provides a directional guideline of getting together small molecular building units into predesigned molecular architecture. These unique advantages coupled with derived covalent strength can be efficiently utilized in applications, including energy conversion and storage, separation, and electronics.

Covalent bonding possesses significant strength which makes it capable to create covalently bonded molecular network of high stability that follows various mode of interactions between versatile range of atoms. Therefore researchers in this field ought to understand and master themselves about the COFs chemistry, so that noble covalent structure can be constructed to enable them to predict the composition and functions of COFs [21–23]. Following the line, experts in organic chemistry have developed the art to monitor the formation of covalent bond in the zero-dimensional (0D) [10]. Subsequently, polymer scientists extended one-dimensional (1D) concept, however, Nobel laureate Roald Hoffmann predicted this concept in two/three-dimensional (2D/3D) as "synthetic wasteland" [24].

Naturally occurring covalently networked organic framework although known, but synthetic analog was first reported by Yaghi et al. in 2005 [25] by connecting the small symmetrical organic building units to yield porous COF [13–15,26,27]. Several examples of natural and synthetic covalent structures in 0D, 1D, 2D, and 3D and several crystalline and amorphous COFs formed by the reversible and irreversible reactions are illustrated in Fig. 9.2A and B, respectively.

In case of higher dimensional COFs, in situ crystallization is essential as different free energies can be generated with multiple covalent bond structure due to the various ways of extension of covalent bonds. However, this concept is associated with a critical problem of crystallization as covalent linking of molecular building units in general, provides amorphous materials. This has been efficiently taken care of introducing COFs by creating the B—O, C—N, B—N, and B—O—Si bonding [29,30]. In this section, the chemistry of COFs in terms of several critical points like molecular building blocks linking into COFs, crystallization, porosity, chemical stability, interlayer stacking, and scalability are discussed.

9.1.3.1 Molecular building blocks linking and crystallization

Beyond the supramolecular chemistry, COFs chemistry is based upon covalent interactions, where two or more building blocks are associated via covalent bonding, resulting extended crystalline structure. As geometry of building block being used in COFs remains unaltered during chemical reaction, the prediction of resulting structure is

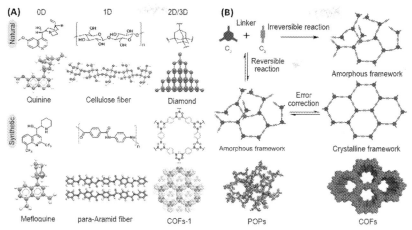

Figure 9.2 (A) Several natural and synthetic structures based on covalent bonding. (B) Schemes of formation of crystalline and amorphous covalent organic frameworks (COFs) via reversible/irreversible mechanisms. *(A) Reprinted with permission from A.P. Cote, A.I. Benin, N.W. Ockwig, M. O'Keeffe, A.J. Matzger, O.M. Yaghi, Porous, crystalline, covalent organic frameworks, Science 310 (2005) 1166−1170 [25]; M. Grujicic, W. Bell, P. Glomski, B. Pandurangan, C.-F. Yen, B. Cheeseman, Filament-level modeling of aramid-based high-performance structural materials, Journal of Materials Engineering and Performance 20 (2011) 1401−1413 [21], Copyright 2005 from publishing group; (B) Reprinted with permission from S.-Y. Ding, W. Wang, Covalent organic frameworks (COFs): from design to applications, Chemical Society Reviews 42 (2013) 548−568 [28], Copyright 2005 from publishing group.*

very much possible. This enables to end-up with a smaller number of predicted structures out of large number of possibilities based on the geometry of the building blocks [11,12]. However, how covalent chemistry could be executed for connecting small molecular units to generate 2D/3D crystalline COFs is still quite complicated. To have an extended crystalline network, the bond formation should be reversible and reaction rate should be on time scale to facilitate the corrections of defects.

The covalent bond formation offers special challenge as they have the tendency to act irreversibly under the mild reaction situation. It is therefore important to identify the conditions favorable for the reversible bond formation beyond extreme temperature and pressure. This goal was achieved by exploiting the condensation reaction with the generation of stoichiometric amount of water as by-product. Conveniently, this approach permits the modulation of the reaction' equilibrium by monitoring the quantity species in the system. For example, COFs were reported for the first time by self-condensation of boronic acid. First crystalline material, COF-1, however, was synthesized by self-condensation of 1,4-phenylenediboronic acid to make a structure of staggered layers with hexagonal pores [27,31]. In addition, molecules with tetrahedral geometry like tetra(4-dihydroxyborylphenyl)methane or its derivatives [8,31] have been utilized to fabricate 3D architectures by condensation method [8,31]. Further,

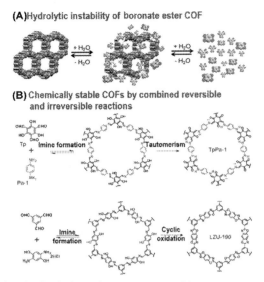

Figure 9.3 (A) Scheme for the hydrolytic decomposition of boronate ester-based covalent organic frameworks (COFs). (B) Scheme for the synthesis of highly stable COFs via reversible/irreversible reactions. *Reprinted with permission from L.M. Lanni, R.W. Tilford, M. Bharathy, J.J. Lavigne, Enhanced hydrolytic stability of self-assembling alkylated two-dimensional covalent organic frameworks, Journal of the American Chemical Society 133 (2011) 13975–13983 [32], Copyright 2005 from publishing group.*

boronic acids can interact with catechols to produce boronate esters (Fig. 9.3A). On the other hand, COFs might also be produced following formation of C—N bonds [33]. For instance, COF-300 was prepared by the condensation of amine linkers and HCHO. Fig. 9.3B shows the synthesis of highly chemically stable COFs by combined reversible and irreversible reactions.

9.1.3.2 Porosity of COFs

The porosity is one of the unique features of COFs, which can be determined by applying gas sorption isotherms and is adjustable as evident by reported COFs with micropores and mesopores. In ORR electrocatalysis, porosity plays a pivotal role by providing significant area for oxygen diffusion; therefore, COFs with high porosity is desirable to achieve expected ORR performance [34]. In case of 2D materials, the porosity arises due to the formation of small channels [35]. The diameter of the channel can be monitored by selecting suitable linkers to produce COFs to have outstanding gas storage capacity. However, 3D COFs are reported to have maximum storage capacity due to high pore volume and surface area as compared to 2D materials [36,37].

9.1.3.3 Chemical stability of COFs
The chemical stability of COFs is of paramount importance and responsible for the high durability of COFs in electrocatalytic devices. Different crystalline COFs can be successfully constructed by following chemical induced reversibility concepts, which in turn imparts uncertainty toward their hydrolytic/chemical stability. COFs are sensitive to hydrolytic decomposition in spite of having high thermal stability and rigidity in various chemical environment, including organic solvents [38]. Water being generated as a product in reversible reaction of COFs formation, the addition of water will favor the backward reaction according to LeChatelier's principle, resulting decomposition of COFs [39]. Therefore due care should be given toward chemical stability of COFs before letting them to place for any practical application [40].

9.1.4 Interlayer stacking in COFs
In spite of the fact that chemically stable imine bonds have high hydrolytic stability as compared to boroxine and boronate ester links, due care ought to be paid during the isolation and purification of 2D imine-linked COFs. To preserve the crystallinity during purification of imine COFs, dry organic solvent was used [41]. Usually, electrostatic repulsion between the C = N unites responsible for weakening of $\pi-\pi$ stacking forces between individual COFs layers makes 2D imine-type COFs sensitive, making COFs layers vulnerable to outdoor provocations [42]. The extensive research efforts have been carried out in the recent years to improve the chemical stability of 2D imine-type COFs with significant crystallinity by strengthening the $\pi-\pi$ stacking forces between the COF layers [40].

9.2 Recent advancements in COF-based electrocatalysts for ORR
Basically, for heterogeneous electrocatalysts, chemical reactions occur at particular active sites, and their activity depends on the inherent activity of individual active sites as well as number of active sites. In aqueous solutions, the desired ORR is tremendously problematic, involving multisteps having various O-intermediates like -O*, -OH*, and -OOH* species [43]. From the point of view of thermodynamics, adsorption, and dissociation energy barriers of -O*, -OH*, and -OOH* intermediate species control the overall kinetics of ORR [44]. Hence, to achieve the satisfactory performance of ORR, electrocatalysts should be able to minimize the energy barriers of above intermediate steps efficiently, accelerating electrons transfer [45,34,35]. Beyond the Pt-based electrocatalysts, COFs materials showed the remarkable ORR activity, selectivity, and significant durability because of their high specific surface area and adjustable pore sizes, high thermal stability, and charge mobility [46,38,47].

COFs permit the precise control of microenvironment of active sites like number of coordination neighboring atoms, and their electronic structure [48]. Hence COFs possesses good stability, incorporating the catalytic active sites into porous crystalline framework and guarantees for high durability toward electrocatalytic reactions in electrolytic conditions. In this section, we focused on recent advances on the fundamental roles of COF-based materials, pristine metal-free COFs, metal-occupied COFs, SACs incorporated COFs, and nanocarbon-supported COFs, as directly or indirectly serving ORR electrocatalytic active sites [49].

9.2.1 Pristine metal-free COFs

Generally, due to limited number of exposed active sites and poor conductivity of pristine COFs as compared to traditional metal-based materials, these materials exhibited poor ORR activity [47,50,45,51]. Therefore COF-based materials with appropriate physicochemical properties and fixed surface composition is of great significance for their practical use. In this regard, many researchers have crafted the electrocatalytic performance of COF-based materials, using various strategies.

Li et al. [52] synthesized two metal-free thiophene-S-COFs (MFTS-COFs) by conducting the Schiff base reaction between 1 2,4,6-tris(4-aminophenyl) benzene (TAPB), 2,5-thiophenedicarboxaldehyde, and 2,2′-bithiophenyl-5,5′-dicarbaldehyde molecular building blocks [Fig. 9.4A(i)]. The transmission electron microscopy (TEM) images for both MFTS-COFs are shown in Fig. 9.4A(ii) and (iii) along with their 2D porous molecular architectures. Further, electrochemical results suggested that MFTS-COFs had significant ORR activity as compared to thiophene-free COF due to the presence of pentacyclic thiophene-sulfur units which acted as ORR active centers [Fig. 9.4A(iv) and (v)]. However, higher S-containing MFTS-COFs showed better ORR activity which could be attributed to its smallest band gap as supported by density-functional theory (DFT) investigations [Fig. 9.4A(vi)] [51,54,55]. The density of states and Gibbs free energy (ΔG) calculations suggested that the thiophene-S-free PDA-TAPB-COF exhibited the broadest band gap around the Fermi level. Therefore these findings demonstrated that rational design and precise synthesis of metal-free COF-based materials could indorse the advancement of novel ORR electrocatalysts.

In another work, Lin et al. [53], introduced an effective mechanochemical strategy, reaction milling, to promote the Schiff-based coupling reaction between melamine and terephthalaldehyde building blocks, producing covalent organic polymer (RM-COP). Here, RM represents the reaction milling, and phosphorus doped RM-COP (RM-COP-PA) materials [Fig. 9.4B(i)]. This strategy involves less hazardous solvents and short reactions time period as compared to conventional solvothermal strategy, and provided optimal crystalline porous materials followed by carbonization at 900°C resulting [Fig. 9.4B(ii)]. Fig. 9.4B(iii) showed the N1s X-ray photoelectron spectroscopy data for RM-COP-PA-900 materials, indicating the four different signals corresponding to the pyridinic N (N1), nitrile N (N2), pyrrolic N (N3), and graphitic N (N4) moieties. However, these findings showed availability of pyridinic N and graphitic N in high

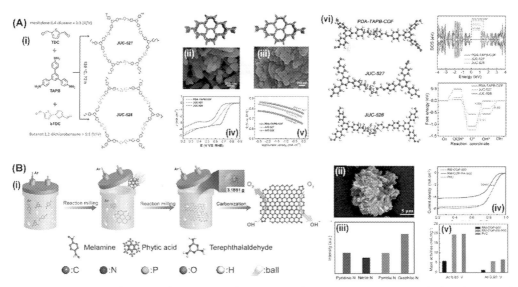

Figure 9.4 (A) (i) Syntheses and structures of JUC-527 and JUC-528, SEM images of (ii) JUC-527 and (iii) JUC-528, (iv) LSV curves (at 1600 rpm) of PDA-TAPB-COF, JUC-527, and JUC-528 in O_2-saturated 0.1M KOH electrolyte, (v) Tafel plots, (vi) DOS and energy diagram. (B) (i) Schematic illustration of preparation of RM-COP-PA-900, (ii) TEM images of RM-COP-PA-900, (iii) different types contents of N in N1s XPS spectra, (iv) LSV curves 1600 rpm in O_2-saturated 0.1M KOH solution for Pt/C, RM-COP-900 and RM-COP-PA-900 for ORR, (v) j_k and mass activities for RM-COP-900, RM-COP-PA-900, and Pt/C at 0.85 V. *COF*, Covalent organic framework; *COP*, covalent organic polymer; *DOS*, density of states; *LSV*, linear sweep voltammetry; *PDA*, terephthaldehyde; *PA*, phytic acid; *SEM*, scanning electron microscopy; *RM*, reaction milling; *TAPB*, 1,3,5-tris(4-aminophenyl)benzene; *TEM*, transmission electron microscopy; *XPS*, X-ray photoelectron spectroscopy. *(A) Reprinted with permission from D. Li, C. Li, L. Zhang, H. Li, L. Zhu, D. Yang, et al., Metal-free thiophene-sulfur covalent organic frameworks: precise and controllable synthesis of catalytic active sites for oxygen reduction, Journal of the American Chemical Society 142 (2020) 8104–8108 [52], Copyright 2020, American Chemical Society; (B) Reprinted with permission from X. Lin, P. Peng, J. Guo, Z. Xiang, Reaction milling for scalable synthesis of N, P-codoped covalent organic polymers for metal-free bifunctional electrocatalysts, Chemical Engineering Journal 358 (2019) 427–434 [53], Copyright 2018, Elsevier.*

amount; therefore these materials showed excellent ORR performance (high ORR activity, better durability, and resistance to crossover effect) in comparison of traditional Pt/C electrocatalyst [Fig. 9.4B(iv) and (v)] [56,57]. Therefore this work opens up a rapid, solvent-free and scalable approach for the further designs of significant ORR electrocatalysts.

9.2.2 Macrocycle-incorporated COFs

Xiang et al. [58] reported a Mn, Fe, and Co metal ions incorporated 2D-COP utilizing the Yamamoto polycondensation chemical reaction [Fig. 9.5A(i)]. An ORR comparison

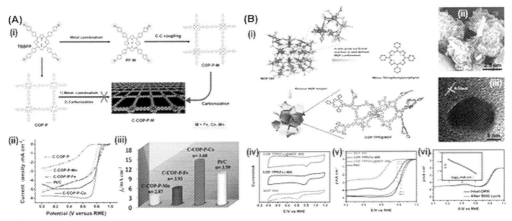

Figure 9.5 (A) (i) Incorporation of Fe, Co, or Mn metals into C-COP, (ii) LSV curves of metal-incorporated C-COP-P-M in O$_2$-saturated 0.1 m KOH at 1600 rpm at 5 mVs^{-1}, (iii) comparison of electrochemical activity given as j$_k$ at 0.75 V versus RHE along with calculated electron transfer numbers. (B) (i) Synthetic route for COP materials within the confined space of MOFs, (ii) SEM and (iii) TEM image of COP material, (iv) CV, (v) LSV, and (vi) stability curves for COP materials. *COP*, covalent organic polymer; *CV*, cyclic voltammetry; *LSV*, linear sweep voltammetry; *MOFs*, metal–organic frameworks; *SEM*, scanning electron microscopy; *RHE*, reversible hydrogen electrode; *RM*, reaction milling; *TEM*, transmission electron microscopy. *Reprinted with permission from Z. Xiang, Y. Xue, D. Cao, L. Huang, J.F. Chen, L. Dai, Highly efficient electrocatalysts for oxygen reduction based on 2D covalent organic polymers complexed with non-precious metals, Angewandte Chemie International Edition 53 (2014) 2433–2437 [58], Copyrights 2014, Wiley-VCH Verlag GmbH & Co. KGaA, Weinheim; (B) Reprinted with permission from J. Guo, Y. Li, Y. Cheng, L. Dai, Z. Xiang, Highly efficient oxygen reduction reaction electrocatalysts synthesized under nanospace confinement of metal–organic framework, ACS Nano 11 (2017) 8379–8386 [59], Copyright 2017, American Chemical Society.*

study between metal-free 2D-COP and Mn, Fe, and Co metal ions incorporated 2D-COP indicates that metal ions incorporated 2D-COP had efficient ORR activity, exhibiting an anodic shift in ORR peak with high current density as compared to metal-free 2D-COP. Further, to achieve their ORR activity, both the COPs were taken under the different carbonization temperatures treatment. It was found that at near 950°C temperature, these COPs achieved significant conductivity, resulting good ORR activity as compared to Pt/C [Fig. 9.5A(ii) and (iii)]. This work opens the best approach to structure 1D and 2D nanoarray of active sites.

Usually, metallomacrocyclic complexes like porphyrin [60], phthalocyanine [61], and their derivatives [62] have been applied as potential models for ORR but such molecules possess poor conductivity and poor stability on electrode surfaces. In this context, one of the most interesting features of COFs is that they can be incorporated with these macrocyclic complexes. For instance, Guo et al. [59] developed a fascinating strategy to fabricate a nonprecious M-N-C material followed by COPs of porphyrin derivative [Fig. 9.5B(i)]. The TEM images showed that the prepared material possessed uniform distribution of metal/nitrogen active sites and high specific surface area [Fig. 9.5B(ii) and (iii)]. This material showed similar activity with traditional Pt/C catalyst in terms of onset and formal potential [Fig. 9.5B(iv)−(vi)].

Further, to demonstrate the ORR activity of COFs with porphyrin moieties, Lin et al., fabricated a series of transition metals, Sc, Ti, V, Cr, Mn, Fe, Co, Ni, Cu, and Zn, incorporated porphyrin-COFs architectures. DFT calculations was conducted to simulate the O-intermediates, involved in ORR, on central metal sites were simulated by hosting dioxygen and H^+ near the central metal sites. The results suggested that both the mechanisms 2e and $4e^-$ occurs on active sites depending on transition metal. The Gibbs free energy indicated that the spontaneously $2e^-$ and 4e transfer were observed in case of Fe and Co metals, while remaining metals exhibited the involvement of high-energy barriers in O-intermediate elementary steps.

9.2.3 COF-derived SAC-based materials

SAC-based electrocatalysts have been recognized as superior materials not only for ORR but also for other electrocatalysis like Oxygen evolution Reaction (OER), Hydrogen evolution reaction (HER), and Carbon dioxide reduction reaction (CO_2RR). Interestingly, their low loading (<5 wt.%) is enough to show good catalytic activity and efficiency. Although, COF-derived SACs are widely studied, but there is a challenge related to the precise identification of real active site in SACs and exact mode of action during the activation of energy rich molecules. This is because at high-temperature treatment, uncontrollable heteroatom doping and unwanted structural destruction take place; therefore advanced in situ techniques are required to characterize the obtained materials and mechanisms.

In another report, Li et al. [63], developed an effective method for the in situ anchoring of metal nanoparticles into M-N-C materials, using high-temperature treatment on metal phthalocyanine-based polymers [Fig. 9.6A(i)]. The TEM images [Fig. 9.6A(ii) and (iii)] showed the uniform distribution of metallic components into porous M-N-C matrix. The obtained M-N-C material exhibited superior ORR performance, in terms of activity, durability, and methanol tolerance in both alkaline as well as in acidic media [Fig. 9.6A(iv) and (v)] as compared to commercial Pt/C catalyst. In addition, this material also exhibited high performance toward cathode reaction in Zn-air battery. The strategy demonstrated in this work can be served as a direction to design efficient electrocatalyst materials with improved ORR performance.

Further, Yi et al. [64], proposed a facile ionothermal-based strategy to fabricate the Fe-SAC on porous porphyrinic triazine-based framework (FeSAs/PTF) with high Fe loading [Fig. 9.6B(i)]. TEM images of this material showed highly hierarchical porosity morphology, having maximum number of $Fe-N_4$ active sites [Fig. 9.6B(ii) and (iii)]. Further, this material displayed satisfactory ORR performance in terms of high

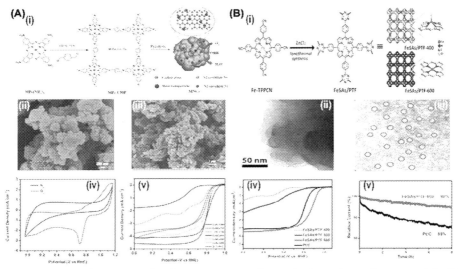

Figure 9.6 (A) (i) Synthetic route for the preparation of MPc-CMPs and metal nanoparticles/MNCs material, SEM images of (ii) CoPc-CMP, (iii) Co-NCs 800, (iv) CV curves of Co-NCs 800 in 0.1M KOH solution saturated by N_2 or O_2 at 50 mVs^{-1}, (v) ORR polarization curves of different samples at a rotation rate of 1600 rpm. (B) (i) Schematic illustration of formation of FeSAs/PTF, (ii) HAADF-STEM images and enlarged images of FeSAs/PTF-600 (single Fe atoms highlighted by circles), (iv) LSVs of FeSAs/PTF-400, -500, -600 and Pt/C at 1600 rpm in O_2-saturated 0.1 5M KOH at 10 mVs^{-1}, (v) current–time chronoamperometry for FeSAs/PTF-600 and Pt/C in an O_2-saturated 0.1M KOH solution. *CV*, Cyclic voltammetry; *HAADF-STEM*, high-angle annular dark field-scanning transmission electron microscopy; *LSVs*, linear sweep voltammetry; *MNCs*, metal nanocomposites; *MPc-CMPs*, metal phthalocyanine-linked conjugated microporous polymers; *ORR*, oxygen reduction reaction; *PTF*, porous thin film; *CoPc-CMPs*, Cobalt phthalocyanine-Conjugated microporous polymers; *Co-NCs*, Cobalt-Nitrogen-doped carbons. (A) Reprinted with permission from Q. Li, Q. Shao, Q. Wu, Q. Duan, Y. Li, H.-g. Wang, In situ anchoring of metal nanoparticles in the N-doped carbon framework derived from conjugated microporous polymers towards an efficient oxygen reduction reaction, Catalysis Science & Technology 8 (2018) 3572–3579 [63], Copyright 2013, RSC; (B) Reprinted with permission from J.-D. Yi, R. Xu, Q. Wu, T. Zhang, K.-T. Zang, J. Luo, et al., Atomically dispersed iron–nitrogen active sites within porphyrinic triazine-based frameworks for oxygen reduction reaction in both alkaline and acidic media, ACS Energy Letters 3 (2018) 883–889 [64], Copyright 2018 American Chemical Society.

activity, durability, and methanol tolerance in both basic and acidic environments [Fig. 9.6B(iv) and (v)]. This work brought new motivations to design highly efficient noble-metal-free catalysts at the atomic scale for ORR.

On the other hand, COFs help to fix the exact location of SACs, therefore SACs incorporated COFs materials can help to resolve above mentioned issues [65]. For example, a series of Pt and Cu-SACs incorporated with covalent triazine frameworks (Pt-CTF and Cu-CTF) was fabricated to demonstrate the ORR performance as well as to realize the active site. In this line, Nakanishi et al. [66] reported Pt-CTF-based materials and the results were found to be satisfactory toward ORR performance.

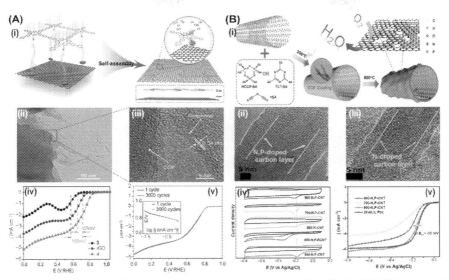

Figure 9.7 (A) (i) Schematic illustration for the preparation of COP/rGO, (ii) HRTEM image of COP/rGO, (iii) an enlarged view of COP/rGO, (iv) LSV curves recorded in O_2-saturated in 0.1M KOH at 1600 rpm, (v) LSV curves of COP/rGO before cycling durability test and after 3000 cycles. (B) (i) Illustration of the preparation of N,P-C material, TEM image of (ii) 800-N,P-CNT, (iii) 800-N-CNT, (iv) CVs of T-N,P-CNT, 800-N-CNT, and 800-N,P-BC in N_2- and O_2-saturated 0.1M KOH solution at 50 mVs^{-1}, (v) LSV curves of 700-, 800-, 900-N,P-CNT, and Pt/C at a rotation rate of 1600 rpm. *COP*, Covalent organic polymer; *CNT*, carbon nanotube; *CV*, cyclic voltammetry; *HRTEM*, high-resolution transmission electron microscopy; *LSV*, linear sweep voltammetry; *rGO*, reduced graphene oxide; *TEM*, transmission electron microscopy. *(A) Reprinted with permission from J. Guo, C.Y. Lin, Z. Xia, Z. Xiang, A. Pyrolysis-Free Covalent organic polymer for oxygen reduction, Angewandte Chemie International Edition 57 (2018) 12567–12572 [68], Copyrights 2018, Wiley-VCH Verlag GmbH & Co. KGaA, Weinheim; (B) Reprinted with permission from Z. Li, W. Zhao, C. Yin, L. Wei, W. Wu, Z. Hu, et al., Synergistic effects between doped nitrogen and phosphorus in metal-free cathode for zinc-air battery from covalent organic frameworks coated CNT, ACS Applied Materials & Interfaces 9 (2017) 44519–44528 [69], Copyright 2017, American Chemical Society.*

Similarly, Cu-CTF-type materials also exhibited remarkable ORR performance in alkaline media [67].

9.2.4 Nanocarbon-supported COFs

Guo et al. [68] prepared a nanocomposite of COP with reduced graphene oxide (rGO) as shown in Fig. 9.7A(i). Due to the incorporation of rGO, the electrical conductivity of COP/rGO prepared nanocomposite was increased significantly as compared to pure COP. The TEM images of the prepared material is shown in Fig. 9.7A(ii) and (iii), indicating the well distribution of active sites. Further, this nanocomposite exhibited the superior ORR activity as compared to pure COP and traditional Pt catalyst, followed by 4e$^-$ transfer mechanism [Fig. 9.7A(iv) and (v)]. Further,

Li et al. [69] prepared a metal-free COF on carbon nanotube (CNT) using hexachlorocyclotriphosphazene and dicyanamide [Fig. 9.7B(i)]. The prepared material contained N-, P-codoped carbon matrix on CNTs surface [Fig. 9.7B(ii) and (iii)]. Moreover, this nanohybrid was found to be highly porous, having strong synergistic effect due to the collaboration between graphitic-N and -P. Further, this material displayed high ORR performance in terms of half-wave potential (-0.162 V), high current density (6.1 mA cm^{-2}) [Fig. 9.7B(iv) and (v)], good stability (83%), and excellent methanol tolerance for ORR in alkaline solution. In addition, the prepared material was successfully applied as a cathode in Zn-air battery, exhibiting high open-circuit potential (1.53 V) and peak power density (0.255 W cm^{-2}).

9.3 Designing approaches of COF-based electrocatalysts

The highly accessible porous materials with variable dimensions are synthesized by combining effective aromatic linkers. To improve the productivity in case of conversion and storage of energy, the development of novel materials with excellent performance is of prior importance among the researchers.

9.3.1 Geometric orientation-based approaches

9.3.1.1 3D-based COFs

COFs relying upon the geometric regularity of the structural building blocks. The 3D-based COFs are basically comprising of various structural building units that consist of different heteroatoms such as boron, nitrogen, carbon, silicon, and try to coordinate with one another through variable functional moieties via means of different types of interactions to extend their 3D spatial arrangements (Fig. 9.8A). Due to the availability of diverse flexible molecular building units, COF materials are available with tremendous opportunities with colossal prospects in underlying model. Interesting, underlying highlights of COFs invest these opportunities with incredible benefits and excellent potential for practical application in various sectors. In case of 3D COFs, the 1D formed section interweave with one another and build long-range formed frameworks that may likewise give transport trails for various applications [31].

9.3.1.2 2D-based COFs

The 2D COFs composed of organic functional moieties that are covalently reinforced with each other and confined in 2D sheets like structure that promote the stacking of different layer that shapes into a layered construction by means of $\pi-\pi$ interaction (Fig. 9.8B). The vast majority of layers are stacked in eclipsed configurations with the arrangement of occasionally adjusted channels rather than into a staggered fashion; however both stacking models delivers comprehensible arrangements. In addition, the vertical column structural portion of 2D COFs generates the desirable shape through

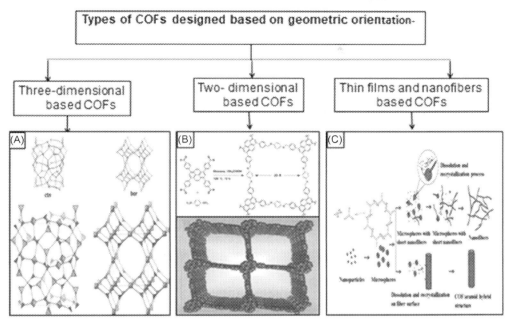

Figure 9.8 (A) The structural units located on the ctn and bor frameworks, as shown in swelled networks; (B) synthesis of ILCOF-1 and space filing representation of the resulting 2D network; (C) formation of COFs nanofibers through dynamic imine chemistry enabled by dissociation—recrystallization process. *COFs*, Covalent organic frameworks; *ILCOF*, imine-linked covalent organic framework. *From S.-Y. Ding, W. Wang, Covalent organic frameworks (COFs): from design to applications, Chemical Society Reviews 42 (2013) 548–568 [28], Copyright 2016, Elsevier.*

the stacking of planar aromatic structural molecular building blocks that helps to developed the desirable polymeric layers that can provide consistent networks to the opening or potential electron transport. Generally, both the types of COFs possess the benefit of modulated skeletal framework, multiple open sites, along with significantly high porosity [70,71].

9.3.1.3 Thin film and nanofiber-based COFs

Surface mediation techniques have been used to construct 2D COFs monolayers or thin films. In this case, when halogenated aromatic monomers react on graphite, Ag or Cu metal under ultrahigh vacuum circumstances, 2D COFs are achieved. The formation of 2D polymers are associated with the dehydration of 1,4-benzenediboronic acid via vapor deposition on well-defined noble surfaces. This concludes that, among Ag(111), Ag(100), and Au(111) surfaces, Ag(111) and Ag(100) are preferable templates than Au(111) and Cu(111) surfaces for this purpose because of their nature and surface diffusion of adsorbed organic precursors. Solvothermal approach was also used to prepare 2D-COF

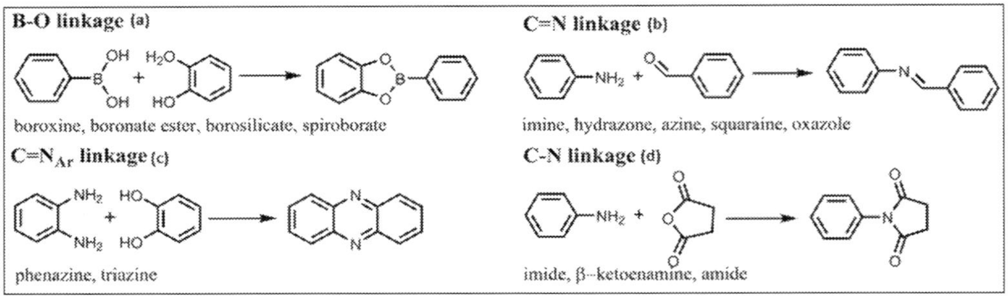

Figure 9.9 Schematic diagram of different reaction schemes that are carried out to obtained Covalent organic framework (COF)-based materials via (a) B-O, (b) C = N, C = N(Aromatic), and (d) C-N linkages. *Reprinted with permission from S.-Y. Ding, W. Wang, Covalent organic frameworks (COFs): from design to applications, Chemical Society Reviews 42 (2013) 548−568 [28]. Copyrights 2018, Wiley-VCH Verlag GmbH & Co. KGaA, Weinheim.*

nanofibers. In this method, dissolution−recrystallization mechanism was used in morphological transformations from microspheres to nanofibers (Fig. 9.8C) [72,73].

9.3.2 Different bonding-based approaches

Boron−carbon-based approach; boron−carbon-based designing approach of COFs is in great demand among the researchers. In this methodology, different molecular structural entities containing different functional groups undergo condensation reactions leading to the formation of ester and anhydride linkage of boron derivatives (Fig. 9.9A). Based on synthesis techniques, this approach can be classified into two categories. Firstly, boron−carbon-based COFs can be generated through the condensation reaction between two similar functional moieties which is termed as self-condensation reaction. For instance, this approach is successfully implemented to obtain staggered layered structure of 2D-COF leading to the formation of planar aromatic six-membered cyclic ring structure consisting of boronate anhydride linkage. Additionally, a tetrahedrally oriented functional entity undergoes self-condensation reaction to obtained 3D boron-based COFs with potentially high surface area. However, these designing approaches are not suitable in moist environmental conditions because of its inefficiency to produce stable boron-based COFs. Although after exposure to air, COF material can retain its porosity and crystalline properties for a limited period of time but its severe loss of structural regularity to some extent [6−8,34,74,75].

9.3.2.1 Carbon−nitrogen-based approach

Another designing approach of COFs material are imine-based containing -C=N- covalent linkage. This approach can be sorted into two categories based on the covalent bond formation between the distinct functional moieties. In the first category, the cocondensation reaction takes place between two different functional entities such as aldehyde and amine functional group that leads to the formation of Schiff base containing nitrogen−carbon double bond (Fig. 9.9B). While in second category, hydrazine-linked COFs materials are successfully obtained through the cocondensation reaction between the aldehyde functional group and hydrazides. However, imine-based COFs are in great demand because of exhibiting combine features of both boron-based COFs and triazine-based COFs. The presence of large number of H bonding among the hydrazone units favored the construction of eclipsed configured structure. Furthermore, nitrogen atom can coordinate with various metal ions. These benefits expand the range of applications of these materials [9−11,76−78].

9.3.2.2 Triazine-based approach (CTFs)

Triazine-based approach is another class of designing technique to synthesized COFs. In these approaches, the molecular entities containing nitrile functional moieties undergoes tricyclization in the presence of Lewis acids such as $ZnCl_2$. Interestingly, the morphology of the desired material can optimize by altering the ratio of $ZnCl_2$ to added monomer. For instance, by increasing the $ZnCl_2$ content, amorphous polymer can be generated delivering significantly higher surface area. Furthermore, these COFs displays comparatively lower crystalline properties but possess excellently higher chemical and thermal stability. However, the nitrogen content can alter the performance rate to significant level. Thus increasing the percent content of N atom enhanced the performance of triazine-based COFs materials especially as a catalytic template [12,13,79,80].

9.3.2.3 Imide-based approach (PICOFs)

Imidization reaction is used to prepare imide-linked COFs and polyimide COFs. Polyimide has some inherent properties like high thermal stability up to 520°C that results in large surface area of PICOFs materials that significantly delivers high potential for the attachment of dye molecules and also plays a vital role in drug delivery system. Currently reported imide-linked COFs displays fluorescence sensing potential which is prepared by solvothermal method [81,36].

9.4 Concluding remarks

To sum-up, COFs have been growing rapidly for last decades as efficient electrocatalysts for the sustainable development of energy sectors. By utilizing the diversity of small molecular building units and covalent bonding construction strategies, COFs can provide exceptional platforms for efficient electrocatalysis due to their predesigned compositions and electronic as well as mechanical properties as compared to traditional materials.

COFs have specific advantages, offering (1) heteroatom incorporation in their porous architecture, (2) incorporation of redox-rich macrocyclic functional moieties into COFs, and (3) unprecedented thermal stability and sufficient electronic character due to the presence of extended $\pi-\pi$ conjugation, for the further enhancement in their electrocatalytic activity. Therefore, considerable progress has been made to design and fabricate various types of COFs like metal-free COFs, metal-occupied COFs, COF-supported SACs and nanocarbon-supported COFs, which demonstrated superior catalytic performance, including free from any methanol-crossover or gas-poisoning effects.

9.4.1 Challenges

In spite of achieving commendable progress in COF-based materials for the applications in energy sectors, some challenges are yet to be addressed: (1) their electrocatalytic performance in acidic situation is not well studied, although unsatisfactory results were obtained in few cases compared to noble metal-based electrocatalysts, (2) stability of COFs needs further improvement before being used in practical applications having harsh experimental conditions, and (3) the realization of real active sites in COF-based materials is still not sufficiently clear, particularly the relation between activity and structure/composition of COF-based electrocatalysts.

9.4.2 Prospects and research directions

To date, the development of COF-based electrocatalysts is still in its early stage and the following perspectives and future directions are listed, which may support to bring COFs to its advanced development stage: (1) several metal-free COFs without carbonization should be fabricated to regularize their catalytic active sites at molecular/atomic levels, (2) fabrication of COFs architectures should have accurately monitored conducting channels, uniform spreading and dopants site, and systematic conjugated network to increase the catalytic performance, (3) COFs should be provided with multitypes of heteroatom or metal ions doping to take benefits of their synergetic effect, and (4) several strategies should be developed to identify the real active sites in COFs and to tune their contents with improved catalytic activity.

References

[1] A. Kumar, Y. Zhang, W. Liu, X. Sun, The chemistry, recent advancements and activity descriptors for macrocycles based electrocatalysts in oxygen reduction reaction, Coordination Chemistry Reviews 402 (2020) 213047.
[2] A. Kumar, V.K. Vashistha, D.K. Das, Recent development on metal phthalocyanines based materials for energy conversion and storage applications, Coordination Chemistry Reviews 431 (2020) 213678.

[3] M. Ma, A. Kumar, D. Wang, Y. Wang, Y. Jia, Y. Zhang, et al., Boosting the bifunctional oxygen electrocatalytic performance of atomically dispersed Fe site via atomic Ni neighboring, Applied Catalysis B: Environmental 274 (2020) 119091.

[4] Y. Wang, A. Kumar, M. Ma, Y. Jia, Y. Wang, Y. Zhang, et al., Hierarchical peony-like FeCo-NC with conductive network and highly active sites as efficient electrocatalyst for rechargeable Zn-air battery, Nano Research 13 (2020) 1090−1099.

[5] T. Zhao, A. Kumar, X. Xiong, M. Ma, Y. Wang, Y. Zhang, et al., Assisting atomic dispersion of Fe in N-doped carbon by aerosil for high-efficiency oxygen reduction, ACS Applied Materials & Interfaces 12 (2020) 25832−25842.

[6] C. Sun, J.A. Alonso, J. Bian, Recent advances in perovskite-type oxides for energy conversion and storage applications, Advanced Energy Materials 11 (2021) 2000459.

[7] K. Huang, Y. Xu, Y. Song, R. Wang, H. Wei, Y. Long, et al., NiPS$_3$ quantum sheets modified nitrogen-doped mesoporous carbon with boosted bifunctional oxygen electrocatalytic performance, Journal of Materials Science & Technology 65 (2021) 1−6.

[8] R. Reeve, P. Christensen, A. Dickinson, A. Hamnett, K. Scott, Methanol-tolerant oxygen reduction catalysts based on transition metal sulfides and their application to the study of methanol permeation, Electrochimica Acta 45 (2000) 4237−4250.

[9] H. Chen, Q.-H. Li, W. Yan, Z.-G. Gu, J. Zhang, Templated synthesis of cobalt subnanoclusters dispersed N/C nanocages from COFs for highly-efficient oxygen reduction reaction, Chemical Engineering Journal 401 (2020) 126149.

[10] S. Kandambeth, K. Dey, R. Banerjee, Covalent organic frameworks: chemistry beyond the structure, Journal of the American Chemical Society 141 (2018) 1807−1822.

[11] W. Chen, L. Wang, D. Mo, F. He, Z. Wen, X. Wu, et al., Modulating benzothiadiazole-based covalent organic frameworks via halogenation for enhanced photocatalytic water splitting, Angewandte Chemie International Edition 59 (2020) 16902−16909.

[12] T. Sun, J. Xie, W. Guo, D.S. Li, Q. Zhang, Covalent−organic frameworks: advanced organic electrode materials for rechargeable batteries, Advanced Energy Materials 10 (2020) 1904199.

[13] M. Li, J. Liu, T. Zhang, X. Song, W. Chen, L. Chen, 2D redox-active covalent organic frameworks for supercapacitors: design, synthesis, and challenges, Small 17 (2021) 2005073.

[14] X. Wang, L. Chen, S.Y. Chong, M.A. Little, Y. Wu, W.-H. Zhu, et al., Sulfone-containing covalent organic frameworks for photocatalytic hydrogen evolution from water, Nature Chemistry 10 (2018) 1180−1189.

[15] J. Larminie, A. Dicks, M.S. McDonald, Fuel Cell Systems Explained, J. Wiley Chichester, 2003.

[16] F. Barbir, PEM Fuel Cells: Theory and Practice, Academic Press, 2012.

[17] G. Sandstede, E. Cairns, V. Bagotsky, K. Wiesener, History of low temperature fuel cells, Handbook of Fuel Cells, Wiley, 2010.

[18] A. Appelqvist, Development of a biocatalytic fuel cell system for low-power electronic applications, Ph.D. thesis, Helsinki University of Technology, 2006.

[19] J. Santana-Casiano, M. Gonzalez-Davila, F.J. Millero, The role of Fe (II) species on the oxidation of Fe (II) in natural waters in the presence of O_2 and H_2O_2, Marine Chemistry 99 (2006) 70−82.

[20] M. Dogru, T. Bein, On the road towards electroactive covalent organic frameworks, Chemical Communications 50 (2014) 5531−5546.

[21] M. Grujicic, W. Bell, P. Glomski, B. Pandurangan, C.-F. Yen, B. Cheeseman, Filament-level modeling of aramid-based high-performance structural materials, Journal of Materials Engineering and Performance 20 (2011) 1401−1413.

[22] R. Dawson, A.I. Cooper, D.J. Adams, Nanoporous organic polymer networks, Progress in Polymer Science 37 (2012) 530−563.

[23] E.J. Corey, B. Czakó, L. Kürti, Molecules and Medicine, John Wiley & Sons, 2007.

[24] R. Hoffmann, How should chemists think? Scientific American 268 (1993) 66−73.

[25] A.P. Cote, A.I. Benin, N.W. Ockwig, M. O'Keeffe, A.J. Matzger, O.M. Yaghi, Porous, crystalline, covalent organic frameworks, Science 310 (2005) 1166−1170.

[26] S.J. Rowan, S.J. Cantrill, G.R. Cousins, J.K. Sanders, J.F. Stoddart, Dynamic covalent chemistry, Angewandte Chemie International Edition 41 (2002) 898−952.

[27] P.J. Waller, F. Gándara, O.M. Yaghi, Chemistry of covalent organic frameworks, Accounts of Chemical Research 48 (2015) 3053–3063.
[28] S.-Y. Ding, W. Wang, Covalent organic frameworks (COFs): from design to applications, Chemical Society Reviews 42 (2013) 548–568.
[29] L. Zhu, Y.-B. Zhang, Crystallization of covalent organic frameworks for gas storage applications, Molecules 22 (2017) 1149.
[30] J. Thote, H. Barike Aiyappa, R. Rahul Kumar, S. Kandambeth, B.P. Biswal, D. Balaji Shinde, et al., Constructing covalent organic frameworks in water via dynamic covalent bonding, IUCrJ 3 (2016) 402–407.
[31] H.M. El-Kaderi, J.R. Hunt, J.L. Mendoza-Cortés, A.P. Côté, R.E. Taylor, M. O'Keeffe, et al., Designed synthesis of 3D covalent organic frameworks, Science 316 (2007) 268–272.
[32] L.M. Lanni, R.W. Tilford, M. Bharathy, J.J. Lavigne, Enhanced hydrolytic stability of self-assembling alkylated two-dimensional covalent organic frameworks, Journal of the American Chemical Society 133 (2011) 13975–13983.
[33] F.J. Uribe-Romo, C.J. Doonan, H. Furukawa, K. Oisaki, O.M. Yaghi, Crystalline covalent organic frameworks with hydrazone linkages, Journal of the American Chemical Society 133 (2011) 11478–11481.
[34] H. Furukawa, O.M. Yaghi, Storage of hydrogen, methane, and carbon dioxide in highly porous covalent organic frameworks for clean energy applications, Journal of the American Chemical Society 131 (2009) 8875–8883.
[35] T. Sick, J.M. Rotter, S. Reuter, S. Kandambeth, N.N. Bach, M. Döblinger, et al., Switching on and off interlayer correlations and porosity in 2D covalent organic frameworks, Journal of the American Chemical Society 141 (2019) 12570–12581.
[36] Q. Fang, J. Wang, S. Gu, R.B. Kaspar, Z. Zhuang, J. Zheng, et al., 3D porous crystalline polyimide covalent organic frameworks for drug delivery, Journal of the American Chemical Society 137 (2015) 8352–8355.
[37] R.-R. Liang, S.-Y. Jiang, A. Ru-Han, X. Zhao, Two-dimensional covalent organic frameworks with hierarchical porosity, Chemical Society Reviews 49 (2020) 3920–3951.
[38] S. Kandambeth, D.B. Shinde, M.K. Panda, B. Lukose, T. Heine, R. Banerjee, Enhancement of chemical stability and crystallinity in porphyrin-containing covalent organic frameworks by intramolecular hydrogen bonds, Angewandte Chemie International Edition 52 (2013) 13052–13056.
[39] M. Dogru, M. Handloser, F. Auras, T. Kunz, D. Medina, A. Hartschuh, et al., A photoconductive thienothiophene-based covalent organic framework showing charge transfer towards included fullerene, Angewandte Chemie International Edition 125 (2013) 2992–2996.
[40] X. Wu, X. Han, Y. Liu, Y. Liu, Y. Cui, Control interlayer stacking and chemical stability of two-dimensional covalent organic frameworks via steric tuning, Journal of the American Chemical Society 140 (2018) 16124–16133.
[41] Y. Fan, Q. Wen, T.G. Zhan, Q.Y. Qi, J.Q. Xu, X. Zhao, A case study on the influence of substitutes on interlayer stacking of 2D covalent organic frameworks, Chemistry—A European Journal 23 (2017) 5668–5672.
[42] C. Kang, Z. Zhang, V. Wee, A.K. Usadi, D.C. Calabro, L.S. Baugh, et al., Interlayer shifting in two-dimensional covalent organic frameworks, Journal of the American Chemical Society 142 (2020) 12995–13002.
[43] Q. Yue, K. Zhang, X. Chen, L. Wang, J. Zhao, J. Liu, et al., Generation of OH radicals in oxygen reduction reaction at Pt–Co nanoparticles supported on graphene in alkaline solutions, Chemical Communications 46 (2010) 3369–3371.
[44] S. Lu, Y. Jin, H. Gu, W. Zhang, Recent development of efficient electrocatalysts derived from porous organic polymers for oxygen reduction reaction, Science China Chemistry 60 (2017) 999–1006.
[45] U. Tylus, Q. Jia, K. Strickland, N. Ramaswamy, A. Serov, P. Atanassov, et al., Elucidating oxygen reduction active sites in pyrolyzed metal–nitrogen coordinated non-precious-metal electrocatalyst systems, The Journal of Physical Chemistry C 118 (2014) 8999–9008.

[46] A.B. Jorge, R. Jervis, A.P. Periasamy, M. Qiao, J. Feng, L.N. Tran, et al., 3D carbon materials for efficient oxygen and hydrogen electrocatalysis, Advanced Energy Materials 10 (2020) 1902494.

[47] H. Zhang, M. Zhu, O.G. Schmidt, S. Chen, K. Zhang, Covalent organic frameworks for efficient energy electrocatalysis: rational design and progress, Advanced Energy and Sustainability Research 2 (2021) 2000090.

[48] K. Xu, Y. Dai, B. Ye, H. Wang, Two dimensional covalent organic framework materials for chemical fixation of carbon dioxide: excellent repeatability and high selectivity, Dalton Transactions 46 (2017) 10780−10785.

[49] H. Wang, Z. Zeng, P. Xu, L. Li, G. Zeng, R. Xiao, et al., Recent progress in covalent organic framework thin films: fabrications, applications and perspectives, Chemical Society Reviews 48 (2019) 488−516.

[50] Y. Yao, Y. Hu, H. Hu, L. Chen, M. Yu, M. Gao, et al., Metal-free catalysts of graphitic carbon nitride−covalent organic frameworks for efficient pollutant destruction in water, Journal of Colloid and Interface Science 554 (2019) 376−387.

[51] R. Ditchfield, W.J. Hehre, J.A. Pople, Self-consistent molecular-orbital methods. IX. An extended Gaussian-type basis for molecular-orbital studies of organic molecules, The Journal of Chemical Physics 54 (1971) 724−728.

[52] D. Li, C. Li, L. Zhang, H. Li, L. Zhu, D. Yang, et al., Metal-free thiophene-sulfur covalent organic frameworks: precise and controllable synthesis of catalytic active sites for oxygen reduction, Journal of the American Chemical Society 142 (2020) 8104−8108.

[53] X. Lin, P. Peng, J. Guo, Z. Xiang, Reaction milling for scalable synthesis of N, P-codoped covalent organic polymers for metal-free bifunctional electrocatalysts, Chemical Engineering Journal 358 (2019) 427−434.

[54] W.J. Hehre, R. Ditchfield, J.A. Pople, Self-consistent molecular orbital methods. XII. Further extensions of Gaussian-type basis sets for use in molecular orbital studies of organic molecules, The Journal of Chemical Physics 56 (1972) 2257−2261.

[55] G. Kresse, J. Furthmüller, Efficient iterative schemes for ab initio total-energy calculations using a plane-wave basis set, Physical Review B 54 (1996) 11169.

[56] C. Tan, X. Cao, X.-J. Wu, Q. He, J. Yang, X. Zhang, et al., Recent advances in ultrathin two-dimensional nanomaterials, Chemical Reviews 117 (2017) 6225−6331.

[57] L. Dai, Y. Xue, L. Qu, H.-J. Choi, J.-B. Baek, Metal-free catalysts for oxygen reduction reaction, Chemical Reviews 115 (2015) 4823−4892.

[58] Z. Xiang, Y. Xue, D. Cao, L. Huang, J.F. Chen, L. Dai, Highly efficient electrocatalysts for oxygen reduction based on 2D covalent organic polymers complexed with non-precious metals, Angewandte Chemie International Edition 53 (2014) 2433−2437.

[59] J. Guo, Y. Li, Y. Cheng, L. Dai, Z. Xiang, Highly efficient oxygen reduction reaction electrocatalysts synthesized under nanospace confinement of metal−organic framework, ACS Nano 11 (2017) 8379−8386.

[60] U.I. Koslowski, I. Abs-Wurmbach, S. Fiechter, P. Bogdanoff, Nature of the catalytic centers of porphyrin-based electrocatalysts for the ORR: a correlation of kinetic current density with the site density of Fe-N_4 centers, The Journal of Physical Chemistry C 112 (2008) 15356−15366.

[61] W. Li, A. Yu, D.C. Higgins, B.G. Llanos, Z. Chen, Biologically inspired highly durable iron phthalocyanine catalysts for oxygen reduction reaction in polymer electrolyte membrane fuel cells, Journal of the American Chemical Society 132 (2010) 17056−17058.

[62] Z. Xu, H. Li, G. Cao, Q. Zhang, K. Li, X. Zhao, Electrochemical performance of carbon nanotube-supported cobalt phthalocyanine and its nitrogen-rich derivatives for oxygen reduction, Journal of Molecular Catalysis A: Chemical 335 (2011) 89−96.

[63] Q. Li, Q. Shao, Q. Wu, Q. Duan, Y. Li, H.-g Wang, In situ anchoring of metal nanoparticles in the N-doped carbon framework derived from conjugated microporous polymers towards an efficient oxygen reduction reaction, Catalysis Science & Technology 8 (2018) 3572−3579.

[64] J.-D. Yi, R. Xu, Q. Wu, T. Zhang, K.-T. Zang, J. Luo, et al., Atomically dispersed iron−nitrogen active sites within porphyrinic triazine-based frameworks for oxygen reduction reaction in both alkaline and acidic media, ACS Energy Letters 3 (2018) 883−889.

[65] K. Kamiya, Development of robust electrocatalysts comprising single-atom sites with designed coordination environments, Electrochemistry 88 (2020) 489−496.

[66] K. Iwase, T. Yoshioka, S. Nakanishi, K. Hashimoto, K. Kamiya, Copper-modified covalent triazine frameworks as non-noble-metal electrocatalysts for oxygen reduction, Angewandte Chemie International Edition 54 (2015) 11068−11072.

[67] R. Gong, L. Yang, S. Qiu, W.-T. Chen, Q. Wang, J. Xie, et al., A nitrogen rich covalent triazine framework as a photocatalyst for hydrogen production, Advances in Polymer Technology 2020 (2020).

[68] J. Guo, C.Y. Lin, Z. Xia, Z. Xiang, A pyrolysis-free covalent organic polymer for oxygen reduction, Angewandte Chemie International Edition 57 (2018) 12567−12572.

[69] Z. Li, W. Zhao, C. Yin, L. Wei, W. Wu, Z. Hu, et al., Synergistic effects between doped nitrogen and phosphorus in metal-free cathode for zinc-air battery from covalent organic frameworks coated CNT, ACS Applied Materials & Interfaces 9 (2017) 44519−44528.

[70] B.J. Smith, W.R. Dichtel, Mechanistic studies of two-dimensional covalent organic frameworks rapidly polymerized from initially homogenous conditions, Journal of the American Chemical Society 136 (2014) 8783−8789.

[71] K.T. Jackson, T.E. Reich, H.M. El-Kaderi, Targeted synthesis of a porous borazine-linked covalent organic framework, Chemical Communications 48 (2012) 8823−8825.

[72] U. Díaz, A. Corma, Ordered covalent organic frameworks, COFs and PAFs. From preparation to application, Coordination Chemistry Reviews 311 (2016) 85−124.

[73] R. Gutzler, H. Walch, G. Eder, S. Kloft, W.M. Heckl, M. Lackinger, Surface mediated synthesis of 2D covalent organic frameworks: 1, 3, 5-tris (4-bromophenyl) benzene on graphite (001), Cu (111), and Ag (110), Chemical Communications (2009) 4456−4458.

[74] N. Huang, X. Ding, J. Kim, H. Ihee, D. Jiang, A photoresponsive smart covalent organic framework, Angewandte Chemie International Edition 127 (2015) 8828−8831.

[75] J.R. Hunt, C.J. Doonan, J.D. LeVangie, A.P. Côté, O.M. Yaghi, Reticular synthesis of covalent organic borosilicate frameworks, Journal of the American Chemical Society 130 (2008) 11872−11873.

[76] W. Luo, Y. Zhu, J. Zhang, J. He, Z. Chi, P.W. Miller, et al., A dynamic covalent imine gel as a luminescent sensor, Chemical Communications 50 (2014) 11942−11945.

[77] C.D. Meyer, C.S. Joiner, J.F. Stoddart, Template-directed synthesis employing reversible imine bond formation, Chemical Society Reviews 36 (2007) 1705−1723.

[78] F.J. Uribe-Romo, J.R. Hunt, H. Furukawa, C. Klock, M. O'Keeffe, O.M. Yaghi, A crystalline imine-linked 3-D porous covalent organic framework, Journal of the American Chemical Society 131 (2009) 4570−4571.

[79] C.E. Chan-Thaw, A. Villa, D. Wang, A. Biroli, G.M. Veith, A. Thomas, et al., PdHx entrapped in covalent triazine framework modulates selectivity in glycerol oxidation [Modulation of palladium activity and stability by a covalent triazine framework], ChemCatChem 7 (2015).

[80] C.E. Chan-Thaw, A. Villa, L. Prati, A. Thomas, Triazine-based polymers as nanostructured supports for the liquid-phase oxidation of alcohols, Chemistry—A European Journal 17 (2011) 1052−1057.

[81] Q. Fang, Z. Zhuang, S. Gu, R.B. Kaspar, J. Zheng, J. Wang, et al., Designed synthesis of large-pore crystalline polyimide covalent organic frameworks, Nature Communications 5 (2014) 1−8.

CHAPTER 10

Nano-inks for fuel cells

Fahimeh Hooriabad Saboor[1] and Tuan Anh Nguyen[2]
[1]Department of Chemical Engineering, University of Mohaghegh Ardabili, Ardabil, Iran
[2]Institute for Tropical Technology, Vietnam Academy of Science and Technology, Hanoi, Vietnam

10.1 Introduction

The chemical energy can be converted to electricity power using a fuel cell that is similar to the operating principle of batteries without needing recharging. Some heat also generates in this conversion without needing direct combustion of fuel. Depends on the chemical character of electrolyte in cell, fuel cells are classified into several types, including alkaline, direct methanol (DMFC), phosphoric acid, sulfuric acid, proton-exchange membrane, molten carbonate, solid oxide (SOFC), and protonic ceramic [1,2]. One of the most promising electricity production devices is SOFCs which generally operate at a temperature ranging from 500°C to 1000°C and have low emission energy conversion [3]. One of the most applicable fuel cells for portable and low-power applications is polymer electrolyte membrane fuel cells (PEMFCs), especially liquid-feed DMFCs with lightweight, easy recharging and storage with no need for hydrogen storage and humidification [4]. The membrane in PEMFCs is coated by a catalyst layer using catalyst ink. The physicochemical properties of the catalyst layer are controlled by catalyst dispersion, viscosity, and drying rate [5]. Microbial fuel cell (MFC) is an emerging system that generates bioelectricity from organic matter of the wastewater using biocatalysts [6].

Various methods have been reported to fabricate fuel cells, such as chemical vapor deposition [7], atomic layer deposition [8], electrochemical vapor deposition [9], screen-printing [10], spin coating or slurry coating [11]. Among various methods, screen-printing is a simple, low cost and flexible technique to fabricate thick-film electronic components such as anode, cathode electrodes, and electrolyte films with a thickness of 10–100 μm in SOFCs [3,12]. A schematic of the main printing techniques is shown in Fig. 10.1 [13]. Several coating/printing techniques are introduced and utilized to deposit layers and patterns in electronic and optoelectronic devices. Some of these techniques include spin coating, blade coating, spray coating, inkjet printing, screen-printing, pad printing, slot-die coating, and gravure coating [14]. The main differences in various printing methods include the formulation of ink, coating/printing time, film/pattern thickness (thick or thin-film), and the resolution and the size of printed features [14]. The inkjet printing

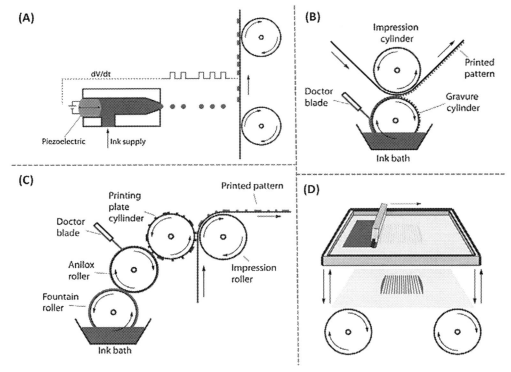

Figure 10.1 Schematic figures of (A) inkjet, (B) gravure, (C) flexographic, (D) screen-printing techniques. *Reproduced with permission from C. Cano-Raya, et al., Chemistry of solid metal-based inks and pastes for printed electronics—a review, Applied Materials Today 15 (2019) 416–430. Copyright 2019 Elsevier.*

method is a simple and controllable technique to apply colloidal inks to form uniform nanostructured ceramic layers in SOFCs [15]. Sobolev et al. [16] reported a highly stable NiO colloidal ink (see Fig. 10.2) with sizes ranging from 7 to 9 nm to fabricate an anode layer of 5 μm in SOFC using commercial printing, then sintering at 900°C. Park et al. [17] introduced and utilized a new electrode fabrication method comprising a micropipette coupled to a peristaltic pump for ink deposition on the membrane surface equipped with a computer-interfaced controller.

A general scheme of the screen-printing process is shown in Fig. 10.3. The screen-printing technique is utilized to produce a uniform pattern on a substrate. A flat substrate is placed under a screen mask, and a printing paste or ink is placed on the mesh of the screen mask. The pattern or thick film is deposited on the substrate by pressing the ink/paste using a polyurethane type squeegee or fill blade [12]. A tunable distance of about 0.5–1 mm between the screen and the substrate is required to fabricate a clear print [14].

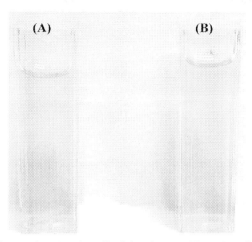

Figure 10.2 (A) As-synthesized NiO ink colloidal solution, (B) NiO colloidal solution after four months of shelf life. *Reproduced with permission from A. Sobolev, P. Stein, K. Borodianskiy, Synthesis and characterization of NiO colloidal ink solution for printing components of solid oxide fuel cells anodes, Ceramics International 46(16, Part A) (2020) 25260–25265. Copyright 2020 Elsevier.*

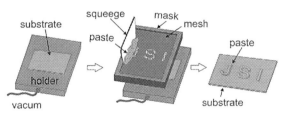

Figure 10.3 Schematic of the screen-printing process. *Reproduced with permission from D. Kuscer, Screen-printing, in: Reference Module in Materials Science and Materials Engineering. 2020, Elsevier. Copyright 2020 Elsevier.*

Here, we briefly represent the rheological characteristics and the influencing parameters of catalyst ink in fuel cells. This chapter focuses on the effect of various parameters including solvents, additives, dispersion method, and composition of catalyst nano-inks to fabricate high performance, uniform, and durable electrodes in several fuel cells such as proton–exchange membrane, SOFC, and MFCs.

10.2 Ink rheological parameters and influencing factors

The ink rheology is a vital factor in controlling the quality of the film and its performance. The steady-state features of screen-printing inks include viscosity, thixotropy, and yield stress. Dynamic features of screen-printing inks include elastic and viscous modulus, damping ratio, and linear viscoelastic region. Several rheological studies have evaluated the ink thixotropic behavior and ink network structure using steady-state and yield stress tests. The viscoelastic and slump characteristics of inks are determined by the oscillation and creep recovery tests, respectively. Solid-like and liquid-like characteristics of inks can be investigated by measuring the elastic modulus and the viscous modulus [3,18–20].

In general, a printing ink consists of a powder solid, a solvent, a binder, and a dispersant. The rheological character of ink depends on the particle size and particle size distribution of solid, which determine the solid content of ink [3,21]. As reported, the high surface area of nano-sized powders resulted in a high viscose ink even at a low solid content of 40%–50%. Low solid content increases the possibility of crack formation, while the inks with high solid content result in a thicker and rougher film. The films prepared using inks with a high solid content of 70% showed an excellent current-voltage characteristic [19,22]. Solvent type is the other crucial rheological factor that affects the solid content of ink [23]. Two common solvents are terpineol ($C_{10}H_{18}O$) and texanol ($C_{12}H_{24}O_3$) that usually utilize a mixture of 94% solvent with a 6% binder [24]. Various binders such as ethylene cellulose, polyvinyl butyral, polymethyl methacrylate, and polyvinyl acetal are usually utilizing in the ink formulation to improve the particle network strength [25–27]. The vital role of dispersant is to tune the viscosity and elastic modulus of inks via modulating interparticle forces between solid particles (i.e., electrostatic or steric hindrance) to prevent the particle agglomeration. Nano-sized powder solids with high surface area require a larger amount of dispersant to be stabilized [28]. A recommended formula to obtain the dispersant content is as follows [3]

$$m_d = m_p(g) \times S_{BET}\left(\frac{m^2}{g}\right) \times A_c\left(\frac{g}{m^2}\right)$$

in which m_d is dispersant amount, A_c is dispersant coverage value, Ac, m_p is the weight of powder and S_{BET} is the specific surface area of powder.

10.3 Recent progress in nano-ink preparation and utilization in fuel cell application

10.3.1 The effect of solvents and additives

The output power of PEMFCs mainly depends on the electrode catalyst layer characteristics such as composition and morphology, which are mainly related to the catalyst ink preparation procedure especially the type of solvent and its dispersion. Solvent properties such as type, boiling point, viscosity, affect the structure and porosity of the catalyst electrode in both supported and unsupported catalyst. It is reported that it depends on the dielectric constant of organic solvents, so the catalyst can be formed as a solution ($\varepsilon > 10$), colloidal ($3 < \varepsilon < 10$), or precipitate ($\varepsilon < 3$). Electrode prepared by a colloidal solution showed a larger electrochemical reaction area and higher cell performance [4,29]. Solvent boiling point and viscosity can affect the porosity of catalyst electrode during ink evaporation step [30]. Alcaide et al. [4] studied the influence of two different solvents, including n-butyl acetate (NBA) and 2-propanol (IPA), on the unsupported PtRu catalyst inks in DMFC applications. The lower polarity of n-butyl acetate resulted in a larger aggregate of the PtRu-Nafion and larger porosity of the catalyst layer formed by the agglomerates. In a single DMFC, PtRu anode catalyst layers prepared using NBA solvent showed higher current densities. Moreover, the proton resistance of the catalyst prepared by NBA is much lower than that of IPA that reveals proper connection of catalyst to Nafion$^@$ ionomers and easier transportation to the membrane [4].

The catalyst ink properties can be changed by applying additives or mixing solvents. Several studies are conducted to evaluate the impact of additives such as polytetrafluoroethylene (PTFE), polyvinylidene fluoride (PVDF), and NH_4HCO_3 on hydrophobic/hydrophilic characters, catalyst dispersion, and cell performance in PEMFCs [5,31–33]. Glycerol, N-methyl-2-pyrrolidone (NMP), and polyethylene glycol (PEG) are some solvents that have been investigated for catalyst ink preparation and their impact on the catalyst-water-ionomer interface, PEMFCs durability, ohmic resistance, the dispersion degree of the catalyst [34–36]. Among the three resistances of mass transfer, charge transfer, and ohmic resistance, the former is more important to control and improve the cell performance. Water flooding is a result of mass transport resistance. Several studies have been studied on lowering the mass transfer resistance via adding pore formers and porous structure of catalyst layer [37,38]. Lee et al. [5] investigated the influence of PEG addition to catalyst ink in a PEM fuel cell. Polyethylene glycol with three different molecular weights from 200–10000 was added to the cathode catalyst ink and showed an increased pore size of the catalyst layer, increased hydrophilicity, and increased performance of current density of 1.5 A cm^{-2} due to a reduced proton transport resistance.

Magnus So et al. [39] studied the agglomeration in different fuel cell catalyst inks (concentrated and diluted inks) applying the Discrete Element Method. They found that the moderate dielectric constant of solvent leads to more stable catalyst ink, while solutions with high or low dielectric constants result in agglomeration. In a solution containing water/NPA, the model predicted higher agglomeration with water content.

The surface tension of the printing ink is a crucial factor to have a uniform catalyst layer that is governed by the surface tension of the solvent. It is reported that the surface tension of the ink needs to be lower than the substrate surface tension to prevent wetting and adhesion problems of the catalyst layer [40]. Bonifácio et al. [40] introduced a one-step screen-printing process to fabricate a catalyst layer with high performance via controlling the surface tension of catalyst ink. The viscosity, total solid content, density, and tack value of the optimized printing ink was 2.75 Pa s, 33.76 wt. %, 1.294 g cm^{-3}, and 92 U.T, respectively.

10.3.2 The effect of the dispersion method

Several studies have been conducted on the influence of dispersion step on the viscosity and rheological properties of catalyst ink [41,42]. Some typical forces that dominate the catalyst dispersion during ink preparation are shown in Fig. 10.4. Du et al. [43] applied two different dispersion methods, including ball milling and sonication. They found that the ionomer solution has a critical role in catalyst dispersion. The ink viscosity characteristics are also affected by the type of dispersion method. Ball milling resulted in a higher viscose ink solution than the sonication method. In conclusion, the catalyst ink dispersed using ball milling was more stable due to the broader size distribution and higher zeta potential than the ink dispersed by sonication.

Pollet et al. [44] investigated the influence of two different dispersion methods, including ultrasonication and mechanically shear-mixed for up to 120 minutes on the catalyst ink properties. Two commercial carbon-supported Pt catalyst ink were studied, and the results showed that ultrasonication for the desired duration is essential for catalyst ink activity. In contrast, sonication for an extended period showed an adverse impact on the ink morphology and composition due to the cavitation phenomena.

10.3.3 The effect of the ink composition

Osmieri et al. [45] investigated the ink composition (water/n-propanol content) of a platinum group metal-free catalysts-based electrode to fabricate a more robust electrode in a PEMFC with high performance. For a Fe−N−C catalyst, the ink containing 50 wt.% H_2O showed improved interactions between ionomer and electrocatalyst and, as a result, a higher performance under low relative humidity. Cleve et al. [46]

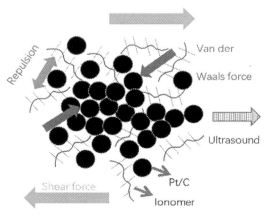

Figure 10.4 Schematic representation of various forces governed during the catalyst ink dispersion. *Reproduced with permission from S. Du, et al., Effects of ionomer and dispersion methods on rheological behavior of proton exchange membrane fuel cell catalyst layer ink, International Journal of Hydrogen Energy 45 (53) (2020) 29430−29441. Copyright 2020 Elsevier.*

investigated the effect of the water content of catalyst ink and n-propanol to water ratio on the microstructural characteristics of the electrode and MEA performance. The MEA comprised a commercial catalyst of 30 wt.% PtCo on HSC, D2020 ionomer, and Nafion 211 membrane. The results revealed an improved oxygen transport, more porous electrode, and smaller-size aggregates were achieved using a water-rich catalyst ink.

The high cost of Pt-based cathodes limits the scalability of MFCs technology. Co and Fe-based materials as low-cost electrocatalysts suffer from cathode flooding, which refers to the water condensation and blocking the catalyst layer's pores, resulting in a decrease in fuel cell performance. Utilizing a microporous layer of conductive ink on the cathode surface can significantly address cathode flooding via oxygen convection [47,48]. Bhowmick et al. [49] applied a cathode comprising Co_3O_4 and Fe_3O_4 blended with conductive ink to improve oxygen reduction reaction at the cathode in dual-chambered aqueous cathode MFCs. Cyclic voltammetry analysis revealed a high reduction current of—30 mA for Co_3O_4 cathode with conductive ink. Cathode with conductive ink containing Co_3O_4 showed a volumetric power density of 6.62 W m^{-3} and a coulombic efficiency (CE) of 22.3%, which proved this cathode can be an efficient alternative to Pt-based cathodes in MFCs. Wu et al. [50] reported a novel and large-size (50 × 60 cm^2) anode for MFCs based on modifying a stainless steel cloth with carbon nanoparticles of Chinese ink. Polypyrrole was used as a building block to prepare the anode with a uniform dispersion of carbon nanoparticles (see Fig. 10.5).

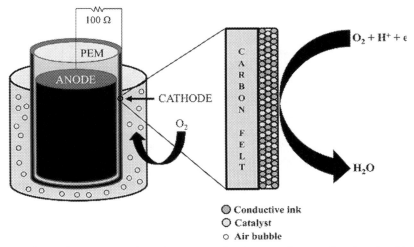

Figure 10.5 Schematic representation of oxygen reduction reaction reaction in microbial fuel cell. *Reproduced with permission from G.D. Bhowmick et al., Improved performance of microbial fuel cell by using conductive ink printed cathode containing Co_3O_4 or Fe_3O_4, Electrochimica Acta 310 (2019) 173−183. Copyright 2019 Elsevier.*

The prepared anode showed the maximum power density of MFC 61.9 mW m^{-2} and the maximum current of 12.4 ± 1.4 mA.

10.4 Conclusion

In this chapter, we briefly discussed the importance of catalyst ink in fuel cell performance. This chapter focused on the effect of various parameters, including solvents, additives, dispersion method, and composition of the catalyst nano-inks to fabricate high performance, uniform, and durable electrodes in several fuel cells, such as proton-exchange membrane, SOFC, and microbial fuel cells. The ink rheology mainly affects the quality of the catalyst layer in fuel cells. Viscosity, thixotropy, yield stress, elastic modulus, viscous modulus, damping ratio, and linear viscoelastic region are some crucial parameters that have been investigated in the rheological studies of printing/coating inks. The rheological character of ink depends on the particle size and particle size distribution of the solid catalyst. In recent years, the catalyst ink preparation procedure, the type of solvent, and its dispersion have been evaluated in several studies. Solvent surface tension, viscosity, and boiling point are controlling factors in the quality of electrodes. In conclusion, printed electronics are an emerging market, and fabrication of high-performance fuel cells based on developing more efficient, stable, durable, low cost, and compatible methods with commercial printing/coating techniques are needed to be continued.

References

[1] A.B. Stambouli, E. Traversa, Solid oxide fuel cells (SOFCs): a review of an environmentally clean and efficient source of energy, Renewable and Sustainable Energy Reviews 6 (5) (2002) 433–455.
[2] N. Sazali, et al., New perspectives on fuel cell technology: a brief review, Membranes 10 (5) (2020) 99.
[3] M.R. Somalu, et al., Screen-printing inks for the fabrication of solid oxide fuel cell films: a review, Renewable and Sustainable Energy Reviews 75 (2017) 426–439.
[4] F. Alcaide, et al., Effect of the solvent in the catalyst ink preparation on the properties and performance of unsupported PtRu catalyst layers in direct methanol fuel cells, Electrochimica Acta 231 (2017) 529–538.
[5] H.-Y. Lee, et al., Reduced mass transport resistance in polymer electrolyte membrane fuel cell by polyethylene glycol addition to catalyst ink, International Journal of Hydrogen Energy 44 (1) (2019) 354–361.
[6] Q. Wen, et al., Improved performance of microbial fuel cell through addition of rhamnolipid, Electrochemistry Communications 12 (12) (2010) 1710–1713.
[7] A.S. Ferlauto, et al., Chemical vapor deposition of multi-walled carbon nanotubes from nickel/yttria-stabilized zirconia catalysts, Applied Physics A 84 (3) (2006) 271–276.
[8] Y. Jee, et al., High performance Bi-layered electrolytes via atomic layer deposition for solid oxide fuel cells, Journal of Power Sources 253 (2014) 114–122.
[9] M. Haldane, T. Etsell, Fabrication of composite SOFC anodes using polarized electrochemical vapor deposition, Fuel Cells Bulletin 2005 (7) (2005) 12–16.

[10] D. Rotureau, et al., Development of a planar SOFC device using screen-printing technology, Journal of the European Ceramic Society 25 (12) (2005) 2633–2636.
[11] Vo, N.X. Phuong, et al., Fabrication and characterization of an anode-side, substrate-supported planar SOFC, Journal of Fuel Cell Science and Technology 8 (2011) 5.
[12] D. Kuscer, Screen printing, Reference Module in Materials Science and Materials Engineering, Elsevier, 2020.
[13] C. Cano-Raya, et al., Chemistry of solid metal-based inks and pastes for printed electronics—a review, Applied Materials Today 15 (2019) 416–430.
[14] I. Verboven, W. Deferme, Printing of flexible light emitting devices: a review on different technologies and devices, printing technologies and state-of-the-art applications and future prospects, Progress in Materials Science 118 (2021) 100760.
[15] A. Matavž, B. Malič, Inkjet printing of functional oxide nanostructures from solution-based inks, Journal of Sol-Gel Science and Technology 87 (1) (2018) 1–21.
[16] A. Sobolev, P. Stein, K. Borodianskiy, Synthesis and characterization of NiO colloidal ink solution for printing components of solid oxide fuel cells anodes, Ceramics International 46 (16, Part A) (2020) 25260–25265.
[17] J. Park, D.J. Myers, Novel platinum group metal-free catalyst ink deposition system for combinatorial polymer electrolyte fuel cell performance evaluation, Journal of Power Sources 480 (2020) 228801.
[18] M.R. Somalu, N.P. Brandon, Rheological studies of nickel/scandia-stabilized-zirconia screen printing inks for solid oxide fuel cell anode fabrication, Journal of the American Ceramic Society 95 (4) (2012) 1220–1228.
[19] R. Rudež, J. Pavlič, S. Bernik, Preparation and influence of highly concentrated screen-printing inks on the development and characteristics of thick-film varistors, Journal of the European Ceramic Society 35 (11) (2015) 3013–3023.
[20] M.G. Rasteiro, E. Antunes, Correlating the rheology of PVC-based pastes with particle characteristics, Particulate Science and Technology 23 (4) (2005) 361–375.
[21] D. Burnat, et al., The rheology of stabilised lanthanum strontium cobaltite ferrite nanopowders in organic medium applicable as screen printed SOFC cathode layers, Fuel Cells 10 (1) (2010) 156–165.
[22] P. Von Dollen, S. Barnett, A study of screen printed yttria-stabilized zirconia layers for solid oxide fuel cells, Journal of the American Ceramic Society 88 (12) (2005) 3361–3368.
[23] M.R. Somalu, et al., The impact of ink rheology on the properties of screen-printed solid oxide fuel cell anodes, International Journal of Hydrogen Energy 38 (16) (2013) 6789–6801.
[24] X. Ge, et al., Screen-printed thin YSZ films used as electrolytes for solid oxide fuel cells, Journal of Power Sources 159 (2) (2006) 1048–1050.
[25] M.R. Somalu, V. Yufit, N.P. Brandon, The effect of solids loading on the screen-printing and properties of nickel/scandia-stabilized-zirconia anodes for solid oxide fuel cells, International Journal of Hydrogen Energy 38 (22) (2013) 9500–9510.
[26] R. Durairaj, et al., Rheological characterisation and printing performance of Sn/Ag/Cu solder pastes, Materials and Design 30 (9) (2009) 3812–3818.
[27] A.C. Nascimento, et al., Correlation between yttria stabilized zirconia particle size and morphological properties of NiO–YSZ films prepared by spray coating process, Ceramics International 35 (8) (2009) 3421–3425.
[28] W.J. Tseng, C.-N. Chen, Effect of polymeric dispersant on rheological behavior of nickel–terpineol suspensions, Materials Science and Engineering: A 347 (1) (2003) 145–153.
[29] M. Uchida, et al., New preparation method for polymer-electrolyte fuel cells, Journal of the Electrochemical Society 142 (2) (1995) 463–468.
[30] M. Chisaka, E. Matsuoka, H. Daiguji, Effect of organic solvents on the pore structure of catalyst layers in polymer electrolyte membrane fuel cells, Journal of the Electrochemical Society 157 (8) (2010) B1218.
[31] G.S. Avcioglu, et al., High performance PEM fuel cell catalyst layers with hydrophobic channels, International Journal of Hydrogen Energy 40 (24) (2015) 7720–7731.

[32] H.-L. Lin, et al., Effect of polyvinylidene difluoride in the catalyst layer on high-temperature PEMFCs, International Journal of Hydrogen Energy 40 (30) (2015) 9400–9409.
[33] J. Zhao, et al., Addition of NH_4HCO_3 as pore-former in membrane electrode assembly for PEMFC, International Journal of Hydrogen Energy 32 (3) (2007) 380–384.
[34] Y.S. Kim, et al., Highly durable fuel cell electrodes based on ionomers dispersed in glycerol, Physical Chemistry Chemical Physics 16 (13) (2014) 5927–5932.
[35] C.-Y. Jung, S.-C. Yi, Improved polarization of mesoporous electrodes of a proton exchange membrane fuel cell using N-methyl-2-pyrrolidinone, Electrochimica Acta 113 (2013) 37–41.
[36] M.-J. Choo, et al., Modulated ionomer distribution in the catalyst layer of polymer electrolyte membrane fuel cells for high temperature operation, ChemSusChem 7 (8) (2014) 2335–2341.
[37] G.-Y. Chen, et al., Gradient design of Pt/C ratio and Nafion content in cathode catalyst layer of PEMFCs, International Journal of Hydrogen Energy 42 (50) (2017) 29960–29965.
[38] H. Li, et al., A review of water flooding issues in the proton exchange membrane fuel cell, Journal of Power Sources 178 (1) (2008) 103–117.
[39] M. So, et al., The effect of solvent and ionomer on agglomeration in fuel cell catalyst inks: simulation by the discrete element method, International Journal of Hydrogen Energy 44 (54) (2019) 28984–28995.
[40] R.N. Bonifácio, et al., Catalyst layer optimization by surface tension control during ink formulation of membrane electrode assemblies in proton exchange membrane fuel cell, Journal of Power Sources 196 (10) (2011) 4680–4685.
[41] K. Woo, et al., Relationship between printability and rheological behavior of ink-jet conductive inks, Ceramics International 39 (6) (2013) 7015–7021.
[42] C. Zhao, C. Yang, An exact solution for electroosmosis of non-Newtonian fluids in microchannels, Journal of Non-Newtonian Fluid Mechanics 166 (17) (2011) 1076–1079.
[43] S. Du, et al., Effects of ionomer and dispersion methods on rheological behavior of proton exchange membrane fuel cell catalyst layer ink, International Journal of Hydrogen Energy 45 (53) (2020) 29430–29441.
[44] B.G. Pollet, J.T.E. Goh, The importance of ultrasonic parameters in the preparation of fuel cell catalyst inks, Electrochimica Acta 128 (2014) 292–303.
[45] L. Osmieri, et al., Utilizing ink composition to tune bulk-electrode gas transport, performance, and operational robustness for a Fe–N–C catalyst in polymer electrolyte fuel cell, Nano Energy 75 (2020) 104943.
[46] T. Van Cleve, et al., Tailoring electrode microstructure via ink content to enable improved rated power performance for platinum cobalt/high surface area carbon based polymer electrolyte fuel cells, Journal of Power Sources 482 (2021) 228889.
[47] C. Santoro, et al., Effects of gas diffusion layer (GDL) and micro porous layer (MPL) on cathode performance in microbial fuel cells (MFCs), International Journal of Hydrogen Energy 36 (20) (2011) 13096–13104.
[48] U. Pasaogullari, C.-Y. Wang, Two-phase transport and the role of micro-porous layer in polymer electrolyte fuel cells, Electrochimica Acta 49 (25) (2004) 4359–4369.
[49] G.D. Bhowmick, et al., Improved performance of microbial fuel cell by using conductive ink printed cathode containing Co_3O_4 or Fe_3O_4, Electrochimica Acta 310 (2019) 173–183.
[50] H. Wu, et al., Stainless steel cloth modified by carbon nanoparticles of Chinese ink as scalable and high-performance anode in microbial fuel cell, Chinese Chemical Letters (2020).

CHAPTER 11

Nanomaterial and nanocatalysts in microbial fuel cells

Sumisha Anappara, Karthick Senthilkumar and Haribabu Krishnan
Department of Chemical Engineering, National Institute of Technology Calicut, Kozhikode, India

11.1 Introduction

Microbial fuel cells (MFCs) are bioelectrochemical devices used for the generation of electricity from biomass. A typical two chamber MFCs consist of an anode and cathode chamber separated by a proton-exchange membrane (PEM). When organic compounds are filled into the anodic chamber, the exoelectrogenic bacteria oxidize the substrates into protons and electrons. The protons pass through the PEM to reach the cathode whereas the electrons are transferred through an external circuit and reduce the electron acceptor at cathode. The electricity produced is equivalent to the amount of electrons flowing through the external circuit [1]. Hence wastewater treatment and simultaneous bioenergy generation are considered as the vital applications of MFC. A typical double-chamber microbial fuel cell (DCMFC) is shown in Fig. 11.1. Taking glucose as the electron donor and O_2 as the electron acceptor, the respective reactions at the anode and cathode and the overall reactions are given as follows [2]:

$$\text{Anode: } C_6H_{12}O_6 + 6H_2O \rightarrow 6CO_2 + 24H^+ + 24e^-$$

$$\text{Cathode: } 6O_2 + 24\,H^+ + 24e^- \rightarrow 12\,H_2O$$

$$\text{Overall reaction: } C_6H_{12}O_6 + 6O_2 \rightarrow 6CO_2 + 6H_2O$$

A basic MFC configuration contains an anode, a cathode and a PEM. The substrates such as glucose, acetate, or lactate are used as the fuel for the biocatalyst growth. The anode chamber is maintained in anaerobic condition whereas the cathode chamber contains oxygen or any other electron acceptors. MFCs have several advantages, viz., low operating temperature, high conversion efficiency with less sludge production; however its large-scale applications are limited because of poor power production and costly components [3]. Generally, carbonaceous materials like carbon cloth (CC), carbon

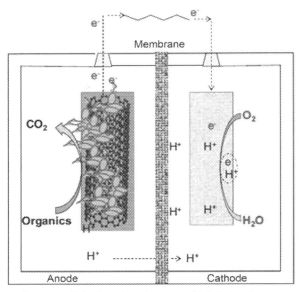

Figure 11.1 Schematic diagram of double-chamber microbial fuel cell.

brush, carbon rod, carbon mesh, carbon paper, granular activated carbon (AC), graphite plate, graphite felt, and reticulated vitreous carbon (RVC) are used as electrode material in MFC. CCs possess high surface area, porosity, and electrical conductivity, but the cost is fairly high. Carbon brush is made of carbon fibers which are twisted on a titanium core. The titanium core increases the electrical conductivity as well as the material cost. Carbon rods hold low surface area and hence it is used as current collector rather than anode. Carbon mesh is a low-cost material having very low conductivity and mechanical strength, that leads to lack of durability under high-flow conditions [4]. Carbon paper is a porous carbonaceous material but costly and fragile. Granular AC is highly porous and has relatively low electrical conductivity [5]. However, it is sometimes used as anode because it is cheaper and shows greater compatibility toward microorganisms. Graphite plate is a simple electrode that holds high electrical conductivity and low cost, but its low surface area results in lower power output in MFC than porous or materials. On the other hand, graphite felt have good conductivity, high surface area, and stability. It also facilitates the biofilm formation. RVC also shows same property as that of graphite felt but the material is relatively fragile and very expensive [6]. Nafion is the widely used PEM in MFCs, but the major problems associated with it are the high cost, substrate loss, and cation transport. Also, it causes oxygen leakage to the anode chamber, which reduces the coulombic efficiency and hinders the growth of anaerobic bacteria [7]. To overcome these drawbacks of electrodes and PEM, proper modifications have to be made to the existing electrodes.

Nowadays, the application of nanomaterials in energy and environment field has drawn great attention due to its physicochemical, thermal and mechanical properties. Nanomaterials possess larger surface area and hence high surface activity than the bulk material. This results in a quick reaction rates and high electrocatalytic activity, which improves the performance of the system. Also, nanomaterials are biologically active than micro or larger particles; therefore it is easy to interact for them with biological system more effectively [4]. The application of different structures of nanomaterials like nanoparticles, nanotubes, nanofiber, nanorods, nanoflowers, etc. in MFC enhances the conductivity, mechanical stability, and electrochemical activity of the electrode materials and also the water retention and mechanical stability of the membranes [5].

11.2 Application of nanomaterial structures in anodes

The anodes in MFC involves both biocatalysis and electrocatalysis; hence the important property of anode is its biocompatibility and extracellular electron transfer capability. The exoelectrogenesis process can be occurring via direct or indirect pathways. The indirect electron transfer between the bacterial cells and the electrode occurs by the addition, or microbial secretion, of electroactive mediator compounds such as quinones and phenazines, that diffusively transport the electrons. In direct or mediatorless electron transfer process, the microorganisms and electrode have direct interaction via the cellular c-type cytochrome proteins, containing heme groups, present in the outer side of cell membrane. C-type cytochromes have been observed in certain microbial species, viz., *Shewanella putrefaciens*, *Geobacter sulferreducens*, etc. [6].

The major losses associated with anode are the activation loss and bacterial metabolism loss due to the excess biofilm formation and biofouling of anode. This leads to the activation overpotential of anode and hence the energy loss. To overcome this, the strategies that can be adapted are; increase of electrode catalysis, addition of mediators in the electrolyte and increase of electrode surface area. However, one important tool to enhance the anode property is to modify the anode with nanomaterials that facilitate the formation of electroactive bacteria and stimulate the extracellular electron transfer process between microbes and the anode (Fig. 11.2) [8]. The materials with nanostructure contains both micropores and nanopores, in which, the microporous structure helps in the biofilm growth, whereas the nanoporous structure helps in achieving efficient electrocatalysis [9]. The various studies on nanostructured anode materials in MFC are discussed further.

11.2.1 Carbon-based nanomaterials

The conventional carbon materials limit its applications in MFC due to the inefficient electronic transmission channel because of their two-dimensional structure and relatively large internal resistance. But, the carbon-based nanomaterials provide both high surface area for bacterial attachment and active sites for electron transfer from

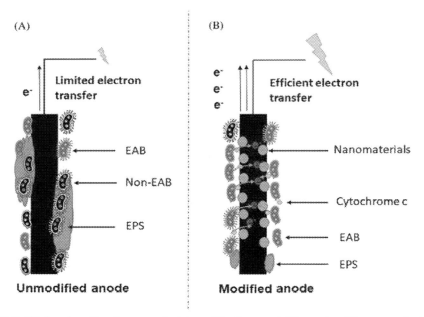

Figure 11.2 (A) Anode without nanomaterial modification; and (B) anode with nanomaterial modification. *Adapted from N. Savla, R. Anand, S. Pandit, R. Prasad, Utilization of nanomaterials as anode modifiers for improving microbial fuel cells performance, Journal of Renewable Materials 8 (12) (2020) 1581–1605.*

extracellular bacteria [9]. The different structures of carbon nanomaterials are carbon nanotubes (CNTs), ACs, diamond films.

CNTs have a fiber structure with high mechanical strength and toughness, large specific surface area, good thermal stability, and conductivity. CNTs are like cylinder rolled up with one-atom-thickness sheet known as graphene. CNTs are classified into single-walled carbon nanotubes (SWCNTs) and multiwalled carbon nanotubes (MWCNTs), depends on the number graphene sheet array (Fig. 11.3). In SWCNT, single sheet of graphene cylinder with honeycomb lattice whereas, in MWCNT, two or more layer of graphene sheets are seen [10].

Liang et al. (2011) [11] used CNT as modifier for carbon paper anode in DCMFC with *G. Sulfurreducens* as biocatalyst and discovered that the anodic biofilm formation was enriched. Also, MFC with modified anode has a low internal resistance of 180 ohms compared to that with bare carbon paper. Roh et al. (2013) [12] also compared the bare carbon paper and CNT-modified carbon paper in DCMFC. The modified carbon paper with CNT, synthesized via layer-by-layer self-assembly method, had very low internal resistance of 258 ohms than that of bare carbon paper, 1163 ohms and produced a maximum power density of 480 mW m^{-2}. Erbay et al. (2015) [13] prepared stainless steel (SS) anode with multiwalled CNT and tested in

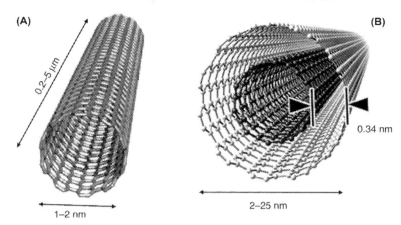

Figure 11.3 Schematic diagram of (A) SWCNT and (B) MWCNT (*MWCNT*, Multiwalled carbon nanotube; *SWCNT*, single-walled carbon nanotube). *Adapted from M.M. Mohideen, Y. Liu, S. Ramakrishna, Recent progress of carbon dots and carbon nanotubes applied in oxygen reduction reaction of fuel cell for transportation, Applied Energy 257 (2020) 114027.*

microfabricated 24-well MFC arrays. The modified anode showed a maximum power density of 3360 mW m^{-2} which is 7.4-fold higher than that of CC electrode. The results also presented the excellent charge transfer characteristics of the electrode due to p-p stacking between the carbon atoms and the microorganisms. The fixed-bed MFC obtained the maximum power density of 680 mW m^{-3} at the organic load of 2.41 g L^{-1} d^{-1}. The efficient MFC reactor could be increased the stability of electricity generation by using repeatedly. However, it is reported that the quantity of CNT used for anode modification should controlled to avoid the cell proliferation and cell death of microorganism due to the CNT cytotoxicity [9].

Single-wall carbon nanohorns (SWCNHs) are newly developed nanomaterials and are self-assembled into "dahlia-like" spherical structures. Yang et al. (2017) [14] used different modifiers for SS anode such as AC, CNT, and SWCNH. Maximum power densities of 327 mW m^{-2} was obtained for SSCNH modified SS followed by CNT (261 mW m^{-2}), AC (244 mW m^{-2}), and bare SS (72 mW m^{-2}).

Another form of carbon-based material is AC, which has porous structure. It is mainly originated from different fruit shells, coals, charcoal, etc. and hence possesses low manufacturing cost unlike other forms of carbon nanostructures. Zhu et al., 2010 [15] studied the performance of activated carbon nanofibers (ACNFs) and treated ACNF with nitric acid and ethylenediamine (EDA) in single-chamber air cathode MFC. The ACNF provides large surface area for the bacterial enrichment and the surface treatment improves the biocatalyst activity. It was reported that EDA-treated ACNF provides higher power density of 2066 mW m^{-2} than nitric acid-treated ACNF (1304 mW m^{-2}) and untreated ACNF (1641 mW m^{-2}).

Diamond is another variation of carbon with high mechanical strength and exhibit applications in electrochemical sensors and electrocatalysis. The amorphous form of nanostructured diamond film has zero toxicity and good biocompatibility. However, due to the high production cost, there are relatively few studies on diamond as anode materials. Wu et al. (2011) [16] analyzed the electrochemical behavior of Shewanella loihica PV-4 on boron-doped diamond (BDD) and nanograss array BDD electrode in a single-chamber microbial fuel cell (SCMFC). The current generated from the nanograss array BDD electrode (1.2 μA cm^{-2}) is much higher than the bare BDD electrode (0.4 μA cm^{-2}). Thus the modification with nanograss array of diamond enhances the biocompatibility and charge transfer capacity.

11.2.2 Nanoconducting polymer

Conductive polymers (CP) have drawn significant interest due to its high electrical conductivity, electrocatalytic activity, easy handling, and thermal and chemical stability. In addition, they possess high stability toward microbial and chemical degradation [17]. The bacterial metabolic products and by-products of the electrocatalytic oxidation process adsorbed on the anode surface, deactivates the anode performance. Hence coating of anode with stable polymers slows down the deactivation [18]. CPs, namely, polyaniline (PANI), polypyrrole (Ppy), polythiophene (PTh), and their composites have been applied in MFC. Conductive polymers with nanostructures possess porous structure, which allowed internal colonization and enhancing the interaction between electrode and biofilms leading to accelerate the electron transfer [19].

PANI is the most extensively studied CP for its excellent environmental stability and low cost. Li et al. (2011) studied the effect of PANI-modified carbon felt (CF) in MFC. The PANI-modified MFCs showed quick increase in voltage during the start-up period, compare with the unmodified CF, and lead to a higher power density by an increase of 35% [20]. In another study, Bin et al. (2011) used modified CC with PANI as anode in a DCMFC. The maximum power density achieved was 1275 mW m^{-2}, which was much higher than that of unmodified CC [21].

Ppy is another attractive CP possesses good stability and biocompatibility. Chi et al. (2013) [22] developed an electropolymerized GF with nano-Ppy. The highest redox currents and lowest ohmic resistance of modified anode enhances the maximum power density of the system by 15% than the control. Chen et al. (2019) [23] fabricated a mini-MFC with Ppy nanowire as anode modifier and produced a maximum voltage of 630 mV. The anode modification with Ppy nanoparticles also accelerated the removal efficiency of metals such as copper in SCMFC [24,25].

Another type of CP is PTh and it is less experimented in MFC due to its complexity in the synthesis and less conductivity compared to PANI and Ppy.

Sumisha et al. (2018) [26] compared the performance of Ppy- and PTh-modified GF in SCMFC and concluded that Ppy-modified anode resulted in maximum power density. Poly(3,4-ethylenedioxythiophene) (PEDOT) is a PTh-derived polymer. Electrospun conducting PEDOT nanofibers as a 3D porous anode was utilized in a miniature MFC by Jiang et al. (2015) [27]. The 3D porous structure of PEDOT helps the microorganism to easily colonize inside the nanofiber. The high electrical conductivity and mean pore size of the nanofiber resulted in a high power density of 423 $\mu W\ cm^{-3}$.

Composite forms of conductive polymers are widely used in MFC to further enhance their properties. The nanocomposites of conductive polymer can be easily formed due to the ease in the doping with other nanomaterials. The nanocomposites reduces the ohmic resistance and enhances the electron transfer rate [9].

Production of MFC anodes with CPs in combination with CNTs and graphene has drawn more attention in this research field. It was found that the multilayer CNT/PANI film, synthesized by a layer-by-layer assembly method, enhances the electroactivity of PANI [28]. In another study, PANI-hybridized 3D graphene exhibited higher performance than a planar carbon electrode in a macroporous and monolithic MFC. The superior performance of modified anode is due to its capability to form three-dimensional interface for the biofilm and to enable electron transfer by high conductive trails [29]. Zhao et al. (2013) [30] discovered a novel electrode contains PANI networks on the graphene nanoribbons (GNRs)-coated carbon paper by electrodeposition method. The power generation obtained was maximum for the composite electrode than the GNR and carbon paper electrodes. The electrical conductivity of GNRs and PANI significantly improved the conductivity of the composite electrode and hence the energy generation.

Ppy is not well-defined in its chemical structure due to the significant amount of α-β' coupling in the polymer side chain, that induce structural disorder and limit its electrochemical response and hence normally applicable in its composite forms [31]. A polypyrrole nanowires coated by graphene oxide (Ppy-NWs/GO) was prepared by one-step electrochemical method and the composite electrode possesses a lower charge transfer resistance due to the synergistic effect of components. The maximum power density obtained was 22.3 $mW\ m^{-2}$, higher than that of Ppy-NWs (15.9 $mW\ m^{-2}$) [32].

Chou et al. (2014) [33] modified a porous melamine sponge anode through PEDOT together with polystyrenesulfonate as an adherence and conductivity enhancer and attached to the composite of reduced graphene oxide-multiwalled carbon nanotube (rGO-MWCNT). The MFC operated with modified sponge showed better performance than the control electrode. PEDOT modified with nickel nanoparticles (Ni-np) and electrochemically reduced graphene (ErGO) was exploited in a glucose/oxygen MFC. The maximum power density of 0.32 $mW\ cm^{-2}$ attained in

PEDOT/ErGO/Ni-np electrode was due to the high biocompatibility between the bacterium and nickel nanoparticles and good electrical conductivity of ErGO [34].

11.2.3 Metal-based nanomaterials

Various metals and metal oxides (MOs) hold higher conductivity than carbon-based materials and are used as modifiers for electrodes to improve the properties such as conductivity, electrocatalytic activity, and biocompatibility. The modification with nano-MOs aids the active sites for microorganisms and act as a bridge for electron transfer between biocatalyst and anode. Recently, many nanostructured MOs have been experimented as the anode modifiers in MFC.

Iron oxides such as goethite [FeO(OH)], magnetite (Fe_3O_4), and hematite (Fe_2O_3) can promote extracellular electron transfer EET either in the form of electrical conduit inside the biofilm and promote the expression of c-type cytochromes or by accumulating at the cell surface; both mechanisms results in the enhanced electricity generation [8,31]. In a study, nanocolloid of hematite was employed as interconnected network for self-assembly of Shewanella and resulted in 50 times maximum power generation compared to the bare, on the basis of inside biofilm mechanism. Whereas, on the basis of interface mechanism; iron oxides can be served as redox couple between Fe (II) and Fe (III) at the interface of the anodic surface [35]. Harshini et al. (2017) [36] used biosynthesized FeO nanoparticles-coated carbon paper in DCMFC. The results concluded that the antibacterial activity of modified anode, due to the presence of nanoparticles, was less effective against the gram positive biocatalyst. Also, the electrode exhibited a good hydrophilic behavior and produced a maximum power density of 140.5 mW m^{-2}. Another important study of iron nanoparticles was carried out by Mohamed et al. (2018) [37]. Fe/Fe_2O_3 nanoparticles were deposited on the surface of different carbonaceous anode materials as an effective catalyst to improve the anode performance of MFC to treat industrial wastewater. The nanostructure facilitated the hydrophilic character of the electrode, the rate of degradation of organic compounds, and the biocompatibility of the electrode surface, and which leads to the decreases in the resistance for electron transfer. Therefore the generated power was considerably increased by 385% compared to bare anode.

Manganese dioxide and ruthenium dioxide nanoparticles possess high stability and low cost. They are able to enhance the electron transfer process by decreasing the internal resistance of modified electrode via the pseudocapacitive properties. In a study, RuO_2 films composed of submicron-/nanosized particles was decorated on CF anode and the maximum power density was increased by 17 times that of CF without MO coating [38]. Nickel oxide (NiO) nanoparticles are also beneficial for the biofilm formation and promote the direct electron transfer. Importance of NiO nanoflaky arrays grown on CC fiber was studied by Qiao et al. (2014) [39]. The unique nanostructure of NiO/CC anode provided maximum power density threefold higher than that of the plain CC anode due to the properties of the modified anode such as the electrocatalytic activity added from the active surface area of electrode, excellent bacterial adhesion, and good interfacial charge transfer.

Generally MOs are playing major role in the enhancement of biofilm formation, while the conductive part of them is weak. Hence carbon-based materials like CNT, graphene, etc. or conductive polymers are usually used to improve the electrical conductivity of the metal-oxide-modified anode [8].

An iron oxide (Fe_3O_4)-based (Fe_3O_4)/CNT nanocomposite electrode was fabricated for MFC as anode. The Fe_3O_4/CNT electrode presented a maximum power density of 830 mW m^{-2}. The high energy generation was due to the synergetic effect of the modifiers. The CNTs strongly attached to the anode surface due to the magnetic properties of Fe_3O_4, which provides high surface area for compact bacterial colonization and good electron transfer between biocatalyst and anode [40]. Manganese ferrite ($MnFe_2O_4$)/PANI-based electrode material have also extensively used in fuel cells. In specific, $MnFe_2O_4$ have good electrochemical properties which can be further enhanced by the addition of PANI. It also possess different valence cations and generally nontoxic in nature, hence have various electrochemical applications [41]. A binder free anode containing biochar, nickel ferrite ($NiFe_2O_4$) and PEDOT was utilized in MFC. The cubic spinel-structured $NiFe_2O_4$ nanorods are homogeneously distributed over the sheet-like amorphous PEDOT. The bioelectrochemical activity of exoelectrogens is improved with the composite electrode and produced a peak power of 1200 mW m^{-2} because of the unique nanostructure, dense biofilm formation, incessant electron pathway, and chemical stability [42].

A unique nanostructured PANI/TiO_2 was developed as a composite anode. TiO_2 is an inorganic mesoporous material with large specific surface area, high porosity, and biocompatibility but due to the poor electrical conductivity of TiO_2, PANI is unified to enhance the conductivity and stability. The modified electrode delivered a maximum power density of 1459 mW m^{-2}, which higher than the TiO_2 and PANI electrode [43].

Multiwalled MnO_2/Ppy/MnO_2 nanotubes (NT-MPMs) was coated on a CC and obtained a maximum power density of 32.7 W cm^{-3}, which is approximately 1.3 times higher than that of the neat CC. The modified CC displays excellent durability and promotes biofilm growth by providing unique active centers and facilitates electron transfer to the electrode [44]. Zhao et al. (2019) [45] developed a cauliflower-like Ppy-MnO_2 modified CC as a capacitive anode for MFC (Fig. 11.4). The 3D-structured composite electrode delivered a power density, which is 3.5-fold higher than that of bare CC. The cauliflower-like structure increases the internal area of the composite and that facilitate the biofilm growth. The composite electrode also possesses long-term cyclabilities with smooth surface and high porosity. Table 11.1 shows the important nanocomposites used in MFC.

Table 11.1 Nanocomposites as anode in MFC.

Sl. no.	Anode	MFC configuration	Power density (mW m^{-2})	Refs.
1.	PANI/CaCO$_3$	DCMFC	1280	[46]
2.	NiO/PANI	DCMFC	1078	[47]
3.	PANI/polypyrrole	DCMFC	2880	[48]
4.	Iodine-doped PTh	DCMFC	760	[24]
5.	Nitrogen-doped/CNT/rGO	DCMFC	1137	[49]
6.	MWCNT/SnO$_2$	DCMFC	1421	[50]
7.	Nanomolybdenum carbide (Mo$_2$C)/CNT	SCMFC	170	[51]
8.	SnO$_2$/rGO	DCMFC	1624	[52]
9.	Cu-doped FeO	DCMFC	161	[53]
10.	Graphene/nickel (G/Ni)	H-shaped MFC	3903	[54]
11.	Gold/poly(e-caprolactone) nanofiber	Microliter size MFC	29	[5]

CNT, Carbon nanotube; *DCMFC*, double-chamber microbial fuel cell; *MFC*, microbial fuel cell; *MWCNT*, multiwalled carbon nanotube; *PANI*, polyaniline; *Pth*, polythiophene; *rGO*, reduced graphene oxide, *SCMFC*, single-chamber microbial fuel cell.

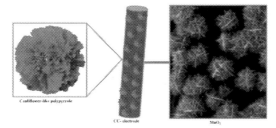

Figure 11.4 Cauliflower-like polypyrrole@MnO$_2$ on CC electrode (*CC*, Carbon cloth). *Adapted from X. Zhao, T. Tian, M. Guo, X. Liu, Cauliflower-like polypyrrole@MnO$_2$ modified carbon cloth as a capacitive anode for high-performance microbial fuel cells, Journal of Chemical Technology and Biotechnology 95 (1) (2019) 163−172.*

11.3 Application of nanomaterial structures in cathodes

Cathode helps in electron transfer, improving electrochemical activity and reduction in reaction kinetics in MFCs. The cathode design is a significant barrier to make an MFC a successful and scalable technology. The chemical reaction at the cathode is hard to develop because the oxygen, electrons, and protons must all meet in a three-phase response at a catalyst. Two different mechanisms usually contribute to the process of oxygen reduction to H$_2$O (four-electron reaction) or H$_2$O$_2$ (two-electron reaction). The electrocatalyst is to

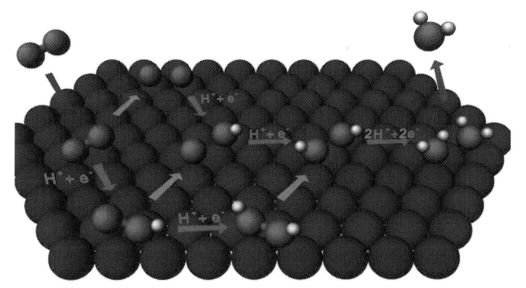

Figure 11.5 Schematic representation of proposed ORR catalysis mechanism on a catalyst surface (ORR, Oxygen reduction reaction) [57]. *Adapted from W. Xia, A. Mahmood, Z. Liang, R. Zou, S. Guo, Earth-abundant nanomaterials for oxygen reduction, Angewandte Chemie International Edition 55 (8) (2016) 2650–2676.*

improve the four-electron oxygen reduction reaction (ORR) kinetic and performance of the MFC [55]. The electrocatalytic study conducted with different noble metals (Pt, Au, Ag, Cu, Pd, etc.) verified that Pt material follows a quasi-4e$^-$ ORR process with less H_2O_2 generation relative to other metals [56] and ORR catalysis mechanism shown in Fig. 11.5. The main problems associated with Pt catalyst for its practical application are the catalytic poisoning and high cost [58]. Therefore different techniques to overcome this issue have been introduced to develop a more effective catalyst.

Due to its high surface area and active sites, the nonprecious nanomaterial exhibited excellent catalytic activity and electrokinetics. To increase the overall performance, ratio between the surface area of catalyst and electron utilization should be optimized. The catalyst surface area has a significant impact on the total power production, if the energy output is below a critical level. Therefore this chapter describes in detail about the nanomaterial-based Pt/C catalyst.

11.3.1 Graphene-based nanomaterial

The recent performance of graphene nanosheets has created new possibilities for the use of two-dimensional (2D) carbon materials as ORR catalysts or as support materials due to their long-term stability, high surface, mechanical strength, and high electrical conductivity [59]. Extensive study was conducted using graphene and metal alloy and

Table 11.2 Summary of graphene-based cathodes and the performances.

Anode coating	Cathode coating	Reactor type	Max power density (W m^{-2})	Refs.
Carbon cloth	Co$_3$O$_4$-Nitrogen doped graphene oxide (NGO)/carbon cloth	Single	0.313	[61]
Carbon brush	Nitrogen-doped carbon nanosheet + graphene/SS	Dual	1.159	[60]
Carbon felt	MnO$_2$ graphene nanosheet/SS	Single	2.083	[62]
Carbon cloth	Graphene + biofilm/carbon cloth	Dual	0.323	[63]
Carbon granules	Co$_3$O$_4$-Nitrogen doped graphene (NG)/glass carbon	Dual	1.340	[64]

SS, Stainless steel.

N-functionalized graphene nanoparticles in MFC cathodes. Qu et al. (2013) examined graphene modified by porous nitrogen-doped carbon nanosheet (PNCN) as metal-free ORR catalyst in SCMFCs and found that PNCN (1.15 W m^{-2}) showed higher catalytic performance over Pt/C. The findings suggest that PN carbon paper is a good alternate catalyst for Pt/C in MFCs [60]. Valipour et al. (2016) evaluated the feasibility of graphene-based nanomaterial as novel cathode catalysts in SCMFCs. GO cathodes MFCs deliver higher power output (1.683 W m^{-2}) which is comparable to Pt/C catalyst [59]. Therefore graphene-based catalyst is promising as a sustainable low-cost green material and provides stable power generation and the long-term operation of MFCs. The graphene material was altered by metal, conductive polymers, and MO and which facilitates the number of active sites of catalysts. Furthermore, the wide surface area of graphene materials will provide more active sites to increase ORR catalysis and Table 11.2 summarize the graphene-based cathodes in MFC. Khilari et al. exposed that electrochemical efficiency of α-MnO$_2$ and graphene loaded α-MnO$_2$ catalyst material in MFC. Graphene loaded catalyst showed comparable performance to Pt/C catalyst in terms of power output and COD removal [65].

11.3.2 Carbon nanotubes and nanofibers

Both CNFs and CNTs revolutionize various research fields in materials science due to their novel properties, such as large surface area and electrical conductivity. Physical and chemical activation can also improve the electrocatalytic properties of CNTs and CNFs [66]. Therefore CNFs and CNTs recommended opening the new approach into nanotechnology in MFC application. As a cathode catalyst in DCMFC, Mahrokh et al. (2017) used MWCNTs (amine functionalized) and obtained a current density of 319.11 mA m^{-2} which is substantially comparable to Pt/C catalysts [67]. Ghasemi et al. (2016) illustrated the use of nanocomposite (CNT/Ppy) as a different cathode catalyst in DCMFC and stated the maximum power density of Ppy (69.12 mW m^{-2}), CNT/Ppy (113.5 mW m^{-2}),

Table 11.3 Summary of the CNT-based cathode performance.

Anode coating/support	Cathode coating/support	Reactor type	Max power density (mW m^{-2})	Refs.
CNTs-textile	CNTs-textile-Pt	Dual	837	[69]
Carbon cloth	Co$_3$O$_4$-NCNTs/carbon cloth	Single	469	[61]
Carbon cloth	Pt-Pd/CNTs	Dual	1274	[70]
Graphite felt	CNTs/SSM	Single	147	[71]
Ppy/AQDS/carbon felt	MnO$_2$/CNTs/stainless steel mesh	Single	2676	[72]
Carbon cloths	Mn-Ppy-CNTs	Single	213	[73]

AQDS, Anthraquinone-2,6-disulfonic disodium salt; *CNT*, carbon nanotube; *NCNT*, nitrogen-doped carbon nanotubes; *Ppy*, polypyrrole; *SSM*, stainless steel mesh.

and Pt (122.85 mW m^{-2}) [68]. The finding explores the use of CNT/Ppy nanocomposite as low-cost cathode catalysts in DCMFCs. Other studies on CNT-based cathode in MFC are listed in Table 11.3. In addition to CNTs, CNFs also displayed interesting activity as ORR cathode catalyst in MFC. Ghasemi et al. (2011) developed activated CNFs as cathode catalysts in DCMFCs and reported a maximum power density of 61.3 mW m^{-2} which was 79% higher than untreated CNFs [74]. Apart from chemical activation; the doping of CNFs also could enhance its electrocatalytic properties. N-doped CNF supported on SS as cathode catalyst in SCMFC was investigated by Chen et al. (2012) and prepared catalyst material shows comparable performance with Pt/C catalyst [75].

11.3.3 Transition metal oxide nanomaterials

Transition metal oxides (TMOs) can show specific attributes like multiple oxidation state, short diffusion path lengths and large surface area that make them the most efficient property materials class covering all areas of solid-state and material science. Metal oxides (MOs) as nanomaterials are highly preferred materials for MFCs due to their excellent electrocatalytic activity, high ORR kinetics, and economic efficiency [76]. Several TMOs have been explored as cathode catalyst materials, such as molybdenum oxide (MoO$_2$), tungsten oxide (WO$_3$), manganese dioxide (MnO$_2$), vanadium oxide (V$_2$O$_5$), iron, nickel, and zinc oxides [55,77]. Ayyaru et al. (2019) recently used vanadium oxide (V$_2$O$_5$) nanorods as a catalyst and showed a maximum power density of 1073 ± 18 mW m^{-2} in SCMFC. This study indicates that V$_2$O$_5$ catalyst efficiency is equivalent to the existing Pt/C catalyst and also economical [78]. Yuan et al. (2015) synthesized MnO$_2$/Ppy/MnO$_2$ multiwalled nanotubes (NT-MPMs) using a cost-effective hydrothermal method, and revealed a maximum power output of 721 mW m^{-2}, which is comparable to the Pt/C catalyst and also possess high electrostability. The NT-MPMs excellent efficiency makes them a potential alternative to Pt/C in MFCs [79]. Bhowmick et al. (2019) used cobalt oxide (Co$_3$O$_4$) and iron oxide (Fe$_3$O$_4$) in MFC as ORR catalysts. They obtained a slightly higher power density of MFC-Co (6.62 W m^{-3}) than MFC-Pt (6.12 W m^{-3}) and the electrochemical analysis also supported the same. The improved efficiency of Co$_3$O$_4$ catalyst suggested an economically efficient substitute to the costly Pt/C-based catalysts in MFCs [80]. Karthick et al. have recently investigated the application of TMO (molybdenum and tungsten oxide)/polpypyrrole composites as cathode catalyst in MFC. The power density and voltage results obtained indicate that the TMO/polpypyrrole composites catalyst is a possible alternative to Pt/C for power generation [81,82].

11.3.4 Conducting polymer nanomaterials

In these nanomaterials, conducting polymers have a wide range of advantages, including structural diversity, easy synthesis, and economic performance because of electric and optical properties. Conducting polymers are promising to use as ORR catalyst in MFCs because they possess weak molecular O—O bond of chemisorbed O$_2$ on the polymer surface and reduces ORR activation barrier [83]. Among different electrical conducting polymers, Ppy, PEDOT, PANI, and poly (3-methyl)thiophene were found to be more dynamic ORR catalyst and also listed in Table 11.4.

Table 11.4 Literature on conducting polymers-based cathode catalyst in MFC.

Cathode composition	Cathode surface area (cm^2)	Cathode loading (mg cm^{-2})	Power density (max.)	Refs.
PANI/CB	7	1	381.8 mW m^{-2}	[84]
Carbon felt/PANI	—	—	199 mW m^{-2}	[85]
PANI/CB/nonpyrolyzed FePc	7	1	630.5 mW m^{-2}	[84]
Mn/Ppy/CNT/CC	28.3	1	169 mW m^{-2}	[73]
PANI/Fe/C/graphite felt	—	—	10,170 mW m^{-3}	[86]
Carbon felt/Poly(aniline-co-2,4-diaminophenol) (PADAP)	—	—	236 mW m^{-2}	[85]
Mn/Ppy/CNT/CC	28.3	2	213 mW m^{-2}	[73]
Prussian blue/PANI/activated carbon	0.785	—	12,820 mW m^{-3}	[87]
Ppy/anthraquinone-2-sulfonate/stainless steel mesh	5.51	—	75 mW m^{-2}	[88]
Pt/C	—	—	9,560 mW m^{-3}	[86]
PANI	12	—	42.4 mW m^{-2}	[89]
Ppy/CB	7	—	401.8 mW m^{-2}	[90]
PANI/V$_2$O$_5$	12	—	79.26 mW m^{-2}	[89]

CB, Carbon black; *CC*, carbon cloth; *CNT*, carbon nanotube; *MFC*, microbial fuel cell; *PANI*, polyaniline; *Ppy*, polypyrrole.

Table 11.5 Comparison of the cost of cathode materials [84,93].

S. no.	Cathode material	Power density (mW m^{-2})	Price (g^{-1})	Power per cost (mW^{-1})
1.	PANI/C/FePc	630.5	10.74	4.11
2.	Pt/C	575.6	29.38	0.55
3.	Ppy/C	401.8	3.39	8.29
4.	FePc/C	336.6	7.42	3.17
5.	PANI/C	381.8	3.39	7.78

PANI, Polyaniline; *Ppy*, polypyrrole.

The polyaniline nanofiber (PANI-NF) as a cathode catalyst in SCMFC has recently been investigated by Ahmed et al. (2012), and is still have less power density (496 mW m^{-2}) than that obtained with existing Pt/C catalysts. Therefore PANI-NF is a promising potential cathode material in MFCs due to its simple synthesis, low cost and porous nature [91]. Sumisha and Haribabu have recently investigated the use of polypyrrole nanoparticle (Ppy-NP)-coated CC as the cathode catalyst for power generation and iron removal in SCMFC. The Ppy/CC (190 ± 4 mW m^{-2}) finding illustrates a significant comparison with Pt/C (278 ± 4 mW m^{-2}) and shows its ability for use as a low-cost cathode catalyst in MFC [26,92]. The comparison of cost of cathode materials is given in Table 11.5.

11.4 Ion-exchange membranes

Ion-exchange membranes (IEM) are classified into five groups as: cation-exchange membranes, mosaic IEMs, anion-exchange membrane, amphoteric IEMs, and bipolar membranes. Some of the major aspects of the membrane material are high proton conductivity, low internal resistance, long-term stability, and high-energy recovery, which make them the right option for use in MFCs. The charge equilibrium between anodic and cathode chamber has a dominated effect on cation transportations of species other than protons through the Nafion since their large concentration K^+, Na^+, Ca^{2+}, Mg^{2+}, and NH^{4+} are significantly higher than proton levels in anolytes and catholytes. For this reason, Naffion and other PEM are employed in MFC. PEM system is capable of affecting the internal resistance, concentration and lack of polarization in an MFC system. While researchers have tried to find lower costs and more stable alternatives but this Nafion remains the excellent PEM membrane on the market. The ratio between PEM surface area and system volume is very critical to ensure optimum power performances. If the power output is below a critical level, the PEM surface region has a significant effect on overall power output. The use of nanomaterial with a really high specific surface area will induce the effect of a greater surface area in the PEM [94].

11.4.1 Perfluorinated membranes

Nafion membranes are proton conductive polymer films composed of polytetrafluoroethylene backbone with ether groups and a sulfonic acid unit as side chains at the end. The high ion density of Nafion groups generates the better ion-rich channels for proton transport and hence widely used in fuel cell applications. The incorporation of nanocomposites in membranes enhances the proton transfer capacity, mechanical strength and morphology. In a study, the performance of self-made CNF/Nafion and ACNF/Nafion nanocomposite membranes were tested and compared in a two cylindrical H-shaped MFC [95]. The nanocomposite in the membranes decreases the pore size and roughness of the membranes, which resulted in the resistance to the oxygen flow from cathode to anode and intern increased the power production to 57.64 mW m^{-2}. The modification with carbon nanomaterials also reduced the fouling of the membrane. Bajestani et al. (2016) [96] studied the effect of Nafion/TiO$_2$ nanocomposite membranes in two cylindrical and H-shaped MFC. The TiO$_2$ nanocomposite membrane prepared with DMF solvent has the best morphology, the highest porosity, and the maximum proton-exchange capacity.

11.4.2 Sulfonated membranes

Nanocomposite membranes based on sulfonated groups are cost effective and resulted in more hydrophilic membranes and generate path for proton transport and also provide negative charge for the enhanced carriage of protons. Polyether ether ketone is a mechanically and thermally stable polymer, which can be easily sulfonated to sulfonated polyether ether ketone (SPEEK). In a study, a newly designed polyimide nanofiber-reinforced/SPEEK composite membrane was used for proton exchange. It showed increased tensile strength and more resistant to fuel and gas crossovers and cost effective than nafion membranes [97]. In another study, nanocomposite membranes modified with SPEEK and sulfonated SiO$_2$ (SiO$_2$-SO$_3$H) were studied in SCMFC. The modification with SiO$_2$-SO$_3$H effectively enhanced the proton conductivity of the membrane. About 7.5% of the modified membrane produced a maximum power density of 1008 mW m^{-2}, which is three times higher than that of Nafion 115 [98].

Sulfonated TiO$_2$ nanoparticles incorporated with polystyrene ethylene butylene polystyrene (S-TiO$_2$/SPSEBS) membranes was utilized in SCMFC by Ayyaru and Dharmalingam (2015) [7]. Interestingly, S-TiO$_2$ composite efficiently improved the proton conductivity of the polymer membrane and the maximum power density obtained was 1345 mW m^{-2} and which is relatively higher than SPSEBS-TiO$_2$ (835 mW m^{-2}) and Nafion membranes (300 mW m^{-2}). As well, the oxygen mass transfer coefficient of the composite membranes decreased with the use of S-TiO$_2$ and hence facilitated the columbic efficiency.

Figure 11.6 TEM image of (A) SGO, (B) SiO$_2$, (C) SGO@SiO$_2$ nanoparticles (*SGO*, Sulfonated graphene oxide; *TEM*, transmission electron microscopy). *Adapted from Q. Xu, L. Wang, C. Li, X. Wang, Study on improvement of the proton conductivity and anti-fouling of proton exchange membrane by doping SGO@SiO2 in microbial fuel cell applications, International Journal of Hydrogen Energy 44 (29) (2019) 15322–15332.*

In a study by Palma et al. (2018), nanocomposites based on polyether sulfone with different amount of Fe$_3$O$_4$ nanoparticles were synthesized and analyzed as PEM in H-type MFC. It was stated that, increasing nanoparticles content can compromise the mechanical properties of membranes leading to a significant brittle behavior, while the tensile strength was found to be suitable for durable MFC operations. The nanocomposite membranes resulted in a good mechanical properties and thermal stability compared to commercial membranes and therefore can be used for longer operations [99]. Furthermore, a study was focused on the development of sulfonated graphene oxide-silicon dioxide (SiO$_2$) (Fig. 11.6) blended with poly(-vinylidene fluoride) grafted sodium styrene sulfonate (PVDF-g-PSSA). The ion-exchange capacity (1.6 meq g^{-1}) and proton conductivity (0.078 S cm^{-1}) of composite membrane was enhanced and found to be higher than the Nafion-117 membrane. Also, the addition of SiO$_2$ improved the hydrophobicity and antifouling characteristics of the membrane and hence resulted in a higher power density of 185 mW m^{-2} [100].

11.4.3 Nonfluorinated and nonsulfonated membranes

The nonfluorinated and nonsulfonated membranes has drawn consideration attention due to its low cost. Polymers like polyvinyl alcohol (PVA) possesses the basic PEM properties such as good mechanical and thermal stabilities, controllable physical properties, and excellent hydrophilic and electrochemical properties [101]. Khilari et al. (2013) [102] studied the feasibility of low cost nanocomposite membrane based on GO and silicotungstic acid cross linked with PVA. The synthesized membrane showed higher proton conductivity and ion-exchange capacity at 0.5 wt.% of GO and later the properties decreased as the GO concentration increased due to the blocking effect of the filler. Hence a maximum power density of 1.48 W m^{-3} was observed for 0.5 wt.% of GO along with 90.03% COD removal. The summary of nanomaterial's based membrane modifications in MFC are listed in Table 11.6.

Table 11.6 Summary of nanomaterial-based membrane modifications in MFCs.

Sl. no.	Membrane type	Membrane	Anode	Cathode	MFC type	Power density	Refs.
1.	Perfluorinated	Nafion/ACNF	carbon paper	carbon paper/Pt	H-type	57.64 mW m^{-2}	[95]
2.	Sulfonated	SPEEK-SiO$_2$	CC	CC/Pt	SCMFC	1008 mW m^{-2}	[98]
3.		PES/Fe$_3$O$_4$	carbon paper	carbon paper	H-type	9.59 mW m^{-2}	[99]
4.		SGO@SiO$_2$/PVDF-g-PSSA	CF	CF	DCMFC	185 mW m^{-2}	[100]
5.		PVA/SS/GO	CC	CC/Pt	SCMFC	193.6 mW m^{-2}	[103]
6.		SPEEK/Cloisite 15 A® Clay/ 2,4,6-triaminopyrimidine	Plain carbon paper	carbon paper/Pt	DCMFC	318 mW m^{-2}	[104]
7.	Nonsulfonated and fluorinated	PVA/silicotungstic acid (STA)/GO	CC	CC/Pt	Cylindrical-type	1.48 W m^{-3}	[102]
8.		CHIT-MWCNT	CC	CC/Pt	SCMFC	46.94 mW m^{-2}	[105]

ACNF, Activated carbon nanofiber; *CC*, carbon cloth; *CNF*, carbon nanofiber; *GO*, graphene oxide; *PES*, polyether sulfone; *PVA*, polyvinyl alcohol; *PVDF-g-PSSA*, poly (vinylidene fluoride) grafted polystyrene sulfonated acid; *SCMFC*, single-chamber microbial fuel cell; *SGO*, sulfonated graphene oxide; *SPEEK*, sulfonated polyether ether ketone; SS, stainless steel.

11.5 Conclusion

This study comprehensively explained the application of different nanostructured materials in MFC as electrode modifiers. The electrodes in MFC have different characteristics and thus the materials and design needed for them also varies. In general, the electrodes acquire the properties such as electrical conductivity, high surface area, mechanical strength, and toughness. In particular, the anode should be biocompatible and should possess long-term stability. The adaptation of nanomaterials in anodes improves the extracellular electron transfer process between microbes and the anode. The nanostructure materials enhance the biofilm growth, and deliver efficient electrocatalysis by providing quicker reaction kinetics at the anode interface. On the other hand, the cathode materials necessitate high catalytic properties to enhance the ORR. Platinum is the best conventional cathode catalyst, but due to its high cost research is ongoing to find alternative for a feasible commercial applications. Nanoparticles of transition metal oxides and its composites with conductive polymers have arisen as promising materials for ORR with low cost. Nanoparticles incorporated membranes shows reduced oxygen transport from cathode to anode, greater proton-exchange and higher coulombic efficiency. Considering all the above, the application of nanomaterials in MFC holds the enhancement of anodic-cathodic catalytic activity and the proton transport, thus improve its performance in terms of power density. Hence the MFC systems can be utilized for large-scale operations.

References

[1] B.E. Logan, C. Murano, K. Scott, N.D. Gray, I.M. Head, Electricity generation from cysteine in a microbial fuel cell, Water Research 39 (5) (2005) 942–952.
[2] V. Gadhamshetty, N. Koratkar, Nano-engineered biocatalyst-electrode structures for next generation microbial fuel cells, Nano Energy 1 (1) (2012) 3–5.
[3] H. Omar, et al., Transition metal nanoparticles doped carbon paper as a cost-effective anode in a microbial fuel cell powered by pure and mixed biocatalyst cultures, International Journal of Hydrogen Energy 43 (2018) 21560–21571.
[4] M. Ghasemi, W. Ramli, W. Daud, S.H.A. Hassan, S. Oh, M. Ismail, Nano-structured carbon as electrode material in microbial fuel cells : a comprehensive review, Journal of Alloys and Compounds 580 (2013) 245–255.
[5] A. Fraiwan, et al., Microbial power-generating capabilities on micro-/nano-structured anodes in micro-sized microbial fuel cells, Fuel Cells 14 (6) (2014) 801–809.
[6] R. Fogel, J.L. Limson, Applications of nanomaterials in microbial fuel cells, Nanomaterials for Fuel Cell Catalysis, Springer, 2016, pp. 551–575.
[7] S. Ayyaru, S. Dharmalingam, A study of influence on nanocomposite membrane of sulfonated TiO_2 and sulfonated polystyrene-ethylene-butylene-polystyrene for microbial fuel cell application, Energy 88 (2015) 202–208.
[8] N. Savla, R. Anand, S. Pandit, R. Prasad, Utilization of nanomaterials as anode modifiers for improving microbial fuel cells performance, Journal of Renewable Materials 8 (12) (2020) 1581–1605.
[9] Y. Liu, X. Zhang, Q. Zhang, C. Li, Microbial fuel cells: nanomaterials based on anode and their application, Energy Technology 8 (9) (2020) 2000206.

[10] M.M. Mohideen, Y. Liu, S. Ramakrishna, Recent progress of carbon dots and carbon nanotubes applied in oxygen reduction reaction of fuel cell for transportation, Applied Energy 257 (2020) 114027.

[11] P. Liang, et al., Carbon nanotube powders as electrode modifier to enhance the activity of anodic biofilm in microbial fuel cells, Biosensors & Bioelectronics 26 (6) (2011) 3000–3004.

[12] S. Roh, Layer-by-layer self-assembled carbon nanotube electrode for microbial fuel cells application, Journal of Nanoscience and Nanotechnology 13 (6) (2013) 6–9.

[13] C. Erbay, et al., Control of geometrical properties of carbon nanotube electrodes towards high-performance microbial fuel cells, Journal of Power Sources 280 (2015) 347–354.

[14] J. Yang, S. Cheng, Y. Sun, C. Li, Improving the power generation of microbial fuel cells by modifying the anode with single-wall carbon nanohorns, Biotechnology Letters 39 (2017) 1515–1520.

[15] N. Zhu, X. Chen, T. Zhang, P. Wu, P. Li, J. Wu, Improved performance of membrane free single-chamber air-cathode microbial fuel cells with nitric acid and ethylenediamine surface modified activated carbon fiber felt anodes, Bioresource Technology 102 (1) (2011) 422–426.

[16] W. Wu, L. Bai, X. Liu, Z. Tang, Z. Gu, Nanograss array boron-doped diamond electrode for enhanced electron transfer from Shewanella loihica PV-4, Electrochemistry Communications 13 (8) (2011) 872–874.

[17] C. Feng, F. Li, H. Liu, X. Lang, S. Fan, A dual-chamber microbial fuel cell with conductive film-modified anode and cathode and its application for the neutral electro-Fenton process, Electrochimica Acta 55 (6) (2010) 2048–2054.

[18] V.L. Beaumont, Investigation of microbial fuel cell performance and microbial community dynamics during acclimation and carbon source pulse tests, M.S. thesis, University of Waterloo, 2007.

[19] H. Liu, B.E. Logan, Electricity generation using an air-cathode single chamber microbial fuel cell in the presence and absence of a proton exchange membrane, Environmental Science & Technology 38 (14) (2004) 4040–4046.

[20] C. Li, L. Zhang, L. Ding, H. Ren, H. Cui, Effect of conductive polymers coated anode on the performance of microbial fuel cells (MFCs) and its biodiversity analysis, Biosensors & Bioelectronics 26 (10) (2011) 4169–4176.

[21] F.J. Hernández-Fernández, et al., Recent progress and perspectives in microbial fuel cells for bioenergy generation and wastewater treatment, Fuel Processing Technology 138 (2015) 284–297.

[22] M. Chi, Graphite felt anode modified by electropolymerization of nano-polypyrrole to improve microbial fuel cell (MFC) production of bioelectricity, Journal of Microbial and Biochemical Technology 01 (S12) (2013) 10–13.

[23] Y. Chen, Z. Zhao, Z. Weng, Performance study of polypyrrole-nanowires based microbial fuel cells, in: 2019 Third International Conference Circuits, System and Simulation, 13–15 June 2019, Nanjing, China, IEEE, pp. 235–238.

[24] A. Sumisha, J. Ashar, A. Asok, S. Karthick, K. Haribabu, Reduction of copper and generation of energy in double chamber microbial fuel cell using *Shewanella putrefaciens*, Separation Science and Technology 55 (13) (2019) 2391–2399.

[25] S. Anappara, A. Kanirudhan, S. Prabakar, H. Krishnan, Energy generation in single chamber microbial fuel cell from pure and mixed culture bacteria by copper reduction, Arabian Journal for Science and Engineering 45 (2020) 7719–7724.

[26] A. Sumisha, K. Haribabu, Modification of graphite felt using nano polypyrrole and polythiophene for microbial fuel cell applications-a comparative study, International Journal of Hydrogen Energy 43 (6) (2018) 3308–3316.

[27] H. Jiang, L.J. Halverson, L. Dong, A miniature microbial fuel cell with conducting nanofibers-based 3D porous biofilm, Journal of Micromechanics and Microengineering 25 (12) (2015) 125017.

[28] K. Dutta, P.P. Kundu, A review on aromatic conducting polymers-based catalyst supporting matrices for application in microbial fuel cells, Polymer Reviews 54 (3) (2014) 401–435.

[29] J. Hou, Z. Liu, P. Zhang, A new method for fabrication of graphene/polyaniline nanocomplex modified microbial fuel cell anodes, Journal of Power Sources 224 (2013) 139–144.

[30] C. Zhao, P.P. Gai, C.H. Liu, X. Wang, H. Xu, J.R. Zhang, et al., Polyaniline networks grown on graphene nanoribbons-coated carbon paper with a synergistic effect for high-performance microbial fuel cells, Journal of Materials Chemistry A 1 (2013) 12587–12594. Available from: https://doi.org/10.1039/C3TA12947K.

[31] Y. Hindatu, M.S.M. Annuar, A.M. Gumel, Mini-review: anode modification for improved performance of microbial fuel cell, Renewable and Sustainable Energy Reviews 73 (2017) 236–248.

[32] X. Li, J. Qian, X. Guo, L. Shi, One-step electrochemically synthesized graphene oxide coated on polypyrrole nanowires as anode for microbial fuel cell, 3 Biotech 8 (2018) 375.

[33] H. Chou, H. Lee, C. Lee, N. Tai, H. Chang, Bioresource technology highly durable anodes of microbial fuel cells using a reduced graphene oxide/carbon nanotube-coated scaffold, Bioresource Technology 169 (2014) 532–536.

[34] L.A. Hernández, G. Riveros, D.M. González, M. Gacitua, M. Angélica, PEDOT/graphene/nickel—nanoparticles composites as electrodes for microbial fuel cells, Journal of Materials Science: Materials in Electronics 30 (2019) 12001–12011.

[35] S. Kato, K. Hashimoto, K. Watanabe, Iron-oxide minerals affect extracellular electron-transfer paths of Geobacter spp, Microbes and Environments 28 (1) (2013) 141–148.

[36] M. Harshiny, N. Samsudeen, R.J. Kameswara, Biosynthesized FeO nanoparticles coated carbon anode for improving the performance of microbial fuel cell, International Journal of Hydrogen Energy 42 (42) (2017) 26488–26495.

[37] H.O. Mohamed, et al., Fe/Fe$_2$O$_3$ nanoparticles as anode catalyst for exclusive power generation and degradation of organic compounds using microbial fuel cell, Chemical Engineering Journal 349 (2018) 800–807.

[38] Z. Lv, D. Xie, X. Yue, C. Feng, C. Wei, Ruthenium oxide-coated carbon felt electrode: a highly active anode for microbial fuel cell applications C1s, Journal of Power Sources 210 (2012) 26–31.

[39] Y. Qiao, X. Wu, C.M. Li, Interfacial electron transfer of *Shewanella putrefaciens* enhanced by nanoflaky nickel oxide array in microbial fuel cells, Journal of Power Sources 266 (2014) 226–231. Available from: https://doi.org/10.1016/j.jpowsour.2014.05.015.

[40] I. Ho, M. Christy, P. Kim, K. Suk, Enhanced electrical contact of microbes using Fe$_3$O$_4$/CNT nanocomposite anode in mediator-less microbial fuel cell, Biosensors & Bioelectronics 58 (2014) 75–80.

[41] R. Kaur, A. Marwah, V.A. Chhabra, K. Kim, S.K. Tripathi, Recent developments on functional nanomaterial-based electrodes for microbial fuel cells, Renewable and Sustainable Energy Reviews 119 (2020) 109551.

[42] N. Senthilkumar, M. Pannipara, A.G. Al-sehemi, G. Gnana, PEDOT/NiFe$_2$O$_4$ nanocomposites on biochar as a free-standing anode for high-performance and durable microbial fuel cells, New Journal of Chemistry 43 (20) (2019).

[43] S.S. Rikame, A.A. Mungray, A.K. Mungray, Modification of anode electrode in microbial fuel cell for electrochemical recovery of energy and copper metal, Electrochimica Acta 275 (2018) 8–17.

[44] H. Yuan, L. Deng, Y. Chen, Y. Yuan, MnO2/Polypyrrole/MnO2 multi-walled-nanotube-modified anode for high-performance microbial fuel cells, Electrochimica Acta 196 (2016) 280–285.

[45] X. Zhao, T. Tian, M. Guo, X. Liu, Cauliflower-like polypyrrole@MnO$_2$ modified carbon cloth as a capacitive anode for high-performance microbial fuel cells, Journal of Chemical Technology and Biotechnology 95 (1) (2019) 163–172.

[46] L. Zou, Y. Qiao, C. Zhong, C.M. Li, Enabling fast electron transfer through both bacterial outer-membrane redox centers and endogenous electron mediators by polyaniline hybridized large-mesoporous carbon anode for high-performance microbial fuel cells, Electrochimica Acta 229 (2017) 31–38.

[47] D. Zhong, X. Liao, Y. Liu, N. Zhong, Y. Xu, Enhanced electricity generation performance and dye wastewater degradation of microbial fuel cell by using a petaline NiO@ polyaniline-carbon felt anode, Bioresource Technology 258 (2018) 125–134.

[48] J.M. Sonawane, S.A. Patil, P.C. Ghosh, S.B. Adeloju, Low-cost stainless-steel wool anodes modified with polyaniline and polypyrrole for high-performance microbial fuel cells, Journal of Power Sources 379 (2018) 103–114.

[49] Y. Hou, et al., Nitrogen-doped graphene/CoNi alloy encased within bamboo-like carbon nanotube hybrids as cathode catalysts in microbial fuel cells, Journal of Power Sources 307 (2016) 561–568.

[50] A. Mehdinia, E. Ziaei, A. Jabbari, Multi-walled carbon nanotube/SnO$_2$ nanocomposite: a novel anode material for microbial fuel cells, Electrochimica Acta 130 (2014) 512–518.

[51] Y. Wang, B. Li, D. Cui, X. Xiang, W. Li, Nano-molybdenum carbide/carbon nanotubes composite as bifunctional anode catalyst for high-performance *Escherichia coli*-based microbial fuel cell, Biosensors & Bioelectronics 51 (2014) 349−355.

[52] A. Mehdinia, E. Ziaei, A. Jabbari, Facile microwave-assisted synthesized reduced graphene oxide/tin oxide nanocomposite and using as anode material of microbial fuel cell to improve power generation, International Journal of Hydrogen Energy 39 (20) (2014) 10724−10730.

[53] A.D. Sekar, T. Jayabalan, H. Muthukumar, N.I. Chandrasekaran, S.N. Mohamed, M. Matheswaran, Enhancing power generation and treatment of dairy waste water in microbial fuel cell using Cu-doped iron oxide nanoparticles decorated anode, Energy (2019) 173−180.

[54] Y. Qiao, X.S. Wu, C.X. Ma, H. He, C.M. Li, A hierarchical porous graphene/nickel anode that simultaneously boosts the bio- and electro-catalysis for high-performance microbial fuel cells, RSC Advances 4 (42) (2014) 21788−21793.

[55] K. Ben, W. Ramli, W. Daud, M. Ghasemi, Non-Pt catalyst as oxygen reduction reaction in microbial fuel cells: a review, International Journal of Hydrogen Energy 39 (10) (2014) 4870−4883.

[56] S. Das, N. Mangwani, Recent developments in microbial fuel cells: a review, Journal of Scientific & Industrial Research 69 (2010) 727−731.

[57] W. Xia, A. Mahmood, Z. Liang, R. Zou, S. Guo, Earth-abundant nanomaterials for oxygen reduction, Angewandte Chemie International Edition 55 (8) (2016) 2650−2676.

[58] H. Wang, A. Bernarda, C. Huang, D. Lee, J. Chang, Micro-sized microbial fuel cell: a mini-review, Bioresource Technology 102 (1) (2011) 235−243.

[59] A. Valipour, S. Ayyaru, Y. Ahn, Application of graphene-based nanomaterials as novel cathode catalysts for improving power generation in single chamber microbial fuel cells, Journal of Power Sources 327 (2016) 548−556.

[60] Q. Wen, S. Wang, J. Yan, L. Cong, Y. Chen, H. Xi, Porous nitrogen-doped carbon nanosheet on graphene as metal-free catalyst for oxygen reduction reaction in air-cathode microbial fuel cells, Bioelectrochemistry (Amsterdam, Netherlands) 95 (2014) 23−28.

[61] T.S. Song, D. Bin Wang, H. Wang, X. Li, Y. Liang, J. Xie, Cobalt oxide/nanocarbon hybrid materials as alternative cathode catalyst for oxygen reduction in microbial fuel cell, International Journal of Hydrogen Energy 40 (10) (2015) 3868−3874.

[62] Q. Wen, et al., MnO_2-graphene hybrid as an alternative cathodic catalyst to platinum in microbial fuel cells, Journal of Power Sources 216 (2012) 187−191.

[63] L. Zhuang, Y. Yuan, G. Yang, S. Zhou, In situ formation of graphene/biofilm composites for enhanced oxygen reduction in biocathode microbial fuel cells, Electrochemistry Communications 21 (1) (2012) 69−72.

[64] Y. Su, et al., A highly efficient catalyst toward oxygen reduction reaction in neutral media for microbial fuel cells, Industrial & Engineering Chemistry Research 52 (18) (2013) 6076−6082.

[65] S. Khilari, S. Pandit, M.M. Ghangrekar, D. Das, D. Pradhan, Graphene supported α-MnO_2 nanotubes as a cathode catalyst for improved power generation and wastewater treatment in single-chambered microbial fuel cells, RSC Advances 3 (21) (2013) 7902−7911.

[66] R.K. Gautam, A. Verma, Electrocatalyst Materials for Oxygen Reduction Reaction in Microbial Fuel Cell, Elsevier B.V, 2019.

[67] L. Mahrokh, H. Ghourchian, K.H. Nealson, M. Mahrokh, An efficient microbial fuel cell using a CNT-RTIL based nanocomposite, Journal of Materials Chemistry A 5 (17) (2017) 7979−7991.

[68] M. Ghasemi, et al., Carbon nanotube/polypyrrole nanocomposite as a novel cathode catalyst and proper alternative for Pt in microbial fuel cell, International Journal of Hydrogen Energy 41 (8) (2016) 4872−4878.

[69] X. Xie, et al., Carbon nanotube-coated macroporous sponge for microbial fuel cell electrodes, Energy & Environmental Science 5 (1) (2012) 5265−5270.

[70] X. Quan, Y. Mei, H. Xu, B. Sun, X. Zhang, Optimization of Pt-Pd alloy catalyst and supporting materials for oxygen reduction in air-cathode microbial fuel cells, Electrochimica Acta 165 (2015) 72−77.

[71] Y. Zhang, J. Sun, Y. Hu, S. Li, Q. Xu, Carbon nanotube-coated stainless steel mesh for enhanced oxygen reduction in biocathode microbial fuel cells, Journal of Power Sources 239 (2013) 169−174.

[72] Y. Chen, et al., Stainless steel mesh coated with MnO$_2$/carbon nanotube and polymethylphenyl siloxane as low-cost and high-performance microbial fuel cell cathode materials, Journal of Power Sources 201 (2012) 136−141.
[73] M. Lu, L. Guo, S. Kharkwal, H. Wu, H.Y. Ng, S.F.Y. Li, Manganese-polypyrrole-carbon nanotube, a new oxygen reduction catalyst for air-cathode microbial fuel cells, Journal of Power Sources 221 (2013) 381−386.
[74] M. Ghasemi, S. Shahgaldi, M. Ismail, B.H. Kim, Z. Yaakob, W.R. Wan Daud, Activated carbon nanofibers as an alternative cathode catalyst to platinum in a two-chamber microbial fuel cell, International Journal of Hydrogen Energy 36 (21) (2011) 13746−13752.
[75] S. Chen, Y. Chen, G. He, S. He, U. Schröder, H. Hou, Stainless steel mesh supported nitrogen-doped carbon nanofibers for binder-free cathode in microbial fuel cells, Biosensors & Bioelectronics 34 (1) (2012) 282−285.
[76] J. Wei, P. Liang, X. Huang, Recent progress in electrodes for microbial fuel cells, Bioresource Technology 102 (20) (2011) 9335−9344.
[77] I. Das, T. Noori, G. Dhar, M.M. Ghangrekar, Synthesis of bimetallic iron ferrite CoO.5ZnO.5Fe$_2$O$_4$ as a superior catalyst for oxygen reduction reaction to replace noble metal catalysts in microbial fuel cell, International Journal of Hydrogen Energy 43 (41) (2018) 19196−19205.
[78] S. Ayyaru, S. Mahalingam, Y. Ahn, A non-noble V$_2$O$_5$ nanorods as an alternative cathode catalyst for microbial fuel cell applications, International Journal of Hydrogen Energy 44 (10) (2019) 4974−4984.
[79] H. Yuan, L. Deng, J. Tang, S. Zhou, Y. Chen, Facile synthesis of MnO$_2$/Polypyrrole/MnO$_2$ multi-walled nanotubes as advanced electrocatalysts for the oxygen reduction reaction, ChemElectroChem 2 (8) (2015) 1152−1158.
[80] G.D. Bhowmick, S. Das, H.K. Verma, B. Neethu, M.M. Ghangrekar, Improved performance of microbial fuel cell by using conductive ink printed cathode containing Co$_3$O$_4$ or Fe$_3$O$_4$, Electrochimica Acta 310 (2019) 173−183.
[81] S. Karthick, A. Sumisha, K. Haribabu, Performance of tungsten oxide/polypyrrole composite as cathode catalyst in single chamber microbial fuel cell, Journal of Environmental Chemical Engineering 8 (6) (2020) 104520.
[82] S. Karthick, K. Haribabu, Bioelectricity generation in a microbial fuel cell using polypyrrole-molybdenum oxide composite as an effective cathode catalyst, Fuel 275 (2020) 117994.
[83] K. Dutta, P.P. Kundu, A review on aromatic conducting polymers-based catalyst supporting matrices for application in microbial fuel cells, Polymer Reviews 54 (3) (2014) 37−41.
[84] Y. Yuan, J. Ahmed, S. Kim, Polyaniline/carbon black composite-supported iron phthalocyanine as an oxygen reduction catalyst for microbial fuel cells, Journal of Power Sources 196 (3) (2011) 1103−1106.
[85] C. Li, L. Ding, H. Cui, L. Zhang, K. Xu, H. Ren, Application of conductive polymers in biocathode of microbial fuel cells and microbial community, Bioresource Technology 116 (2012) 459−465.
[86] B. Lai, P. Wang, H. Li, Z. Du, L. Wang, S. Bi, Calcined polyaniline-iron composite as a high efficient cathodic catalyst in microbial fuel cells, Bioresource Technology 131 (2013) 321−324.
[87] L. Fu, S.J. You, G.Q. Zhang, F.L. Yang, X.H. Fang, Z. Gong, PB/PANI-modified electrode used as a novel oxygen reduction cathode in microbial fuel cell, Biosensors & Bioelectronics 26 (5) (2011) 1975−1979.
[88] C. Feng, Q. Wan, Z. Lv, X. Yue, Y. Chen, C. Wei, One-step fabrication of membraneless microbial fuel cell cathode by electropolymerization of polypyrrole onto stainless steel mesh, Biosensors & Bioelectronics 26 (9) (2011) 3953−3957.
[89] K.B. Ghoreishi, M. Ghasemi, M. Rahimnejad, Development and application of vanadium oxide/polyaniline composite as a novel cathode catalyst in microbial fuel cell, International Journal of Energy Research 38 (1) (2014) 70−77.
[90] Y. Yuan, S. Zhou, L. Zhuang, Polypyrrole/carbon black composite as a novel oxygen reduction catalyst for microbial fuel cells, Journal of Power Sources 195 (11) (2010) 3490−3493.
[91] J. Ahmed, H.J. Kim, S. Kim, Polyaniline nanofiber/carbon black composite as oxygen reduction catalyst for air cathode microbial fuel cells, Journal of the Electrochemical Society 159 (5) (2012) B497−B501.

[92] A. Sumisha, K. Haribabu, Nanostructured polypyrrole as cathode catalyst for Fe(III) removal in single chamber microbial fuel cell, Biotechnology and Bioprocess Engineering 25 (1) (2020) 78−85.

[93] Y. Han, Y. Furukawa, Conducting polyaniline and biofuel cell, International Journal of Green Energy 3 (1) (2006) 17−23.

[94] A. Rahimpour, M. Shabani, H. Younesi, M. Ponti, M. Rahimnejad, A. Akbar, A critical review on recent proton exchange membranes applied in microbial fuel cells for renewable energy recovery, Journal of Cleaner Production 264 (2020) 121446.

[95] M. Ghasemi, S. Shahgaldi, M. Ismail, Z. Yaakob, New generation of carbon nanocomposite proton exchange membranes in microbial fuel cell systems, Chemical Engineering Journal 184 (2012) 82−89.

[96] M. Bazrgar, S.A. Mousavi, Effect of casting solvent on the characteristics of Nafion/TiO_2 nanocomposite membranes for microbial fuel cell application, International Journal of Hydrogen Energy 41 (1) (2015) 476−482.

[97] K. Chae, et al., Sulfonated polyether ether ketone (SPEEK)-based composite proton exchange membrane reinforced with nanofibers for microbial electrolysis cells, Chemical Engineering Journal 254 (2014) 393−398.

[98] A. Sivasankaran, D. Sangeetha, Influence of sulfonated SiO_2 in sulfonated polyether ether ketone nanocomposite membrane in microbial fuel cell, Fuel 159 (2015) 689−696.

[99] L. Di, et al., Synthesis, characterization and performance evaluation of Fe_3O_4/PES nano composite membranes for microbial fuel cell, European Polymer Journal 99 (2018) 222−229.

[100] Q. Xu, L. Wang, C. Li, X. Wang, Study on improvement of the proton conductivity and antifouling of proton exchange membrane by doping SGO@SiO_2 in microbial fuel cell applications, International Journal of Hydrogen Energy 44 (29) (2019) 15322−15332.

[101] G. Kuppurangam, G. Selvaraj, Chapter 8, An overview of current trends in emergence of nanomaterials for sustainable microbial fuel cells, Emerging Nanostructured Materials for Energy and Environmental Science, Springer, 2019.

[102] S. Khilari, S. Pandit, M.M. Ghangrekar, D. Pradhan, D. Das, Graphene oxide-impregnated PVA-STA composite polymer electrolyte membrane separator for power generation in a single-chambered microbial fuel cell, Industrial & Engineering Chemistry Research 52 (33) (2013) 11597−11606.

[103] R. Rudra, V. Kumar, N. Pramanik, P. Paban, Graphite oxide incorporated crosslinked polyvinyl alcohol and sulfonated styrene nanocomposite membrane as separating barrier in single chambered microbial fuel cell, Journal of Power Sources 341 (2017) 285−293.

[104] H. Ilbeygi, et al., Power generation and wastewater treatment using a novel SPEEK nanocomposite membrane in a dual chamber microbial fuel cell, International Journal of Hydrogen Energy 40 (1) (2014) 477−487.

[105] P.N. Venkatesan, S. Dharmalingam, Characterization and performance study on chitosan-functionalized multi walled carbon nano tube as separator in microbial fuel cell, Journal of Membrane Science 435 (2013) 92−98.

CHAPTER 12

Nanomembranes in fuel cells

Yunfeng Zhang
Sustainable Energy Laboratory, China University of Geosciences Wuhan, Wuhan, China

12.1 The introduction of nanomembranes in fuel cells

12.1.1 Description of nanomembranes

12.1.1.1 Proton-exchange membranes

In recent years, the problem of nonrenewable energy exhaustion and environmental pollution has become increasingly serious, the development of new clean energy has become a top priority for scientists [1,2]. As a device capable of converting reactive chemical energy into electric energy without generating pollution, fuel cells now have great potential for development in electric vehicles, electronic equipment, rail transit, and many aspects of people's production and life [3–6]. According to different electrolyte types, fuel cells can be divided into: (1) alkaline fuel cells, (2) phosphoric acid fuel cells, (3) proton-exchange membrane fuel cells (PEMFCs), (4) molten carbonate fuel cells, and (5) solid oxide fuel cells [7,8]. According to the different operating temperature, it can be divided into: low-temperature fuel cell, medium temperature fuel cell, and high-temperature fuel cell. Due to the difference of battery operation characteristics, different types of fuel cell applications are also completely different. PEMFC has attracted wide attention due to its advantages such as excellent energy-conversion efficiency, remarkable power density, high power-to-weight ratio, fast start-up at low temperature, distinguished assemblability, and low pollution degree [9,10]. It is considered as a green, clean, efficient, and environmentally friendly power source with broad development prospects [11]. In the context of sustainable and renewable energy, PEMFC has a broad development prospect [12–15].

The main function of proton-exchange membrane (PEM) in PEMFCs is to separate fuel/oxidant and avoid direct contact between them. Highly efficient proton conduction constitutes the current path; it also prevents short circuits in batteries by preventing electron transfer [16] The membrane electrode assembly (MEA) composed of PEM, catalyst layer, and gas diffusion layer is an important place where chemical energy is converted into electrical energy, which is the heart of the whole battery (Fig. 12.1). The whole battery reaction, namely, the generation, dispersion, and consumption of protons and electrons, takes place within MEA. Therefore MEA performance directly affects the performance and service life of the battery [17,18]. In general, the membrane will be subjected to thermal, mechanical,

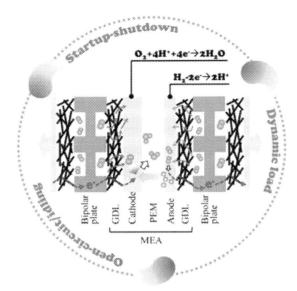

Figure 12.1 PEM fuel cell composition schematic and operating principle, and three operating conditions governing fuel cell aging [1]. *PEM*, Proton-exchange membrane.

and chemical aging. Thermal load accelerates the decomposition of molecular structure, of which hot spot induced degradation is typical. Mechanical aging refers to the occurrence of cracks and pinholes in the film due to mechanical factors such as frequent shrinkage and expansion during the dry-wet cycle. Attacks by free radicals lead to chemical degradation.

12.1.1.2 Characteristic parameters of nanomembranes

As an important part of MEA, degradation of PEM leads directly to fuel and oxidant penetration, which leads to mixed polarization and ultimately damage of fuel cell. Therefore an ideal PEMs must have the following characteristics: (1) extremely poor fuel permeability; (2) excellent proton conductivity, supernal current density, few electron conductivity, and low resistance loss; (3) a degree of chemical stability and mechanical strength; (4) distinguished electrolyte delivery to maintain uniform contact with the electrolyte to prevent local drying; and (5) prominent assemblability and low production costs [19,20].

PEM is generally divided into Nafion membrane, nonfluorine PEM, and composite PEM, [21] which produced by the US Dupont perfluorinated sulfonic PEM (Nafion series) with high mechanical strength, high proton conductivity, good chemical stability, and oxidation resistance become the only commercial production of PEM [22,23]. The common perfluorosulfonic acid-type PEM is composed of the aliphatic perfluorinated main chain and the side chain connected with ether. The end of the

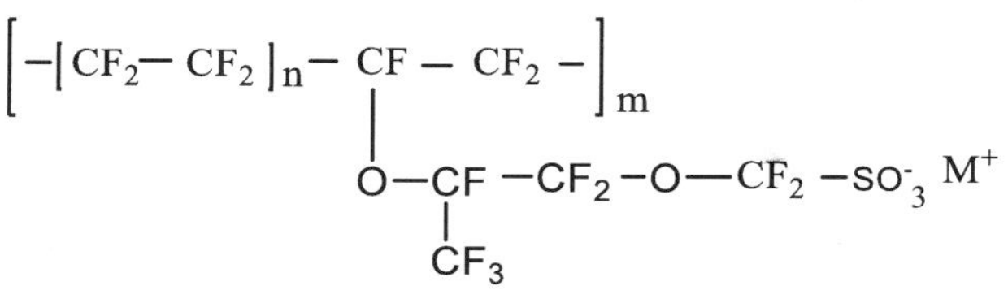

$$\left[-\left|CF_2-CF_2\right|_n - \underset{\underset{CF_3}{|}}{\underset{|}{CF}} - CF_2 - \right]_m$$
$$ O-CF-CF_2-O-CF_2-SO_3^- \ M^+$$

Figure 12.2 General structure diagram of perfluorosulfonic acid PEMs polymer molecules [22]. *PEMs*, Proton-exchange membranes.

side chain is connected with the sulfonic acid group as the cation-exchange site (Fig. 12.2). The formation of this structure enables Nafion membranes to have long-term stability under oxidation and reduction conditions [24].

Due to the strong electronegativity of fluorine, sulfonic acid groups are highly acidic and can be completely dissociated in water, thus showing good proton conductivity [25]. In addition, polytetrafluoroethylene has strong hydrophobicity, sulfonic acid group has hydrophilicity, and the flexible branched chain makes the molecule have obvious microphase separation structure. The sulfonic acid groups on the flexible branch chain tend to gather together to form ion-rich regions, and these ion-rich regions are connected with each other to form favorable channels for proton transmission, thus forming high proton conductivity. The connected hydrophilic domains are responsible for proton and water transport, while the hydrophobic domains provide morphological stability for the polymer and prevent it from dissolving in water [26].

However, perfluorinated sulfonic acid (PFSA) membranes also have many fatal disadvantages, for example, in direct methanol fuel cells (DMFCs), methanol penetration is serious due to the poor barrier of Nafion membranes to methanol [27,28]. In addition, the further development of PFSA membranes is seriously hindered by the disadvantages such as high production cost, easy chemical degradation, poor high-temperature performance, long synthesis route, and strong dependence of proton conduction on water [18,29]. Therefore the development of Nafion membrane substitutes and the modification of PFSA membrane to improve performance and reduce cost have become the main tasks of PEM research [24].

12.1.1.3 Classification of nanomembranes

Nanofilms have attracted much attention in recent years because of their ability to design continuous and efficient nanoproton transport channels. Nanofilms are generally divided into two types: (1) polymer membranes with unique structures and special functional groups. Zhang et al. [30] reviewed various ways to build continuous and efficient nanoproton transport channels by designing polymer structures. There have been many reports on how to improve the overall performance of PEM by using these methods. (2) Organic−inorganic composite membranes are prepared by filling the polymer matrix with nanomaterials or blending the two. In terms of morphology, nanomaterials include nanoparticles, nanofilms, nanoblocks, etc. In terms of the types of materials, it can be divided into: (1) nanometal/nonmetallic oxide materials: titanium dioxide, silicon dioxide, alumina, etc. Sulfonated block copolymers were added with titanium dioxide nanoparticles by Ortiz-Negrón et al. [31] to improve the mechanical and thermal stability of PEMs, while affecting their transport performance in DMFCs. (2) Carbon nanomaterials: nanofibers, nanotubes, graphite, etc. The interface compatibility between the sulfonated poly(ether ether ketone) (SPEEK) and the sulfonated carbon nanofibers (SCNFs) was improved by Liu and collaborators, and the SPEEK nanofibers were prepared by electrostatic spinning method [32]. In addition, the carbon nanofiber (CNF) felt was successfully cut into short lengths to facilitate the dispersion of SCNFs in the composite membrane. (3) Natural nanomineral materials: montmorillonite (MMT), kaolin, etc. Kotal and Bhowmick [33] reviewed the various modified natural and synthetic clays that were used in most of the polymer science and nanotechnology applications. The modified clays that were discussed for their preparation, structure, and properties were MMT, hectorite, kaolinite, sepiolite, laponite, saponite, rectorite, bentonite, vermiculite, biedellite, and chlorite. (4) Metal-organic frameworks (MOFs) materials. A new Ni-MOFs/polyacrylonitrile light flexible nanofiber was successfully prepared by Bai's group by electrostatic spinning method [34]. Nanofibers are composed of one-dimensional (1D) imidazole proton conduction channels with high proton conductivity, which proton conductivity achieve 6.04×10^{-5} S cm^{-1} at 363K and 90% relative humidity (RH). (5) Heteropoly acid (HPA), etc. Ramesh Prabhu et al. [35] prepared a PEM for fuel cells by cross-linking sulfonic succinic acid with modified chitosan/poly(ethylene oxide) (PEO)-doped HPA with different weight percentages. The membrane was characterized by XRD (X-ray diffractometer), SEM (scanning electron microscope), AFM (atomic force microscope), and impedance analysis. The results shows that the structure of the composite membranes becomes more compact compared with undoped membranes due to interaction between the sulfonic acid groups and amine groups. This could lead to increase the conductivity and reduce the excess swelling of the membrane. There have been many reports on the preparation of organic–inorganic nanocomposite membranes with silicon dioxide, titanium dioxide, MMT, and carbon nanomaterials as fillers [36–40].

12.1.2 Proton transport channels
12.1.2.1 Proton transfer mechanism
Proton conductivity is a basic index to measure proton transport efficiency in PEM, and proton-conduction mechanism is the explanation of proton transport mode [19]. Proton transport on both sides of PEM in fuel cells is generally divided into vehicle mechanism (diffusion mechanism) with low mobility and hopping mechanism

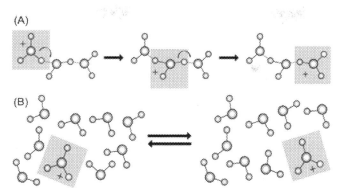

Figure 12.3 Schematic diagram of proton transfer mechanism: (A) Grotthuss mechanism and (B) vehicle mechanism [7].

(the Grotthuss mechanism) with high mobility [26,41−44]. In the vehicle mechanism, dissociated protons have strong affinity with water molecules. Therefore water molecules can be used as a carrier to form hydronium ions (H_3O^+), Zundel cations ($H_5O_2^+$), and characteristic cations ($H_9O_4^+$) with protons [45]. Under the action of electroosmosis, concentration difference and pressure difference, through the diffusion of these hydronium in the free water area, proton migration can be achieved. However, the situation is slightly more complicated in the Grotthuss mechanism, hydrogen bonds will be formed between proton and carrier [21,30]. When Nafion is hydrated, the sulfonic acid groups on the side chains dissociate producing transportable proton and sulfonate anion groups. The sulfonated anion group accumulates to form the hydrophilic water, and the proton jumps from a sulfonated anion to a nearby water molecule or sulfonated anion, constantly jumping to realize the migration from anode to cathode (Fig. 12.3) [7,46].

Studies have shown that at a higher hydration degree, the proton Grotthuss mechanism becomes more important. When the temperature rises, the diffusion coefficient of proton hopping motion increases faster than that of vehicle motion. Proton jumping mechanism is the dominant mechanism for longer time and distance transport events between sulfonate groups, while vehicular motion controls the shorter time and shorter distance transmission behavior [47].

12.1.2.2 Proton transport channel models

The high proton conductivity of Nafion membrane is not only related to its unique proton conduction mechanism, but also closely related to the structure of its proton transmission channel. In recent years, there have been many reports on the proton transport channel model in Nafion membrane [48−50]. Although the specific shape of hydrated Nafion is still under debate, there are currently three widely accepted classical models: spherical, cylindrical, and layered microstructure. In traditional water-assisted PEM, its hydrophilic proton conduction functional groups, such as sulfonic acid groups, can form interconnected proton-conduction channels with water molecules. These channels provide pathways for protons to diffuse or migrate through bulk water or hydrogen bonds formed among water molecules or between water and the -SO$_3$H groups (Fig. 12.4) [30]. Ion transport in Nafion perfluorinated membranes is controlled by osmosis, which means that the connectivity of ion clusters is crucial (Fig. 12.5) [51]. Here, the cluster network model deserves to be mentioned, which has existed for many

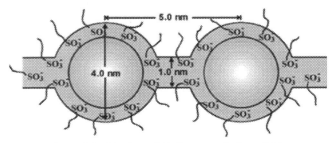

Figure 12.4 Cluster network model of hydrated Nafion morphology [51].

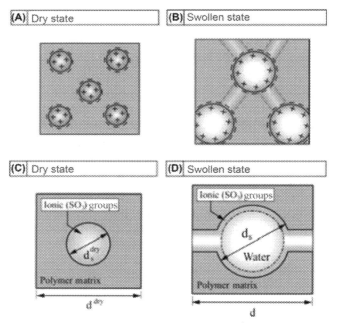

Figure 12.5 Cluster network models and corresponding geometric representations of perfluorinated sulfonic acid films in (A) dry and (B) expansive states, including a polymer matrix of a spherical cluster, are shown in two-dimensional views in (C) dry and (D) expansive states [52].

years. The Nafion's cluster network model was revealed by Hsu and his colleagues through means of mechanical and dielectric relaxation, nuclear magnetic resonance, infrared spectroscopy, transport experiments, electron microscopy and X-ray, small-angle X-ray experiments, etc. [53].

For the cylindrical proton transport channel model, the small-angle scattering data of hydrated Nafion were simulated by Schmidt-Rohr, K's team introducing an algorithm (Fig. 12.4) [54]. The characteristic "ionomer peak" arises from long parallel but otherwise randomly packed water channels surrounded by partially hydrophilic side branches, forming inverted-micelle cylinders. The cylinder model can explain the important characteristics of Nafion, including the rapid diffusion of water and protons in Nafion and its continuous diffusion at low temperatures, as well as many aspects that spherical clusters fail to explain [55]. Kusoglu et al. [52] proposed a representative volume element based on the description of the membrane nanostructure of PFSA in the literature. Based on the mechanical model, the functional relationship between young's modulus of membrane tension and water volume fraction of PFSA at different temperatures was estimated. The results show that the water-filled cylindrical channel with spherical clusters is more effective than that with spherical clusters alone (Fig. 12.6).

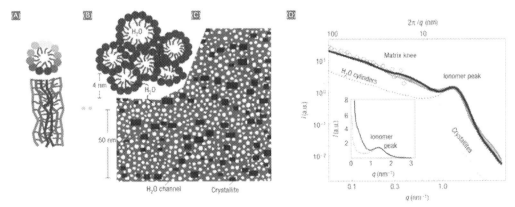

Figure 12.6 Parallel water-channel (inverted-micelle cylinder) model of Nafion. (A) Two views of an inverted-micelle cylinder, with the polymer backbones on the outside and the ionic side groups lining the water channel. Shading is used to distinguish chains in front and in the back. (B) Schematic diagram of the approximately hexagonal packing of several inverted-micelle cylinders. (C) Cross-sections through the cylindrical water channels (white) and the Nafion crystallites (black) in the noncrystalline Nafion matrix (dark gray), as used in the simulation of the small-angle scattering curves in (D). (D) Small-angle scattering data (circles) of Rubatat et al. [56] in a log(*I*) versus log(*q*) plot for Nafion at 20 vol.% of H_2O, and our simulated curve from the model shown in (C) (solid line). The inset shows the ionomer peak in a linear plot of *I*(*q*). Simulated scattering curves from the water channels and the crystallites by themselves (in a structureless matrix) are shown dashed and dotted, respectively [54].

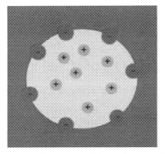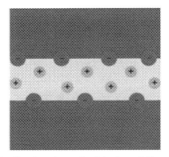

Figure 12.7 Schematic diagram of ion charge distribution in cylindrical and flat water structures. Note that a flat water structure corresponds to more negative electrostatic energy [57].

As for the proton transmission channel with layered structure, Laurent Rubatat et al. [56] from the contrast variation method combined with the SANS technique, the structure between 10 and 1000 Å of the Nafion can be described in term of elongated polymeric aggregates connected at larger scale to form a film with good mechanical properties [57,58]. Rubatat and coworkers showed through experiments and calculations that, in the case of Nafion, water was not confined within a spherical cavity but between fibrous objects [59]. In addition, the morphology of Nafion is not fixed. Liu et al. [60] found through comparison of experimental results that Nafion may change morphically during hydration, that is, with the increase of hydration degree and the expansion of hydrophilic regions, Nafion may change from stratiform or cylindrical shape (Fig. 12.7).

12.1.2.3 Modification strategies of proton transport channel

Outstanding proton transfer efficiency is one of the basic requirements of an ideal PEM. Therefore many researchers have been working on proton transfer channels for many years. The Nafion membrane is modified to construct a continuous and efficient nanoproton transport channel, usually through the following methods:

1. By changing the structure of the polymer and introducing some special groups to help form an interconnected ion cluster to promote proton transfer [30]. Such as: (1) Aliphatic and aromatic organic hybrid, introduction of aliphatic fragments into fully aromatic PEMs has been shown to facilitate proton conduction in aqueous micelles [61]. (2) Change the side chain length of the polymer, the chemical structure of Nafion contains hydrophilic sulfonic acid groups attached to a long and flexible side chain, sufficient to form microphase separation between the sulfonic group in the long side chain and the main chain of hydrophobic polymerization [62−64]. Therefore PEMs with long side chain sulfonic acid group has received extensive attention. (3) Block copolymers, self-assembled block copolymers can form definite nanostructures whose morphology and domain size are adjustable on the nanoscale [30,65]. Self-assembled ordered nanostructures of materials are

conducive to proton transport under various conditions, which is required by the strong performance of fuel cells. Ion-containing block copolymers are expected to be the next generation of PEMs in hydrogen and methanol fuel cells. (4) The introduction of locally dense sulfonated fragments in PEMs contributes to the formation of a continuous and efficient proton transport channel [66−68]. The polarity difference between the local dense sulfonated unit and the hydrophobic unit of the polymer is very great, thus forming a clear phase separation structure and improving the proton conduction efficiency. (5) Polymer-cross-linked PEMs can improve the mechanical stability and water stability of the membrane, while maintaining acceptable proton conductivity under low WU and low humidity conditions [30]. Common cross-linked PEMs include self-cross-linked network and semiinterpenetrating polymer-cross-linked network. In the process of heat treatment, a small part of sulfonic acid groups on the membrane, as cross-linking groups, react with the activated benzene ring on the membrane to produce self-cross-linking. Such thermal cross-linking makes the cross-linking membrane maintain high proton conductivity and breaking elongation [69]. For semiinterpenetrating polymer networks (IPNs) polymer cross-linking networks, membrane expansion can be effectively solved because one polymer chain is trapped in the cross-linking network of another polymer [70,71]. In addition, the molecular scale dispersion of sulfonated polymer chains in semi-IPNs cross-linking networks can ensure the uniform distribution of proton carriers.

2. Using nanometer material filling the preparation of organic and inorganic composite membrane, is the most direct way to build effective ion conductive path [72]. Nanomaterials enhance the water retention capacity of PEM by virtue of their small size and large specific surface area, thus improving proton conductivity and expanding the operating temperature range of PEMFCs. Due to the high specific surface area of nanoparticles, the interface interaction between polymer matrix and nanoparticles can become very high, which has a great impact on the properties of the polymer itself [73]. In the organic−inorganic hybrid nanocomposite PEM, the organic polymer can provide mechanical strength and proton carrier, while the inorganic nanomodule can help to establish the connection between the heterogeneous components through the interaction between the heterogeneous components (polymer matrix, inorganic filler, water, and proton) [30]. The introduction of inorganic nanometer components can significantly improve the mechanical strength, proton conductivity and dimensional stability of PEMs. Zhang's group prepared sulfonated titanium dioxide (STiO$_2$) by reaction of titanium dioxide (TiO$_2$) with 1,3-propanesulfonic acid. Novel STiO$_2$ incorporated sulfonated poly(aryl ether sulfone) (SPAES) nanocomposite PEMs were made by solution casting. Characterization test results showed that TiO$_2$ composite film has better stability, higher proton conductivity, lower methanol permeability and better relative selectivity than Nafion film [74].

Carbon nanotubes (CNTs) have become a new type of advanced inorganic filler due to their unique properties such as high tensile modulus and strength, large surface volume ratio, high flexibility, excellent electrochemical activity, and potential applications in the field of energy materials. Gong and colleagues first functionalized CNTs with inorganic proton conductor boron phosphate (BPO$_4$) by the easily degradable polydopamine-assisted sol-gel method to obtain BPO$_4$@CNTs [75]. The new additive was then used to modify the SPEEK. Polydopamine coating layer can act as a special adhesive to homogeneously adhere BPO$_4$ nanoparticles on CNTs, which not only reduces the risk of short-circuiting, but also creates new proton conductive pathways in the composite membrane. The comprehensive characterization showed that compared with pure SPEEK, the proton conductivity, thermal stability, tensile property, and dimensional stability of composite PEMs were significantly improved.

Graphene oxide (GO) is usually prepared by the strong oxidant graphite oxide, while usually retaining the layered structure of the parent graphite, providing a large surface area. This property, combined with its insulating properties, makes GO have great potential in PEMs applications of PEMFC. However, due to the lack of a proton-exchange group, pure GO did not improve the performance of PEMs as much as expected. Sulfonate functionalized GO may be a better solution. When the content of sulfonated GO sulfonate group is high, not only the water retention of PEM can be improved, but also its proton conductivity can be improved. Jiang and his partners modified SPEEK using sodium dodecylbenzene sulfonate (SDBS) adsorbed GO as a filler. The addition of SDBS-GO greatly improves the ion-exchange capacity (IEC), water absorption and proton conductivity, while reducing the methanol permeability through SPEEK membranes [76].

MOFs has a large specific surface area and can accommodate a large number of water molecules and acid groups, providing sufficient receptors for proton transfer. A large number of hydrogen bonds in MOFs can interact with polymer matrix to form a tighter hydrogen bond network and additional proton transport channels. In addition, the small aperture of MOFs can hinder the diffusion of fuel and oxidant and improve the selectivity [77–80]. In recent years, MOFs modified PEMs has shown excellent performance and great advantages in commercial applications, and there are many related research reports. Tsai et al. [81] first reported the preparation of novel Nafion-based composite membranes (PEM-1 and PEM-2) using two 1D channel microporous MOFs, CPO-27 (Mg), and MIG-53 (Al) as fillers. The results show that the water absorption, proton conductivity and power density of the composite membrane are greatly improved. HPA and polyoxic metal acid are a class of transition metal oxide clusters with diverse structures. They are characterized by ultraacidity, controllable oxidizability, high proton conductivity, etc. which make them increasingly attractive in recent decades. The influence of HPA on improving proton conductivity and chemical stability of Nafion has been confirmed [82,83].

12.2 Polymer-based nanomembranes
12.2.1 Polymer membranes with tailored main chains

As one of the core components of PEMFC, PEM, known as the heart of PEMFC, determines the power of the battery. It has the function of isolating anode and cathode, providing proton transmission channel and preventing fuel penetration, and has the advantages of high proton conductivity and good mechanical properties. Proton conductivity is one of the key properties of PEM.

There are two ways to improve proton conductivity. The first is to increase the content of proton groups. Polymers with high IEC values usually show high proton conductivity. However, the dimensional stability of high IEC PEM is not good [84–88]. High IEC values often have negative effects on PEMs performance, such as dimensional stability, water resistance, and mechanical strength. The second is to establish an effective proton transmission channel, so as to improve the proton transmission efficiency (under low IEC values, the proton conductivity of polymer films is low; under a higher IEC value, although the proton conductivity of the polymer can be effectively improved, it is easy to cause excessive swelling of the membrane, resulting in a sharp decline in the dimensional stability). The former increases the proton conductivity by increasing the proton-exchange capacity, but also causes excessive water absorption of the film, resulting in serious swelling of the film and even water dissolution. With a certain proton-exchange capacity, the latter separates the hydrophilic and hydrophobic phases in PEM so that the hydrophilic proton groups gather effectively and form a continuous and efficient proton transfer channel.

The polymer backbone structure has an important influence on the properties of PEM, such as microscopic phase separation, proton conductivity, water absorption and mechanical strength, etc. By optimizing the design of the backbone structure, PEM with excellent performance can be obtained. The following part mainly introduces the construction of proton transport channel and its influence on PEM performance from the aspects of stiffness and flexibility of main chain segment, random and block structure, hydrophilic and hydrophobic branch chain, etc.

12.2.1.1 Flexible and rigid chain segments

Polymerization backbone structure can improve the performance of PEM by introducing rigid and flexible chain segments. Rigid polymer chain segment is mainly the aromatic hydrocarbon, it has rigid skeleton of benzene ring, the main chain of the rigid section of PEM low water swelling and high mechanical strength, but the whole aromatic hydrocarbons of the preparation of polymer membrane brittleness is bigger, to improve the flexibility of the main chain can easily promote the transfer of ionic groups, thereby enhancing the nanophase separation of membrane morphology. Lee et al. reported that fluranyl polyether (aryl ethersulfone) contains locally and densely

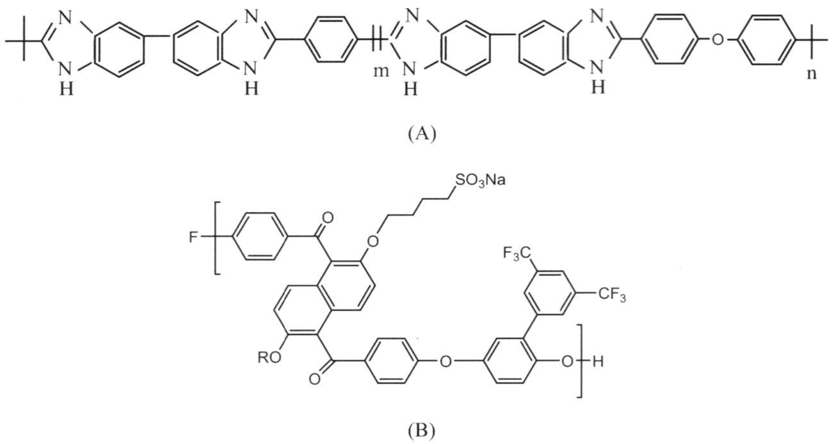

Figure 12.8 (A) The structural formula of polybenzimidazoles and (B) SDF-PAEK.

distributed flexible butyl sulfonic acid suspension units. Unique molecular design enables different phase separation and enhanced proton conductivity [62]. As described as Higashihara's group observed that the new poly(benzene) with flexible alkyl chain segment had obvious phase separation. At 80°C and 30% RH, the membrane proton conductivity of 2.93 mmol g^{-1} in IEC was 9.1 mS cm^{-1}, higher than that of Nafion membrane under the same condition [89]. Moreover, the introduction of the flexible chain segment can also enhance the mobility of the hydrophilic chain segment and the hydrophobic chain segment, promote the microphase separation of the thin film, facilitate the formation of hydrophilic clusters, effectively aggregate the hydrophilic proton groups to form a continuous and efficient proton transfer channel, and improve the proton conductivity. Wang et al. [90] synthesized a series of segmental block polybenzimidazoles (PBIs) (Fig. 12.8A) with different molar ratios and similar molecular weights. Due to the combination of rigid and flexible segments in block copolymers, block copolymer films show obvious nanometer phase separation structures. At a low phosphoric acid doping level (0.1 S cm^{-1} at 180°C), high proton conductivity of the blocking membrane can be obtained. Li and Bu [91] synthesized a novel SDF-PAEK polymer (Fig. 12.8B), which has a rigid hydrophobic main chain, and prepared a series of SDF-PAEK-Nafion-X films by doping the new SDF-PAEK skeleton molecules into the Nafion film, as a new PEM material suitable for DMFC. As skeleton molecules of Nafion membrane, short high-density trifluoromethyl side chain and long flexible aliphatic side group side chain can promote PEM phase separation, facilitate the formation of hydrophilic clusters, and improve proton conductivity. However, there is intermolecular incompatibility between SDF-PAEK containing rigid groups and Nafion molecules based on flexible chains. With the increase of the composition and incompatibility of SDF-PAEK, some small dispersed ion clusters are separated from each other, and the extension of the distance between two adjacent proton transport groups will shorten the hydrogen bond life in the transport channel of the fluorination system [92]. Therefore when the temperature increases, the membrane with the internal sulfonic acid groups closer together will have higher conductivity. Different proportions of SDF-PAEK were added to the Nafion membrane and its conductivity was tested, and it was found that the proton conductivity of the SDF-PAEK composite membrane was the highest at 15 wt.%. The material exhibited the highest proton conductivity of 0.197 S cm^{-1} at 80°C. By summarizing the above work, it can be seen that reasonable regulation of the rigid flexibility of the main chain structure can improve the proton conductivity and flexibility of the film, while the flexible alkyl chain will reduce the chemical stability of the film. Referring to Nafion's flexible fluorocarbon structure, fluorinated alkyl chain can be introduced into the rigid chain to overcome this problem.

12.2.1.2 Regular and atactic chain segments

The rigid flexible structure of the main chain can affect the performance of PEM, and the random and block structure of the chain segment can affect the microscopic phase separation, water absorption and proton conductivity of PEM. Block copolymers have ordered chain structure arrangement, and the microstructure and performance of PEM can be adjusted through structural design. After forming the film, self-assembled ordered nanostructures can be formed to promote proton long-distance transmission and improve proton conductivity, which is less dependent on temperature and humidity than random copolymers. Moreover, block copolymerization can obtain stereo-controlled polymers, and it is found that polymers with such molecular structure tend to form hydrophilic/hydrophobic microphase separation. By means of this microphase separation, the membrane has a wider proton transport channel and is very stable under hydrolysis and oxidation [93,94]. At the same IEC level, the proton conductivity of multiblock copolymer films is higher than that of random copolymer films. As described as Pan's group [95] synthesized sulfonated polybenzimidazole-polyimide (PI) block copolymers (Fig. 12.9A) through high-temperature polycondensation. Change the length of the PI chain to obtain a series of block copolymers with various block lengths. All the block copolymers exhibit good thermal stability, dimensional stability, mechanical strength, and proton conductivity. Compared to the random sulfonated PI-containing benzimidazole membranes with the same degree of sulfonation, the membranes prepared from the block copolymers show higher proton conductivities. The proton conductivities of the block copolymer membranes range from 6.2×10^{-4} to 1.1×10^{-2} S cm^{-1} at 105°C. The block copolymer membrane doped with phosphoric acid exhibits proton conductivity higher than 0.2 S cm^{-1} at 160°C, indicating its potential applications in PEMFCs operated under high-temperature and low-humidity conditions. Shuntaro Amari et al. [96] proposed a novel and simple molecular design concept for aromatic random copolymers based on

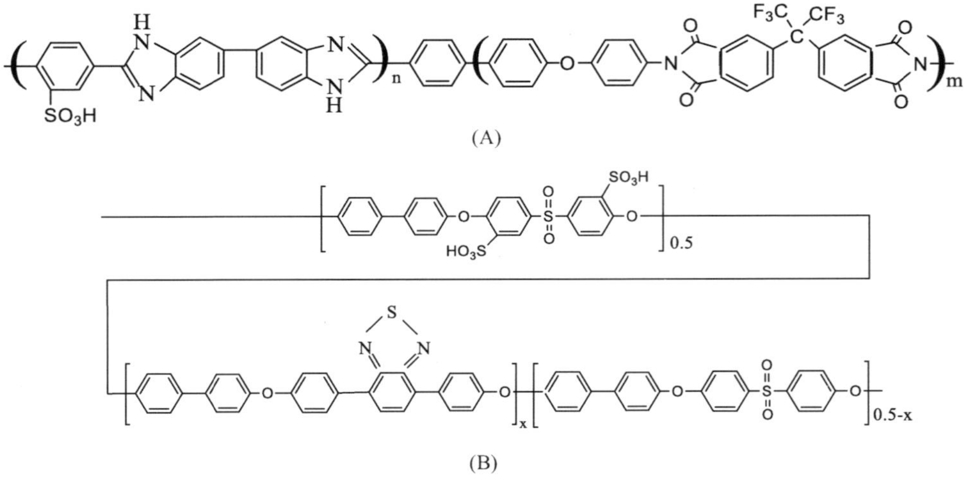

Figure 12.9 (A) The structural formula of SPBIb-PIs and (B) SBT. *SBT*, Sulfonated benzothiadiazole; *SPBIb-PIs*, sulfonated polybenzimidazole-polyimide block copolymers.

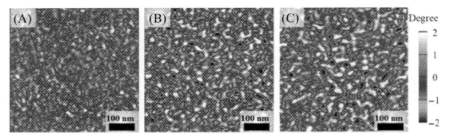

Figure 12.10 TM-AFM phase images for the (A) SPES membrane, (B) the 0.1% SBT membrane, and (C) 1% SBT membrane (scan size: 500 nm) [96]. *SBT*, Sulfonated benzothiadiazole; *SPES*, sulfonated poly(ether sulfone); *TM-AFM*, tapping-mode atomic force microscopy.

sulfonated benzothiadiazole (SBT) units (Fig. 12.9B) to induce nanostructuring. Aromatic random copolymers with different content of SBT (0.1% SBT and 1% SBT) were synthesized. Because the sulfonic acid group density in the SBT unit is high, proton conductivity can be enhanced by increasing the SBT unit content in the polymer. Using AFM (Fig. 12.10), it is confirmed that introducing a small amount of the SBT unit in the sulfonated poly(arylene ether sulfone) membrane induced significant hydrophilic and hydrophobic domain aggregation. Furthermore, these morphological transformations provided high density of proton carrier density in membrane and enabled effective proton conductivity even under low-humidity condition (2.0 mS cm^{-1} at 30% RH). These results indicate that the SBT unit is a key trigger for constructing favorable nanostructure for ion transport and improving ion conductivity in the PEM. Zhang et al. [97] used random with low sulfonation degree and block sulfonated poly(ether sulfone)s (SPES) as the

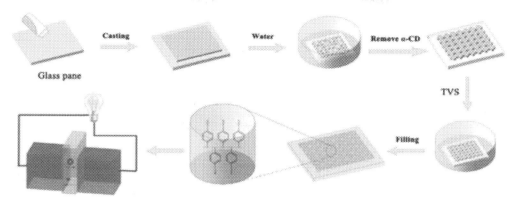

Figure 12.11 Schematic illustration of the preparation procedure of the pore-filling composite membrane [97].

porous membrane matrix to prepare a new block polymer porous filler membrane, which was used to improve proton conductivity in DMFC applications. The composite films prepared by random polymer and block polymer were compared. The pore-filling membrane with block polymers shows higher proton conductivity (0.083 S cm^{-1} at 80°C) than that prepared by random polymers (0.072 S cm^{-1} at 80°C). Compared with Nafion 117, the composite membrane exhibits higher dimensional stability (6.6%) and methanol resistance. This is due to the well-developed conduction channels formed by the aggregation of the polymer chains with sulfonic acid groups (Fig. 12.11). Therefore by summarizing the above work, it can be concluded that, compared with random sulfonated films, block copolymer films have higher proton conductivity, block copolymers show outstanding structural ordering and performance regulation, and have a potential application prospect.

12.2.2 Polymer membranes with functional side chains
12.2.2.1 Hydrophilic side chains
For the structural design of polymer PEM, it cannot only change the backbone structure, but also improve the PEM performance by grafting functional branches on the polymer skeleton. For the side chain, because in the side chain of overhang of sulfonic acid group is easy to gather to form hydrophilic ionic domain, is advantageous to the proton transport, so they usually have higher proton conductivity than the main chain in the process of polymer film, the main chain of the branched chain mobility than the skeleton, so through the reasonable design of the branched chain to control the hydrophilic/hydrophobic phase distribution, can optimize the performance of the PEM.

For sulfonated aromatic polymers, high degree of sulfonation is a necessary condition for obtaining high proton conductivity. But the biggest problem is that sulfonated aromatics polymers with high degree of sulfonation have poor mechanical properties and

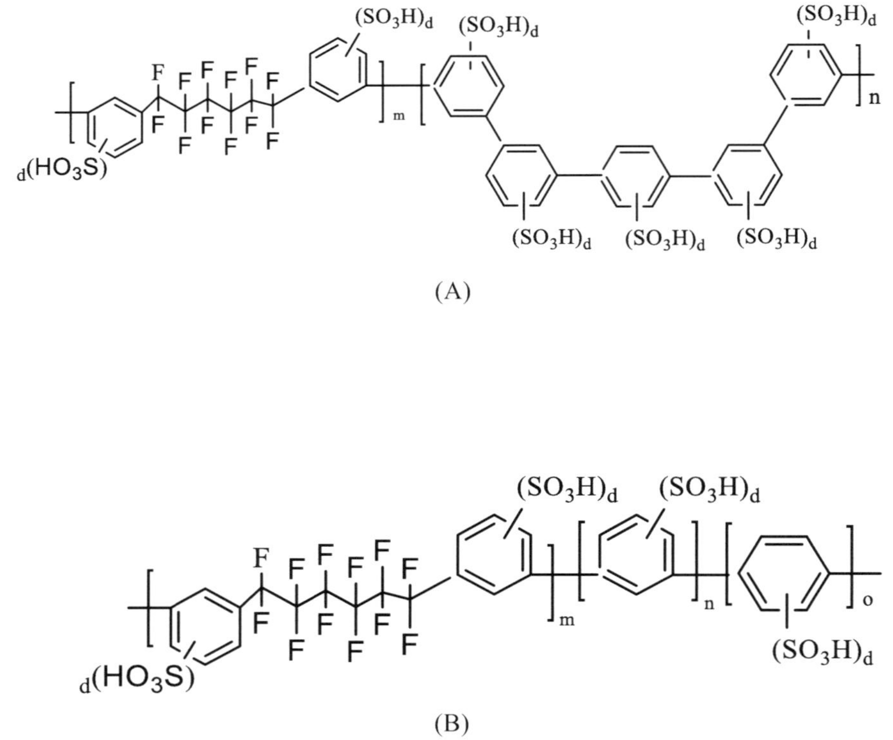

Figure 12.12 The structural formula of (A) SPAF-QP and (B) SPAF-MM.

poor dimensional stability [98–100]. The reason is that the sulfonic acid group is always on the backbone of the polymer. Thus hydrophilic and hydrophobic phase separation structures cannot be easily formed [101]. To solve this problem, side-chain polymers have been widely studied. In the side chain membrane, the hydrophilic sulfonic acid group is located on the side chain, away from the main chain of hydrophobic polymerization [102–107]. As described as Long's group [108] synthesized a poly(arylene perfluoroalkylene)-based ionomers containing densely sulfonated quinquephenylene groups in the main chain (SPAF-QP) (Fig. 12.12A). The combination of highly hydrophobic perfluoroalkylene and highly hydrophilic sulfonated quinquephenylene units is designed to achieve high proton conductivity with reasonable IEC values (1.83–1.97 meq g^{-1}). Compared with a new type of sulfonated poly(arylene peruoroalkylene) (SPAF-MM) (Fig. 12.12B) membrane with fewer sulfonated hydrophilic groups, SPAF-QP membrane has obvious hydrophilic/hydrophobic differences in polymer components, resulting in obvious phase separation morphology, high proton conductivity and compatibility with the catalyst layer, thus the proton conductivity of SPAF-QP membrane is significantly improved. The proton conductivity of SPAF-QP membrane is obviously better than that of SPAF-MM

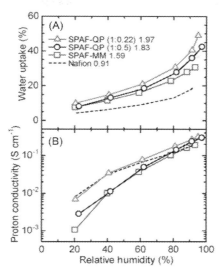

Figure 12.13 Relative humidity dependence of (A) water uptake and (B) proton conductivity at 80°C.

membrane, especially at low humidity (up to 6.9 S cm^{-1} at 80°C and 20% RH) (Fig. 12.13). The selected SPAF-QP (1:0.22) membrane exhibits much better fuel cell performance than the SPAF-MM membrane and comparable performance to the Nafion NRE 211 membrane even under low-humidified conditions.

12.2.2.2 Hydrophobic side chains

In addition to grafting hydrophilic branch chain on the main chain to improve the performance of PEM, the hydrophobic branch chain can also be grafted on the polymerization main chain with its hydrophobicity and mobility, and a continuous ionic conductive channel can be constructed in the film to obtain high ionic conductivity. Moreover, the graft hydrophobic branch chain can obtain lower water absorption rate and swelling rate under the same IEC conditions. Wang et al. [62] the first use of 9,9-double (3-4-hydroxy phenyl-) phenyl fluorine (Fig. 12.14) and 4,4'-(six fluorine and propyl) phenol, 1,4-(4-fluorine benzene formyl) benzene was synthesized via aromatic nucleophilic polycondensation of fluorene and new polymer poly(aryl ether ketone benzene) side base structure (4-PAEK-xx), further after mild sulfonation reaction, and the preparation of a series of hydrophobic side chain type poly(aryl ether ketone) PEM (4-SPAEK-xx). The results showed that the PEM had moderate water absorption and low swelling rates, which ranged from 21% to 51.2% and 7.4% to 17.2%, respectively at 80°C. This type of polyaryl ether ketone PEM shows good ionic conductivity. At 80°C, the ionic conductivity of the 4-PAEK-xx membrane ranges from 115 to 171 mS cm^{-1}, and the ionic conductivity of the 4-PAEK-xx membrane has exceeded that of commercial

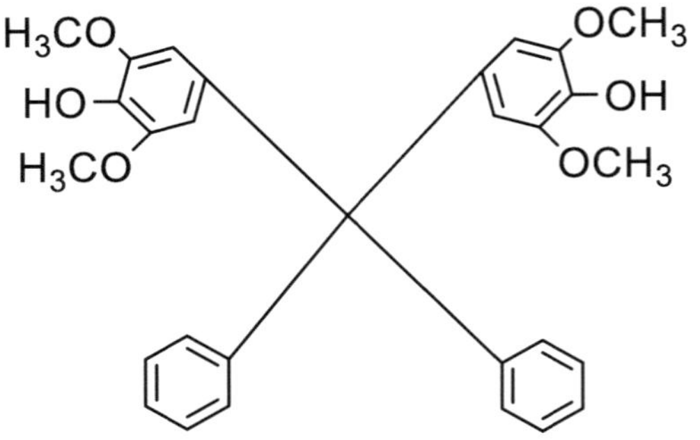

DMHF

Figure 12.14 The structural formula of 4-PAEK-xx.

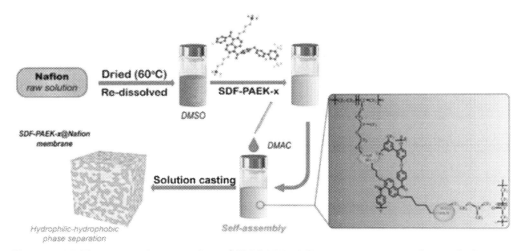

Figure 12.15 The preparation procedure of SDF-PAEK@Nafion-x composite membranes [91].

Nafion membrane. Li and Bu [91] synthesized a novel SDF-PAEK polymer with a rigid hydrophobic backbone, a short high-density trifluoromethyl side chain and a long flexible aliphatic side group side chain as the skeleton molecules of Nafion membrane (Fig. 12.15). Due to the unique molecular interaction selectivity during membrane formation, SDF-PAEK automatically matches Nafion molecular conformation through self-assembly. A series of SDF-PAEK-Nafion-X films were prepared by doping the novel SDF-PAEK skeleton molecules into the Nafion films, which were used as new PEM materials suitable for DMFC applications. Compared with recasting Nafion, the cost of the composite membrane is significantly reduced, as its composition can be up to 20%. Transmission

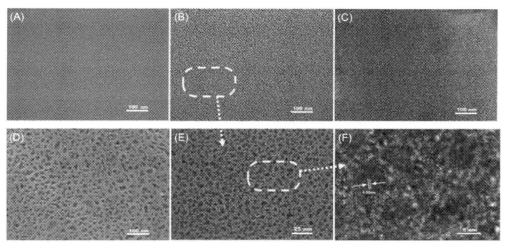

Figure 12.16 TEM images of SDF-PAEK@Nafion composite membranes: (A) SDF-PAF@Nafion-10%; (B, E, and F) SDF-PAEK@Nafion-15%; (C) SDF-PAEK@Nafion-20%; (D) recast Nafion [91].

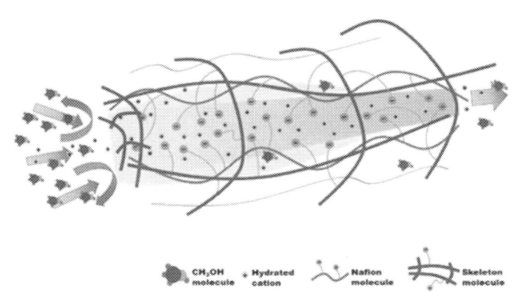

Figure 12.17 Schematic diagram for the proton transport channel of the prepared membranes [91].

electron microscope (TEM) analysis (Fig. 12.16) showed the existence and distribution of ion cluster microphase lattice and double-continuous proton transport channels. The unique skeleton molecules of SDF-PAEK can promote the phase separation between hydrophilic and hydrophobic aggregates, thus forming nanowires and promoting proton transport at the same time (Fig. 12.17). Proton transport channels caused by hydrophilic and hydrophobic phase separation were finally distributed, which significantly promoted proton transport and significantly enhanced proton conductivity. At 80°C, the proton conductivity of SDF-PAEK@Nafion-15% was 0.197 S cm^{-1}, while the methanol permeability value decreased by half compared with that of cast Nafion. Therefore the introduction of hydrophobic chain segments into the molecular structure design can promote the phase separation between hydrophilic and hydrophobic aggregates, thus forming nanochannels, and promote proton transport and improve proton conductivity.

By grafting functional branchlets on the main chain of polymerization, the high mobility of hydrophilic or hydrophobic branchlets can effectively promote the microphase separation of PEM and form a continuous and efficient proton-conducting channel. However, functional branched chains usually increase the free volume in the film, leading to increased water absorption and swelling, and reducing the size and mechanical stability of PEM. Generally, the main chain enables PEM to have high mechanical and chemical stability, and the graft side chain can provide functional groups. Therefore obtaining a branch chain PEM with good performance is to optimize the ratio between the main chain and the side chain of polymerization.

12.2.3 Polymer membranes with special groups

12.2.3.1 Large and rigid groups

The introduction of large volume rigid groups (naphthalene, fluorene) can promote the formation of ion clusters, because the planar parts will accumulate in the polymer matrix, which is conducive to the generation of obvious phase separation. Meanwhile, the introduction of free volume can solve the problem of excessive swelling of the film due to water absorption [109,110]. Membrane with similar IEC can increase the conductivity of proton by increasing the free volume of membrane. Zhang [111] designed and synthesized sulfonated PIs by naphthalene diimide and sulfonation units (Fig. 12.18A). This structure is composed of aliphatic block alternately, allowing the accumulation of naphthalene diimide part and sulfonic acid groups gathered at the same time, giving high conductivity and excellent hydrolysis stability, and the PEM showed low swelling. Lee's group [112] synthesized a new series of proton-conducting polymer electrolyte membranes based on sulfonated poly(arylene ether)s with fluorene (Fig. 12.18B). And the proton conductivity is as high as 260 mS cm^{-1}, with low thickness expansion rate and strong mechanical properties. Yuan and coworkers [113] synthesized a novel rigid sulfonated polyfluorene (aryl ether ketone sulfone) from 3,3-disulfonic acid-4,4-dichlorodiphenylsulfone and 9,9-bis (4-hydroxyphenyl) fluorene via a typical polycondensation reaction. The proton conductivity of SPAEKS-50 is 66.5 mS cm^{-1}, which is slightly higher than that of Nafion 117 under the same conditions. Compared with Nafion 117 membrane, all membranes showed lower methanol permeability and higher proton selectivity. The hard fluorene in SPEEKs is expected to improve the dimensional stability

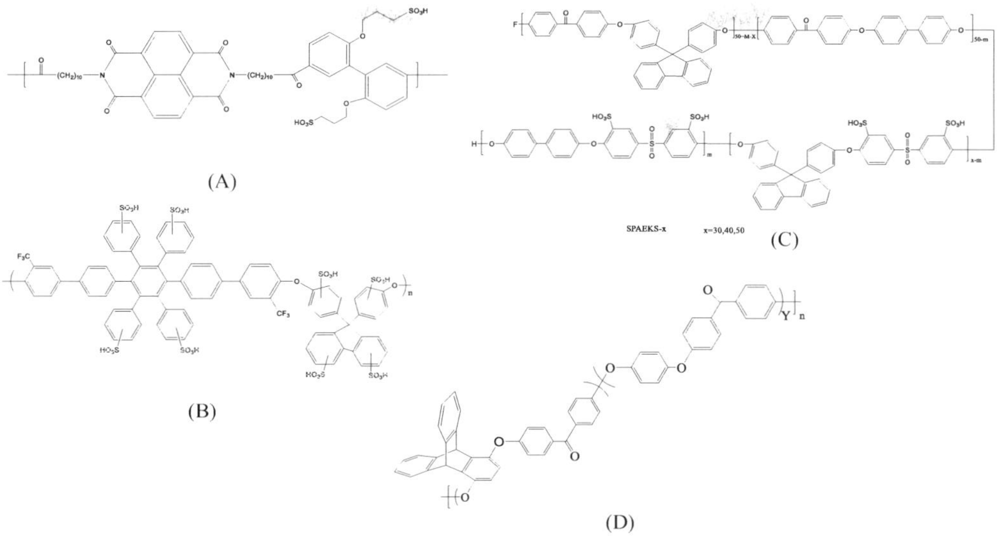

Figure 12.18 Chemical structures of some PEMs: (A) SPI [111], (B) SP4-X [112], (C) SPAEKs [113], and (D) TripPEEK [114]. *PEMs*, Proton-exchange membranes; *SPAEKs*, sulfonated poly(aryl ether ketones); *SPI*, sulfonated polyimides.

and thermal stability and ensure the long-term stability of SPAEKs films. Therefore the introduction of large rigid groups can increase the free volume of the membrane and thus increase the proton conductivity (Fig. 12.18C). As described as Timothy's group, [114] a trimethyl ether ketone (triptycene) based on triadiene was synthesized and sulfonated to form a proton-exchange system. The free volume of triphenyl was increased due to the incorporation of triphenyl. The proton conductivity was 334 mS cm^{-1} at 85°C (Fig. 12.18D).

12.2.3.2 Amino groups

The introduction of basic groups (-NH$_2$ and -NH-) can achieve ionic cross-linking in the membrane [100]. Joseph and collaborator [115] synthesized a sulfonated polysulfone (SPAES-50) and an ion-cross-linked blend membrane containing a commercially available PBI derivative (PBI-OO) with an activated benzene ring in the main chain. The Friedel-Crafts reaction was activated by heating the film. Due to excessive PBI-OO, sulfonic acid groups preferentially react with PBI-OO, leading to covalent cross-linking (Fig. 12.19A). In fuel cell tests, they showed for the first time that a sulfoxide cross-linking blend membrane could be used in fuel cells. Guan et al. [116] used vinyl imidazole as a cross-linking agent for vinyl benzyl substituted PBI to prepare water-free seed conducting membranes (Fig. 12.19B). The results showed that the cross-linking degree of the membrane increased with the increase of cross-linking agent

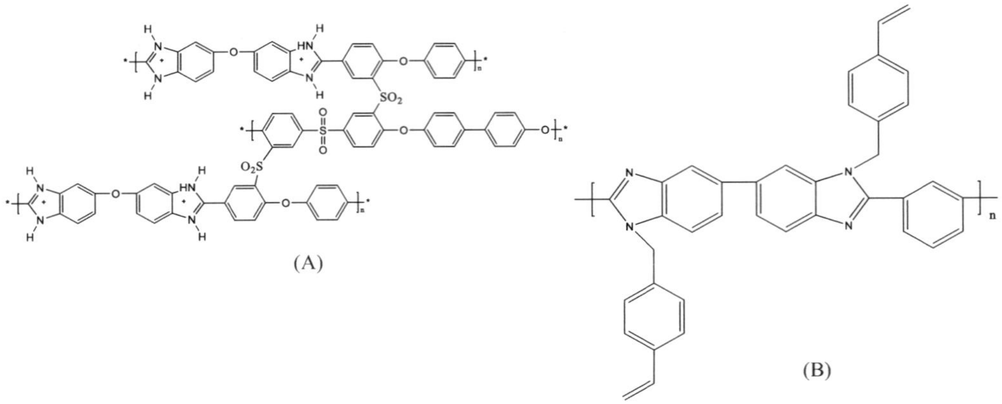

Figure 12.19 (A) The cross-linking reaction [115] and (B) synthesis of vinyl benzyl substituted polybenzimidazole [116].

content, and the acid doping ability decreased with the increase of PBI cross-linking degree. N-vinyl imidazole introduced by PBI plays an important role in improving the performance of PBI membrane. It can improve the alkaline site of the membrane, retain free radicals, slow down the free radical reaction in the process of degradation, and improve its oxidation stability.

12.2.3.3 Densely sulfonated units

At present, the most advanced PEMs are perfluorosulfonic acid membranes and the trademarks are Nafion, Flemion, and Aquivion. Although they have high proton conductivity and excellent chemical stability, they have limited performance at high temperatures, and they are expensive to manufacture. Therefore sulfonated aromatic polymers with high thermal stability and high mechanical strength have attracted extensive attention as alternative high-performance PEMs. The most widely reported aromatic PEMs include sulfonated derivatives of polybenzene, poly(aryl ether ketone), poly(aryl ether sulfone), poly(aryl sulfoxide), etc. The sulfonic groups in these polymers are located on the backbone, and the rigid multiaromatic backbone prevents the formation of continuous ion transport channels. Therefore these sulfonated polymers can only obtain appropriate conductivity at high IEC and high water content, resulting in large size changes and poor mechanical properties [26,86]. A feasible strategy to improve the performance of aromatic PEM is to significantly separate hydrophilic sulfonic group from hydrophobic polymerization backbone by locating the sulfonic group on the flexible side chain, so as to improve the nanometer phase separation between hydrophilic and hydrophobic regions [117]. The results also show that these side-chain-sulfonated polymers exhibit good proton conductivity at relatively low water content. The functionalization of aromatic backbone with sulfonic acid groups usually gives them good solubility and considerable proton conductivity.

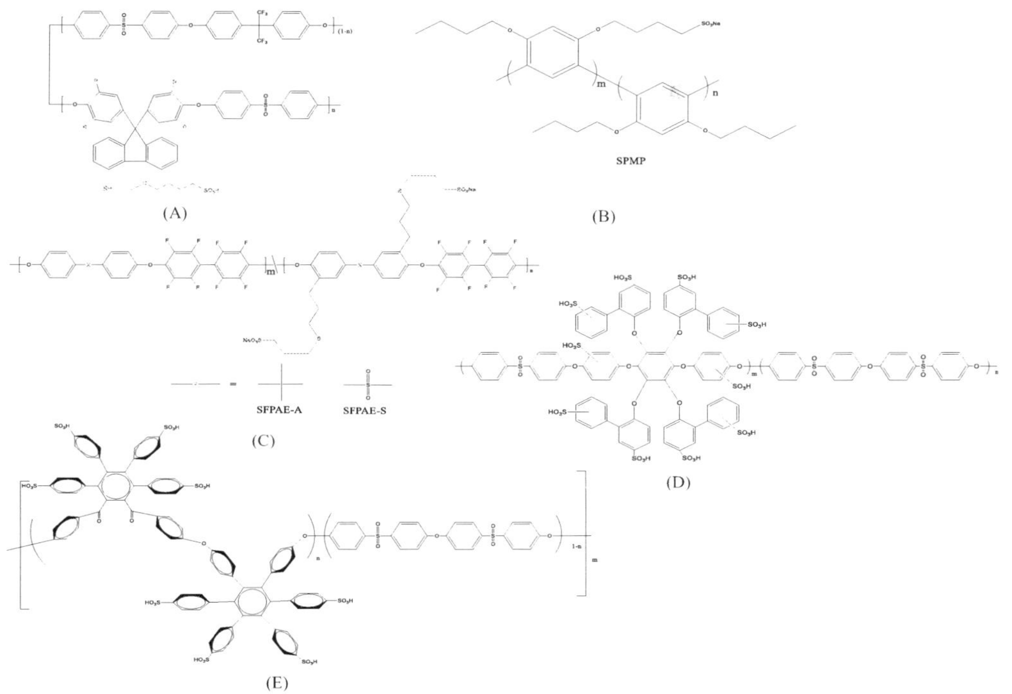

Figure 12.20 (A) The respective side-chain type sulfonated copolymers (SPAES) [62], (B) the structure of SPMPs [89], (C) synthesis of DAFPAEs and SFPAEs [118], (D) the structure of locally sulfonated poly(ether sulfone)s with highly sulfonated units [119], and (E) the structure of MSBP [120]. *SPAES*, Sulfonated poly(aryl ether sulfone); *SFPAEs*, sulfonated fluorinated poly(arylene ether)s; *SPMPs*, sulfonated poly(m-phenylene)s.

Currently, several strategies have been developed to enhance the phase separation of sulfonated aromatic polymers. A simple approach is to design sulfonic groups like Nafion on the flexible side chain, so that the sulfonic groups can easily assemble into hydrophilic phases, while the polymer backbone forms a hydrophobic phase. For example, Wang and coworkers [62] reported a poly(aryl ether sulfone) based on fluorene (Fig. 12.20A), which contains locally dense flexible butyl sulfonic acid side unit. This unique molecular design enables unique phase separation (ion domain size: 2–5 nm under TEM) and enhanced proton conductivity (61–209 mS cm^{-1} to 30°C under hydration). Zhang et al. [89] also observed significant phase separation in new polyphenols with flexible alkyl sulfonated side chains, and successfully obtained a series of poly(m-phenyl), SPMP, via long alkyl side chains and sulfonic acid (Fig. 12.20B). The proton conductivity of C-SPMP8-2 membrane with IEC ratio of 2.93 mmol g^{-1} at 80°C and 30% RH was 9.1 mS cm^{-1}, higher than that of Nafion under the same conditions. These previous works demonstrated the feasibility of promoting phase separation in PEMs by designing sulfonic acid groups at the end of long side chains of polymers. Chen and coworkers [118] successfully synthesized two partial fluorinated polyaryl ethers SFPAE-A-70 and SFPAE-S-70 containing long alkyl sulfonate side chains (Fig. 12.20C). The low polymerization temperature avoids the rearrangement of allyl groups in the polymer, resulting in the formation of longer alkyl side chains after the mercaptolene reaction than previously reported. Both TEM and SAXS characterization revealed obvious hydrophilic and hydrophobic phase separation in sulfonated polysulfones. SFPAE-A-70 and SFPAE-S-70 exhibited proton conductivity similar to Nafion 212 at all test temperatures, and all sulfonated polyaryl ether ketones were thermally and mechanically stable because of their rigid aromatic backbone. Compared with Nafion 212 battery, SFPAE-S-70 battery has higher coulomb efficiency and energy efficiency. Matsumoto and collaborator [119] successfully synthesized a new local sulfonated polyethersulfone for fuel cells (Fig. 12.20D). Each repeating unit contains 10 sulfonic acid groups. Due to the high contrast of polarity between the highly sulfonated unit and the hydrophobic polyethersulfone unit, the definite phase separation structure and the well-connected proton transport channel were observed. The proton conductivity in a wide range of RH is comparable to that of Nafion 117 at low humidity. Dong et al. [120] synthesized a polysulfonated block copolymer (Fig. 12.20E). Sulfonic acid on the side chain will provide better aggregation effect and hydrophilic/hydrophobic separation, improving battery performance. At 25°C, the water absorption rate is 9.6%–42.5%, at 80°C, the water absorption rate is 14.7%–90.8%. The obtained IEC is 0.79–1.87 (mmol g^{-1}). The results show that the morphology of polymer matrix has great influence on the performance of the cell, and the membrane with good hydrophilic–hydrophobic phase separation is very important in fuel cell applications. Pang and Han [121] synthesized a new hard carbazole derivative and a dense sulfonated carbazole derivative, PEM, which can form clusters of sulfonic acid clusters, forming continuous wide dimensional nanochannels. All membranes are stable at 250°C. These membranes also exhibit good dimensional stability and excellent proton conductivity. Spesk-cz-15 has a very high proton conductivity (165.3 mS cm^{-1} at 80°C).

Kyu and collaborator [122] synthesized poly(aryl ether ketone) block copolymers with hydrophilic precursors of different lengths. SPEAK's ATM Phas image shows the apparent connection and separation coexistence of proton channels (Fig. 12.21).

From the point of view of molecular design, the sulfonated block polymers composed of hydrophilic sulfonated and hydrophobic segments have excellent phase separation properties and connect the proton transport channels. Dense sulfonated structure is a new method to prepare high-performance PEMS. Dense sulfonic groups can form large ion clusters, which is favorable for phase separation and Proton Transport. It is a promising method to improve the properties of PEMs by introducing sulfonic acid groups into the grafted side chains of polymer backbone to form obvious phase separation morphology.

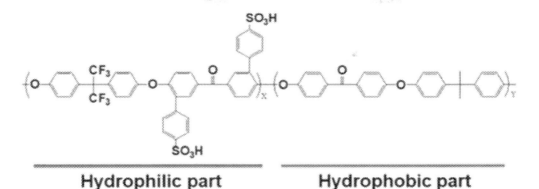

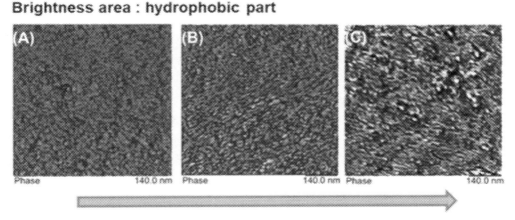

Figure 12.21 AFM phase image of the SPAEK membranes under 100% RH (700 nm × 700 nm): (A) SPAEK (X4.8Y8.8), (B) SPAEK (X7.5Y8.8), and (C) SPAEK (X.1Y8.8). *AFM*, Atomic force microscope; *RH*, relative humidity; *SPAEK*, sulfonated poly(aryl ether ketones).

Khabibullin and coworkers [123] used silicon dioxide as a rigid substrate, to change the content of sulfonic group in the silica gel membrane, different proportions of 3-sulfopropylmethacrylate (SPM) and 2-ethoxy ethyl methacrylate (EEMA) were copolymerized in the pores of the membrane, the proton conductivity of the films with different sulfonic groups was measured (Fig. 12.22). As can be seen from the diagram, the proton conductivity of the porous membrane varies in an S shape with respect to the Mole% of the monomer, with three distinct regions. First, there is a low conductivity region corresponding to 20%—40% sulfonic group content. Increasing the content of sulfonic group to 50% can increase the proton

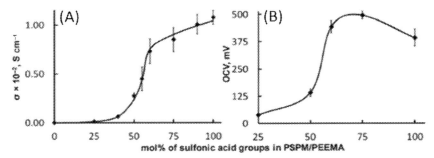

Figure 12.22 Plot of room temperature proton conductivity for self-assembled membranes (A) and room temperature open circuit voltage for pressed membranes (B) as a function of the SPM content in the pore-filling copolymer [123].

conductivity. In the small range of 50%—75% sulfonic acid, the proton conductivity increases sharply. Finally, the third region is saturated with proton conductivity, where the sulfonic group content increases from 75% to 100%, resulting in only a 20% increase in proton conductivity. In most PEMs, acidic groups form ion-rich clusters that must be interconnected to provide proton conductivity. This is promoted by water, which is prone to form hydrophilic ion pathways, the number of which increases with the increase of the sulfonic group. From the proton conductivity data in the diagram, it can be concluded that when the sulfonated monomers account for 50%—60% of the copolymer, the ionic-rich clusters are connected and the proton conductivity increases rapidly.

Sulfonic acid group content (mol%)	σ (mS cm^{-1})	Methanol uptake (%)	Open circuit potential (OCP (mV))
0	Negligible	N/A	N/A
25	0.12 ± 0.01	N/A	40 ± 5
40	0.64 ± 0.12	N/A	N/A
50	2.7 ± 0.3	3	140 ± 20
55	4.5 ± 1.2	N/A	N/A
60	7.3 ± 1.3	5	440 ± 30
75	8.5 ± 1.2	13	495 ± 20
90	10.0 ± 1.0	N/A	N/A
100	10.7 ± 0.7	17	390 ± 40

Abbreviation: N/A, Not available.

12.2.4 Cross-linking membranes

The stability of sulfonated aromatic film in water is still a key point to be noted. The film is dry and fragile, and the stability and mechanical properties of the film are reduced in the hydrated state. A large number of researchers have proposed the method of cross-linking membrane to improve the stability of the membrane under dry and hydrated conditions. Chemical cross-linking is an important technology to improve the physical and chemical properties of polymers. This method can limit the excessive swelling of PEM in water and improve its mechanical properties [124]. There are two main ways to achieve cross-linking. One is ionic cross-linking, which is obtained by preparing acidic/alkaline polymer blends. The other is covalent cross-linking, which can be achieved by heat treatment or by forming an amide-type connection through the imidazole group of the polymer. There are two types of PEM of cross-linking: full cross-linking type and semiinterpenetrating cross-linking type, which are, respectively, introduced in the following two aspects.

12.2.4.1 Semiinterpenetrating

Semiinterpenetrating cross-linking polymers are cross-linked networks of two or more polymers containing linear polymers interspersed with other or more polymers [125,126]. Compared with Nafion, the acidity of aromatic sulfuric acid on the sulfonated aromatic polymer is relatively low. When IEC is the same as Nafion, the proton conductivity of the sulfonated aromatic polymer membrane is far lower than that of Nafion membrane, and to obtain higher proton conductivity, a higher IEC is required, which leads to excessive swelling of the membrane. To have higher proton conductivity and higher mechanical properties, semicross-linking has been developed [127]. Linear polymer chains on molecular scale dispersed in half interpenetrating network, make the proton carrier uniform distribution. At the same time, half interpenetrating cross-linked can reduce the water absorbing capacity and swelling ratio of PEM, and the narrow hydrophilic channel reduces the methanol permeability [128]. In the research of the Liu's group [70], the cross-linking network based on bromomethylated poly(phenylene oxide) (BPPO) and binary amine [2,2′-(ethylenedioxy) bis (ethylamine), EBEA] was introduced into SPEEK. Heat treatment made the reaction of benzyl bromide of BPPO and amino of EBEA to form Cr-BPPO covalent cross-linking network (Fig. 12.23). There was also ion bond interaction between -SO$_3$H and -NH- in the thin film. The results showed that the content of the cross-linked network had an important influence on the film performance. With the increase of the content of the cross-linked network, the dimensional stability of the film improved, and the area swelling of the PEM with a cross-linked network of 40 wt.% in water at 20°C and 80°C hardly changed. In addition, the methanol permeability of the thin film was significantly reduced, and the PEM with a cross-linking network of 20 wt.% had the highest relative selectivity (25.57×10^4 S cm^{-3}), which was five times higher than Nafion 117.

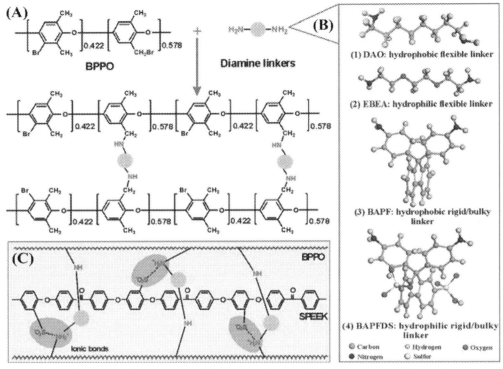

Figure 12.23 Schematic illustration of fabricating the semi-IPN membranes, (A) the formation of the cross-linking networks through the alkylation reaction between BPPO and diamine linkers, (B) four different diamine cross-linkers, and (C) the constructed semi-IPNs containing ionic bonds [70]. *BPPO*, bromomethylated poly(phenylene oxide); *IPNs*, interpenetrating polymer networks.

12.2.4.2 Fully cross-linking

The full cross-linked polymers have a three-dimensional (3D) network structure. Unlike semiinterpenetrating cross-linked networks, it does not contain linear polymers. SPEEK has been the focus of attention to replace Nafion membrane. Sun and coworkers [129] synthesized chloromethyl SPEEK (SPEEK-Cl) and prepared cross-linked membrane under mild conditions through the Friedel-Crafts alkylation reaction, which not only retained the sulfonic acid groups on the polymer skeleton, but also improved mechanical strength, dimensional stability, proton conductivity and power density (Fig. 12.24). Kim et al. [130] developed a PEM for a cross-linked copolymer containing polybenzoxazine and PBI (Fig. 12.25A). The proton conductivity is improved and the battery performance is excellent. Li and collaborators [131] successfully designed and synthesized the cross-linked full aromatic sulfonated polyamide (Cr-FA-SPA) (Fig. 12.26). Then Cr-FA-SPA was mixed with SPEEK to prepare a series of Cr-FA-SPA content

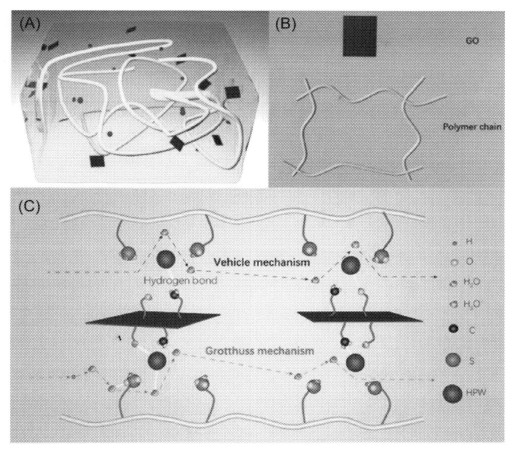

Figure 12.24 The cross-section of a assumed composite membranes (A), self-cross-linking polymer chains and graphene oxide model (B), and schematic representation of proton transfer in self-cross-linking hybrid membrane (C) [129].

SPEEK/Cr-FA-SPA composite films (Fig. 12.25B). The morphology of SPEEK/Cr-FA-SPA membrane was analyzed by SEM, and related properties such as water absorption, size swelling, proton conductivity, and methanol permeability were systematically studied. The surface of the membrane presented uniform and smooth morphology due to the good compatibility between SPEEK and Cr-FA-SPA. After blending with Cr-FA-SPA, the size expansion of pure SPEEK membrane at high temperature was significantly reduced. When the content of Cr-FA-SPA in SPEEK/Cr-FA-SPA membrane increased to 40 wt.%, the methanol permeability was 2 orders of magnitude lower than that of Nafion 117 membrane. Table 12.1 summarizes the values of proton conductivity for various polymeric materials of PEM.

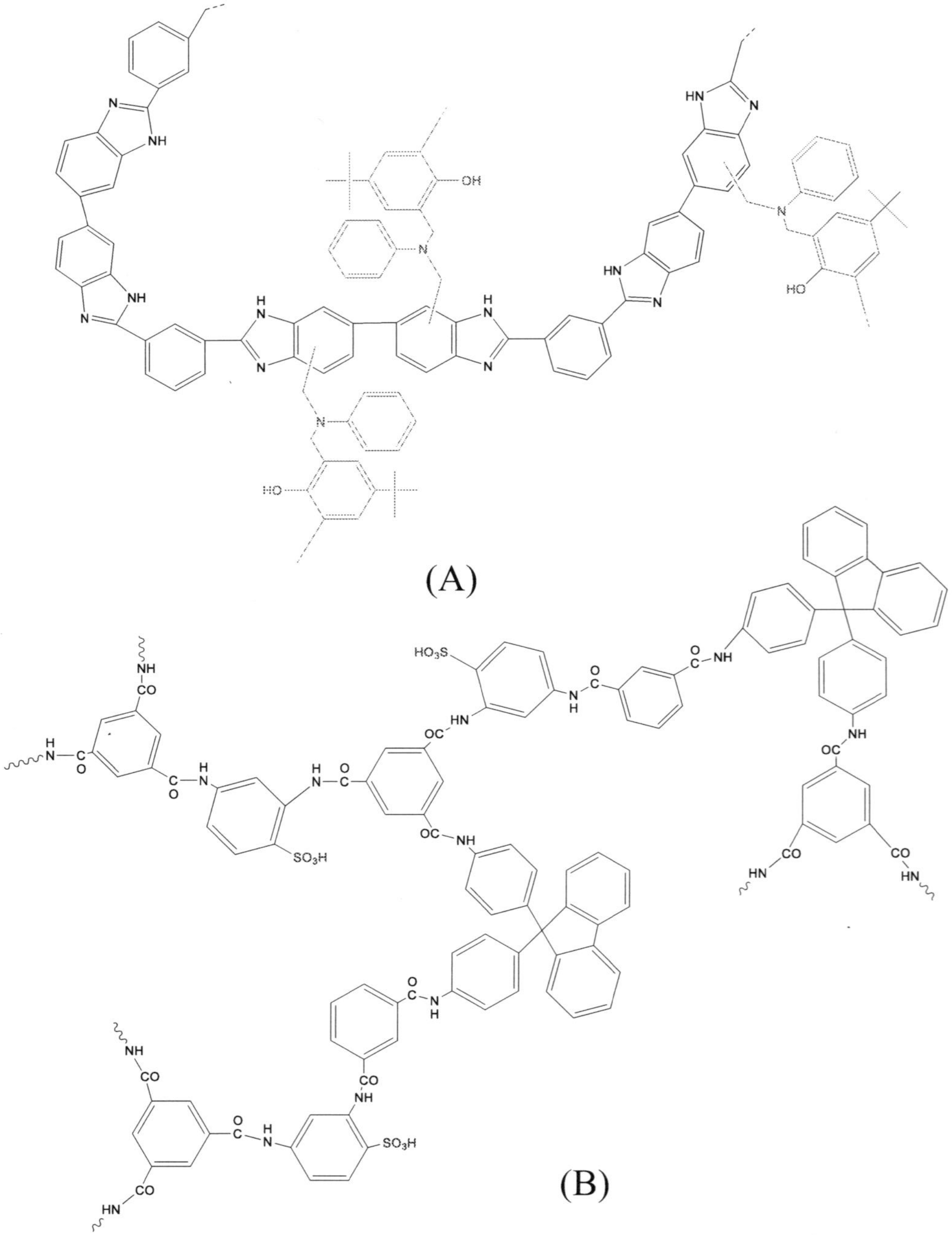

Figure 12.25 (A) Possible chemical structure of P(pBUa-co-BI)s [130], (B) the structure of Cr-FA-SPA [131]. *Cr-FA-SPA*, Cross-linked full aromatic sulfonated polyamide.

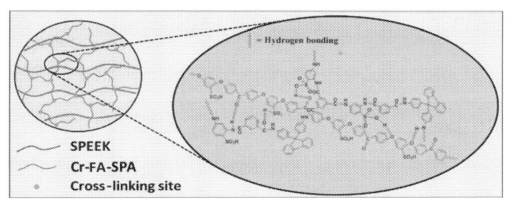

Figure 12.26 Schematic representation of the hydrogen bonding interactions between SPEEK and Cr-FA-SPA [131]. *Cr-FA-SPA*, Cross-linked full aromatic sulfonated polyamide; *SPEEK*, sulfonated poly(ether ether ketone).

Table 12.1 Values of proton conductivity for various polymeric materials of PEM.

Basic material	Proton conductivity (S cm^{-1})	References
Block polybenzimidazole	0.100 (180°C)	[90]
SDF-PAEK polymer	0.197 (80°C)	[91]
Sulfonated polybenzimidazole-polyimide block copolymers	0.060 (160°C)	[95]
Random copolymers based on sulfonated benzothiadiazole units	0.300 (80°C)	[96]
The pore-filling membrane with random polymers	0.072 (80°C)	[97]
The pore-filling membrane with block polymers	0.083 (80°C)	[97]
Poly(arylene perfluoroalkylene)-based ionomers	6.900 (80°C)	[108]
Sulfonated polyimides	0.114 (30°C)	[111]
Poly(arylene ether)	0.260 (80°C)	[112]
Trimethyl ether ketone (triptycene)	0.334 (85°C)	[114]
Polybenzimidazole	0.100 (150°C)	[116]
Fluorene-based poly(arylene ether sulfone)	0.146 (30°C)	[62]
Sulfonated poly(m-phenylene)s	0.151 (80°C)	[89]
Diallyl-containing bisphenol monomers	0.090 (25°C)	[118]
Multisulfonated block copolymers	0.131 (80°C)	[120]
Densely sulfonated rigid structure	0.165 (80°C)	[121]
Sulfonated poly(arylene ether ketone)	0.131 (90°C)	[122]
Semiinterpenetrating polymer networks	0.069 (30°C)	[70]
Self-cross-linked sulfonated polyether ether ketone membrane	0.119 (80°C)	[129]
H$_3$PO$_4$-doped cross-linked benzoxazine-benzimidazole copolymer membrane	0.120 (150°C)	[130]
Sulfonated poly(ether ether ketone)	0.110 (80°C)	[131]

12.3 Hybrid membranes containing nanofillers
12.3.1 Effect of nanofillers

Recently, modification of polymeric membranes by nanoparticles is in the center of attention. Incorporation of nanomaterials with different types of materials could produce synergistic effects. In hybrid organic—inorganic composite membranes, the inorganic part usually offers good thermal and mechanical stability, while the organic functional groups such as -SO$_3$H gives specific chemical reactivity [132]. In this idea, the nanocomposite membrane can aggregate the merits of each component to achieve high proton conductivity and suitable mechanical property at the same time [133].

Incorporating hydrophilic inorganic nanoparticles has been generally accepted as an effective approach to improving the proton conductivity and stability of a PEM at the same time. Zero-dimensional (0D) nanomaterials (e.g., titanium oxide), 1D nanomaterials [e.g., single/multiwalled carbon nanotubes (SWCNTs/MWCNTs)], and two-dimensional (2D) nanomaterials (e.g., GO and graphitic carbon nitride nanosheets) have been used as nanofillers [134]. For instance, 1D nanomaterials MWCNTs possess a large surface/volume ratio and excellent thermal stability [135]. Comparing to 0D and 1D nanomaterials, 2D nanomaterials are considered as the most ideal materials owing to their larger specific surface area and higher aspect ratio, which facilitate the building of continuous long-range proton conductive nanochannels [134,136].

The selection of fillers is the biggest challenge in the manufacturing process of mixed matrix membranes. The structure, molecular sieving effect and the interaction between the membrane and the penetrant have great influence on the properties of the fillers. The utilization of the inorganic fillers could enhance the interfacial interaction between the polymer and fillers [137]. Several fillers have been added to the electrolyte, such as SiO$_2$, TiO$_2$, and ZrO$_2$ particles to reduce alcohol crossover and to enhance thermal stability and water retention. But in general, these composite electrolytes exhibit lower proton conductivity. To overcome these problems, various types of clay minerals and other clay-related materials, such as MMTs, bentonite, have been explored as alternative fillers to minimize the loss of proton conduction.

The incorporation of these fillers in the polymer matrix produces the improvement of membrane properties, such as proton conductivity and thermal stability at high temperatures [138]. Noto et al. [139] prepared two families of hybrid Nafion-based inorganic-organic proton-conducting membranes using two different types of nanofillers, respectively. TG measurements showed that the addition of nanofiller stabilizes the -SO$_3$H groups of the Nafion host polymer. In contrast, the interactions between the nanofiller and the polymer matrix stabilize the perfluoroether side chains of the Nafion host polymer in the hybrid membranes. Nafion nanocomposite membranes based on Fe$_2$TiO$_5$ perovskite were prepared by dispersion of Fe$_2$TiO$_5$ nanoparticles within the commercial Nafion membranes [140]. Incorporation of Fe$_2$TiO$_5$ nanoparticles in Nafion matrix improved the thermal stability of Nafion membranes. In another study, a polymer nanocomposite containing inorganic nanofillers can alleviate the performance constraints of SPEEK by enhancing the membrane's adsorption of water molecules and restraining polymeric chain alignment [141]. Moreover, the incorporation of inorganic fillers, for example, SiO_2, TiO_2, and ZrO_2 with Nafion and other polymers such as polyvinylidene fluoride (PVDF) and polypyrrole, are used to increase resistance in fuel permeation [23]. This work has demonstrated that addition of sulfonated SiO_2 nanoparticles to PBI can effectively increase the proton conductivity, improve the mechanical strength, and reduce the methanol permeability of the membranes [142].

12.3.2 Nanoparticles
12.3.2.1 Nanometal oxides
In recent years, great progress has been made in the development of membrane materials for various applications, which also increases the demand for modified new membranes [143]. The synergistic effect of organic/inorganic hybrid materials provides a good way for the synthesis of novel membranes with excellent performance, which is also the main choice for the current membrane selection trend [16,144]. Among them, adding inorganic nanoparticles such as TiO_2, Al_2O_3, SiO_2, and ZrO_2 to improve the performance of the membrane is one of the commonly used modification methods [145]. The following results can be achieved: (1) improve the thermal and dimensional stability of the membrane; (2) improve proton conductivity; (3) reduce methanol transport channels, thereby reducing methanol permeability of the membrane [146].

Proton conductivity and power density can be improved by grafting sulfonated polymer onto TiO_2 nanoparticles. Compared with the pure SPEEK membrane, the mechanical stability and dimensional stability of the nanocomposite membrane were also improved. The maximum power density of the membrane containing a certain amount of poly(2-acrylamido-2-methylpropane sulfonic acid) (PAMPS)-g-TiO_2 and poly(styrene sulfonic acid) (PSSA)-g-TiO_2 nanoparticles was 80°C obtained 283 and 245 mW cm^{-2}, respectively [147]. A series of nanocomposite PEMs were prepared by blending various titanium oxide/graphitized carbon nitride (TiO_2/g-C_3N_4) nanocomposites with sulfonated polyethersulfone (SPAES). The introduction of nanocomposites makes the membrane have obvious proton conductivity, mechanical stability and dimensional stability. Especially, the SPAES-TiO_2/g-C_3N_4-1.0 nanocomposite membrane demonstrates the prominent proton conductivity of 325.3 mS cm^{-1} at 80°C [148]. Miao et al. [149] found that adding TiO_2 particles into poly (vinylidene fluoride-co-hexafluoropropylene) (PVDF-HFP) polymer electrolyte matrix can improve the ionic conductivity and the stability of electrolyte/electrode interface. In another study, STiO_2 was successfully prepared by condensation reaction, and was uniformly interspersed into the PVDF-g-PSSA (PPSSA), thus the S-TiO_2/PPSSA composite PEMs were synthesized (Fig. 12.27A). The reasonable addition of S-TiO_2 improved the physicochemical and electrochemical properties of the membranes. The S-TiO_2/PPSSA

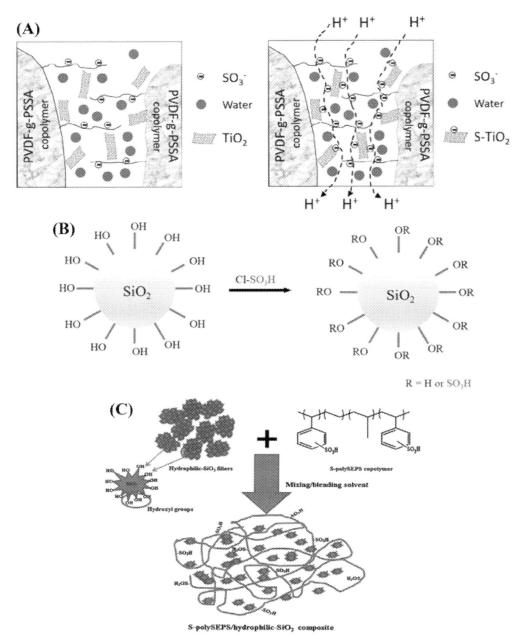

Figure 12.27 (A) Schematic diagram of the pathways of proton transport in composite membrane [150], (B) preparation of functionalized SiO$_2$(f-SiO$_2$) inorganic nanofillers [151], and (C) schematic illustration for preparation of the S-poly-SEPS/hydrophilic-SiO$_2$ composite membranes [152]. *S-poly-SEPS*, Sulfonated polystyrene-block-poly(ethyl-ran-propylene)-block-polystyrene.

composite membrane has higher proton conductivity (0.067 S cm^{-1}) than pristine PPSSA membrane [150]. TiO$_2$ nanoparticles investigated for the fuel cell performance were modified by sulfonation and surface coated using polyaniline (PANI). Fabricated membranes were characterized and analyzed. The highest proton conductivity was 2.30×10^{-4} S cm^{-1} with SPES/STiO$_2$-PANI (0.5%) composite membrane [153]. Elakkiya et al. synthesized PEM from polyethersulfone (PES), SPEEK, and nanoparticles by phase inversion technology. Metal oxide nanoparticles such as Fe$_3$O$_4$, TiO$_2$, and MoO$_3$ were added to polymer blends (PES and SPEEK), respectively. Compared with the original PES membrane, the polymer composite membrane showed excellent performance in IEC and proton conductivity [154]. Nafion composite membranes containing Fe$_3$O$_4$-sulfonated graphene oxide (SGO) with different wt.% were successfully fabricated by solution casting method. The combined action of Fe$_3$O$_4$-GO improves the performance of the cell under high temperature and low humidity, compared with the PEMFC of the pristine Nafion membrane. The proton conductivity of Nafion/Fe$_3$O$_4$-SGO (3 wt.%) composite membrane at 120°C under 20% RH was 11.62 mS cm^{-1} [155]. MoO$_3$ is a kind of excellent conductive material, which also exhibits excellent physicochemical properties for their use in energy related fields [156]. Branchi et al. [157] reported on the synthesis and characterization of sulfated Al$_2$O$_3$ and of composite membranes, prepared by dispersing the functionalized oxide in Nafion, acting as electrolytes in PEMFCs. Recently, the high electrochemical performance of fuel cells at low temperature or low humidity has become the target of rapid commercial research considering the operating temperature of fuel cells [151]. Based on this, Lee et al. [151] studied the composite membranes of SPAEK block copolymers blended with inorganic nanofiller (SiO$_2$ or f-SiO$_2$) with different particle sizes to improve the proton conductivity and fuel cell performance under low RH. Test results show that the electrochemical performance and proton conductivity of the composite membrane were improved under low RH (Fig. 12.27B). Yi et al. [158] prepared Nafion/SiO$_2$ composite membrane by in-situ sol-gel method. By controlling the size of SiO$_2$ nanoparticles, the effects of the size of SiO$_2$ nanoparticles on the membrane properties and cell performance were analyzed. In another study, Han [152] and collaborators fabricated a composite membrane by direct-mixing hydrophilic-SiO$_2$ particles with sulfonated polystyrene-block-poly(ethyl-ran-propylene)-block-polystyrene (S-poly-SEPS) in the blending solvent (Fig. 12.27C). This research showed that the composite membrane with hydrophilic SiO$_2$ has better ionic conductivity and better cell performance than Nafion. SiO$_2$ has played an important role in improving the performance of fuel cells by enhancing the membrane properties. In another work, Ying et al. [159] describes the interactions between SiO$_2$ and different types of polymer matrices in fuel cells and how they improve fuel cell performance. Taghizadeh et al. [160] synthesized ZrO$_2$ nanoparticles via ultrasonic-assisted method and the

Nafion/ZrO$_2$ nanocomposite membranes with different ZrO$_2$ contents were prepared by the solution casting procedure. Compared with pure Nafion membrane, the nanocomposite membrane exhibited more chemical stability.

12.3.2.2 Metal-organic frameworks

MOFs is a class of crystalline hybrid materials, composed of metal centers or clusters and organic connectors, with unique chemical versatility, combined with designable frames and unprecedented large and permanent internal pores [160]. Since the chemical structures of the metal centers and the linkers can be designed by synthetic tailoring to obtain the desired properties for a certain application (Fig. 12.28A), MOFs have recently attracted significant attention [161,164]. Ahmadian-Alam et al. prepared a novel ternary composite polymer electrolyte membrane using sulfonated PSU, MOF, and SiO$_2$ nanoparticles as carriers. The embedding of MOF/Si nanoparticles significantly improved the mechanical and thermal properties of the prepared membranes. According to the result of proton conductivity, proton transport of the prepared nanocomposite membrane is improved, and the power density is higher than the original PSU blend membrane [165].

As the carrier of proton-conducting materials, MOFs have attracted much attention because it has higher surface area than traditional fillers, which can allow encapsulation of proton transfer material [166,167]. In addition, from a material design perspective, the high porosity of MOFs can form pathways through which protons can be carried [168]. Erkartal [169] and coworkers demonstrated a novel ternary composite membrane consisting of poly(vinyl alcohol) (PVA), PAMPS, zeolitic imidazolate framework-8 (ZIF-8), which is prepared by physical blending and casting methods. In this study, it is clearly shown that the enhanced effects of ZIF-8 nanoparticles on proton conductivity in the membrane. Rao [78] and coworkers designed different composite PEMs filled with UiO-66-SO$_3$H and UiO-66-NH$_2$ as MOFs by single doping and codoping of two MOFs, respectively (Fig. 12.28B). It was found that codoping of these two MOFs with suitable sizes was more conducive to the proton conductivity enhancement of the composite PEM. Besides, the methanol trapping effect of MOF pores made the methanol permeability of the codoped PEM decrease greatly. Mukhopadhyay et al. [170] prepared an aryl ether-type polybenzimidazole (OPBI) based nanocomposite membranes of postsynthetic modification (PSM)-1 and PSM-2 MOFs with different weight percentages by solution casting blend method. This work clearly demonstrates the benefits of using rationally designed PSM-1 and PSM-2 MOFs as nanofiller to prepare the nanocomposite membranes, which can perform excellent proton conduction in a wide range of temperature. Shimizu and coworkers developed two MOFs with promising proton conductivity at high RH (Fig. 12.28C and D) [162]. As reported as Yang's group, a chemically stable and structurally flexible MOF with a 3D frame structure of a 1D channel was prepared, in which the high-density sulfonic acid sites are arranged on the channel

Nanomembranes in fuel cells 321

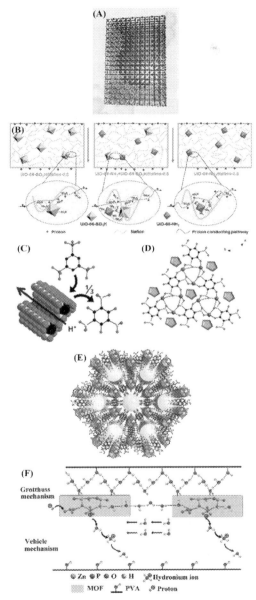

Figure 12.28 (A) A multivariate MOF (MTV-MOF) [161], (B) transport mechanism diagram of Nafion/UiO-66-NH$_2$, Nafion/UiO-66-SO$_3$H, and Nafion/UiO-66-NH$_2$ + UiO-66-SO$_3$H [78], (C) the space-filling cross-section structure of β-PCMOF2 with its one-dimensional proton conduction channel, (D) a two-dimensional crystal layout of β-PCMOF2 [162], (E) structure of But-8(M) (M = Cr, Al) and the ion exchange in But-8(Cr) [163], and (F) the possible mechanisms of the proton transport in JUC-200 and JUC-200@PVA composite membrane [80]. *MOF*, Metal-organic framework; *PCMOF*, proton conducting MOF; *PVA*, polyvinyl alcohol.

surface for proton conduction (Fig. 12.28E). This MOF achieved proton conductivity of 1.27×10^{-1} S cm^{-1} (at 80°C under 100% RH), and can also maintain moderately high proton conductivities under a wide range of RH and temperature [163]. It is found that the open framework structure of MOFs improved the performance of the composite membrane by loading particles with special functions into the matrix as proton carriers [77]. A hexaphosphate ester-based 3D MOF and its composite membrane with PVA were prepared (Fig. 12.28F). The composite membrane showed high stabilities against water and acid, and exhibited a proton conductivity of 1.62×10^{-3} S cm^{-1} at 80 in water [80].

12.3.2.3 Mineral materials

Clay/polymer nanocomposites are the typical examples of nanotechnology. The nanoparticles or layered materials were dispersed in the polymeric matrix, which changes the structure and improves the performance of the matrix. The interaction between polymer (organic system) and nanoclay (inorganic system) at molecular level is the main reason for the improvement of mechanical and electrical properties of composites [171,172]. The intercalated or exfoliated MMT nanoclay plays a significant role for ionic conduction in the electrolytes [173].

Smectite-type clays have a layered structure. The addition of MMT in the composite membrane can assist the proton transfer and also improve the IEC. PEM was prepared by solution casting with chitosan, PEO/modified MMT as the main materials. According to the test results, proton conductivities of the blend membranes are about 2.99×10^{-2} S cm^{-1} at room temperature (Fig. 12.29). And all the membranes are homogenous that could be revealed from the XRD studies [171]. PI PEMs doped with different MMT content were prepared by Abidin and team using solution casting technology [174]. The SEM images and XRD results showed that the MMT nanoclay particles are uniformly distributed in the membrane. At 70°C, the proton conductivity of a certain membrane was 7.1×10^{-2} S cm^{-1}. MMT is probably one of the most widely

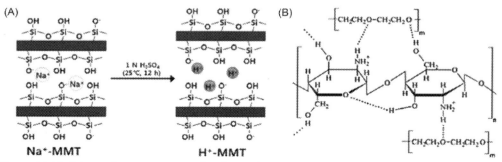

Figure 12.29 Schematic representation for the (A) functionalization of layered silicates and (B) chitosan/PEO blend [171]. *PEO*, Polyethylene oxide.

used natural clay minerals due to its unique combination of swelling, intercalation, and ion-exchange properties [138]. Ahmed et al. [175] used low sulfonate polyether ether ketone (SPEEK) mixed with different proportions of sodium rich Smectite clay as the electrolyte membrane of fuel cell. The composite membranes proved to have a higher thermal stability than SPEEK. The proton conductivity of the SPEEK/clay composite membranes increased compared to the SPEEK membrane, especially at low humidity. A sulfonated MMT/sulfonated poly(biphenyl ether sulfone)/polytetrafluoroethylene (SMMT/SPSU-BP/PTFE) composite membrane for fuel cells was prepared by Xing and collaborators [176]. The organic modification of MMT can improve its capability of dispersing, adsorption and nanocomposite in the organic systems. Alonso et al. [177] reported a series of Nafion-clay nanocomposite membranes, which were prepared by dispersion of clay nanoparticles into Nafion. The nanocomposite membrane maintained a high proton conductivity while significantly reducing the permeability of methanol. In addition, the membranes are much stiffer and can withstand higher temperatures than pure Nafion. On the side, an inorganic-organic composite membrane consisting of MMT and PBI, containing ether units (PBI-O), was obtained by Wang and collaborators [178]. Compared to pure PBI-O membranes, the conductivity of 5%-MMT-PBI-O composite membranes were improved nearly 400% without sacrificing other properties. Moreover, a series of composite membranes for fuel cells were prepared by combining the organic modified MMTs with SPEEK polymer. The thermal stability of the SPEEK membrane was improved by adding modified MMTs. Besides, the addition of DMDO-MMT increased the proton conductivity of the composite membrane [179]. Using kaolinite and sepiolite clay as additives, Aygün et al. prepared SPEEK polymer-based thermally cross-linked polymer membranes by sol-gel synthesis. The addition of sepiolite and kaolinite not only improved the thermal stability of the synthetic membranes, but also increased the proton conductivity from 0.172 to 0.268 S cm^{-1} when adding 9% kaolinite, and to 0.329 S cm^{-1} at 80 when adding 9% sepiolite [180].

12.3.3 Nanocarbon materials

Carbon nanomaterials are promising materials in many technologies, such as seawater desalination, fouling resistance, reverse osmosis and nanofiltration, water purification and flame retardancy. Nanocarbon materials are widely used in electrochemical, carbon nanomaterials can be used as additive of PEM fuel cell, such as CNFs and CNTs, graphene, such as material, due to their excellent mechanical strength and good methanol blocking rate, can enhance the proton conductivity of PEM and lower PEM methanol permeability. Among carbon nanomaterials, CNTs and go are the two most widely studied carbon nanomaterials. At present, carbon nanomaterials have been proved to be one of the most promising materials for improving the performance of PEM in fuel cells.

12.3.3.1 Nanofibers

CNFs are widely used in PEMFC, lithium ion batteries and other electrochemical energy storage devices due to their excellent mechanical properties, large specific surface area, high aspect ratio, good electrical conductivity and thermal conductivity, and difficulty in agglomeration. Different composite films can be prepared by hybridization of CNFs and various organic/inorganic materials. It is found that adding CNFs can improve the mechanical strength, proton conductivity, water absorption, methanol permeability, and thermal stability of composite films. Preparation of CNFs usually have chemical vapor deposition method, arc method, laser sputtering method, etc., but the conditions of these methods are more demanding, and product impurity is more, so far the most commonly used and most effective method is the electrostatic spinning technology, through the preparation of electrostatic spinning can be simple and quick for CNF (CCNF), and the carbon fiber is longer, less impurity, specific surface area.

In this research, Lee et al. [181] used coaxial electrostatic spinning technology to prepare a new core-shell SiO_2@CNF (SiO_2@C) composite membrane. Through the characterization of the material, found SiO_2@C core-shell structure of the specific surface area, electrical conductivity, mechanical strength, porosity and water absorption were better than that of pure CNFs, according to the battery performance test results show that in the 50°C—80°C temperature range and 15% RH, using SiO_2@C material of PEMFC power density was significantly higher than that of pure CNF porous membrane, and 66%—302% higher than traditional hydrophobic carbon black powder porous membrane. Zhang et al. [182] used a simple method on the basis of the spinning preparation for CNF and activated carbon nanofibers (ACNFs) load sulfonated poly real SPEEK membrane (Fig. 12.30), through a series of tests, the results show that the CNFs load film significantly increased along with the introduction of the film's mechanical properties, water absorption and proton conductivity. The composite membrane containing 0.47 wt.% CNF showed proton conductivity of 0.0311 S cm^{-1} at 20°C and 100% RH, which was higher than that of pure SPEEK membrane (0.0201 S cm^{-1}).

Di et al. [183] developed a simple method based on the technology of power spinning preparation of the polyacrylonitrile nanofibers, the carbide as precursor of continuous CNF, then will CNF incorporating SPEEK in the preparation of the nanometer composite membrane (Fig. 12.31), by characterization of the composite membrane, found that the composite membrane has low methanol transmittance and good proton conductivity. The composite membrane containing 0.51 wt.% continuous CNF showed a proton conductivity of 0.041 S cm^{-1}. The composite membrane containing 2.52 wt.% continuous CNF was 1.5 times more selective than the pure SPEEK membrane. SCNF/SPEEK composite membranes with different weight ratios were successfully prepared by solution casting in the study of Liu's group [32] (Fig. 12.32), and a series of characterization and tests were carried out. It was found that the proton conductivity and methanol permeability of the composite membrane were greatly

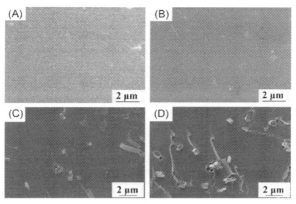

Figure 12.30 SEM micrographs of composite membranes, surfaces of (A) CNF/SPEEK and (B) ACNF/SPEEK; cross-sections of (C) CNF/SPEEK and (D) ACNF/SPEEK [182]. *ACNF*, Activated carbon nanofiber; *CNF*, carbon nanofiber; *SEM*, scanning electron microscope; *SPEEK*, sulfonated poly(ether ether ketone).

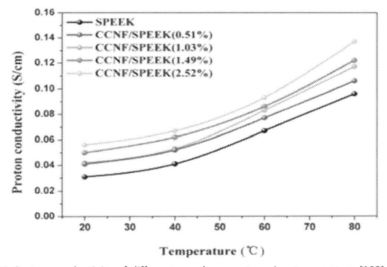

Figure 12.31 Proton conductivity of different membranes at varying temperature [183].

improved. In particular, the composite membrane with a content of 1 wt.% SCNF has the highest proton conductivity, with a proton conductivity of 0.128 S cm^{-1} at 60°C (Fig. 12.33), which is comparable to Nafion 117, and the methanol permeability of the composite membrane is much lower than that of Nafion 117, only for 5.12×10^{-7} cm^2 s^{-1}. This study shows that the composite membrane has good mechanical and electrochemical properties and is very suitable for PEMFC.

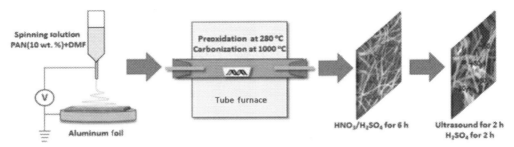

Figure 12.32 Procedure for preparation of SCNFs [32]. *SCNFs*, Sulfonated carbon nanofibers.

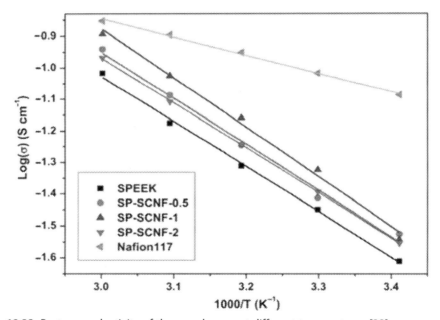

Figure 12.33 Proton conductivity of the membranes at different temperatures [32].

12.3.3.2 Nanotubes

CNT, a tubular structure with a diameter in the nanometer and a length in the micron range, can be used as an additive material for PEM in fuel cells due to their unique structure and physical properties as well as good electrical and thermal conductivity. CNT can be divided into SWCNTs and MWCNTs according to their different forms of existence. SWCNTs are more ordered and have better flexibility and electrical conductivity than MWCNTs. On the other hand, had lower electronic conductivity and higher surface defects [184]. Because of this property, MWCNTs are more widely used in PEM than SWCNTs. The tensile strength of CNTs is about 50 times that of steel. Therefore the mechanical strength of PEM can be improved by adding CNTs.

In the study of Altaf [185], novel sulfonated polysulfone (SPS)-based composite polymer PEM filled with polydopamine-modified carbon nanotubes (PD-CNTs) have been prepared using the phase inversion technique (Fig. 12.34). The composite membrane was characterized by various test methods, it was found that its water absorption and proton conductivity were significantly improved. Compared with the original SPS membrane, the proton conductivity of PD-CNTs/SPS composite membrane increased by 43%, and the proton conductivity of the composite membrane at 80°C was 0.1216 S cm^{-1}. Compared with recast Nafion membrane (23.00 × 10^{-7} S cm^{-1}), methanol permeability of composite membrane (5.68 × 10^{-7} S cm^{-1}) was significantly reduced. Therefore the composite membrane as PEM has great potential in the application of DMFC. Hasani-sadrabadi et al. [186] modified CNTs with chitosan and then dissolved them in Nafion membrane matrix, which significantly improved membrane selectivity and proton conductivity, and significantly reduced the methanol transmittance, thus increasing the fuel cell efficiency to about 16%, while the fuel cell efficiency of commercial Nafion 117 membrane was only 11%. Moreover, the generating capacity of the fuel cell using the composite membrane reached 110 mW cm^{-2}, more than twice that of the commercial Nafion 117 membrane fuel cell (47 mW cm^{-2}).

Molla-Abbasi et al. [187] use of Nafion, four ethoxy silane modified CNTs and phosphotungstic acid modified CNTs synthesized a new type of composite membrane

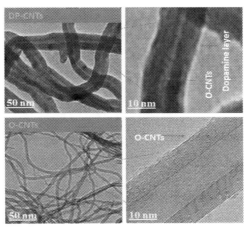

Figure 12.34 TEM images of polydopamine-modified CNTs and oxidized-CNTs [185]. *CNTs*, Carbon nanotubes; *TEM*, transmission electron microscope.

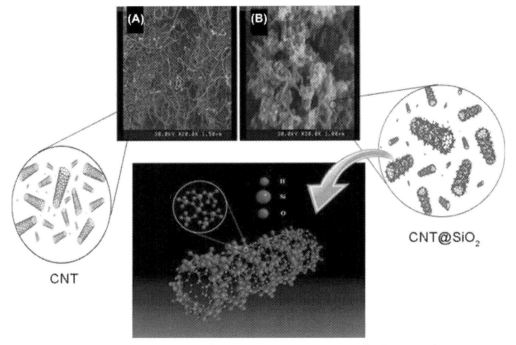

Figure 12.35 Field emission scanning electron microscope (FESEM) of (A) CNT and (B) CNT@SiO$_2$ [187]. *CNT*, Carbon nanotube.

(Fig. 12.35). And by test shows that the surface modification of CNTs and phosphotungstic acid group compound can improve the performance of DMFCs and the methanol permeability of the composite membrane is 2.63×10^{-7} cm^2 s^{-1}, compared with recast Nafion membrane (2.25×10^{-6} cm^2 s^{-1}), the selectivity of the prepared nanocomposite membrane (330,700 S s cm^{-3}) was significantly higher than that of recast Nafion membrane (38,222 S s cm^{-3}). Kim et al. [188] successfully prepared the amine-functionalized carbon nanotube (ANCT)/SPEEK composite membrane by embedding ANCT into the SPEEK matrix (Fig. 12.36). It was found that the mechanical strength and thermal stability of the composite membrane were greatly improved compared with the original SPEEK membrane. When the load current density is 101.2 mA cm^{-2}, the peak power density of the composite membrane doped by 1.5 wt.% ACNF is 51.1 mW cm^{-2} at 60°C in 20% RH.

Yin et al. [189] successfully prepare multilayer single-walled sulfonated CNTs (Su-CNTs)/Nafion composite membranes by layer assembly. Through research, they have found that proton conductivity and mechanical strength of composite membranes are enhanced. In particular, the proton conductivity of the 80-layer single-walled

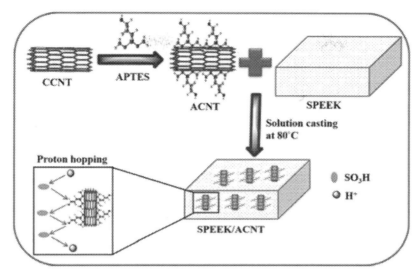

Figure 12.36 Preparation process of SPEEK/ACNT composite membrane and its proton hopping [188]. *ACNT*, Activated carbon nanotube; *SPEEK*, sulfonated poly(ether ether ketone).

Su-CNTs/Nafion composite membrane is 0.33 S cm^{-1} at 152°C. Moreover, it also shows good electrochemical stability and is a PEM with great potential. Gahlot et al. [190] were prepared using the solution casting method, such as electric orientation functionalized CNT and SPEEK hybrid membrane (Fig. 12.37), the functionalization of CNT is done through sulfonation and carboxyl, the preparation of composite membrane ion selective, electrical conductivity and methanol permeability resistance, thermal stability and mechanical strength are strengthened (Fig. 12.38), and the composite film under high temperature and good proton conductivity, these properties make it can be high temperature range of the PEM fuel cell materials.

Li et al. [191] fixed phosphotungstic acid molecules on CNT through the electrostatic self-assembly of polydiallyl dimethyl ammonium chloride (PDDA), and then added these particles into PVA to prepare methanol blocking films for DMFCs. Under the condition of 60°C, the proton conductivity of the prepared membrane was 9.4 S cm^{-1}, much higher than that of the PVA membrane. Moreover, it showed better proton conductivity stability than PVA membrane in 120-hour test. In addition, the methanol permeability of the composite membrane was about 40% lower than that of the PVA membrane (6.1 × 10^{-7} cm^2 S^{-1}), and the selectivity of protons to methanol was also improved due to the enhanced stability after the addition of CNTs. The maximum power density of a single DMFC based on composite film reached 16 mW cm^{-2} at 60°C.

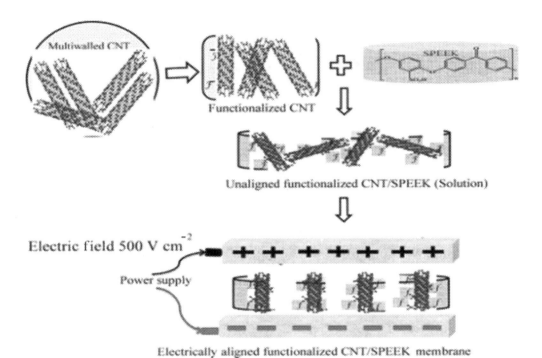

Figure 12.37 Schematic representation of electric field aligned CNT/SPEEK composite membrane [190]. *CNTs*, Carbon nanotubes; *SPEEK*, sulfonated poly(ether ether ketone).

Figure 12.38 TEM of (A) randomly distributed, (B and C) electrically aligned SPEEK/f CNT (S-sCNT-5) composite membranes (arrows show the direction of CNT alignment) [190]. *CNTs*, Carbon nanotubes; *SPEEK*, sulfonated poly(ether ether ketone); *TEM*, transmission electron microscope.

12.3.3.3 Graphite

In addition to CNTs, one of the most widely used carbon nanomaterials in PEM is GO, which can be prepared by mechanical dissection from graphite and subsequent oxidation treatment. Because GO contains hydrophilic carboxylic acid groups, it can be dispersed in a variety of solvents [192–194]. GO has good insulation performance, large specific surface area and proton conductivity. The single GO is not satisfactory in some performance, so GO needs to be functionalized to improve its performance. The compatibility between GO and the main polymer matrix improves the mechanical properties of the composite film. In addition, GO has excellent methanol barrier property, so it can be used as a nanofiller for DMFC.

Qu et al. [195] doped GO grafted with (3-amino-propyl) triethoxy silane into the SPEEK matrix by solution casting method to prepare the composite membrane, and characterized the composite membrane. It was found that the GO treated with silane coupling agent improved its stability and compatibility in SPEEK and reduced the agglomeration phenomenon. Both the hydroscopicity and proton conductivity of the composite membrane were improved. At 120°C, the proton conductivity of the SPEEK membrane containing 2 wt.% amine-functionalized GO (AGO) reached 11.32 mS cm^{-1}, which was 2.45 times that of the SPEEK membrane. Cai et al. [196] use ^{60}Co γ-ray radiation Go to prepare the new functional GO, and then use the solvent casting out of polybenzimidazole/radiation grafting graphene oxide (PBI company/RGO) composite membrane, and phosphoric acid (PA) for doping to improve the proton conductivity of composite membrane. With the increase of RGO content, the tensile strength (27.3—38.5 MPa) of the PBI/RGO/PA composite membrane first increased and then decreased, which was significantly higher than other PA-doped PBI base membranes. Under the condition of no humidity and 170°C, the proton conductivity of the PBI/RGO/PA composite membrane reached 28.0 mS cm^{-1}, which was 72% higher than that of the PBI/PA membrane.

Through hydrogen bond elimination reaction, Cao et al. [197] achieved sulfonated graphene (SG) and SPEEK cross-linking through hydrogen bond elimination reaction, and SG/SPEEK composite membranes were prepared (Fig. 12.39). With the increase of doping amount, the composite membranes showed good proton conductivity. The proton conductivity of the composite membrane loaded with 2 wt.% SG at 54°C was 0.063 S cm^{-1}, which was 1.54 times that of the SPEEK membrane (0.041 S cm^{-1}). The methanol permeability of the composite membrane loaded with 2 wt.% SG was 1.834×10^{-9} cm^2 S^{-1}, which was also lower than that of the SPEEK membrane (4.537×10^{-9} cm^2 S^{-1}). SG/SPEEK composite membranes also have better water absorption and thermal stability than GO/SPEEK and SPEEK membranes. In this study, Yusoff et al. [198] successfully prepared the SGO/PBI composite membranes with different weight ratios (Fig. 12.40), and found that the proton conductivity of the composite membranes was higher than that of the pure PBI membranes. The composite membrane containing 2 wt.% SGO displayed the highest proton conductivity, with a value of 9.14 mS cm^{-1} at 25°C, and it increased to 29.30 mS cm^{-1} at 150°C. In the PEMFC test, compared with a pristine PBI membrane, the maximum power density was increased by 40% with the use of the composite membrane with 2 wt.% SGO.

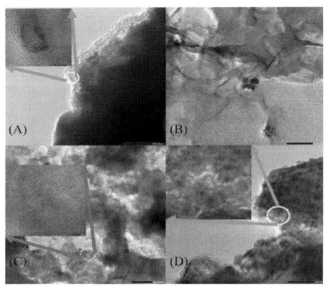

Figure 12.39 TEM of (A) GO, (B) SG, (C) G-1, and (D) S-1 [197]. *GO*, Graphene oxide; *SG*, Sulfonated graphene; *TEM*, transmission electron microscope.

Lee et al. [199] used GO and sulfonated poly(arylene thioether sulfone)-grafted graphene oxide (SATS-GO) as raw materials to prepare sulfonated poly(arylene ether sulfone) (SPAES) composite membrane and used it in PEMFC (Fig. 12.41). The mechanical strength, chemical stability, and proton conductivity of the composite membrane are greatly improved compared with those before. The composite membrane containing 2 wt.% SATS-GO has the best membrane performance. Kausar et al. [200] through the polycondensation reaction prepare of ethylenediamine-tetraacetic dianhydride, 4-aminophenyl sulfone, and poly(ethylene glycol) bis(amine) to prepare poly(imide-ethylene glycol) (PIEG), then doping GO to prepare undoped and acid-doped composite films with good thermal and mechanical stability. When the content of the GO from 1 to 5 wt.%, found not-doping PIEG/GO the tensile strength of the composite membrane increased from 59.7 to 65.9 MPa, thermal performance has been significantly enhanced, and the acid-doped composite film under the RH of 94%, with the increase of GO content, ionic conductivity (2.4–2.9 mmol g^{-1}) and proton conductivity (1.8–2.7 S cm^{-1}) also improved. In this research, sulfonated poly (arylene ether sulfone) (SPAES) composite membranes having GO and sulfonated polytriazole–grafted graphene oxide (SPTA-GO) were prepared by Han's group [201] (Fig. 12.42). All the SPAES-based composite membranes with GO and SPTA-GO revealed better physical/chemical stabilities and mechanical strength than the pristine SPAES membrane. The maximum power density of the MEA employing

Figure 12.40 SEM images for (A) commercial graphite, (B) GO, and (C) SGO at a magnification of 5000; and TEM images for (D) graphite, (E) GO, and (F) SGO at a magnification of 17,000 [198]. *GO*, Graphene oxide; *SEM*, scanning electron microscope; *SGO*, sulfonated graphene oxide; *TEM*, transmission electron microscope.

the SPAES composite membrane with SPTA-GO reached up to 1.58 W cm^{-2} at 80°C and 100% RH condition, which is larger than that employing commercial Nafion 212 (1.33 W cm^{-2}).

12.3.3.4 Fullerenes

Carbon nanomaterials with CNT and GO and so on due to the low solubility of fullerenes, poor compatibility, and easy to the gathered themselves together, and

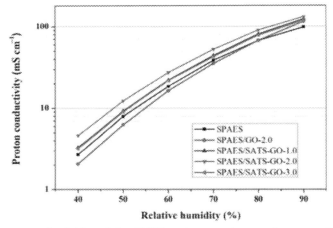

Figure 12.41 Proton conductivities of the SPAES and composite membranes at 80°C as a function of relative humidity [199]. *SPAES*, Sulfonated poly(aryl ether sulfone).

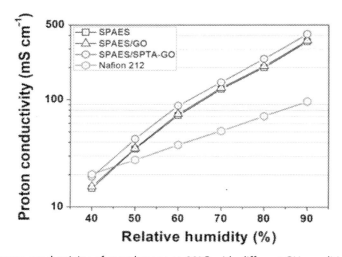

Figure 12.42 Proton conductivity of membranes at 80°C with different RH conditions (40%—90%) [201]. *RH*, relative humidity.

therefore limits the fullerene materials in the application of PEM, but fullerenes have many unique properties, such as high electron affinity, high volume density of functional groups, and radical scavenging properties, high mechanical properties of fullerenes radical scavenging performance can improve the oxidation stability of the composite membrane [202]. Therefore fullerenes can be functionalized to improve

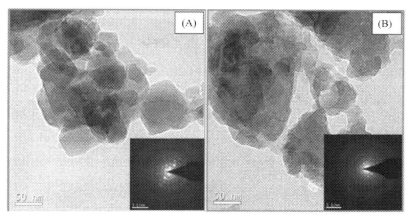

Figure 12.43 TEM morphology of (A) pristine fullerene and (B) FF [205]. *TEM*, Transmission electron microscope.

its performance. Compared with the nonfunctionalized fullerenes nanocomposite membranes, the functionalized fullerenes can significantly improve the proton conductivity and tensile strength of the membranes, and they can be well dispersed in the membranes, so they can be well applied in PEM.

Neelakandan et al. [203] added sulfonated fullerenes to the highly branched sulfonated polyether ketone sulfone membrane, which improved the methanol barrier of the composite membrane and did not reduce the proton conductivity. The maximum proton conductivity of the sulfonated graphene membranes with appropriate load was 0.332 S cm^{-1} at 80°C, and the maximum power density of the composite membranes at 60°C was 74.38 mW cm^{-2}, which was about 30% higher than the Nafion 212 membrane (51.78 mW cm^{-2}) under the same conditions. Rikame [204] prepared by solution casting method such as the phosphorylated fullerene/sulfonated polyvinyl alcohol (PFSP) composite membrane, and found that the composite film has excellent performance in fuel cells, the proton conductivity of the composite membrane preparation of 11.7 × 10^{-2} S cm^{-1}, 120% water imbibition, IEC of 1.67 meq g^{-1}, the obtained power density is 499.1 mW m^{-2}, and the membrane is expected to be used in fuel cells.

Rambabu et al. [205] functionalized fullerenes by using the 4-benzenesulfonic acid precursor generated by the diazoylation reaction of aminobenzonic acid, and doped the functionalized fullerenes (FF) into Nafion ionomers by solution casting method to form a composite membrane (Fig. 12.43), and characterized the physical properties of the fullerenes. In composite membrane because of sulfonic acid group and show good proton conductivity, compared with the recasting of Nafion membrane, composite

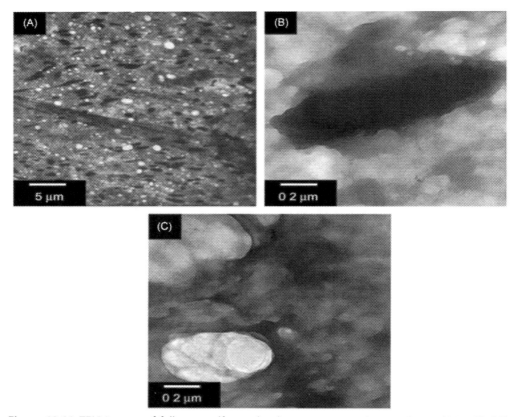

Figure 12.44 TEM images of fullerene-sulfonated polystyrene composite membrane (1.4% Flu-PS): (A) bulk of prepared membrane; (B) fullerene region of (A) with high magnification; (C) micropore region of (A) with high magnification [206]. *TEM*, Transmission electron microscope.

membrane has higher selectivity of electrochemical, so can increase the output power of the fuel cell, load 1 wt.% functionalization of graphene composite membrane in 2M methanol solution can be 146 mW cm^{-2} peak power density, higher than the Nafion 117. Saga et al. [206] with sulfonated polystyrene and fullerene as raw materials, using the solution casting method out of the fullerene-sulfonated polystyrene composite film (Fig. 12.44), composite film could improve the antioxidant capacity, also reduces the methanol permeability, with 1.4 wt.% of the composite membrane under the current density of 200 mA cm^{-2} in the fuel cell power density reached 47 mW cm^{-2}, through the study found that by raising the fullerene at the nanoscale dispersion and reduce the membrane body of microporous composite membrane would generate higher mechanical strength. Table 12.2 summarizes the values of proton conductivity for various nanofillers in PEM.

Table 12.2 Values of proton conductivity for various nanofillers in PEM.

Basic material	Proton conductivity (S cm^{-1})	References
SPAES-TiO$_2$/g-C$_3$N$_4$	0.325 (80°C)	[148]
PVDF-HFP-TiO$_2$	1.66 × 10^{-3} (25°C)	[149]
S-TiO$_2$/PPSSA	0.067 (30°C)	[150]
SPES/STiO$_2$-PANI	2.30 × 10^{-4}	[153]
PES/SPEEK/Fe$_3$O$_4$, PES/SPEEK/TiO$_2$, and PES/SPEEK/MoO$_3$	3.57, 4.57, 2.67, × 10^{-4}	[154]
Nafion/Fe$_3$O$_4$-SGO	0.012 (120°C)	[155]
SPAEK/SiO$_2$ or f-SiO$_2$	0.159–0.199 (90°C, 100% RH)	[151]
SiO$_2$/S-poly-SEPS	0.150 (90°C)	[152]
Nafion/ZrO$_2$	0.016 (25°C)	[160]
Sulfonated PSU/MOF/SiO$_2$	0.017 (70°C)	[165]
PVA:PAMPS:ZIF-8	0.134 (80°C)	[169]
UiO-66-SO$_3$H and UiO-66-NH$_2$	0.256 (90°C, 95% RH)	[78]
PSM-1-OPBI, PSM-2-OPBI	0.29, 0.308 (160°C)	[170]
BUT-8(Cr)A	0.127 (80°C)	[163]
Hexaphosphate ester@PVA	1.62 × 10^{-3} (80°C)	[80]
Chitosan/PEO/modified MMT	2.99 × 10^{-2} (25°C)	[171]
PI/MMT	0.071 (70°C)	[174]
SPEEK/clay	0.020 (140°C)	[175]
SMMT/SPSU-BP/PTFE	0.073 (25°C)	[176]
Kaolinite and sepiolite clay/SPEEK	0.329 (80°C)	[180]
Activated CNF-supported SPEEK	0.031 (20°C)	[182]
Continuous CNF/SPEEK	0.041 (25°C)	[183]
Sulfonated CNFs/SPEEK	0.128 (60°C)	[32]
Polydopamine-modified CNTs/SPS	0.122 (80°C)	[185]
Nafion/chitosan-wrapped CNT	0.104 (25°C)	[186]
Nafion/PWA/CNTs	0.087 (25°C)	[187]
Amine-functionalized CNT/SPEEK	0.153 (90°C)	[188]
Sulfonated CNTs/Nafion	0.330 (152°C)	[189]
APTES/SPEEK/GO	0.011 (120°C)	[195]
PBI/RGO/PA	0.028 (170°C)	[196]
Sulfonated graphene/SPEEK	0.063 (54°C)	[197]
PBI-sulfonated GO	0.029 (150°C)	[198]
SPAES/SATS-GO	0.131 (80°C)	[199]
SPAES/SPTA-GO	0.413 (80°C)	[201]
BSPAEKS/s-Fu	0.332 (80°C)	[203]
Phosphorylated fullerene/PFSP	0.117 (50°C)	[204]
Nafion-FF	0.097 (60°C)	[205]

APTES, 3-Amino-propyl triethoxy silane; *CNF*, carbon nanofiber; *CNTs*, carbon nanotubes; *GO*, graphene oxide; *MMT*, montmorillonite; *MOF*, metal-organic framework; *PA*, phosphoric acid; *PAMPS*, poly(2-acrylamido-2-methylpropane sulfonic acid); *PANI*, polyaniline; *PBI*, polybenzimidazole; *PEM*, proton-exchange membrane; *PEO*, poly(ethylene oxide); *PES*, polyethersulfone; *PFSP*, sulfonated polyvinyl alcohol; *PI*, polyimide; *PSM*, postsynthetic modification; *PSU*, poly(biphenyl ether sulfone); *PTFE*, polytetrafluoroethylene; *PVA*, poly(vinyl alcohol); *PVDF*, polyvinylidene fluoride; *PWA*, phosphotungstic acid; *RGO*, radiation grafting graphene oxide; *SATS*, sulfonated poly(arylene thioether sulfone); *SGO*, sulfonated graphene oxide; *SMMT*, sulfonated montmorillonite; *SPAEK*, sulfonated poly(aryl ether ketones); *SPAES*, sulfonated poly(aryl ether sulfone); *SPEEK*, sulfonated poly(ether ether ketone); *SPES*, sulfonated poly(ether sulfones); *S-poly-SEPS*, sulfonated polystyrene-block-poly(ethyl-ran-propylene)-block-polystyrene; *SPS*, sulfonated polysulfone; *SPSU-BP*, sulfonated poly(biphenyl ether sulfone); *SPTA*, sulfonated polytriazole; *ZIF-8*, zeolitic imidazolate framework-8; *BSPAEKS*, branched sulfonated poly(arylene ether ketone sulfone); *s-Fu*, sulfonated fullerenes.

References

[1] P. Ren, et al., Degradation mechanisms of proton exchange membrane fuel cell under typical automotive operating conditions, Progress in Energy and Combustion Science 80 (2020) 100859.
[2] Y. Wang, et al., Materials, technological status, and fundamentals of PEM fuel cells—a review, Materials Today 32 (2020) 178—203.
[3] H. Chen, et al., The reactant starvation of the proton exchange membrane fuel cells for vehicular applications: a review, Energy Conversion and Management 182 (2019) 282—298.
[4] L. Du, et al., Low-PGM and PGM-free catalysts for proton exchange membrane fuel cells: stability challenges and material solutions, Advanced Materials 33 (6) (2020) e1908232.
[5] X.-X. Xie, et al., Proton conductive carboxylate-based metal—organic frameworks, Coordination Chemistry Reviews 403 (2020) 213100.
[6] M.M. Barzegari, et al., Performance prediction and analysis of a dead-end PEMFC stack using data-driven dynamic model, Energy 188 (2019) 116049.
[7] J. Escorihuela, et al., Proton conductivity of composite polyelectrolyte membranes with metal-organic frameworks for fuel cell applications, Advanced Materials Interfaces 6 (2) (2019) 1801146.
[8] W.R.W. Daud, et al., PEM fuel cell system control: a review, Renewable Energy 113 (2017) 620—638.
[9] M. Akel, et al., Nano hexagonal boron nitride-Nafion composite membranes for proton exchange membrane fuel cells, Polymer Composites 37 (2) (2016) 422—428.
[10] F.C. Teixeira, et al., Nafion phosphonic acid composite membranes for proton exchange membranes fuel cells, Applied Surface Science 487 (2019) 889—897.
[11] F. Salimi Nanadegani, E. Nemati Lay, B. Sunden, Computational analysis of the impact of a micro porous layer (MPL) on the characteristics of a high temperature PEMFC, Electrochimica Acta 333 (2020) 135552.
[12] S. Chugh, et al., Experimental and modelling studies of low temperature PEMFC performance, International Journal of Hydrogen Energy 45 (15) (2020) 8866—8874.
[13] P. Liu, et al., Experimental investigation on the voltage uniformity for a PEMFC stack with different dynamic loading strategies, International Journal of Hydrogen Energy 45 (50) (2020) 26490—26500.
[14] E.M. Izurieta, et al., Process intensification through the use of multifunctional reactors for PEMFC grade hydrogen production: process design and simulation, Chemical Engineering and Processing—Process Intensification 147 (2020) 107711.
[15] E. Ogungbemi, et al., Review of operating condition, design parameters and material properties for proton exchange membrane fuel cells, International Journal of Energy Research 45 (2) (2021) 1227—1245.
[16] P. Rodríguez-Garnica, et al., Silica based hybrid organic-inorganic materials for PEMFC application, International Journal of Hydrogen Energy 45 (33) (2020) 16698—16707.
[17] A.C. Bhosale, P.C. Ghosh, L. Assaud, Preparation methods of membrane electrode assemblies for proton exchange membrane fuel cells and unitized regenerative fuel cells: a review, Renewable and Sustainable Energy Reviews 133 (2020) 110286.
[18] S.R. Raja Rafidah, et al., Recent progress in the development of aromatic polymer-based proton exchange membranes for fuel cell applications, Polymers (Basel) 12 (5) (2020) 1061.
[19] Y. Hu, et al., Improving the overall characteristics of proton exchange membranes via nanophase separation technologies: a progress review, Fuel Cells 17 (1) (2017) 3—17.
[20] X. Li, et al., Molecular dynamics simulation study of a polynorbornene-based polymer: a prediction of proton exchange membrane design and performance, International Journal of Hydrogen Energy 41 (36) (2016) 16254—16263.
[21] S.J. Peighambardoust, S. Rowshanzamir, M. Amjadi, Review of the proton exchange membranes for fuel cell applications, International Journal of Hydrogen Energy 35 (17) (2010) 9349—9384.
[22] F.A. Zakil, S.K. Kamarudin, S. Basri, Modified Nafion membranes for direct alcohol fuel cells: an overview, Renewable and Sustainable Energy Reviews 65 (2016) 841—852.

[23] D. Dhanapal, et al., A review on sulfonated polymer composite/organic-inorganic hybrid membranes to address methanol barrier issue for methanol fuel cells, Nanomaterials (Basel) 9 (5) (2019) 668.
[24] C.Y. Wong, et al., Additives in proton exchange membranes for low- and high-temperature fuel cell applications: a review, International Journal of Hydrogen Energy 44 (12) (2019) 6116−6135.
[25] H. Ahmad, et al., Overview of hybrid membranes for direct-methanol fuel-cell applications, International Journal of Hydrogen Energy 35 (5) (2010) 2160−2175.
[26] S.J. Paddison, Proton conduction mechanisms at low degrees of hydration in sulfonic acid−based polymer electrolyte membranes, Annual Review of Materials Research 33 (1) (2003) 289−319.
[27] G. Xu, et al., Performance dependence of swelling-filling treated Nafion membrane on nanostructure of macromolecular filler, Journal of Membrane Science 534 (2017) 68−72.
[28] Y.-L. Liu, Developments of highly proton-conductive sulfonated polymers for proton exchange membrane fuel cells, Polymer Chemistry 3 (6) (2012) 1373−1383.
[29] H. Ito, et al., Properties of Nafion membranes under PEM water electrolysis conditions, International Journal of Hydrogen Energy 36 (17) (2011) 10527−10540.
[30] Y. Zhang, et al., Recent developments on alternative proton exchange membranes: strategies for systematic performance improvement, Energy Technology 3 (7) (2015) 675−691.
[31] A. Ortiz-Negrón, D. Suleiman, The effect of TiO_2 nanoparticles on the properties of sulfonated block copolymers, Journal of Applied Polymer Science 132 (41) (2015).
[32] X. Liu, et al., Electrospun multifunctional sulfonated carbon nanofibers for design and fabrication of SPEEK composite proton exchange membranes for direct methanol fuel cell application, International Journal of Hydrogen Energy 42 (15) (2017) 10275−10284.
[33] M. Kotal, A.K. Bhowmick, Polymer nanocomposites from modified clays: recent advances and challenges, Progress in Polymer Science 51 (2015) 127−187.
[34] Z. Bai, et al., Enhanced proton conduction of imidazole localized in one-dimensional Ni-metal-organic framework nanofibers, Nanotechnology 31 (12) (2020) 125702.
[35] J. Kalaiselvimary, M. Ramesh Prabhu, Fabrications and investigation of physicochemical and electrochemical properties of heteropoly acid-doped sulfonated chitosan-based polymer electrolyte membranes for fuel cell applications, Polymer Bulletin 76 (3) (2018) 1401−1422.
[36] R.P. Pandey, et al., Graphene oxide based nanohybrid proton exchange membranes for fuel cell applications: an overview, Advances in Colloid and Interface Science 240 (2017) 15−30.
[37] M. Tohidian, et al., Sulfonated aromatic polymers and organically modified montmorillonite nanocomposite membranes for fuel cells applications, Journal of Macromolecular Science, Part B 52 (11) (2013) 1578−1590.
[38] K.Y. Mudi, et al., Development of carbon nanofibers/Pt nanocomposites for fuel cell application, Arabian Journal for Science and Engineering 45 (9) (2020) 7329−7346.
[39] H. Zarrin, et al., Functionalized graphene oxide nanocomposite membrane for low humidity and high temperature proton exchange membrane fuel cells, The Journal of Physical Chemistry C 115 (42) (2011) 20774−20781.
[40] I. Petreanu, et al., Corrigendum to "Synthesis and testing of a composite membrane based on sulfonated polyphenylene oxide and silica compounds as proton exchange membrane for PEM fuel cells" Mater. Res. Bull., 96, (December (Part 3)), (2017), 136−142, Materials Research Bulletin 108 (2018) 281.
[41] S. Feng, J. Savage, G.A. Voth, Effects of polymer morphology on proton solvation and transport in proton-exchange membranes, The Journal of Physical Chemistry C 116 (36) (2012) 19104−19116.
[42] E. Spohr, P. Commer, A.A. Kornyshev, Enhancing proton mobility in polymer electrolyte membranes: lessons from molecular dynamics simulations, The Journal of Physical Chemistry B 106 (41) (2002) 10560−10569.
[43] M.N. Tsampas, S. Brosda, C.G. Vayenas, Electrochemical impedance spectroscopy of, fully hydrated Nafion membranes at high and low hydrogen partial pressures, Electrochimica Acta 56 (28) (2011) 10582−10592.
[44] J. Savage, Y.-L.S. Tse, G.A. Voth, Proton transport mechanism of perfluorosulfonic acid membranes, The Journal of Physical Chemistry C 118 (31) (2014) 17436−17445.

[45] M.K. Petersen, G.A. Voth, Characterization of the solvation and transport of the hydrated proton in the perfluorosulfonic acid membrane Nafion, The Journal of Physical Chemistry B 110 (37) (2006) 18594–18600.
[46] N. Agmon, The Grotthuss mechanism, Chemical Physics Letters 244 (5) (1995) 456–462.
[47] S. Feng, G.A. Voth, Proton solvation and transport in hydrated Nafion, The Journal of Physical Chemistry B 115 (19) (2011) 5903–5912.
[48] K.D. Kreuer, On the development of proton conducting polymer membranes for hydrogen and methanol fuel cells, Journal of Membrane Science 185 (1) (2001) 29–39.
[49] W.Y. Hsu, T.D. Gierke, Elastic theory for ionic clustering in perfluorinated ionomers, Macromolecules 15 (1) (1982) 101–105.
[50] J. Rozière, D.J. Jones, Non-fluorinated polymer materials for proton exchange membrane fuel cells, Annual Review of Materials Research 33 (1) (2003) 503–555.
[51] K.A. Mauritz, R.B. Moore, State of understanding of Nafion, Chemical Reviews 104 (10) (2004) 4535–4586.
[52] A. Kusoglu, et al., Micromechanics model based on the nanostructure of PFSA membranes, Journal of Polymer Science Part B: Polymer Physics 46 (22) (2008) 2404–2417.
[53] W.Y. Hsu, T.D. Gierke, Ion transport and clustering in Nafion perfluorinated membranes, Journal of Membrane Science 13 (3) (1983) 307–326.
[54] K. Schmidt-Rohr, Q. Chen, Parallel cylindrical water nanochannels in Nafion fuel-cell membranes, Nature Mater 7 (1) (2008) 75–83.
[55] X. Kong, K. Schmidt-Rohr, Water–polymer interfacial area in Nafion: comparison with structural models, Polymer 52 (9) (2011) 1971–1974.
[56] L. Rubatat, et al., Evidence of elongated polymeric aggregates in Nafion, Macromolecules 35 (10) (2002) 4050–4055.
[57] K.-D. Kreuer, G. Portale, A critical revision of the nano-morphology of proton conducting ionomers and polyelectrolytes for fuel cell applications, Advanced Functional Materials 23 (43) (2013) 5390–5397.
[58] H.G. Haubold, et al., Nano structure of NAFION: a SAXS study, Electrochimica Acta 46 (10) (2001) 1559–1563.
[59] L. Rubatat, G. Gebel, O. Diat, Fibrillar structure of Nafion: matching Fourier and real space studies of corresponding films and solutions, Macromolecules 37 (20) (2004) 7772–7783.
[60] S. Liu, J. Savage, G.A. Voth, Mesoscale study of proton transport in proton exchange membranes: role of morphology, The Journal of Physical Chemistry C 119 (4) (2015) 1753–1762.
[61] N. Asano, et al., Aliphatic/aromatic polyimide ionomers as a proton conductive membrane for fuel cell applications, Journal of the American Chemical Society 128 (5) (2006) 1762–1769.
[62] C. Wang, et al., Fluorene-based poly(arylene ether sulfone)s containing clustered flexible pendant sulfonic acids as proton exchange membranes, Macromolecules 44 (18) (2011) 7296–7306.
[63] C. Wang, et al., Poly(arylene ether sulfone) proton exchange membranes with flexible acid side chains, Journal of Membrane Science 405–406 (2012) 68–78.
[64] Y. Zhang, et al., Fabrication of a polymer electrolyte membrane with uneven side chains for enhancing proton conductivity, RSC Advances 6 (83) (2016) 79593–79601.
[65] Y.A. Elabd, M.A. Hickner, Block copolymers for fuel cells, Macromolecules 44 (1) (2011) 1–11.
[66] K. Matsumoto, T. Higashihara, M. Ueda, Locally and densely sulfonated poly(ether sulfone)s as proton exchange membrane, Macromolecules 42 (4) (2009) 1161–1166.
[67] S.Y. Lee, et al., Morphological transformation during cross-linking of a highly sulfonated poly(phenylene sulfide nitrile) random copolymer, Energy & Environmental Science 5 (12) (2012) 9795–9802.
[68] C. Li, et al., Fabrication of sulfonated poly (ether ether ketone)/sulfonated fully aromatic polyamide composite membranes for direct methanol fuel cells (DMFCs), Energy Technology 7 (1) (2019) 71–79.
[69] L. Fu, G. Xiao, D. Yan, High performance sulfonated poly(arylene ether phosphine oxide) membranes by self-protected cross-linking for fuel cells, Journal of Materials Chemistry 22 (27) (2012) 13714–13722.

[70] X. Liu, et al., Semi-interpenetrating polymer network membranes from SPEEK and BPPO for high concentration DMFC, ACS Applied Energy Materials 1 (10) (2018) 5463–5473.

[71] X. Liu, et al., Semi-interpenetrating polymer networks toward sulfonated poly(ether ether ketone) membranes for high concentration direct methanol fuel cell, Chinese Chemical Letters 30 (2) (2019) 299–304.

[72] N. Cele, S.S. Ray, Recent progress on Nafion-based nanocomposite membranes for fuel cell applications, Macromolecular Materials and Engineering 294 (11) (2009) 719–738.

[73] G. Alberti, et al., Preparation of nano-structured polymeric proton conducting membranes for use in fuel cells, Annals of the New York Academy of Sciences 984 (1) (2003) 208–225.

[74] X. Zhang, et al., Preparation of sulfonated polysulfone/sulfonated titanium dioxide hybrid membranes for DMFC applications, Journal of Applied Polymer Science 137 (32) (2020) 48938.

[75] C. Gong, et al., A new strategy for designing high-performance sulfonated poly(ether ether ketone) polymer electrolyte membranes using inorganic proton conductor-functionalized carbon nanotubes, Journal of Power Sources 325 (2016) 453–464.

[76] Z. Jiang, et al., Composite membranes based on sulfonated poly(ether ether ketone) and SDBS-adsorbed graphene oxide for direct methanol fuel cells, Journal of Materials Chemistry 22 (47) (2012) 24862–24869.

[77] Q. Liu, et al., Metal organic frameworks modified proton exchange membranes for fuel cells, Frontiers in Chemistry 8 (2020) 694.

[78] Z. Rao, B. Tang, P. Wu, Proton conductivity of proton exchange membrane synergistically promoted by different functionalized metal-organic frameworks, ACS Applied Materials & Interfaces 9 (27) (2017) 22597–22603.

[79] G. Huang, et al., Universal, controllable, large-scale and facile fabrication of nano-MOFs tightly-bonded on flexible substrate, Chemical Engineering Journal 395 (2020) 125181.

[80] K. Cai, et al., An acid-stable hexaphosphate ester based metal–organic framework and its polymer composite as proton exchange membrane, Journal of Materials Chemistry A 5 (25) (2017) 12943–12950.

[81] C.-H. Tsai, et al., Enhancing performance of Nafion®-based PEMFC by 1-D channel metal-organic frameworks as PEM filler, International Journal of Hydrogen Energy 39 (28) (2014) 15696–15705.

[82] H. Wu, et al., High performance proton-conducting composite based on vanadium-substituted Dawson-type heteropoly acid for proton exchange membranes, Composites Science and Technology 162 (2018) 1–6.

[83] A. Amiri, et al., Preparation and characterization of magnetic Wells-Dawson heteropoly acid nanoparticles for magnetic solid-phase extraction of aromatic amines in water samples, Journal of Chromatography A 1483 (2017) 64–70.

[84] H. Yao, et al., Highly sulfonated co-polyimides containing hydrophobic cross-linked networks as proton exchange membranes, Polymer Chemistry 7 (29) (2016) 4728–4735.

[85] L. Christiani, et al., Evaluation of proton conductivity of sulfonated polyimide/dihydroxy naphthalene charge-transfer complex hybrid membranes, Journal of Polymer Science Part A: Polymer Chemistry 52 (20) (2014) 2991–2997.

[86] H. Pu, et al., Studies on anhydrous proton conducting membranes based on imidazole derivatives and sulfonated polyimide, Electrochimica Acta 54 (9) (2009) 2603–2609.

[87] G. Meyer, et al., Degradation of sulfonated polyimide membranes in fuel cell conditions, Journal of Power Sources 157 (1) (2006) 293–301.

[88] Y. Yin, et al., Sulfonated polyimides with flexible aliphatic side chains for polymer electrolyte fuel cells, Journal of Membrane Science 367 (1–2) (2011) 211–219.

[89] X. Zhang, et al., Polymer electrolyte membranes based on poly(m-phenylene)s with sulfonic acid via long alkyl side chains, Polymer Chemistry 4 (4) (2013) 1235–1242.

[90] C. Li, et al., Single-ion conducting electrolyte based on electrospun nanofibers for high-performance lithium batteries, Advanced Energy Materials 9 (10) (2019) 1803422.

[91] J. Li, F. Bu, C. Ru, H. Jiang, Y. Duan, Y. Sun, X. Pu, L. Shang, X. Li, C. Zhao, et al., Enhancing the selectivity of Nafion membrane by incorporating a novel functional skeleton molecule to improve the performance of direct methanol fuel cells, J. Mater. Chem. A 8 (2020) 196–206.

[92] Y. Song, et al., Structural characteristics of hydrated protons in ion conductive channels: synergistic effect of the sulfonate group and fluorine studied by molecular dynamics simulation, The Journal of Physical Chemistry C 122 (4) (2018) 1982–1989.
[93] C. Zhao, et al., Block sulfonated poly(ether ether ketone)s (SPEEK) ionomers with high ion-exchange capacities for proton exchange membranes, Journal of Power Sources 162 (2) (2006) 1003–1009.
[94] Z.P. Guan, et al., Synthesis and characterization of poly(aryl ether ketone) ionomers with sulfonic acid groups on pendant aliphatic chains for proton-exchange membrane fuel cells, European Polymer Journal 46 (1) (2010) 81–91.
[95] H. Pan, et al., Preparation and properties of sulfonated polybenzimidazole-polyimide block copolymers as electrolyte membranes, Ionics 24 (6) (2017) 1629–1638.
[96] S. Amari, et al., Effect of a sulfonated benzothiadiazole unit on the morphology and ion conduction behavior of a polymer electrolyte membrane, Industrial & Engineering Chemistry Research 57 (47) (2018) 16095–16102.
[97] C. Zhang, et al., Novel pore-filling membrane based on block sulfonated poly (ether sulphone) with enhanced proton conductivity and methanol resistance for direct methanol fuel cells, Electrochimica Acta 307 (2019) 188–196.
[98] J. Xu, et al., Direct polymerization of novel functional sulfonated poly(arylene ether ketone sulfone)/sulfonated poly(vinyl alcohol) with high selectivity for fuel cells, RSC Advances 6 (33) (2016) 27725–27737.
[99] J. Xu, et al., Synthesis and properties of a novel sulfonated poly(arylene ether ketone sulfone) membrane with a high β-value for direct methanol fuel cell applications, Electrochimica Acta 146 (2014) 688–696.
[100] J. Xu, et al., Construction of a new continuous proton transport channel through a covalent cross-linking reaction between carboxyl and amino groups, International Journal of Hydrogen Energy 38 (24) (2013) 10092–10103.
[101] N. Li, et al., Polymer electrolyte membranes derived from new sulfone monomers with pendent sulfonic acid groups, Macromolecules 43 (23) (2010) 9810–9820.
[102] Q. Zhang, et al., Poly(arylene ether) electrolyte membranes bearing aliphatic-chain-linked sulfophenyl pendant groups, Journal of Membrane Science 428 (2013) 629–638.
[103] C. Wang, et al., Erratum to "Poly(arylene ether sulfone) proton exchange membranes with flexible acid side chains" [J. Membr. Sci. 405–406 (2012) 68–78], Journal of Membrane Science 375 (2012) 421–422.
[104] C. Wang, et al., Proton-conducting membranes from poly(ether sulfone)s grafted with sulfoalkylamine, Journal of Membrane Science 427 (2013) 443–450.
[105] H. Xie, et al., Synthesis and properties of highly branched star-shaped sulfonated block polymers with sulfoalkyl pendant groups for use as proton exchange membranes, Journal of Membrane Science 497 (2016) 55–66.
[106] J. Pang, et al., Low water swelling and high proton conducting sulfonated poly(arylene ether) with pendant sulfoalkyl groups for proton exchange membranes, Macromolecular Rapid Communications 28 (24) (2007) 2332–2338.
[107] B. Lafitte, P. Jannasch, Proton-conducting aromatic polymers carrying hypersulfonated side chains for fuel cell applications, Advanced Functional Materials 17 (15) (2007) 2823–2834.
[108] Z. Long, J. Miyake, K. Miyatake, Proton exchange membranes containing densely sulfonated quinquephenylene groups for high performance and durable fuel cells, Journal of Materials Chemistry A 8 (24) (2020) 12134–12140.
[109] P.Y. Xu, et al., Effect of fluorene groups on the properties of multiblock poly(arylene ether sulfone)s-based anion-exchange membranes, ACS Applied Materials & Interfaces 6 (9) (2014) 6776–6785.
[110] K. Miyatake, B. Bae, M. Watanabe, Fluorene-containing cardo polymers as ion conductive membranes for fuel cells, Polymer Chemistry 2 (9) (2011) 1919–1929.
[111] Z. Zhang, T. Xu, Proton-conductive polyimides consisting of naphthalenediimide and sulfonated units alternately segmented by long aliphatic spacers, Journal of Materials Chemistry A 2 (30) (2014) 11583–11585.

[112] H.-F. Lee, et al., Synthesis of highly sulfonated polyarylene ethers containing alternating aromatic units, Materials Today Communications 3 (2015) 114–121.
[113] C. Yuan, Y. Wang, Synthesis and characterization of a novel sulfonated poly (aryl ether ketone sulfone) containing rigid fluorene group for DMFCs applications, Colloid and Polymer Science 299 (2021) 93–104.
[114] L.C.H. Moh, et al., Free volume enhanced proton exchange membranes from sulfonated triptycene poly(ether ketone), Journal of Membrane Science 549 (2018) 236–243.
[115] D. Joseph, et al., Thermal crosslinking of PBI/sulfonated polysulfone based blend membranes, Journal of Materials Chemistry A 5 (1) (2017) 409–417.
[116] Y.S. Guan, et al., Preparation and characterisation of proton exchange membranes based on crosslinked polybenzimidazole and phosphoric acid, Fuel Cells 10 (6) (2010) 973–982.
[117] M. Rikukawa, K. Sanui, Proton-conducting polymer electrolyte membranes based on hydrocarbon polymers, Progress in Polymer Science 25 (10) (2000) 1463–1502.
[118] X. Chen, et al., Partially fluorinated poly(arylene ether)s bearing long alkyl sulfonate side chains for stable and highly conductive proton exchange membranes, Journal of Membrane Science 549 (2018) 12–22.
[119] K. Matsumoto, T. Higashihara, M. Ueda, Locally sulfonated poly(ether sulfone)s with highly sulfonated units as proton exchange membrane, Journal of Polymer Science Part A: Polymer Chemistry 47 (13) (2009) 3444–3453.
[120] D.W. Seo, et al., Preparation and characterization of block copolymers containing multi-sulfonated unit for proton exchange membrane fuel cell, Electrochimica Acta 86 (2012) 352–359.
[121] X. Han, et al., Synthesis and properties of novel poly(arylene ether)s with densely sulfonated units based on carbazole derivative, Journal of Membrane Science 589 (2019) 117230.
[122] K.H. Lee, et al., Enhanced performance of a sulfonated poly(arylene ether ketone) block copolymer bearing pendant sulfonic acid groups for polymer electrolyte membrane fuel cells operating at 80% relative humidity, ACS Applied Materials & Interfaces 10 (24) (2018) 20835–20844.
[123] A. Khabibullin, I. ZharovIlya Zharov, Fuel Cell Membranes Based on Polymer-Modified Silica Colloidal Crystals and Glasses: Proton Conductivity and Fuel Cell Performance, Mater. Res. Soc. Symp. Proc. 1502 (2013) 2294–2304, doi:10.1557/opl.2013.568.
[124] A.N. Ponomarev, et al., A new synthesis approach for proton exchange membranes based on ultra-high-molecular-weight polyethylene, Journal of Applied Polymer Science 137 (47) (2020) 49563.
[125] L. Chikh, V. Delhorbe, O. Fichet, (Semi-)Interpenetrating polymer networks as fuel cell membranes, Journal of Membrane Science 368 (1–2) (2011) 1–17.
[126] C. Chen, B. Chen, R. Hong, Preparation and properties of alkaline anion exchange membrane with semi-interpenetrating polymer networks based on poly(vinylidene fluoride-co-hexafluoropropylene), Journal of Applied Polymer Science 135 (5) (2018) 45775.
[127] V. Delhorbe, et al., Influence of the membrane treatment on structure and properties of sulfonated poly(etheretherketone) semi-interpenetrating polymer network, Journal of Membrane Science 427 (2013) 283–292.
[128] H. Yu, et al., Improved chemical stability and proton selectivity of semi-interpenetrating polymer network amphoteric membrane for vanadium redox flow battery application, Journal of Applied Polymer Science 138 (6) (2020) 49803.
[129] F. Sun, et al., Friedel-Crafts self-crosslinking of sulfonated poly(etheretherketone) composite proton exchange membrane doped with phosphotungstic acid and carbon-based nanomaterials for fuel cell applications, Journal of Membrane Science 611 (2020) 118381.
[130] S.-K. Kim, et al., Cross-linked benzoxazine–benzimidazole copolymer electrolyte membranes for fuel cells at elevated temperature, Macromolecules 45 (3) (2012) 1438–1446.
[131] C. Li, et al., Cross-linked fully aromatic sulfonated polyamide as a highly efficiency polymeric filler in SPEEK membrane for high methanol concentration direct methanol fuel cells, Journal of Materials Science 53 (7) (2017) 5501–5510.
[132] K. Selvakumar, S. Rajendran, M. Ramesh Prabhu, Influence of barium zirconate on SPEEK-based polymer electrolytes for PEM fuel cell applications, Ionics 25 (5) (2018) 2243–2253.

[133] Y. Lu, et al., Sulfonated graphitic carbon nitride nanosheets as proton conductor for constructing long-range ionic channels proton exchange membrane, Journal of Membrane Science 601 (2020) 117908.

[134] Y. Liu, et al., Immobilized phosphotungstic acid for the construction of proton exchange nanocomposite membranes with excellent stability and fuel cell performance, International Journal of Hydrogen Energy 45 (35) (2020) 17782−17794.

[135] Y. Zhang, et al., Sulfonated poly(ether ether ketone)-based hybrid membranes containing polydopamine-decorated multiwalled carbon nanotubes with acid-base pairs for all vanadium redox flow battery, Journal of Membrane Science 564 (2018) 916−925.

[136] W. Chen, et al., $SO_4^{(2-)}/SnO_2$ solid superacid granular stacked one-dimensional hollow nanofiber for a highly conductive proton-exchange membrane, ACS Applied Materials & Interfaces 12 (36) (2020) 40740−40748.

[137] M. Taufiq Musa, N. Shaari, S.K. Kamarudin, Carbon nanotube, graphene oxide and montmorillonite as conductive fillers in polymer electrolyte membrane for fuel cell: an overview, International Journal of Energy Research 45 (2) (2021) 1309−1346.

[138] K. Charradi, et al., Silica/montmorillonite nanoarchitectures and layered double hydroxide-SPEEK based composite membranes for fuel cells applications, Applied Clay Science 174 (2019) 77−85.

[139] V. Di Noto, et al., Inorganic−organic membranes based on Nafion, $[(ZrO_2) \cdot (HfO_2)_{0.25}]$ and $[(SiO_2) \cdot (HfO_2)_{0.28}]$. Part I: synthesis, thermal stability and performance in a single PEMFC, International Journal of Hydrogen Energy 37 (7) (2012) 6199−6214.

[140] K. Hooshyari, et al., Fabrication $BaZrO_3$/PBI-based nanocomposite as a new proton conducting membrane for high temperature proton exchange membrane fuel cells, Journal of Power Sources 276 (2015) 62−72.

[141] G. Gnana kumar, A. Manthiram, Sulfonated polyether ether ketone/strontium zirconite@TiO_2 nanocomposite membranes for direct methanol fuel cells, Journal of Materials Chemistry A 5 (38) (2017) 20497−20504.

[142] Suryani, et al., Polybenzimidazole (PBI)-functionalized silica nanoparticles modified PBI nanocomposite membranes for proton exchange membranes fuel cells, Journal of Membrane Science 403−404 (2012) 1−7.

[143] E. Bet-moushoul, et al., TiO_2 nanocomposite based polymeric membranes: a review on performance improvement for various applications in chemical engineering processes, Chemical Engineering Journal 283 (2016) 29−46.

[144] I.A. Anu Karthi Swaghatha, L. Cindrella, Self-humidifying novel chitosan-geopolymer hybrid membrane for fuel cell applications, Carbohydrate Polymers 223 (2019) 115073.

[145] L. Mazzapioda, S. Panero, M.A. Navarra, Polymer electrolyte membranes based on Nafion and a superacidic inorganic additive for fuel cell applications, Polymers (Basel) 11 (5) (2019) 914.

[146] J. Wang, L. Wang, Preparation and properties of organic−inorganic alkaline hybrid membranes for direct methanol fuel cell application, Solid State Ionics 255 (2014) 96−103.

[147] P. Salarizadeh, et al., Novel proton exchange membranes based on proton conductive sulfonated PAMPS/PSSA-TiO_2 hybrid nanoparticles and sulfonated poly (ether ether ketone) for PEMFC, International Journal of Hydrogen Energy 44 (5) (2019) 3099−3114.

[148] P.B. Ingabire, et al., Titanium oxide/graphitic carbon nitride nanocomposites as fillers for enhancing the performance of SPAES membranes for fuel cells, Journal of Industrial and Engineering Chemistry 91 (2020) 213−222.

[149] R. Miao, et al., PVDF-HFP-based porous polymer electrolyte membranes for lithium-ion batteries, Journal of Power Sources 184 (2) (2008) 420−426.

[150] C. Li, et al., Synthesis, characterization and application of S-TiO_2/PVDF-g-PSSA composite membrane for improved performance in MFCs, Fuel 264 (2020) 116847.

[151] K.H. Lee, et al., Effect of functionalized SiO_2 toward proton conductivity of composite membranes for PEMFC application, International Journal of Energy Research 43 (10) (2019) 5333−5345.

[152] S.-Y. Jang, S.-H. Han, Sulfonated polySEPS/hydrophilic-SiO_2 composite membranes for polymer electrolyte membranes (PEMs), Journal of Industrial and Engineering Chemistry 23 (2015) 285−289.

[153] S. Elakkiya, et al., Polyaniline coated sulfonated TiO_2 nanoparticles for effective application in proton conductive polymer membrane fuel cell, European Polymer Journal 112 (2019) 696–703.

[154] S. Elakkiya, et al., Enhancement of fuel cell properties in polyethersulfone and sulfonated poly (ether ether ketone) membranes using metal oxide nanoparticles for proton exchange membrane fuel cell, International Journal of Hydrogen Energy 43 (47) (2018) 21750–21759.

[155] M. Vinothkannan, et al., Sulfonated graphene oxide/Nafion composite membranes for high temperature and low humidity proton exchange membrane fuel cells, RSC Advances 8 (14) (2018) 7494–7508.

[156] A. Manivel, et al., Synthesis of MoO_3 nanoparticles for azo dye degradation by catalytic ozonation, Materials Research Bulletin 62 (2015) 184–191.

[157] M. Branchi, et al., Functionalized Al_2O_3 particles as additives in proton-conducting polymer electrolyte membranes for fuel cell applications, International Journal of Hydrogen Energy 40 (42) (2015) 14757–14767.

[158] C.-C. Ke, et al., Preparation and properties of Nafion/SiO_2 composite membrane derived via in situ sol-gel reaction: size controlling and size effects of SiO_2 nano-particles, Polymers for Advanced Technologies 23 (1) (2012) 92–98.

[159] Y.P. Ying, S.K. Kamarudin, M.S. Masdar, Silica-related membranes in fuel cell applications: an overview, International Journal of Hydrogen Energy 43 (33) (2018) 16068–16084.

[160] M.T. Taghizadeh, M. Vatanparast, Ultrasonic-assisted synthesis of ZrO_2 nanoparticles and their application to improve the chemical stability of Nafion membrane in proton exchange membrane (PEM) fuel cells, Journal of Colloid and Interface Science 483 (2016) 1–10.

[161] H. Furukawa, et al., The chemistry and applications of metal-organic frameworks, Science 341 (6149) (2013) 1230444.

[162] S. Kim, et al., Achieving superprotonic conduction in metal-organic frameworks through iterative design advances, Journal of the American Chemical Society 140 (3) (2018) 1077–1082.

[163] F. Yang, et al., A flexible metal–organic framework with a high density of sulfonic acid sites for proton conduction, Nature Energy 2 (11) (2017) 877–883.

[164] N. Stock, S. Biswas, Synthesis of metal-organic frameworks (MOFs): routes to various MOF topologies, morphologies, and composites, Chemical Reviews 112 (2) (2012) 933–969.

[165] L. Ahmadian-Alam, H. Mahdavi, A novel polysulfone-based ternary nanocomposite membrane consisting of metal-organic framework and silica nanoparticles: As proton exchange membrane for polymer electrolyte fuel cells, Renewable Energy 126 (2018) 630–639.

[166] S. Horike, D. Umeyama, S. Kitagawa, Ion conductivity and transport by porous coordination polymers and metal–organic frameworks, Accounts of Chemical Research 46 (11) (2013) 2376–2384.

[167] B.Y. Xia, et al., A metal–organic framework-derived bifunctional oxygen electrocatalyst, Nature Energy 1 (1) (2016) 15006.

[168] Y. Ye, et al., High anhydrous proton conductivity of imidazole-loaded mesoporous polyimides over a wide range from subzero to moderate temperature, Journal of the American Chemical Society 137 (2) (2015) 913–918.

[169] M. Erkartal, et al., Proton conducting poly(vinyl alcohol) (PVA)/poly(2-acrylamido-2-methylpropane sulfonic acid) (PAMPS)/zeolitic imidazolate framework (ZIF) ternary composite membrane, Journal of Membrane Science 499 (2016) 156–163.

[170] S. Mukhopadhyay, et al., Fabricating a MOF material with polybenzimidazole into an efficient proton exchange membrane, ACS Applied Energy Materials 3 (8) (2020) 7964–7977.

[171] J. Kalaiselvimary, et al., Effect of surface-modified montmorillonite incorporated biopolymer membranes for PEM fuel cell applications, Polymer Composites 40 (S1) (2019) E301–E311.

[172] A. Jalali, et al., Polymer/clay nanocomposites based on aliphatic-aromatic polyamide containing thiazole ring and new organomodified nanosilicate layers: synthesis, characterization, and thermal properties study, Polymer Composites 39 (S3) (2018) E1407–E1420.

[173] M.J. Koh, et al., Preparation and characterization of porous PVdF-HFP/clay nanocomposite membranes, Journal of Materials Science & Technology 26 (7) (2010) 633–638.

[174] K.S. Abidin, et al., Role of structural modifications of montmorillonite, electrical properties effect, physical behavior of nanocomposite proton conducting membranes for direct methanol fuel cell applications, Materials Science-Poland 35 (4) (2018) 707–716.

[175] Z. Ahmed, et al., Physicochemical characterization of low sulfonated polyether ether ketone/Smectite clay composite for proton exchange membrane fuel cells, Journal of Applied Polymer Science 138 (1) (2020) e49634.

[176] D. Xing, et al., Preparation and characterization of a modified montmorillonite/sulfonated polyphenylether sulfone/PTFE composite membrane, International Journal of Hydrogen Energy 36 (3) (2011) 2177—2183.

[177] R. Herrera Alonso, et al., Nafion—clay nanocomposite membranes: morphology and properties, Polymer 50 (11) (2009) 2402—2410.

[178] F. Wang, D. Wang, H. Zhu, Montmorillonite—polybenzimidazole inorganic-organic composite membrane with electric field-aligned proton transport channel for high temperature proton exchange membranes, Polymer-Plastics Technology and Engineering 57 (17) (2018) 1752—1759.

[179] H. Doğan, et al., Organo-montmorillonites and sulfonated PEEK nanocomposite membranes for fuel cell applications, Applied Clay Science 52 (3) (2011) 285—294.

[180] A. Çal ı, A. Şahin, İ. Ar, Incorporating sepiolite and kaolinite to improve the performance of SPEEK composite membranes for proton exchange membrane fuel cells, The Canadian Journal of Chemical Engineering 98 (4) (2019) 892—904.

[181] H.F. Lee, P.C. Wang, Y.W. Chen-Yang, An electrospun hygroscopic and electron-conductive core-shell silica@carbon nanofiber for microporous layer in proton-exchange membrane fuel cell, Journal of Solid State Electrochemistry 23 (3) (2019) 971—984.

[182] B. Zhang, et al., Carbonaceous nanofiber-supported sulfonated poly (ether ether ketone) membranes for fuel cell applications, Materials Letters 115 (2014) 248—251.

[183] Y.B. Di, et al., Preparation and characterization of continuous carbon nanofiber-supported SPEEK composite membranes for fuel cell application, RSC Advances 4 (94) (2014) 52001—52007.

[184] G. Rambabu, S.D. Bhat, F.M.L. Figueiredo, Carbon nanocomposite membrane electrolytes for direct methanol fuel cells—a concise review, Nanomaterials 9 (9) (2019) 1292.

[185] F. Altaf, et al., Synthesis and applicability study of novel poly(dopamine)-modified carbon nanotubes based polymer electrolyte membranes for direct methanol fuel cell, Journal of Environmental Chemical Engineering 8 (5) (2020) 104118.

[186] M.M. Hasani-Sadrabadi, et al., Nafion/chitosan-wrapped CNT nanocomposite membrane for high-performance direct methanol fuel cells, RSC Advances 3 (20) (2013) 7337—7346.

[187] P. Molla-Abbasi, K. Janghorban, M.S. Asgari, A novel heteropolyacid-doped carbon nanotubes/Nafion nanocomposite membrane for high performance proton-exchange methanol fuel cell applications, Iranian Polymer Journal 27 (2) (2018) 77—86.

[188] A.R. Kim, et al., Amine functionalized carbon nanotube (ACNT) filled in sulfonated poly (ether ether ketone) membrane: effects of ACNT in improving polymer electrolyte fuel cell performance under reduced relative humidity, Composites Part B: Engineering 188 (2020) 107890.

[189] C.S. Yin, et al., Lateral-aligned sulfonated carbon-nanotubes/Nafion composite membranes with high proton conductivity and improved mechanical properties, Journal of Membrane Science 591 (2019) 117356.

[190] S. Gahlot, V. Kulshrestha, Dramatic improvement in water retention and proton conductivity in electrically aligned functionalized CNT/SPEEK nanohybrid PEM, ACS Applied Materials & Interfaces 7 (1) (2015) 264—272.

[191] Y.H. Li, et al., A poly(vinyl alcohol)-based composite membrane with immobilized phosphotungstic acid molecules for direct methanol fuel cells, Electrochimica Acta 224 (2017) 369—377.

[192] J.R. Potts, et al., Graphene-based polymer nanocomposites, Polymer 52 (1) (2011) 5—25.

[193] Y.S. Ye, et al., A new graphene-modified protic ionic liquid-based composite membrane for solid polymer electrolytes, Journal of Materials Chemistry 21 (28) (2011) 10448—10453.

[194] K. Feng, B.B. Tang, P.Y. Wu, "Evaporating" graphene oxide sheets (GOSs) for rolled up GOSs and its applications in proton exchange membrane fuel cell, ACS Applied Materials & Interfaces 5 (4) (2013) 1481—1488.

[195] S.G. Qu, et al., Enhanced proton conductivity of sulfonated poly(ether ether ketone) membranes at elevated temperature by incorporating (3-aminopropyl)triethoxysilane-grafted graphene oxide, Korean Journal of Chemical Engineering 36 (12) (2019) 2125—2132.

[196] Y.B. Cai, Z.Y. Yue, S.A. Xu, A novel polybenzimidazole composite modified by sulfonated graphene oxide for high temperature proton exchange membrane fuel cells in anhydrous atmosphere, Journal of Applied Polymer Science 134 (25) (2017).
[197] N. Cao, et al., Synthesis and characterization of sulfonated graphene oxide reinforced sulfonated poly (ether ether ketone) (SPEEK) composites for proton exchange membrane materials, Materials 11 (4) (2018) 516.
[198] Y.N. Yusoff, et al., Sulfonated graphene oxide as an inorganic filler in promoting the properties of a polybenzimidazole membrane as a high temperature proton exchange membrane, International Journal of Hydrogen Energy 45 (51) (2020) 27510−27526.
[199] H. Lee, et al., Highly sulfonated polymer-grafted graphene oxide composite membranes for proton exchange membrane fuel cells, Journal of Industrial and Engineering Chemistry 74 (2019) 223−232.
[200] A. Kausar, Study on poly(imide-ethylene glycol) and graphene oxide-based hybrid proton exchange membrane, International Journal of Polymer Analysis and Characterization 21 (6) (2016) 537−547.
[201] J. Han, et al., Sulfonated poly(arylene ether sulfone) composite membrane having sulfonated polytriazole grafted graphene oxide for high-performance proton exchange membrane fuel cells, Journal of Membrane Science 612 (2020) 118428.
[202] H.B. Wang, et al., Fabrication of new fullerene composite membranes and their application in proton exchange membrane fuel cells, Journal of Membrane Science 289 (1−2) (2007) 277−283.
[203] S. Neelakandan, et al., Highly branched poly(arylene ether)/surface functionalized fullerene-based composite membrane electrolyte for DMFC applications, International Journal of Energy Research 43 (8) (2019) 3756−3767.
[204] S.S. Rikame, A.A. Mungray, A.K. Mungray, Synthesis, characterization and application of phosphorylated fullerene/sulfonated polyvinyl alcohol (PFSP) composite cation exchange membrane for copper removal, Separation and Purification Technology 177 (2017) 29−39.
[205] G. Rambabu, N. Nagaraju, S.D. Bhat, Functionalized fullerene embedded in Nafion matrix: a modified composite membrane electrolyte for direct methanol fuel cells, Chemical Engineering Journal 306 (2016) 43−52.
[206] S. Saga, et al., Polyelectrolyte membranes based on hydrocarbon polymer containing fullerene, Journal of Power Sources 176 (1) (2008) 16−22.

CHAPTER 13

Shape-controlled metal nanoparticles for fuel cells applications

Ajit Behera
Department of Metallurgical & Materials Engineering, National Institute of Technology, Rourkela, Rourkela, India

13.1 Introduction

The development of advanced electrocatalysts with higher properties in terms of selectivity, activity, and stability for use in electrochemical applications, such as energy storage and conversion devices, has become the focus of interest. It should be noted that the properties of nanostructured metal mainly based on their size and shape of nanoparticle (NP). The resulting electrocatalytic activity (Fig. 13.1) is the sum of the specific contributions of each type of surface site present on the NP [1]. Therefore controlling the size and shape of metal NPs has always been a very important research goal, because this target can optimize their performance. In recent decades, strategies and schemes for the preparation of metal NPs of controlled size and shape, such as spherical, octahedron, cubic, cubic octahedron, rod, wire, decahedron, and 20 facets and plates has been developed [2–4]. Furthermore, even shapes that are thermodynamically unpopular with conventional polyhedra are often synthesized, including high-index facets, concave surfaces, and branching structures. The size of the NPs in the catalyst is between 1 and 20 nm. It is widely accepted that in a typical fuel cell, only 25%–35% of the total available Pt is effective [5]. This is due to the fact that Pt NPs used as fuel cell catalysts are irregular particles with a wide particle size distribution and also in aggregate form. These large particles and aggregates are less active in the electrocatalysis of fuel cells because their surface area is reduced or their active sites are chemically or physically blocked [6]. All of the aforementioned problems are largely related to the synthesis conditions, such as pH, temperature, and the type of support material used. Therefore it is important to select the correct reaction conditions to achieve maximum Pt utilization. Therefore the utilization rate of Pt can be improved by controlling the shape of the NPs [7]. Therefore the use of metal NPs in a controlled manner has proven to be one of the most interesting options for improving the electrocatalytic performance of many nanomaterials. Metal NPs with controlled shapes play a fundamental role in chemical reactions because they can regulate the electronic, optical, magnetic, and catalytic properties of functional materials.

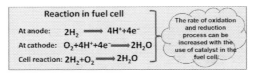

Figure 13.1 Reactions in the fuel cell and importance of the catalyst.

13.2 What is shape-controlled catalyst

The activity and selectivity of the catalyst particles strongly depend on the larger relative surface area, the orientation of the atoms on the surface of the catalyst particles and on the shape setting of the NPs. This clearly indicates the catalytic activity mainly based on the nanoform of the catalyst. For fuel cells, the orientation of the most commonly used Pt catalyst can be easily obtained. It is also known that there are different facets of Pt in different proportions, shape and size of the NPs. For the dehydrocyclization of n-heptane, the activity of the hexagonal structure of Pt (111) is five times higher than that of cubic NPs (100) [8]. The oxygen reduction reaction activity of cubic Pt particles is four times higher than that of polyhedra or truncated cubes. Pt NP with preferential crystalline orientation (111) shows better performance than polycrystalline Pt particles. The cubic octahedral particles contain (100) and (111) faces. The electrochemical stability and activity of this Pt-based catalytic system are closely related to the surface energy of several aspects. The degradation of cubic particles with higher surface energy during the oxidation of methanol is much faster than the degradation of octahedral particles with lower surface energy [9]. Therefore managing the dimensionality of the NPs can not only increase the activity but also increase the stability of the particles, thereby reducing the cost and improving the life of the fuel cell. The catalysts used in various fuel cells (Table 13.1) have been converted into NPs in a controllable way to improve electrochemical reactions.

13.3 Platinum catalyst

Platinum is a precious stable electrode metal and has excellent unique properties as a CO oxidation catalyst in catalytic converters, H_2/O_2 reactions in fuel cells, HNO_3 production and oil cracking. A lot of investigation has been done to reduce the size of the Pt catalyst, thus maximizing the surface area [13]. The researchers also highlighted the selectivity of the improved Pt nanocatalysts that would be useful for exposing glass faces that are inherently more active for specific processes. Truncated cubes, tetrahedra, and octahedra have been developed for Pt, and numbers of work has been put into synthesis, which involves H_2 and capping agent $[(C_3H_3NaO_2)_n]$ as a to reduce the K_2PtCl_4 [14]. Initially, Pt tetrahedron changed to truncated octahedron and again

Table 13.1 Type of catalyst in various fuel cells [10–12].

	Polymer electrolyte fuel cells (PEFCs)	Alkaline fuel cell (AFC)	Phosphoric acid fuel cells (PAFCs)	Molten carbonate fuel cell (MCFC)	Solid oxide fuel cell (SOFC)
Electrolyte	Hydrated polymeric ion-exchange membrane	Mobilized or immobilized potassium hydroxide in asbestos matrix	Immobilized liquid phosphoric acid in SiC	Immobilized liquid molten carbonate in LiAlO$_2$	Perovskites (Ceramics)
Electrodes	Carbon	Transition metals	Carbon	Nickel and nickel oxide	Perovskite and perovskite/metal cermets
Catalyst	Nano-Pt	Nano-Pt	Nano-Pt	Nanoscaled electrode material	Nanoscaled electrode material
Charge carrier	H$^+$	OH$^+$	H$^+$	CO$_3^{-2}$	O^{-2}

changed to cube when Pt atom preferentially deposited on {111} plane. Preferential addition of Pt atom as well as the presence of capping agent leads to the development of preferential facet. Pt nanocubes and cubicoctahedra can be prepared by reduction of K$_2$PtCl$_4$ with the presence of H$_2$ that escaped from NaBH$_4$. Here, tetradecyl-trimethyl-ammonium-bromide is used as a surfactant. And again, using ascorbic acid reductant gives rise the generation of Pt nanodendritic crystals [15].

Pt nanocrystals can be produced by the polyol reduction technique in presence of polyvinyl pyrrolidone (PVP) and Ag$^+$ as ion mediators. Ag$^+$ helps the growth of nanocrystals parallel to the <100> axis and form the Pt octahedral [16]. In the presence of gold, Pt octahedra grows further to give rise the branched structures. This phenomenon is observed only in low concentrations of seeds and high concentrations of Pt atoms. To comply with these standards, two strategies have been adopted. One system is the oxidative attack of the Fe(III) species that combined with the polyol reduction process to keep the initial Pt-seed concentration lower. Therefore by continuously blowing N$_2$ during the reaction, oxidative attack is prevented, to generate and maintain a high concentration of Pt atoms. In this way, the octahedral seeds of Pt can become tetrapods due to excessive growth in the corners {111}. The second system is based on H$_2$ reduction and uses a high concentration of Pt precursor to ensure the necessary overgrowth [17]. El-Sayed and his colleagues used tetrahedral Pt seeds to grow single-crystal star-shaped Pt nanostructures up to 30%. Monocrystalline Pt nanowires can be produced with the same procedure as the polyol synthesis for the Pt tetrapods. The reaction can be carried out in air rather than nitrogen. In this case, the reduction of the precursor remains slow throughout the reaction. They grow parallel to the <111> axis, as do the ultrafine gold nanowires produced by the reduction of the oleylamine-AuCl complex [18]. Although autocatalytic nucleation may be involved, the growth mechanism of Pt nanowires along the <111> axis is still unknown. Slow reduction in the solution causes low Pt concentration and difficult for more nucleation site. [PtCl$_4$]$_2$- or [PtCl$_2$ (H$_2$O)$_2$] are the two precursor forms dimer that initiated the long-chain Pt complexes and after their reduction, Pt nanowires form. The growth of Pt nanowire is different from that of the Au and Pd nanowire. Ceramic nanofibers, silica/polymer beads, wire mesh, silicon wafers, etc. are the media can be used for Pt nanowires. Furthermore, nanobars and Pt nanowires were synthesized using the gamma irradiation method [19] and the Pd-seeding process, respectively. Finally, in related work, interconnected Pt nanowires were synthesized in a two-phase water-chloroform system containing cetyltrimethyl-ammonium bromide (CTAB). The reduction of the Pt precursor and the growth of the nanocrystals are limited to the worm-shaped microcellular network that forms in the chloroform droplets [20].

In addition to conventional nanocrystals, unconventional Pt nanocrystals covered by high refractive index facets can be produced. In principle, possible unconventional forms are: {hk0} covered tetrahedron, {hkk} covered trapezium, {hhl} covered triooctahedron, and {hkl} covered hexahedron (h > k > l). It is emphasized that the important role of the surface covered by the gaseous substance can be to guide or control the shape of the nanocrystal. Since the high refractive index facets contain many pendent bonds and atomic steps, the Pt atoms on the surface can easily form Pt-O bonds, which can protect those facets. Conversely, the planes {111} and {100} are so compact that oxygen must diffuse into or out of the crystal lattice [21]. As a result, these ordered surface structures will be destroyed.

Pt generally tends to take on a single-crystal seed-based form. But, it is extremely difficult to obtain a Pt plate structure using a strategy based on dynamic control. For example, when we use PVP as a mild reducing agent such as in the synthesis of Ag and Pd nanoplates, only 15% of the products consist of plate structures and they are all very small (edges <10 nm). This is due to the expected greater internal deformation of the Pt twin structure. However, recently, we have shown that using Pd nanoplates as seeds for heteroepitaxial deposition of Pt, Pt nanoplates with larger lateral dimensions (edge > 50 nm). In this method, Pd nanoplates are synthesized by reducing the Na$_2$PdCl$_4$ precursors using PVP as a reducing agent and then used as seeds for the nucleation of Pt formed by reducing H$_2$PtCl$_6$ by citric acid [22]. The product consists of a core of hexagonal or triangular Pd nanoplates and a shell consisting of thin, uniform Pt layers, showing that Pt grows epitaxially on the Pd surface layer by layer. The close network correspondence between Pd/Pt and the low reduction rate linked with the moderate reducing capacity of C H O , helps to obtain a compliant epitaxial growth of the Pt layers on Pd nanoplates [23].

13.4 Fe, Co, and Ni catalysts

Catalysts of Fe, Co, and Ni are the ferromagnetic materials. It is very difficult to synthesize their nanocrystals, keeping in focus the magnetic moments of the entire nanocrystal that align when a magnetic field is applied. Due to its potential superparamagnetism, its synthesis is also very interesting. Suitable precursors are used to reduce for the preparation of these materials. The resultant nanocrystal shape regulated by the rate of reduction, by the concentration of the precursor, and by the selected capping agent. However, the solvents and precursors commonly used in these systems are very different from the solvents and precursors used to prepare Pt, Pd, Au, and Ag nanocrystals, so it is difficult to directly compare them [24]. In a well-controlled inert environment, Fe, Co, and Ni oxidize easily, so the resulting nanocrystals are often coated with their natural oxides. Usually, Fe(CO)$_5$ is act as a precursor for the synthesis of Fe nanocrystals. Many methods have been recorded to produce Fe nanocrystals with sizes between 2 and 20 nm [25]. In addition to Fe(CO)$_5$, Fe[N(SiMe$_3$)$_2$]$_2$ (where Me = CH$_3$) has been shown to be a good precursor of Fe NPs. Fe nanocube (7 nm size) can be prepared using Fe[N(SiMe$_3$)$_2$]$_2$ reducing agent, in addition of hexadecylamine and oleic acid/hexa-decyl-ammonium-chloride as capping agent in trimethylbenzene at 150°C [26]. The precursor Fe(II)oleate complexes can be used to produce the Fe nanocubes of edge length 20 nm. By thermal decomposition technique, organometallic compounds helps to reduce the iron containing salt to give Fe nanocrystal. By this procedure, production of Fe nanowires is possible in the introduction of ligand (phenanthroline) [27].

In contrast, production of Co nanocrystal is depending on the phase change irrespective of the size change. For example, ε-phase having space group p413 is metastable at ambient temperature and becomes highly reaction-sensitive when exposed to 500°C that forms FCC-phase. Solution-based technique can be used to prepare the ε-Co NP, by adding CoCl$_2$ contained octylether solution with perhydrogen solution at a temperature of 200°C whereas hcp-Co NPs can be formed at 300°C [28]. Monodispersed Co NPs in the ε-phase can be prepared by rapid pyrolysis of Co$_2$(CO)$_8$ using CoCl$_2$ reducing agent and surfactants such as oleic acid, lauric acid [29]. In case of Co NPs production, temperature plays a key role to control the phase. Nonspherical Co nanocrystals are mostly hcp-Co. For example, hcp-Co nanodiscs can be produced by reducing Co$_2$(CO)$_8$ with linear alkylamines. The alkylamine amino group oppose the growth on {001} plane and there is the development of disk shape of Co nanocrystal. It is observed that, alkyl chain also an important parameter to alter the shape of nanotubes [30].

FCC-phase Ni nanocrystals attracted attention for their magnetic properties. The typical process for the synthesis of Ni nanocrystals can be produce using the precursor of Ni(COD)$_2$(COD = 1,5-cyclooctadiene) that reduce up to 60°C temperature. Beside the nonspherical form, hexagonal and triangular Ni nanoplates have recently been prepared. In this process, Fe(CO)$_5$ was added to moderate the reduction rate of Ni formate. Since Ni is an FCC metal, this slow reduction process can be considered the reason for the typical controlled kinetic synthesis of plate-shaped morphology [31].

13.5 Various functionality of shape-selected nanoparticles

Various properties after the selection of shape control parameters are discussed further.

13.5.1 Activity enhancement

The most impressive example of activity enhanced through shape control is the realization of the Pt$_3$Ni plan (111). The surface oxygen reduction reaction activity of Pt$_3$Ni (111) is found to be about 90 times that of the commercially available Pt/C. Monocrystalline surface with different structures and compositions has been prepared and found that the surface of Pt$_3$Ni (111) has the top skin layer complete with Ni bottom layer that regulates the electronic structure of Pt, thus weakening the O$_2$ absorption force and leading to oxygen reduction reactions. Thus using NPs in a controlled way is possible to create the surface of Pt$_3$Ni(111) [32]. The mass activity of commercially available Pt/C is known to be ~ 0.1 A mgPt^{-1}. By using the gaseous reducing agent, octahedron from Pt$_3$Ni synthesized with a mass activity of 0.44 A mgPt^{-1}. The 24 icosahedra of Pt$_3$Ni have (111) crystalline planes in the same research group and their mass activity is greater, which is 0.62 A mgPt^{-1}. The octahedron synthesis of PtNi reportedly has an activity of 1.45 A mgPt^{-1}. The size, exact composition and cleanliness of the surface will affect the dough activity, but it is clear that shape control has greatly improved the dough activity of the Pt$_3$Ni nanoframes [33]. The massive activity of Pt in the oxygen reduction reaction, which is the bottleneck reaction of fuel cell commercialization.

Shaped Pd NPs are commonly used as electrocatalytic catalysts for the oxidation of formic acid. The faceted Pd cube (100) is much more active than the faceted Pd octahedron (111) and the surface of the Pd cube is more prone to surface oxidation. Concave dipyramidal Pd nanocrystals combined with (110) facets show increased activity when Pd is grown locally in excess in Pt cubes. The PtPd cube shows a higher rotation frequency than the PtPd octahedron. NPs with high-index surfaces have been reported to have specific enhanced activities for various electrocatalytic reactions [34]. However, NPs are generally very large (greater than 50 nm) and many atoms within the particles are blocked. Therefore the overall activity of the slurry is generally worse than that of commercially available catalysts.

Shaped metal oxide NPs are also used to enhance the activity of the catalytic reaction. Once the Au is deposited, the cerium oxide shape is used in the gas exchange reaction of the water (CO + H$_2$O-CO$_2$ + H$_2$). The oxidation state of the deposited Au is significantly different depending on the shape of the ceria. Au oxide deposited on the (110) faceted ceria nanocrystals exhibits a higher CO conversion rate than the Au metal deposited on the (100) faceted ceria nanocrystals. Similarly, Au or Co deposited on (110) faceted ceria NPs showed improved activity for steam reforming of methanol or ethanol, respectively. The oxidation of soot on the ceria catalyst also showed shape dependence [35].

Co_3O_4 nanrods with (110) facets show high CO oxidation activity even at $-77°C$. Although for Co_3O_4 nanrods, the CO conversion rate at room temperature is 100%, for NPs of formless Co_3O_4, the conversion rate is much lower (less than 40%). Cubic Cu_2O nanocrystals with (100) faces, octahedra with (111) faces and rhombic dodecahedra with (110) faces were also synthesized and used for the photocatalytic degradation of methyl orange. The rhombohedral dodecahedron exhibits much greater photocatalytic activity than the cube [36]. The bandgap energy of Cu_2O materials can be controlled using different ways. When comparing Cu_2O cubes, cylinders, and spheres, the band gap energy of these cubes is approximately 1 eV lower than that of spherical materials. Cu_2O with high refractive index facets shows greater activity for the oxidation of CO in the gas phase. In the photocatalytic reaction under various conditions, the decahedral TiO_2 particles with crystalline faces (110) are greater than or at least equivalent to $P_2 5$. High aspect ratio TiO_2 nanorods are more conducive to H_2 production through water splitting. MgO nanopowders or nanosheets show a high catalytic activity for the degradation of volatile organic compounds or the condensation of benzaldehyde and acetophenone. MgO cubes or ZnO nanobelts are also used as supports for Au or Pd deposition and exhibit high activity for CO oxidation [37].

13.5.2 Selectivity enhancement

The catalytic application of shaped NPs mainly focuses on enhancing activity, but recently the use of shaped NPs to control selectivity has received increasing attention. The high selectivity will minimize unwanted products, thus reducing the separation and purification load. The reaction mechanism was studied on the single-crystal surface, indicating that the surface structure will be a key factor in increasing selectivity. The size of the NPs was generally adjusted to control the surface structure and improve selectivity. Pt cubes with (100) faces and Pt cubes with (100) and (111) faces are first reported to have evident gas phase reactions such as benzene hydrogenation [38]. The hydrogenolysis reaction of methyl-cyclopentane was tested and different selectivities are there depending on the shape of the NPs used. For the hydrogenation of cyclic chemicals such as furan and pyrrole, Ct or C—O bonds are easier to break on the Pt (100) surface than on the Pt (111) surface. However, the Pt (111) plan is more advantageous for the isomerization of trans-2-butene to cis-2-butene (in food chemistry) [39]. Hydrogenolysis of methyl-cyclopentane on the Pt (100) side of the Pt nanocubes will produce more cleavage compounds, while the hexane will be produced on the Pt (111) side of the octahedral Pt NPs (110) cyclohexane. The electrocatalytic hydrogenation of ketene also showed shape dependence, resulting in the production of partially hydrogenated cyclohexanone in cubes of Pt [40].

Hydrogen can be more readily supplied to the reactants adsorbed on the surface of Pd (100) by decomposition of the Pd hydride, and the ethylene produced on the surface will more easily leave the surface of the Pd (100). Therefore Pd cubes with an average particle size of 12.8 nm are more selective for ethylene than Pd spheres of the same size, while the amount of ethane produced is the smallest [41]. These Pd nanocubes have a higher selectivity than conventional Pd catalysts. Under the same conversion level, its particle size is much smaller, only 5 nm. When well-defined ZnO crystals are used together with NiMo as a cocatalyst for the combined reforming-hydrogenolysis reaction of glycerol, a larger portion of (100) nonpolar facets is more effective for the selective production of 1,2-propanediol [42].

13.5.3 Optical behavior

Under the irradiation of light, free electrons in the metal are driven by the alternating electric field and oscillate together in phase with the incident light. This collective oscillation is called surface plasmon resonance, which can effectively scatter and absorb light under resonant conditions and give metals (especially Ag and Au) luminous colloidal colors [43]. Related to electronic oscillation is that when excited by light, surface charge polarization also occurs on the metal surface. Unlike bulk metals, the induced charges of NPs cannot travel along a flat surface in the form of waves, but are confined and concentrated on the surface of the particle, in this case it is called plasmon resonance local surface area (LSPR). LSPR generates a strong local electric field within a few nanometers of the particle surface. As a direct application of LSPR, this near-field effect improves the Raman diffusion cross section of the molecules adsorbed on the surface of the NPs, thus providing a "fingerprint" spectrum rich in chemical information [44].

For nearly a hundred years, people have known that the shape of nanocrystals affects their interaction with light. To demonstrate how shape affects light absorption and scattering, a series of simulated spectra for Ag nanospheres were obtained using Mie theory or the discrete dipole approximation method for other Ag nanocrystals. The simplest is also the most symmetrical: a sphere. The dipole resonance results from the polarization of the electron density throughout the sphere, such that the sphere itself has a dipole moment, which reverses its sign at the same frequency as the incident light. Weak quadrupolar resonance is characterized by two parallel dipoles of opposite signs, which is caused by the inhomogeneity of the incident light throughout the sphere due to the loss of energy [45]. In principle, the charge separation caused by the electron density in relation to the polarization of the positive ion lattice provides the main restoring force of the electronic oscillations and largely determines the frequency and intensity of the resonant peak of a given metal. Therefore changes in the size, shape, and dielectric environment of the nanocrystals can affect the surface polarization and change the resonance spectral profile. For simple geometric shapes, both the sharpness of the corners and the symmetry of the shape can change the polarization of the surface, thus changing the LSPR peak. Taking into account the sharpness of the corners, recent near-field calculations around the NPs indicate that surface charges accumulate in the corners. As a result of this improved charge separation, resistance to electronic oscillations will be reduced. For example, compared to the 40 nm spheres, the strongest peak is red shifted by approximately 100 nm. Redshifts are also observed in tetrahedral and octahedra, because the angles of the tetrahedron are the sharpest, so the resonance peak of the redshift is the largest. However, changing the degree of sharpness of the corners cannot take into account all the spectral characteristics [46].

13.6 Summary

To enhance the catalytic effects in the fuel cell, this age gifted the shape controlling technology. This chapter discussed how the shape is controlled for a catalyst for a fuel cell. The most used platinum catalyst has been discussed along with Fe, Co, and Ni catalysts. Various functionality of shape-controlled NPs such as activity, selectivity, and the optical index has been discussed. More research is required to enhance the discussed activity of various materials which is to be used in future fuel cell.

References

[1] C.L. Bentley, M. Kang, P.R. Unwin, Nanoscale surface structure—activity in electrochemistry and electrocatalysis, Journal of the American Chemical Society 141 (6) (2019) 2179−2193. Available from: https://doi.org/10.1021/jacs.8b09828.

[2] J. Xiao, Q. Kuang, S. Yang, et al., Surface structure dependent electrocatalytic activity of Co_3O_4 anchored on graphene sheets toward oxygen reduction reaction, Scientific Reports 3 (2013) 2300. Available from: https://doi.org/10.1038/srep02300.

[3] Z.W.S. Yang, W. Wu, Shape control of inorganic nanoparticles from solution, Nanoscale 8 (2016) 1237−1259. Available from: https://doi.org/10.1039/C5NR07681A.

[4] K. An, G.A. Somorjai, Size and shape control of metal nanoparticles for reaction selectivity in catalysis, ChemCatChem 4 (10) (2012) 1512−1524. Available from: https://doi.org/10.1002/cctc.201200229.

[5] M. Bandapati, S. Goel, B. Krishnamurthy, Platinum utilization in proton exchange membrane fuel cell and direct methanol fuel cell—review, Bioelectrochemistry & Fuel Cells 9 (4) (2019). Available from: https://doi.org/10.5599/jese.665.

[6] J.E. Newton, J.A. Preece, N.V. Rees, S.L. Horswell, Nanoparticle catalysts for proton exchange membrane fuel cells: can surfactant effects be beneficial for electrocatalysis? Physical Chemistry Chemical Physics 16 (2014) 11435−11446. Available from: https://doi.org/10.1039/C4CP00991F.

[7] M.S. Jameel, A.A. Aziz, M.A. Dheyab, Green synthesis: proposed mechanism and factors influencing the synthesis of platinum nanoparticles, Green Processing and Synthesis 9 (1) (2020) 386−398. Available from: https://doi.org/10.1515/gps-2020-0041.
[8] M.J. Lundwall, S.M. McClure, D.W. Goodman, Probing terrace and step sites on Pt nanoparticles using CO and ethylene, Journal of Physical Chemistry C 114 (17) (2010) 7904−7912. Available from: https://doi.org/10.1021/jp9119292.
[9] I. Khan, K. Saeed, I. Khan, Nanoparticles: properties, applications and toxicities, Arabian Journal of Chemistry 12 (7) (2019) 908−931. Available from: https://doi.org/10.1016/j.arabjc.2017.05.011.
[10] Y. Qiao, C.M. Li, Nanostructured catalysts in fuel cells, Journal of Materials Chemistry 21 (2011) 4027−4036. Available from: https://doi.org/10.1039/C0JM02871A.
[11] C. Song, Fuel processing for low-temperature and high-temperature fuel cells: challenges, and opportunities for sustainable development in the 21st century, Catalysis Today 77 (1−2) (2002) 17−49. Available from: https://doi.org/10.1016/S0920-5861(02)00231-6.
[12] L. Carrette, A. Friedrich, U. Stimming, Fuel cells: principles, types, fuels, and applications, ChemPhyChem 1 (4) (2000) 162−193. https://doi.org/10.1002/1439-7641(20001215)1:4 < 162::AID-CPHC162 > 3.0.CO;2-Z.
[13] S. Guo, S. Dong, E. Wang, Three-dimensional Pt-on-Pd bimetallic nanodendrites supported on graphene nanosheet: facile synthesis and used as an advanced nanoelectrocatalyst for methanol oxidation, ACS Nano 4 (1) (2010) 547−555. Available from: https://doi.org/10.1021/nn9014483.
[14] J. Chen, B. Lim, E.P. Lee, Y. Xia, Shape-controlled synthesis of platinum nanocrystals for catalytic and electrocatalytic applications, Nano Today 4 (1) (2009) 81−95. Available from: https://doi.org/10.1016/j.nantod.2008.09.002.
[15] E. Antolini, J. Perez, The renaissance of unsupported nanostructured catalysts for low-temperature fuel cells: from the size to the shape of metal nanostructures, Journal of Materials Science 46 (2011) 4435−4457. Available from: https://doi.org/10.1007/s10853-011-5499-3.
[16] J. McGinty, N. Yazdanpanah, C. Price, J.H. ter Horst, J. Sefcik, Chapter 1: Nucleation and crystal growth in continuous crystallization, The Handbook of Continuous Crystallization, Royal Society of Chemistry, 2020, pp. 1−50. Available from: http://doi.org/10.1039/9781788013581-00001.
[17] M. Jiang, B. Lim, J. Tao, P.H.C. Camargo, C.M.Y. Zhu, Y. Xia, Epitaxial overgrowth of platinum on palladium nanocrystals, Nanoscale 2 (2010) 2406−2411. Available from: https://doi.org/10.1039/C0NR00324G.
[18] A. Loubat, M. Impéror-Clerc, B. Pansu, F. Meneau, B. Raquet, G. Viau, et al., Growth and self-assembly of ultrathin Au nanowires into expanded hexagonal superlattice studied by in situ SAXS, Langmuir 30 (14) (2014) 4005−4012. Available from: https://doi.org/10.1021/la500549z.
[19] T.K.L. Nguyen, N.D. Nguyen, V.P. Dang, D.T. Phan, T.H. Tran, Q.H. Nguyen, Synthesis of platinum nanoparticles by gamma Co-60 ray irradiation method using chitosan as stabilizer, Advances in Materials Science and Engineering 2019 (2019) 1−5. Available from: https://doi.org/10.1155/2019/9624374.
[20] Y. Song, R.M. Garcia, R.M. Dorin, H. Wang, Y. Qiu, E.N. Coker, et al., Synthesis of platinum nanowire networks using a soft template, Nano Letters 7 (12) (2007) 3650−3655. Available from: https://doi.org/10.1021/nl0719123.
[21] Y. Ma, Q. Kuang, Z. Jiang, Z. Xie, R. Huang, L. Zheng, Synthesis of trisoctahedral gold nanocrystals with exposed high-index facets by a facile chemical method, Angewandte Chemie International Edition 47 (46) (2008) 8901−8904. Available from: https://doi.org/10.1002/anie.200802750.
[22] B. Lim, J. Wang, P.H.C. Camargo, M. Jiang, M.J. Kim, Y. Xia, Facile synthesis of bimetallic nanoplates consisting of Pd cores and Pt shells through seeded epitaxial growth, Nano Letters 8 (8) (2008) 2535−2540. Available from: https://doi.org/10.1021/nl8016434.
[23] X. Huang, Z. Zeng, S. Bao, et al., Solution-phase epitaxial growth of noble metal nanostructures on dispersible single-layer molybdenum disulfide nanosheets, Nature Communications 4 (2013) 1444. Available from: https://doi.org/10.1038/ncomms2472.
[24] O. Pascu, S. Marre, C. Aymonier, Creation of interfaces in composite/hybrid nanostructured materials using supercritical fluids, Nanotechnology Reviews 4 (6) (2015) 487−515. Available from: https://doi.org/10.1515/ntrev-2015-0009.

[25] J. Watt, G.C. Bleier, M.J. Austin, S.A. Ivanov, D.L. Huber, Non-volatile iron carbonyls as versatile precursors for the synthesis of iron-containing nanoparticles, Nanoscale 9 (20) (2017) 6632–6637. Available from: https://doi.org/10.1039/c7nr01028a. PMID: 28304414.

[26] W.J. Huang, C.C. Chen, S.W. Chao, S.S. Lee, F.L. Hsu, Y.L. Lu, et al., Synthesis of N-hydroxycinnamides capped with a naturally occurring moiety as inhibitors of histone deacetylase, ChemMedChem 5 (4) (2010) 598–607. Available from: https://doi.org/10.1002/cmdc.200900494.

[27] Y. Zhang, M.K. Ram, E.K. Stefanakos, D.Y. Goswami, Synthesis, characterization, and applications of ZnO nanowires, Journal of Nanomaterials 2012 (2012) 1–22. Available from: https://doi.org/10.1155/2012/624520.

[28] R.J Joseyphus, T. Matsumoto, H. Takahashi, D. Codama, K. Tohoji, B. Jeyadevan, Designed synthesis of cobalt and its alloys by polyol process, Journal of Solid State Chemistry 180 (11) (2007) 3008–3018. Available from: https://doi.org/10.1016/j.jssc.2007.07.024.

[29] E. Zacharaki, M. Kalyva, H. Fjellvåg, A.O. Sjåstad, Burst nucleation by hot injection for size controlled synthesis of ε-cobalt nanoparticles, Chemistry Central Journal 10 (1) (2016). Available from: https://doi.org/10.1186/s13065-016-0156-1.

[30] Y. Xia, Y. Xiong, B. Lim, S.E. Skrabalak, Shape-controlled synthesis of metal nanocrystals: simple chemistry meets complex physics? Angewandte Chemie International Edition 48 (1) (2009) 60–103. Available from: https://doi.org/10.1002/anie.200802248.

[31] A. Berlie, I. Terrya, M. Szablewsk, Controlling nickel nanoparticle size in an organic/metal–organic matrix through the use of different solvents, Nanoscale 5 (2013) 12212–12223. Available from: https://doi.org/10.1039/C3NR04883G.

[32] S. Duan, Z. Du, H. Fan, R. Wang, Nanostructure optimization of platinum-based nanomaterials for catalytic applications, Nanomaterials (Basel) 8 (11) (2018) 949. Available from: https://doi.org/10.3390/nano8110949.

[33] S. Chen, Z. Niu, C. Xie, M. Gao, M. Lai, M. Li, et al., Effects of catalyst processing on the activity and stability of Pt − Ni nanoframe electrocatalysts, ACS Nano 12 (2018) 8697–8705. Available from: https://doi.org/10.1021/acsnano.8b04674.

[34] L.-N. Zhou, X.-T. Zhang, Z.-H. Wang, S. Guo, Y.-J. Li, Cubic superstructures composed of PtPd alloy nanocubes and their enhanced electrocatalysis for methanol oxidation, Chemical Communications 52 (2016) 12737–12740. Available from: https://doi.org/10.1039/C6CC07338G.

[35] C. Sun, H. Liab, L. Chen, Nanostructured ceria-based materials: synthesis, properties, and applications, Energy & Environmental Science 5 (2012) 8475–8505. Available from: https://doi.org/10.1039/C2EE22310D.

[36] W.-C. Huang, L.-M. Lyu, Y.-C. Yang, M.H. Huang, Synthesis of Cu_2O nanocrystals from cubic to rhombic dodecahedral structures and their comparative photocatalytic activity, Journal of the American Chemical Society 134 (2) (2012) 1261–1267. Available from: https://doi.org/10.1021/ja209662v.

[37] G. Glaspell, H.M.A. Hassan, A. Elzatahry, et al., Nanocatalysis on supported oxides for CO oxidation, Topics in Catalysis 47 (2008) 22–31. Available from: https://doi.org/10.1007/s11244-007-9036-1.

[38] S. Dey, G.C. Dhal, Applications of rhodium and ruthenium catalysts for CO oxidation: an overview, Polytechnica 3 (2020) 26–42. Available from: https://doi.org/10.1007/s41050-020-00023-5.

[39] J. Li, P. Fleurat-Lessard, F. Zaera, F. Delbecq, Switch in relative stability between cis and trans 2-butene on Pt(111) as a function of experimental conditions: a density functional theory study, ACS Catalysis 8 (4) (2018) 3067–3075. Available from: https://doi.org/10.1021/acscatal.8b00544.

[40] Y. Zhang, X.C.F. Shi, Youquan Deng, Nano-gold catalysis in fine chemical synthesis, Chemical Reviews. 112 (4) (2012) 2467–2505. Available from: https://doi.org/10.1021/cr200260m.

[41] A.O. Ibhadon, S.K. Kansal, The reduction of alkynes over Pd-based catalyst materials—a pathway to chemical synthesis, Journal of Chemical Engineering & Process Technology 9 (2) (2018). Available from: https://doi.org/10.4172/2157-7048.1000376.

[42] X. Li, C. Zhang, H. Cheng, W. Lin, P. Chang, B. Zhang, et al., A study on the oxygen vacancies in ZnPd/ZnO-Al and their promoting role in glycerol hydrogenolysis, ChemCatChem 7 (8) (2015) 1322–1328. Available from: https://doi.org/10.1002/cctc.201403036.

[43] V.G. Kravets, A.V. Kabashin, W.L. Barnes, A.N. Grigorenko, Plasmonic surface lattice resonances: a review of properties and applications, Chemical Reviews 118 (12) (2018) 5912−5951. Available from: https://doi.org/10.1021/acs.chemrev.8b00243.

[44] X.-J. Chen, G. Cabello, D.-Y. Wu, Z.-Q. Tian, Surface-enhanced Raman spectroscopy toward application in plasmonic photocatalysis on metal nanostructures, Journal of Photochemistry and Photobiology C: Photochemistry Reviews 21 (2014) 54−80. Available from: https://doi.org/10.1016/j.jphotochemrev.2014.10.003.

[45] C. Liu, L. Chen, T. Wu, Y. Liu, R. Ma, J. Li, et al., Characteristics of electric quadrupole and magnetic quadrupole coupling in a symmetric silicon structure, New Journal of Physics 22 (2020) 023018, February. Available from: https://doi.org/10.1088/1367-2630/ab6cde.

[46] J. Massoudi, M. Smari, K. Nouric, E. Dhahri, K. Khirounid, S. Bertainae, et al., Magnetic and spectroscopic properties of Ni−Zn−Al ferrite spinel: from the nanoscale to microscale, RSC Advances 10 (2020) 34556−34580. Available from: https://doi.org/10.1039/D0RA05522K.

CHAPTER 14

Fuel cells recycling

Ajit Behera
Department of Metallurgical & Materials Engineering, National Institute of Technology, Rourkela, Rourkela, India

14.1 Introduction to fuel cells and recycling

Fuel cell functions same as batteries, but they do not discharge or required to be recharged. As long as they provide fuel, they will generate electricity and heat. It was only in 1960 that the nascent government organization National Aeronautics and Space Administration (NASA) began to find experimental power sources to expand the space missions [1,2] and fuel cells were in a state of obscurity. Through research and development sponsored by the NASA and private companies, fuel cells are expected to become an alternative to internal combustion engines and redesign public utilities by making energy cleaner, more cheaper, and more portable [3]. A fuel cell consists of two electrodes, anode (or negative electrode) and cathode (or positive electrode), placed in between the electrolyte. Examples of fuel cells include alkaline fuel cells, hydrogen-oxygen fuel cells, etc. Fuel cells generate electricity through electrochemical reactions, called reverse electrolysis [4]. The reactant is continuously supplied to the electrode from the container. Unlike conventional batteries, when chemicals are consumed, the fuel cell does not need to be discharged. They are free from polluting emission because the only reaction product is water (used in hydrogen and oxygen fuel cells). The reaction in the fuel cell combines H_2 and O_2 to form water vapor (H_2O), heat, and electricity. Fuel cell can use the three by-products of the reverse electrolysis reaction. The residual heat can be used to heat and cool the spaces. Water vapor can be captured and used as a raw material to supplement hydrogen and electricity can be transmitted to an external loop, where any electrical equipment can be used [5]. Today, fuel cells are used in a wide aspect of applications, from refueling at homes and businesses, servicing key facilities such as hospitals, supermarkets and data centers, cell phones, laptops, to driving various vehicles. The main key driver driving demand for fuel cells is the estimated cost, for domestic fuel cell systems, the price ranges from approximately $35,000 to $100,000 or more. The domestic plant will involve the source of all elements from the hydrogen generator to the storage tank. Below is a calculation example used to determine the cost of running a 1 kW hydrogen fuel cell. In this calculation, a compressed hydrogen (normal temperature and pressure) cylinder with a hydrogen volume of

Table 14.1 Cost of operation for a 1 kW fuel cell on hydrogen.

Compressed hydrogen cylinder contains 200 ft³ of hydrogen (NTP) or 5667.37 L.
The cost of the hydrogen cylinder will be measured at $100 + $14/month (rental) + $20 (delivered) of the 1 kW fuel cell system and the hydrogen utilized is 13 standard liters/min. Operational hour at 1 kW = 5667.37/13 = 7.26 h Cost measurement:

Case 1	Case 2
When 01 cylinder is procured/month, the cost of running a 1 kW fuel cell for 7.26 h is $134. On a kilowatt, this equates to $18.40 per hour.	As the number of hydrogen cylinders used increases, the cost will decrease. Here, assume 05 cylinder of hydrogen are utilized/month, the run time is 7.26 × 5 = 36.33 h. The cost of the gas cylinder is (5 × $100) + (5 × $14) + $20 = $590. In kilowatt units, this equates to $16.24 per hour.

Therefore fuel cells are also scalable: It indicates the individual fuel cells can be connected together to make a stack. Again, these stacks are joined to generate larger systems. The size and power of fuel cell systems vary widely, from replacing internal combustion engines in electric vehicles to large multimegawatt devices that supply power directly to the public grid.

NTP, Normal temperature and pressure.

200 ft³ is used at normal atmospheric condition, and the purity of the hydrogen utilized is 99.99999%. Table 14.1 shows the typical running cost of a 1 kW hydrogen fuel cell [6,7].

The second key performance factor is durability, which meets application expectations based on the life span of the fuel cell system. Under actual operating conditions, the lifespan targets for stationary fuel cells and transportation are 40,000 hours and 5000 hours, respectively [8]. After shelf life, the main concern is how to minimize waste-solid materials and chemical materials in micro and nanoforms. Once the life of the fuel cell has expired, the remaining components are the truly environmentally problematic materials, triggering the recycling of various technologies, so that every part of the fuel cell can be recycled.

14.2 Nanomaterials used in fuel cells

In current practice, nanoscale materials are preferred rather than microscale materials. One opportunity to improve cells through nanoscience/technology is to use unique nanoscale science to produce materials with improved properties and/or reduced costs.

For example, the electrochemical behavior of nanostructured α-Fe$_2$O$_3$ was significantly improved compared to microsamples of the same material, where the amount of intercalated lithium increased from 0.03 Li per formulation (micro) to 0.47 Li per formulation (nanometer). In another example, electrochemically inert Li$_2$MnO$_3$ is activated creating nanostructured materials [9]. For each type of fuel cell, all potential nanomaterials used in fuel cell components are discussed further.

14.2.1 Nanomaterials in the polymer electrolyte membrane or proton-exchange membrane

The main component of the proton-exchange membrane water electrolysis (PEMWE) and proton-exchange membrane fuel (PEMF) cells is the membrane electrode assembly (MEA), which includes two essential layers (anode and cathode) separated by a polymer membrane electrolyte. The development of proton conduction membranes based on sulfonated polymers (in the 1960s) initiated world research work on polymer electrolytic fuel cells, which are considered to have potential, high efficiency, environmental compatibility, and mobility. This is the first innovation in conversion device technology [10]. There are three main functionalities of MEA: reactant barrier, electrical insulator, and proton conductor. Current and commercial membrane materials are perfluorinated materials that suffer from dehydration and temperature limitations. However, for most applications, PEMF cell batteries should operate at temperatures above 1000°C. including the inorganic fillers with polymer electrolyte membrane (PEM) enhances the range of operating temperature, reduce the dehydration and enhance the mechanical properties of PEM. [11]. It is foreseeable that the technical upgradation of fuel cell will be based on nanotechnology. Among them, the main methods are nanoscale catalysts, nanocomposite membranes, and nanostructured MEAs. Methods based on nanotechnology have several advantages. Nanostructured catalysts with platinum nanoparticles are able to increase the activity of the catalyst. Likewise, nanoporous carbon can be added as support material to manage the particle size as well as the catalyst distribution. Especially when oxygen-reducing promoters (e.g., Ru-N complexes) and highly metal-interactive elements (e.g., S) are incorporated into the nanoporous carbon, the catalyst is utilized to improve activity and stability [12]. Hydrocarbon nanocomposite membranes of exfoliated clay, especially in direct methanol fuel cells, have high ion conductivity and low permeability to methanol due to their sufficiently high degree of sulfonation. High surface density nanostructured MEA can reduce ohmic loss and activation loss in the polarization process and improve fuel efficiency, which can increase overall fuel cell efficiency by more than 40% [13].

The size of the silica nanoparticles in the Nafion silica nanocomposites will influence the enhancement of the properties of the nanocomposites (e.g., water absorption, conductivity, and resistance to protons). Pt-based materials are the primary electrocatalysts utilized in PEM stacks. PEMWE involves the use of metal constituents instead of C-based components. The main materials utilized in PEMWE for flow fields and separators are based on titanium alloys or coated-stainless steel [14]. Among PEMWE, Ir is preferable to Ru and Pt due to its adequate catalytic behavior and greater resistance to corrosion in oxygen evolution reactions. The MEA is stacked in series with multiple bipolar plates interconnected until the required stack power is reached. The bipolar plate acts as a current collector between the batteries from one end to the other. They usually consist of graphite, carbon-polymer compounds, and polymers. Aluminum and metal alloys (e.g., CrNi-steel) are also favorable for bipolar plates due to their high electrical conductivity and mechanical properties. But, due to corrosives, bipolar metal sheets must be modified or surface coated [15].

Carbon nanotube (CNT) can act as an alternative catalyst material in PEM fuel cells due to its encouraging properties. CNT provides a higher surface area and chemical stability that can improve the electrocatalytic activity and stability of Pt-catalyst particles through their active cooperation. The typical surface modification of the Pt catalyst with CNT need a complicated sequence to follow, for example, wet chemical techniques, such as oxidation and purification of the CNT surface and solution reduction of the Pt precursor [16]. Recently, multimetal research has been conducted on the design and nanoengineering of size, shape, and composition parameters at the nanoscale of multimetal and platinum-based gold-based nanoparticles and catalysts [17]. Pt- and Au-based polymetallic nanoparticles incudes transition metals such as Ni, Fe, Ti, Co, V, Cu, Cr, W, Zr, Pd, Ag, Ir, and Rh.

14.2.2 Nanomaterials in the molten carbonate fuel cells

Molten carbonate fuel cell (MCFC) is considered a promising alternative energy source for power generation systems due to its high electrochemical conversion efficiency, fuel flexibility, environmental friendliness, and heat-power cogeneration. Today, MCFC is ready to enter the commercialization phase. For commercialization, the MCFC stack runtime is expected to be 40,000 hours. One necessary factor for long-term performance of MCFC is ensuring the matrix stability, which can be used as a reservoir for liquid electrolytes and plays a great character as an inhibitor between the cathode and anode [18]. Also, due to the direct reaction between hydrogen and oxygen (rather than an electrochemical reaction), cracks formed in the matrix during battery operation can greatly reduce battery performance. The decrease in battery performance also prevents the MCFC battery from running more than 40,000 hours for a long time. To avoid the formation of cracks during battery operation, it is important to increase the substrate strength. Some research has performed, for example, adding the sintering aids such as boron oxide (B_2O_3) and aluminum acetylacetonate [$(C_5H_7O_2)_3Al$] to strengthen the $LiAlO_2$ matrix [19]. The development of electrode materials for MCFCs is a key issue, as due to the specific operating environment of these batteries, a balance between mechanical and electrochemical properties is required. Regarding the anode, one possible strategy is to use improved nanoparticles in the electrode manufacturing process. Candidate nanomaterials include, but are not limited to, TiO_2, CeO_2, ZrO_2, Al, Ti, Mg, etc. [20]. The use of carbide oxide nanoparticles allows for better functionality and potential prospects regarding the use of the third alloying metal. However, although nanocerium oxide has the highest mechanical properties, its polarization and power curve are also higher, so the performance of nanozirconia is slightly better than nanocerium oxide. The MCFC anode material must have high creep resistance and good mechanical properties to withstand high temperatures and corrosive environments.

14.2.3 Materials in phosphoric acid fuel cell

Phosphoric acid fuel cell (PAFC) is a fuel cell that utilizes liquid phosphoric acid as an electrolyte. This is the first commercial fuel cell. They were developed in the mid-1960s and have been field tested since the 1970s and have achieved gradual improvements in various factors (such as, stability, performance, and cost) by incorporating various materials. These features make PAFC an ideal choice for first stationary applications. With different concentrations of H_3PO_4 and by observing the kinetics of the oxygen reduction reaction (ORR), the effect of phosphate anion poisoning on the Pt-type cathode catalyst of phosphoric acid was investigated. The toxic effect of phosphate on ORR is much less severe in PtNi/C than in Pt/C. Indeed, the potential of the phosphate anion adsorbed on PtNi is greater than that of PtNi, which inhibits the activation of water by OH and OH to O conversion, thus providing more positions for ORR. Due to the presence of phosphate anion at the superior site of PtNi/C, OH is now unable to occupy these sites due to steric hindrance or low utilization of the superior site. Hence it appears that the remaining sites (twofold and threefold) for making molecular oxygen undergo ORR [21].

14.2.4 Materials in solid oxide fuel cells

Compared to most existing fuel cells, solid oxide fuel cell (SOFC) shows great versatility in fuels. Due to this flexibility, hydrocarbons are allowed as a fuel and can tolerate the existence of carbon monoxide. A higher working temperature (650°C–1000°C) can also recover thermal energy for coproduction. The ionic conductivity of 0.1 S cm^{-1}, which is a key requirement, limits conventional SOFC electrolytic material. Additionally, SOFC produces zero or very low SO_x release and avoids the noise normally associated with traditional power harvest systems. SOFC made up of dense ceramic membrane (O_2 ion transporter) sandwiched between two porous electrodes [22]. The membrane can rest on a porous anode/cathode (5–50 mm) and act as a structural component having approximately 100 mm thickness. Yttria–stabilized zirconia (YSZ) is the most advanced electrolyte in SOFC due to its long-term and high-temperature stability. A single SOFC cell is joined in series by the help of bipolar plates as interconnects. Perovskite-type oxides are the preferable cathode materials and lanthanum chromite (LaCrO$_3$) is the most common interconnector material. When working temperature is less than 700°C and cost is the dominating factor, then porous stainless steel is the better choice for substrate. Typical SOFC electrolytic materials have shortcomings in monophasic materials and biphasic materials and nanocomposite materials need to be designed and developed. Interphase handling of nanocomposites can overcome the challenges of SOFC, thereby increasing and improving the electrical conductivity of the material. The stationary SOFC system successfully combines the carbon neutral design concept with the MgH$_2$ material to produce a low heat value of 82% and, in all available cases, when the total energy production is reduced to a minimum. Minimum the efficiency, energy production becomes 68.6% provides heat to the exhaust gases and to the system to heat the tank [23]. To develop magnesium-based hydrogen storage materials, various nanoprocessing technologies have been widely used to synthesize magnesium-based materials with small particles and crystallite sizes, resulting in good hydrogen storage kinetics but poor thermal conductivity.

14.3 Recycling processes

After the useful life of the fuel cell, it is more crucial to recycle it to protect the environment from waste. Table 14.2 lists the parts and materials available in a typical fuel cell. In this case, the exact composition of the material depends on the corresponding processing and can vary between the brand and the life of the battery pack.

Various advanced processes to recycle the fuel cell matter have been discussed further.

Table 14.2 Separation of general materials from PEM fuel cells.

Steps	Separation action	Fraction produced
1	Tie rods and casing removal	Connecting elements (high-grade steel)
	End plates removal	Connecting elements (high-grade steel)
2		Insulation (plastics)
		End plates (high-grade steel or Al alloy)
	Stack's individual layers removal,	Seal (plastics)
	Bipolar plates, gas, seal, diffusion layer, and membrane electrode removal	Bipolar plates (polymeric binders, graphite, separating agents, additives)
3		Gas diffusion layer (nonwoven fabric, carbon paper)
		Membrane electrode assembly (polymeric membrane coated with graphite and Pt)

14.3.1 Hydrometallurgical process

The hydrometallurgical approaches imply the dissipation of target elements from solid substrates by caustic or acid attack. This stage is usually followed by precipitation separation, solvent-extraction, distillation, ion-exchange, cementation, or filtration. The major advantage of this method includes higher selectivity toward metals, lower energy expenditure and the ability to recover reagents. However, the hydrometallurgical process implies the requirement for mechanical pretreatment to enhance the activity of exposed surface to the reagents, the production of large quantities of solution and wastewater that can be corrosive and/or toxic. The hydrometallurgical process of recovering Pt-based metal catalysts from PEMF cell and PEMWE generally involves subprocesses of pretreatment, leaching, separation, and purification [24]. To increase the overall technical and economic efficiency of the procedure, multiple threads may be involved, for example, solvent regeneration. The efficiency of leaching, separation, and precipitation is high, which can improve the overall recovery rate. The leaching agent is a solution of a strong acid and an oxidizing agent. Aquaregia is a solution of HCl and HNO_3 (3:1 molar ratio) is generally used to recover platinum-based metals from catalysts used on carbon supports. However, due to the high acid concentration at pH < 1, leaching will create a rather severe environment. After leaching, in the filtration stage, carbon particles remove from the Pt-based metal solution. The metal anions of the platinum group can be separated in several ways. For Pt, there are two options: one is an anion-exchange resin processes followed by desorption of the resin; the second is liquid-liquid extraction followed by Pt separation. To precipitate platinum as $(NH_4)_2PtCl_6$, the Pt-rich stream in the separation technique is processed with NH_4Cl. Finally, the flux is filtered and the platinum is recovered as a solid [25]. It is possible to add an ignition phase at 350°C to obtain high purity Pt (99.9%). Experimental results show that the Pt-recovery rate reaches 76%. In other way, $(NH_4)_2PtCl_6$ can be used with C powder to produce advanced electrode ink. This technique is also applicable to the recovery of Ir and Ru. Similarly, by altering the lechant, this series of processes can be used for Ni recovery in SOFC, Raney-Ni and silver recovery in AWE [26].

14.3.2 Pyrohydrometallurgical process

In the PEM systems, high-temperature hydrometallurgical process involve with the calcination step, in which the gas diffusion layer, membrane and electrodes are incinerated. The generated ash products are treated by acid dissolution and then the Pt-based metals are recovered by precipitation. Compared to hydrometallurgical treatment, the high-temperature hydrometallurgical process does not require mechanical pretreatment and has high-recovery performance, however needs higher energy [27]. The number of steps and easy procedures associated with hydrometallurgical processing favor its applicability. In unfavorable conditions such as higher operating temperature and pH can cause safety and environmental problems. The process consists of three phases: (1) drying and calcination, (2) leaching and purification, and (3) reduction and filtration. The first step is to dry the MEA at 80°C for 3 hours. The dried MEA was then calcined at 600°C for 6 hours to remove the remaining carbon and volatiles. The ash produced by calcination was leached using HNO_3 at 65°C (Fig. 14.1). In next purification stage, the solution was heated to 110°C in addition to HCl. Then use deionized water and NaOH to dissolve the resulting H_2PtCl_6 solid at pH 3—4. Add formic acid to reduce Pt. At end of the step, the platinum black was recovered by drying and vacuum filtration [28].

14.3.3 Hydrothermal process

Hydrothermal treatment associates treating waste with steam at higher temperatures and pressures. Fig. 14.2 shows the hydrothermal recovery process of YSZ polycrystals. Due to its complexity, this process can take a relatively long time. Use 200°C—240°C water to decompose YSZ contained materials. The powder was collected by filtration and then dried at 50°C. Then the dry powder was ground and sieved until the size was less than 100 mm. The screened micropowder is pressed uniaxially and then cold isostatic pressed. The resulting material was sintered at a temperature of 1400°C—1600°C for 2 hours to get the recycled zirconia [29].

14.3.4 Selective electrochemical dissolution

This process can recover the carbon support and catalyst from the catalyst coated membrane of the MEA (Fig. 14.3). This process depends on the electrochemical dissolution of materials under various voltage and pH. The first step is the electrochemical cleaning of the catalyst coating membrane. Then purge the system with O_2 and

Figure 14.1 Pyrohydrometallurgical Pt-recovery process.

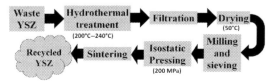

Figure 14.2 Yttria-stabilized zirconia recovery by hydrothermal process.

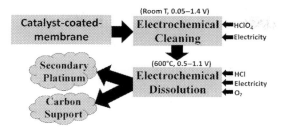

Figure 14.3 Selective electrochemical dissolution of catalyst-coated membrane.

change the voltage and pH. Dissolution process carried out at potential below 1.2 V to avoid the corrosion of the carbon support and allowing its recovery. By changing the pH and voltage conditions, this method can be extended to the recovery of other materials in other fuel cell and hydrogen devices [30]. This technology can recover high purity catalysts under relatively mild pH, temperature, and voltage conditions, providing adequate technical environmental performance. Since the carbon surface of the holder shows changes that affect its function, it can be reused directly, so it must first be renewed.

14.3.5 Transient dissolution through potential alteration

This method can simultaneously recover Pt-based and carbon-based materials. It focuses on transients caused by periodic changes in the oxidation state of the Pt surface. In the transition phase, the presence of chloride can stabilize Pt ions. O_3 and CO are used as oxidizing and reducing agents, respectively. Oxidation at Pt surface changes periodically until the Pt dissolves completely. The main advantages of this process are avoiding the external potential, higher rate of recovery and periodic working conditions. By adjusting the operating conditions (e.g., number of cycles and duration), this method can be used to recover other metals in PEMWE and AWE, such as Ru, Pd, and Ir [31].

14.3.6 Membrane and Pt-recovery acid process

The acidic method is used to recover Pt-based materials. In this process strong acid is used oxidize the carbon support followed by separation steps such as filtration and centrifugal separation. This approach can efficiently recover platinum group metals and ionomers of the membrane coated with catalyst (Fig. 14.4). One of the main features of this method is that it can regenerate the membrane of a new battery cell with adequate electrochemical performance. Compare with traditional techniques (such as hydrometallurgy and thermal hydrometallurgy), this new method has advantages in improving the recovery rate of 1Pt-based materials and reducing harmful emissions [32]. However, the harsh pH conditions and the relatively complex chains and long processes represent the main limitations of the technique.

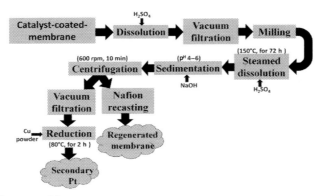

Figure 14.4 Membrane and Pt-recovery process.

14.3.7 Alcohol solvent process

This process uses an alcohol solvent process to recover ionomers and precious metals from the PEM unit. In this case a pretreatment is carried out so that, in addition to the catalyst, the polymer resin can be subsequently recovered. The first step associated with delamination of the MEA layer by the help of alkyl solution. The microwave heater raises the temperature until a range suitable for dissolution is reached. The particle size of the polymer in the dispersion is a key parameter for subsequent filter press and ultrafiltration separation. The microwave heater regulates the parameter by varying the time period and temperature. In general, the technique is characterized by mild pH, temperature, and voltage conditions. However, this recycling process raises concerns about its economic viability. In this sense, solvent recovery can improve technical environmental performance and, at the same time, can significantly reduce costs [33].

14.4 Microbial fuel cell recycling mechanism

Humans can utilize the natural process of microorganisms by using microbial fuel cells (MFCs) and integrated photomicrobial fuel cells containing electrodes, thus maximizing the oxidation rate of microorganisms by controlling two types of microorganisms microbes (choice of algae and bacteria) and the applied physicochemical conditions [34]. It focuses on the natural recycling of natural elements by microorganisms, the productivity of bacteria and microalgae as fuels, the decomposition and use of MFCs to supplement the main biomass production (at the cathode) and the decomposition process and transformation by heterotrophic microbes (at the anode). It shows the potential use of the photomicrobial fuel cell as a complete recycling machine in the future. It has advantages over all other biorecycling systems, including fast recycling rate, water production, carbon dioxide removal, oxygen release, and generation (not electricity). The MFC originated from the discovery and invention of Potter1 in 1911 and has a history of over 100 years. However, due to the extremely low level of energy (nW) produced, fuel cells were considered the curiosity of the laboratory at the time. In 1931 Cohen created a semifuel cell for the batch culture of microorganisms, when connected in series; the fuel half-cell can generate a voltage greater than 35 V, which means that the half-cell equivalent is 75—80 MFC [35].

14.5 Summary

The fuel cell system is a clean, efficient, reliable, and quiet energy source. But at the end of their life cycle, fuel cell materials can become an environmental problem. This chapter looks at the various processes for recycling and reusing fuel cell materials.

References

[1] See: <fuelcellstore.com/education>, 2021 (accessed 01.03.21).
[2] E.M. Cohn, NASA's fuel cell program, Advances in Chemistry 47 (1969) 1—8. Available from: http://doi.org/10.1021/ba-1965-0047.ch001.
[3] I. Staffell, D. Scamman, A.V. Abad, P. Balcombe, E. Paul, The role of hydrogen and fuel cells in the global energy system, Energy & Environmental Science 12 (2019) 463—491. Available from: https://doi.org/10.1039/C8EE01157E.
[4] M.A. Laguna-Bercero, Recent advances in high temperature electrolysis using solid oxide fuel cells: a review, Journal of Power Sources 203 (2012) 4—16. Available from: https://doi.org/10.1016/j.jpowsour.2011.12.019.
[5] A. Kirubakaran, S. Jain, R.K. Nema, A review on fuel cell technologies and power electronic interface, Renewable and Sustainable Energy Reviews 13 (9) (2009) 2430—2440. Available from: https://doi.org/10.1016/j.rser.2009.04.004.
[6] T. Elmer, M. Worall, S. Wu, S.B. Riffat, Fuel cell technology for domestic built environment applications: state of-the-art review, Renewable and Sustainable Energy Reviews 42 (2015) 913—931. Available from: https://doi.org/10.1016/j.rser.2014.10.080.
[7] See: <https://www.fuelcellstore.com/hydrogen-education>, 2021 (accessed 03.03.21).
[8] See: <https://www.energy.gov/eere/fuelcells/durability-working-group>, 2021 (accessed 03.02.21).
[9] G. Liu, S. Zhang, One-step synthesis of low-cost and high active Li_2MnO_3 cathode materials, International Journal of Electrochemical Science 11 (2016) 5545—5551. Available from: https://doi.org/10.20964/2016.07.35.
[10] P. Kallem, N. Yanar, H. Choi, Nanofiber-based proton exchange membranes: development of aligned electrospun nanofibers for polymer electrolyte fuel cell applications, ACS Sustainable Chemistry & Engineering 7 (2) (2019) 1808—1825. Available from: https://doi.org/10.1021/acssuschemeng.8b03601.
[11] B.P. Tripathi, M. Kumar, V.K. Shahi, Highly stable proton conducting nanocomposite polymer electrolyte membrane (PEM) prepared by pore modifications: an extremely low methanol permeable PEM, Journal of Membrane Science 327 (1—2) (2009) 145—154. Available from: https://doi.org/10.1016/j.memsci.2008.11.014.
[12] G. Rao, Y. Gogotsi, K. Swider-Lyons, Nanomaterials for polymer electrolyte membrane fuel cells; materials challenges facing electrical energy storage, in: 2009 MRS Fall Meeting, Boston, MA, United States, 2010. doi:10.2172/984665.

[13] D. Hao, J. Shen, Y. Hou, Y. Zhou, H. Wang, An improved empirical fuel cell polarization curve model based on review analysis, International Journal of Chemical Engineering 2016 (2016) 1−10. Available from: https://doi.org/10.1155/2016/4109204.

[14] S. Lædre, O.E. Kongstein, A. Oedegaard, H. Karoliussen, F. Seland, Materials for proton exchange membrane water electrolyzer bipolar plates, International Journal of Hydrogen Energy 42 (5) (2017) 2713−2723. Available from: https://doi.org/10.1016/j.ijhydene.2016.11.106.

[15] S. Karimi, N. Fraser, B. Roberts, F.R. Foulkes, A review of metallic bipolar plates for proton exchange membrane fuel cells: materials and fabrication methods, Advances in Materials Science and Engineering 2012 (2012) 1−22. Available from: https://doi.org/10.1155/2012/828070.

[16] X. Hu, S. Dong, Metal nanomaterials and carbon nanotubes—synthesis, functionalization and potential applications towards electrochemistry, Journal of Materials Chemistry 18 (2008) 1279−1295. Available from: https://doi.org/10.1039/B713255G.

[17] N.V. Long, C.M. Thi, Y. Yong, M. Nogami, M. Ohtaki, Platinum and palladium nano-structured catalysts for polymer electrolyte fuel cells and direct methanol fuel cells, Journal of Nanoscience and Nanotechnology 13 (7) (2013) 4799−4824. Available from: https://doi.org/10.1166/jnn.2013.7570. PMID: 23901503.

[18] J. Asenbauer, T. Eisenmann, M. Kuenzel, A. Kazzazi, Z. Chen, D. Bresser, The success story of graphite as a lithium-ion anode material—fundamentals, remaining challenges, and recent developments including silicon (oxide) composites, Sustainable Energy Fuels 4 (2020) 5387−5416. Available from: https://doi.org/10.1039/D0SE00175A.

[19] H.-S. Kim, S.-A. Song, S.-C. Jang, D.-N. Park, H.-C. Ham, S.-P. Yoon, et al., Enhancement of mechanical strength using nano aluminum reinforced matrix for molten carbonate fuel cell, Transactions of the Korean Hydrogen and New Energy Society 23 (2) (2012) 143−149. Available from: https://doi.org/10.7316/khnes.2012.23.2.143.

[20] G. Accardo, D. Frattini, S.P. Yoon, H.C. Ham, S.W. Nam, Performance and properties of anodes reinforced with metal oxide nanoparticles for molten carbonate fuel cells, Journal of Power Sources 370 (2017) 52−60. Available from: https://doi.org/10.1016/j.jpowsour.2017.10.015.

[21] N. Sammes, R. Bove, K. Stahl, Phosphoric acid fuel cells: fundamentals and applications, Current Opinion in Solid State and Materials Science 8 (5) (2004) 372−378. Available from: https://doi.org/10.1016/j.cossms.2005.01.001.

[22] S.C. Singhal, Solid oxide fuel cells for power generation, Wires Energy and Environment 3 (2) (2014) 179−194. Available from: https://doi.org/10.1002/wene.96.

[23] M. Powell, K. Meinhardt, V. Sprenkle, L. Chick, G. McVay, Demonstration of a highly efficient solid oxide fuel cell power system using adiabatic steam reforming and anode gas recirculation, Journal of Power Sources 205 (2012) 377−384. Available from: https://doi.org/10.1016/j.jpowsour.2012.01.098.

[24] A. Valente, D. Iribarren, J. Dufour, End of life of fuel cells and hydrogen products: from technologies to strategies, International Journal of Hydrogen Energy 44 (38) (2019) 20965−20977. Available from: https://doi.org/10.1016/j.ijhydene.2019.01.110.

[25] B.R. Reddy, B. Raju, J.Y. Lee, H.K. Park, Process for the separation and recovery of palladium and platinum from spent automobile catalyst leach liquor using LIX 841 and Alamine 336, Journal of Hazardous Materials 180 (1−3) (2010) 253−258. Available from: https://doi.org/10.1016/j.jhazmat.2010.04.022.

[26] A. Faes, A. Hessler-Wyser, A. Zryd, J. Van Herle, A review of RedOx cycling of solid oxide fuel cells anode, Membranes (Basel) 2 (3) (2012) 585−664. Available from: https://doi.org/10.3390/membranes2030585.

[27] M.K. Jha, A. Kumari, R. Panda, J.R. Kumar, K. Yoo, J.Y. Lee, Review on hydrometallurgical recovery of rare earth metals, Hydrometallurgy 165 (1) (2016) 2−26. Available from: https://doi.org/10.1016/j.hydromet.2016.01.035.

[28] R.R. Srivastava, M. Kim, J. Lee, et al., Resource recycling of superalloys and hydrometallurgical challenges, Journal of Materials Science 49 (2014) 4671−4686. Available from: https://doi.org/10.1007/s10853-014-8219-y.

[29] A.M. Férriz, A. Bernad, M. Mori, S. Fiorot, End-of-life of fuel cell and hydrogen products: a state of the art, International Journal of Hydrogen Energy 44 (25) (2019) 12872−12879. Available from: https://doi.org/10.1016/j.ijhydene.2018.09.176.
[30] J.A. Moghaddam, M.J. Parnian, S. Rowshanzamir, Preparation, characterization, and electrochemical properties investigation of recycled proton exchange membrane for fuel cell applications, Energy 161 (2018) 699−709. Available from: https://doi.org/10.1016/j.energy.2018.07.123.
[31] A. Pavlišič, P. Jovanovič, V.S. Šelih, M. Šala, N. Hodnik, M. Gaberšček, Platinum dissolution and redeposition from Pt/C fuel cell electrocatalyst at potential cycling, Journal of the Electrochemical Society 165 (6) (2018).
[32] F. Xu, S. Mu, M. Pan, Recycling of membrane electrode assembly of PEMFC by acid processing, International Journal of Hydrogen Energy 35 (7) (2010) 2976−2979. Available from: https://doi.org/10.1016/j.ijhydene.2009.05.087.
[33] H.-C. Cheng, F.-S. Wang, Optimal process/solvent design for ethanol extractive fermentation with cell recycling, Biochemical Engineering Journal 41 (3) (2008) 258−265. Available from: https://doi.org/10.1016/j.bej.2008.05.004.
[34] J. Greenman, I. Gajda, I. Ieropoulos, Microbial fuel cells (MFC) and microalgae; photo microbial fuel cell (PMFC) as complete recycling machines, Sustainable Energy Fuels 3 (2019) 2546−2560. Available from: https://doi.org/10.1039/C9SE00354A.
[35] C. Munoz-Cupa, Y. Hu, C. Xu, A. Bassi, An overview of microbial fuel cell usage in wastewater treatment, resource recovery and energy production, Science of the Total Environment 754 (2021) 142429. Available from: https://doi.org/10.1016/j.scitotenv.2020.142429.

CHAPTER 15

Micro/nanostructures for biofilm establishment in microbial fuel cells

Linbin Hu[1,2], Jun Li[1,2], Qian Fu[1,2], Liang Zhang[1,2], Xun Zhu[1,2] and Qiang Liao[1,2]

[1]Key Laboratory of Low-grade Energy Utilization Technologies and Systems, Chongqing University, Ministry of Education, Chongqing, China
[2]Institute of Engineering Thermophysics, School of Energy and Power Engineering, Chongqing University, Chongqing, China

15.1 Introduction

In the past decades, the increasingly environmental pollution and energy shortage call for the development renewable technology for clean energy production and environmental remediation [1–3]. Among these, microbial fuel cell (MFC) technology that can recover bioelectricity from the chemical energy in organic wastewater by the metabolic activity of electrochemical active bacteria (EABs) have attracted worldwide attention [4,5]. The working principle of a typical MFC is shown in Fig. 15.1, the MFC system maintains the mass and energy balance. In detail, the EABs in electroactive biofilms attached to the anode oxidizes organic matters in wastewater to generate electrons and protons. The produced electrons are then transferred to the anode, and pass through the external circuit to the cathode due to the potential difference between the anode and cathode, thus generating current. The protons produced from the anode reaction diffuses to the cathode due to the concentration difference and participates in the cathode reaction with the reception of electrons and electron acceptor [6]. As an example, the reactions for the MFC using acetate and oxygen as the substrate and cathodic electron acceptor at operating conditions of pH 7, 5 mM HCO_3^-, 1 g L^{-1} acetate, and 25°C are as follows [Eqs. (15.1)–(15.3)]:

Anode reaction:

$$CH_3COO^- + 4H_2O \rightarrow 2HCO_3^- + 9H^+ + 8e^- \quad E_{anode} = -0.300 \text{ V} \quad (15.1)$$

Cathode reaction:

$$O_2 + 4e^- + 4H^+ \rightarrow 2H_2O \quad E_{cathode} = 0.805 \text{ V} \quad (15.2)$$

Overall reaction:

$$CH_3COO^- + 2O_2 \rightarrow 2HCO_3^- + H^+ \quad E_{cell} = 1.105 \text{ V} \quad (15.3)$$

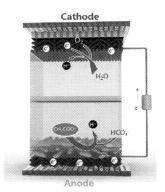

Figure 15.1 Schematic diagram of a typical MFC (*MFC*, Microbial fuel cell).

As predicted by Nernst equation, the theoretical voltage that can be generated from a MFC is 1.105 V. However, in practical application, the operating voltage of an MFC is only about $0.3 \sim 0.4$ V [7,8], which is much lower than the theoretical one. During the past decades, tremendous efforts have been devoted to the optimization of MFC design, the synthesis of cathode materials, the preparation of membrane/separator, and the understanding of the fundamentals issues of microbial metabolism, aiming to improve the MFC performance. However, the power generation is till insufficient as compared to the theoretical volumetric power density of the cell (53 kW m^{-3}) for industrial application [7].

As an essential part of an MFCs, the anode material have a significant impact on the MFC performance because it is the solid matrix for biofilm establishment, electron transfer and substrate degradation. In general, the establishment of biofilm on the electrode surface involves the following procedures: (1) adhesion of individual bacteria on the electrode surface; (2) formation of mono/multilayers of bacteria; (3) construction of polymerized scaffold of biofilm; and (4) mature biofilm establishment with 3D porous structure [9]. It is well accepted that modifying electrode materials by introducing functional groups or nanostructures on the electrode surface usually lead to a better biocompatibility, a higher electrical conductivity, and a larger specific surface area that can promote the extracellular electron transfer (EET) and biofilm establishment on the electrode surface, and consequently resulting in a higher anode performance [10,11]. However, EABs are living colloidal particles that range in size from about 0.5 um to several micrometers in length and ~ 1 um in diameter. The size incompatibility between the anode porous structure and EABs usually causes blockage of substrate and product transfer inside the electrode, inducing proton diffusion limitation and insufficient substrate supply in the interior of the electrode, eventually inhibiting the internal microbial colonization of the anode and seriously limiting the anode performance [12,13].

Although many reviews on the anode materials [14,15], strain type [16,17], electrons transfer mechanisms [18—20], and actual application [21,22] have been published previously, there are very few reviews concerning the influence of electrode structure on biofilm establishment. Recently, significant progress have been achieved on the fabrication and application of micro- and nanostructured electrode for the facilitation of EAB biofilm establishment and the improvement of the MFC performance. Hence this chapter reviews the up-to-date development of electrode with micro/nanostructures that have been used for the MFC anode over the past decades. Specifically, (1) provides a detailed discussion on how the nanostructure affecting the bacteria—surface interaction; (2) comprehensively reviews the electrodes with microstructures for anode for the facilitated biofilm establishment; and (3) discusses the critical challenges for the practical application of the electrodes with micro/nanostructures in MFC anodes.

15.2 Nanostructure for promoting electrochemical active bacteria/electrode interaction

In a MFC, anode serves as the substratum of EAB biofilm and is in direct contact with EAB. Therefore the nanostructure and functional groups on its surface play a crucial role on the lag time during start-up, the active biomass and EET efficiency during stable operation by influencing the roughness, substrate/product transfer, and EET pathway of the electrode [23,24]. In general, the anode nanostructure can be mainly prepared by (1) modifying the electrode with carbon nanomaterials; (2) coating with conductive polymers; and (3) loading with metal oxides nanoparticles. In the following section, the recent progress of the above-mentioned categories will be discussed in detail.

15.2.1 Carbon nanomaterials decoration

Carbon nanomaterials, such as carbon black, carbon nanotubes (CNTs), and graphene, have been widely employed in anode fabrication due to their large specific surface area and excellent electrochemical properties and biocompatibility [25,26]. Typical anodes coated with carbon nanomaterials used for biofilm were shown in Fig. 15.2 and listed in Table 15.1. Among various commercial carbon nanomaterials, carbon black is one of the most cost effective one (Fig. 15.2A and B). It is a conductive powdery nanomaterial with a large surface area of $10-3000 \text{ m}^2 \text{ g}^{-1}$ and can be easily produced from incomplete combustion or thermal decomposition of hydrocarbons such as coal and petroleum. Previous studies demonstrated that the modification of commonly used electrode materials, such as 3D macroporous carbon fiber electrode [34] and stainless steel mesh [49] by carbon black increased the roughness, electrical conductivity, and biocompatibility of the electrode, allowing a faster microbial adhesion and more active biomass on the electrode surface, leading to an the enhanced EET rate [36] and thus a better anode performance. In addition, carbon black can effectively reduce the charge transfer resistance

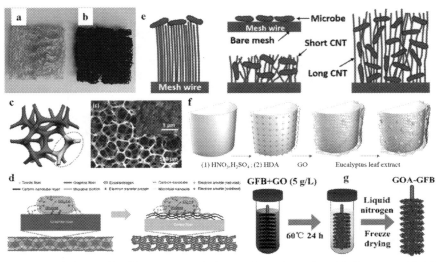

Figure 15.2 Digital photo of loofah sponge (A) before and (B) after carbon black coating. (C) Schematic diagram (left) and SEM image (right) of CNT-sponge revealing the macroporous structure and continuous 3D CNT coating. (D) Schematic of the electrode configuration and electron-transfer mechanisms for the CNT-textile anode (right), compared with the widely used carbon cloth anode (left). (E) Schematic diagram of the interactions between bacteria and the CNT-stainless steel mesh electrode. (F) Schematic diagram of the synthesis of the EL-RGO/CFP hybrid anode. (G) Schematic diagram for the fabrication process of GOA-GFB electrode (*CNT*, Carbon nanotube; *SEM*, scanning electron microscopy; *EL-RGO/CFP*, eucalyptus leaves-reduced graphene oxide/carbon fiber paper; *GOA-GFB*, graphene oxide aerogel-graphite fiber brush). *Reprinted with permission from J. Zheng, C. Cheng, J. Zhang, X. Wu, Appropriate mechanical strength of carbon black-decorated loofah sponge as anode material in microbial fuel cells, International Journal of Hydrogen Energy 41 (2016) 23156–23163, https://doi.org/10.1016/j.ijhydene.2016.11.003 [27]. Copyright 2016. Elsevier. Reprinted with permission from X. Xie, et al., Carbon nanotube-coated macroporous sponge for microbial fuel cell electrodes, Energy & Environmental Science 5 (2012) 5265–5270, https://doi.org/10.1039/c1ee02122b [28]. Copyright 2012. The Royal Society of Chemistry. Reprinted with permission from X. Xie, et al., Three-dimensional carbon nanotube-textile anode for high-performance microbial fuel cells, Nano Letters 11 (2011) 291–296, https://doi.org/10.1021/nl103905t [29]. Copyright 2010. American Chemical Society. Reprinted with permission from C. Erbay, et al., Control of geometrical properties of carbon nanotube electrodes towards high-performance microbial fuel cells, Journal of Power Sources 280 (2015) 347–354, https://doi.org/10.1016/j.jpowsour.2015.01.065 [30]. Copyright 2015. Elsevier. Adapted and reprinted with permission from S. Zhou, M. Lin, Z. Zhuang, P. Liu, Z. Chen, Biosynthetic graphene enhanced extracellular electron transfer for high performance anode in microbial fuel cell, Chemosphere 232 (2019) 396–402, https://doi.org/10.1016/j.chemosphere.2019.05.191 [31]. Copyright 2019. Elsevier. Adapted and reprinted with permission from X. Yang, et al., Eighteen-month assessment of 3D graphene oxide aerogel-modified 3D graphite fiber brush electrode as a high-performance microbial fuel cell anode, Electrochimica Acta 210, (2016) 846–853, https://doi.org/10.1016/j.electacta.2016.05.215 [32]. Copyright 2016. Elsevier.*

and enhance the EET. Pretreatment of carbon black with oxidation can also effectively enhance the concentration of oxygen-containing functional groups on the electrode surface, which enhances the biocompatibility of electrodes and facilitates the adhesion and

Table 15.1 Summary of different carbon nanomaterials-modified electrodes as MFC anodes (MFC, Microbial fuel cell).

Carbon nanomaterial	Anode base material	Substrate	Modification method	Max power/current density	Anode surface characterization	Optimization effect	References
Carbon black	Carbon veil/cloth	Urine	Silk screen	60.7/ 50.6 mW m^{-2}	High roughness and surface area	220%/180%	[33]
Carbon black	Carbon fiber mat	Acetate	Electrospun	21 A m^{-2}	High porosity and conductivity	124%	[34]
Carbon black	Stainless steel mesh	Acetate	Binder-free dipping/drying	3215 mW m^{-2}	Good adhesion		[35]
Carbon black and coke	Polyvinyl alcohol	Toluene	Mixing and drying	72 mW m^{-2}	High porosity and conductivity	278%	[36]
Carbon black	Loofah sponge	Acetate	Dipping and drying	61.7 W m^{-3}	Good biocompatibility	117%	[27]
CNT	Carbon paper	Glucose	Self-assembly	480 mW m^{-2}	High specific surface area and conductivity	148%	[37]
CNT	Porous textile	Glucose	Dipping and drying	1098 mW m^{-2}	Two-scale porous and nanowires	168%	[29]
CNT	Sponge	Domestic wastewater	Dipping and drying	1240 mW m^{-2}	Uniform macroporous and high conductivity	250% (compared to previous study)	[28]
MWCNT	Stainless steel mesh	Acetate	Chemical vapor deposition	3360 mW m^{-2}	High specific surface area and conductivity	740% (compared to bare carbon cloth)	[30]
CNT	Gold-coated glass slides	Acetate	Spraying/drying	3320 mW m^{-3}	High conductivity and biocompatibility	850% (compared to previous study)	[38]
MWCNT	Biofilm	Acetate	Adsorption–filtration	1130 mW m^{-2}	High porosity and conductivity	159%	[39]
Graphene	Stainless steel mesh	Glucose	Pressing and drying	2668 mW m^{-2}	Good biocompatibility	1800%	[40]
Graphene	Carbon paper	Glucose	Layer-by-layer assembly	368 mW m^{-2}	High roughness and surface area	151% (performance)69% (start-up time)	[41]
Graphene	Carbon cloth	Acetate	Dipping and drying	1018 mW m^{-2}	Good biocompatibility and high roughness	214%	[42]
Graphene	Stainless steel felt	Acetate	Dipping and drying	2142 mW m^{-2}	Good biocompatibility and reactivity	167% (compared to modification by CNT)	[43]
Graphene	Carbon fiber paper	Glucose	Self-assembly	1158 mW m^{-2}	High hydrophilic and specific surface area	170%	[31]

(Continued)

Table 15.1 (Continued)

Carbon nanomaterial	Anode base material	Substrate	Modification method	Max power/current density	Anode surface characterization	Optimization effect	References
Graphene	FeS$_2$ nanoparticles	Acetate	Freeze-drying	3220 mW m^{-2}	Good adhesion	142% (performance) 10% (start-up time)	[44]
GO	Graphite fiber brush	Lactate	Vacuum freeze-drying	53.8 W m^{-3}	High specific surface area	414 times	[32]
rGO	Carbon cloth	Acetate	Evaporating and compression	~2400 mW m^{-2}	High stability and larger surface area	216% (compared to modification by AC)	[45]
Graphene/PANI	Titanium suboxide	Glucose	Electrodeposite	2073 mW m^{-2}	Good biocompatibility and conductivity	1270% (compared to modification by CC)	[46]
Graphene/gold nanoparticles	Carbon brush	Acetate	Layer-by-layer assembly	33.7 W m^{-3}	Good biocompatibility and conductivity	320% (performance) 25% (start-up time)	[47]
rGO/PPy	Carbon cloth	Glucose	Dipping and drying	1068 mW m^{-2}	High conductivity and surface area	283%	[48]

CNT, Carbon nanotube; *GO*, graphene oxide; *MWCNT*, multiwalled carbon nanotube; *rGO*, reduced graphene oxide; *PANI*, polyaniline; *PPy*, polypyrrole; *AC*, activated carbon; *CC*, carbon cloth.

biofilm formation of EAB [27]. However, upon deposition in the anode, carbon black nanoparticles usually aggregated into a rich microporous structure, which hindered substrate supply, product removal, and bacterial adhesion inside the electrode. Furthermore, the low electrical conductivity of the amorphous carbons containing in carbon black also impeded the further improvement of anode performance. As a consequence, advanced carbon nanomaterials like CNTs, graphene, and its derivatives have been applied to increase the anode performance.

CNT is a 1D tubular carbon nanomaterial with unique hollow nanofiber structure, high mechanical strength, large specific surface area, and excellent electrical conductivity [24,50]. Like carbon black, anode modification by CNT lead to an increased specific surface area and more rough surface of the electrode, benefitting a faster bacterial adhesion, a more active biomass on the anode, and a strengthened EET efficiency between the EAB and electrode, and therefore the higher performance. Xie et al. [28,29] fabricated CNT-textile composite electrodes by depositing a layer of microporous CNTs on the surface of a macroporous textile (Fig. 15.2C). The macroporous (∼100 um) inside the composite electrodes can provide sufficient space for bacterial colonization, and the microporous produced by CNT modification not only increase the specific surface area of textiles but also enhance the affinity of bacterial colonization on the electrode surface, thus enhancing the EET efficiency. Moreover, as shown in Fig. 15.2D, the loose network distribution of CNTs can allow bacteria to penetrate into its interior and promote bacterial colonization inside it [14,37,50]. In addition, carbon cloth [51], graphite felt [52], reticulated vitreous carbon (RVC) [53], stainless steel [43], and other materials are often used for surface modification by CNTs to form composite electrodes (Fig. 15.2E). It has been also revealed that the properties of CNTs, such as length, stacking density, and surface properties can have a significant impact on the performance of MFC anodes [30,38]. For example, a longer length of CNTs can induce more EET sites for EAB, whereas a smaller stacking density facilitates the insertion of bacteria inside the CNT networks. Interestingly, to further enhance the electron transfer within the biofilm, a hybrid biofilms can be produced by inserting CNTs into the EAB biofilm. The high conductivity of CNT and the porous structure in the bacterial-multiwalled carbon nanotube (MWCNT) hybrid system allowed a more effective electron transport and substrate diffusion within the biofilm, and thus facilitated biofilm formation [39]. Therefore compared to the MFC with natural growth biofilm, MFC with hybrid biofilm of 40 um thickness achieved 58.8% higher power density and 53.8% reduction in start-up time. It should be noted that, however, CNTs have potential toxic effects on bacteria to some extent, which can inhibit the growth of bacteria or even cause the death of bacterial death [54–56]. Although there is no reports of such toxic effects of CNT on EAB in MFC anodes, the amount of CNT during surface modification needs to be strictly controlled.

Graphene is a 2D honeycomb crystalline structure made of tightly packed sp^2 hybridized carbon atoms [24,57,58]. It poses excellent properties such as high mechanical strength, excellent physicochemical properties, and electrochemical catalytic activity, as well as a high specific surface area of $2630\ m^2\ g^{-1}$ [59,60]. It has been demonstrated that graphene can accelerate the bacterial growth and act as an intermediary molecule to enhance EET efficiency by promoting the expression of microbial signaling molecules [40,61,62]. Both carbon materials and stainless steels can be used as the substratum for graphene modification. Zhang et al. [40] developed a graphene-modified stainless steel mesh anode. The maximum power density of the cell reached $2668\ mW\ m^{-2}$, which is 18 times higher than that of the MFC with unmodified one. Guo et al. [41] proposed a facile layer-by-layer assembly method to modify carbon paper by graphene and found that the increased surface roughness of the modified electrode induced an improved growth and colonization of electroactive bacteria on its surface. In addition, carbon cloth and stainless steel fiber felts are also often used as substrates for the modification of graphene [42]. The thin wrinkles and folds formed on the electrode surface greatly improve its surface roughness and biocompatibility, promoting the colonization of bacterial and reducing the overpotential during MFC operation. In another study, graphene-modified carbon paper anodes were prepared by plant-mediated bioreduction combined with self-assembly [31]. The biomolecules adsorbed on the graphene surface can effectively change the surface properties of graphene, by rendering it hydrophilic and biocompatible, which effectively enhances the biofilm-electrode interaction, and results in a faster EET rates and higher energy conversion efficiency.

The successful case of graphene motivated researchers to investigate the feasibility of using graphene derivatives for anode modification (Fig. 15.2G). Yang et al. [32] constructed 3D-3D composite electrodes by applying graphene oxide aerogel on the surface of graphite fiber brushes using vacuum freeze-drying method. The microorganism-reduced graphene oxide formed on the electrode surface could obtain a better affinity with EAB, promoting bacterial colonization and achieving a maximum power density of $53.8\ W\ m^{-3}$. Although graphene and its derivatives can strengthen biofilm-electrode interactions by increasing the specific surface area, roughness, and biocompatibility of the electrodes [32,45], the electrical conductivity of graphene oxide remains unsatisfyingly low due to the disruption of the conjugated sp^2 hybridization network of its hydrophilic functional groups. Reducing graphene oxide by removing the oxygen-containing functional groups can be an option to enhance the electrical conductivity, but the irreversible agglomeration formed during the reduction would decrease the specific surface area of these materials [48,50]. Therefore combining graphene and its derivatives with other nanomaterials with high electrical conductivity or high specific surface area (e.g., carbon nanomaterials, metal/metal oxides, conductive polymers) can be a solution [46,47].

For example, a method by combining graphene with polyaniline (PANI) on titanium suboxides was proposed to fabricate anode. The synergistic effect of the graphene and PANI on titanium suboxides optimized the surface properties of the electrode, leading to a 12.7 times higher MFC performance [46]. Similar to CNTs, graphene is also toxic to microorganisms due to the irreversible damage on cell membrane resulting from its sharp edges. Therefore the loading of graphene and its derivatives should be optimized before it was applied to the anode modification [63,64].

15.2.2 Conductive polymer coating

Conductive polymers are the materials with loosely held electrons in their π-conjugated backbones [65,66]. Recently, these materials have earned much attention in the improvement of anode performance because they possess the unique properties, such as high conductivity, high biocompatibility resulting from high hydrophilicity, and strong corrosion resistance. The commonly used conductive polymers for anode modification are PANI, polypyrrole (PPy), poly(aniline-co-o-aminophenol), poly(3,4-ethylene dioxythiophene) (PEDOT), etc. (Fig. 15.3, Table 15.2). Sonawane et al. [67] modified a highly uniform PANI coating on the surface of stainless steel plate by galvanostatic polymerization, which resulted in a significant enhancement of the hydrophilicity of the electrode surface, effectively promoting the growth of EAB (Fig. 15.3A and B). Wang et al. [68] prepared a PANI-coated graphite-felt anode by cyclic voltammetry. It was found that the electrical resistance of the modified anode dramatically decreased and the start-up period was reduced from 140 to 78 hours. In another study, the PANI was anchored on the carbon cloth surface by in situ polymerization [72]. It was found that decorating carbon cloth electrode with PANI increased the electrochemical surface by two times, allowing more EAB adhesion on the electrode surface. Pu et al. [69] tried to coat PPy on the stainless steel surface by in situ electrochemical deposition to improve the corrosion resistance and biocompatibility of the electrode. They found that the conductivity and surface roughness of the electrodes were enhanced. Due to the presence of PPy film on the electrode surface, the maximum power density of the MFC was 29 times higher than that with bare stainless steel electrode (Fig. 15.3C and D). Similar results were also observed for stainless steel wool [74], nickel foam [82,83], and copper wire mesh [84]. Apart from aforementioned properties, an extra beneficial role of conductive polymers for the improvement of anode performance was found by Kang et al. [78]. They found that coating graphite plate, carbon cloth, and graphite-felt with PEDOT enhanced the attached active biomass due to the electrostatic force between the positively charged PEDOT polymer backbone and the negatively charged microbial cell walls.

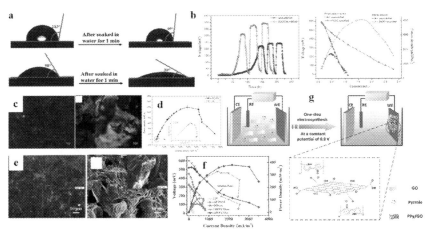

Figure 15.3 (A) Contact angle variation measurement for pristine SS-P (top) and PANI-modified SS-P (bottom). (B) Comparison of start-up curves (left) and power density curves (right) for MFCs with/without PANI-modified graphite-felt anodes. (C) Inverted fluorescence microscopy of microorganisms on anode of bare SS (left) and PPy/SS (right). (D) Power density output of MFCs with bare SS and PPy/SS anode. (E) SEM images of GP (left) and GO/PANI$_{OS}$ (right). (F) Polarization curve of MFCs with anode modified by different methods. (G) Schematic diagram of the one-step electrosynthesis of PPy/GO composites via electropolymerization (*GO*, Graphene oxide; *MFCs*, microbial fuel cell; *PANI*, polyaniline; *PPy*, polypyrrole, *SEM*, scanning electron microscopy; *SS-P*, stainless steel-plate; *SS*, stainless steel; *GP*, graphite paper). Reprinted with permission from J. M. Sonawane, S. Al-Saadi, R. K. Singh Raman, P. C. Ghosh, S. B. Adeloju, Exploring the use of polyaniline-modified stainless steel plates as low-cost, high-performance anodes for microbial fuel cells, Electrochimica Acta 268 (2018) 484–493, https://doi.org/10.1016/j.electacta.2018.01.163 [67]. Copyright 2018. Elsevier. Reprinted with permission from P. Wang, H. R. Li, Z. W. Du, Polyaniline synthesis by cyclic voltammetry for anodic modification in microbial fuel cells, International Journal of Electrochemical Science 9 (2014) 2038–2046 [68]. Copyright 2014. Electrochemical Science Group. Reprinted with permission from K.-B. Pu, et al., Polypyrrole modified stainless steel as high performance anode of microbial fuel cell, Biochemical Engineering Journal 132 (2018) 255–261, https://doi.org/10.1016/j.bej.2018.01.018 [69]. Copyright 2018. Elsevier. Reprinted with permission from D. Z. Sun, et al., In-situ growth of graphene/polyaniline for synergistic improvement of extracellular electron transfer in bioelectrochemical systems, Biosensors & Bioelectronics 87 (2017) 195–202, https://doi.org/10.1016/j.bios.2016.08.037 [70]. Copyright 2016. Elsevier. Adapted and reprinted with permission from Z. Lv, et al., One-step electrosynthesis of polypyrrole/graphene oxide composites for microbial fuel cell application, Electrochimica Acta 111 (2013) 366–373, https://doi.org/10.1016/j.electacta.2013.08.022 [71]. Copyright 2013. Elsevier.

According to the literature [70], conductive polymers can participate the cytochrome based direct electron transfer. Thus compositing conductive polymers with other nanomaterials could be an effective way to further upgrade the EET efficiency between EAB and electrode. As indicated in Fig. 15.3C–F, Sun et al. [70] proposed a graphite paper anode coated with graphene and PANI, and found that the maximum power density is 3.4 times and 5.7 times higher than that with PANI and graphene alone. Similarly, Lv et al. [71] constructed composite electrodes with uniform PPy and graphene oxide on graphite-felt by one-step electrosynthesis (Fig. 15.3G). The specific

Table 15.2 Summary of different conductive polymers-modified electrodes as MFC anodes.

Conductive polymers	Anode base material	Substrate	Modification method	Max power/current density	Anode surface characterization	Optimization effect	References
PANI	Stainless steel plate	Acetate	Galvanostatic polymerization	780 mW m^{-2}	High conductivity and stability	780%	[67]
PANI	Graphite-felt	Acetate	Electrochemical synthesize	4 W m^{-3}	Good adhesion	235% (performance) 56% (start-up time)	[68]
PANI	Carbon cloth	Lactate	In situ polymerizatidekhon	388.6 mW m^{-2}	Good biocompatibility and high surface area	650%	[72]
PANI	Graphene foam	Lactate	Dipping and drying	768 mW m^{-2}	High specific surface area and EET rate	486% (compared to bare carbon cloth)	[73]
PANI	Carbon cloth	Lactate	Dipping and drying	323 mW m^{-2}	High specific surface area and EET rate	204%	[73]
PANI	Stainless steel wool	Acetate	Constant current	2880 mW m^{-2}	High hydrophilic and roughness	227%	[74]
PPy	Self-assembly	Lactate	Self-degraded template	612 mW m^{-2}	Large surface area and open porous structure	665% (compared to bare carbon paper)	[75]
PPy	Stainless steel	Acetate	Electrochemical deposition	1190.9 mW m^{-2}	High roughness and conductivity	2900%	[69]
PPy	Carbon dots	Acetate and lactate	Electrochemical polymerization	291.4 mW m^{-2}	High electroactive surface area	212%	[76]
PPy	Stainless steel wool	Acetate	Electropolymerization	1870 mW m^{-2}	High hydrophilic and roughness	147%	[74]
PPy	CNT	Glucose	In situ chemical polymerization	228 mW m^{-2}	High biocompatibility and conductivity	—	[77]
PEDOT	Graphite plate	Acetate	Coating and drying	640 mW m^{-2}	Good adhesion and high roughness	400%	[78]
PEDOT	Carbon cloth	Acetate	Coating and drying	790 mW m^{-2}	Good adhesion and high roughness	226%	[78]

(Continued)

Table 15.2 (Continued)

Conductive polymers	Anode base material	Substrate	Modification method	Max power/current density	Anode surface characterization	Optimization effect	References
PEDOT	Graphite–felt	Acetate	Dipping and drying	1620 mW m^{-2}	Good adhesion and high roughness	312%	[78]
Graphene/PANI	Graphite paper	Acetate	Chemical synthesize	4442 mW m^{-2}	High EET efficiency	800%	[70]
PPy/GO	Graphite–felt	Lactate	Electropolymerization	1326 mW m^{-2}	High biocompatibility and stability	799%	[71]
MWCNT/PANI	Carbon felt	Sediment	Electrochemical deposition	527 mW m^{-2}	High wettability and good adhesion	400%	[79]
PPy/PDA	EAB	Acetate	Mixture and centrifugation	2520.2 mW m^{-2}	High conductivity and good adhesion	1100%	[80]
PANI/CNT	Graphite–felt	Acetate	Polymerization and deposition	257 mW m^{-2}	High hydrophilic and surface area	343%	[52]
PANI/Graphene	Carbon cloth	Acetate	Dipping and drying	884 mW m^{-2}	High EET efficiency and conductivity	190%	[81]
PANI/TiO$_2$	Nickel foam	Glucose	Pressing and drying	1495 mW m^{-2}	High specific surface area	200% (compared to previous study)	[82]
PANI/TC/Chit-nickel foam	Nickel foam	Bad wine	Coating and drying	18.8 W m^{-3}	High conductivity and specific surface area	227%	[83]

EAB, Electrochemical active bacterium; *MWCNT*, multiwalled carbon nanotube; *PANI*, polyaniline; *PPy*, polypyrrole; *PEDOT*, poly(3,4-ethylene dioxythiophene); *PDA*, polydopamine; *TC*, titanium carbide; *Chit*, chitosan.

surface area, conductivity, biocompatibility, and stability of the electrode remarkably exceeded those of the electrodes modified with PPy or graphene alone. The performance of the cell using composite electrode was improved by nearly eight times compared to bare graphite-felt anode. Besides, a composite anode prepared by electrochemical deposition of MWCNT and PANI on the surface of carbon felt [85] resulted in nine times higher amount of active biomass than that of the unmodified electrode.

15.2.3 Other types of modifications

Apart from the above methods, biofilm establishment can also be promoted by incorporating metal or metal oxide nanoparticles (e.g., Mn^{4+}, $Fe^{2+/3+}$, and TiO_2) [86—88]. Bouabdalaoui et al. [87] prepared FePS/graphite composite electrodes by adding electroactive iron to graphite. The MFC with composite anode eliminated the start-up delay time due to the optimized electrode interface. Moreover, it has been reported that ferric ions and magnetite nanoparticles can also play an enhanced role in the interfacial electron transfer [44,89]. Ying et al. [88] modified TiO_2 thin films on stainless steel mesh as MFC anodes with a greatly improved electrode biocompatibility. The prepared electrode exhibited a biofilm thickness of 28.3 um with a high microbial protein density of 2431 ug prot cm^{-2}, which was 12 times higher than that of the untreated anode. Meanwhile, the modification of TiO_2 also enhanced the EET rate at the electrode interface, and the charge transfer resistance of the electrode was as low as 3.55 Ω. Adjusting the surface properties of carbon-based materials by strong acid or ammonia treatment to introduce functional groups on the electrode surface is also an effective way to enhance biofilm formation. Cheng et al. [90] modified carbon cloth anode with ammonia at 700°C. The ammonia treatment increased the surface charge of the carbon cloth anode, enhancing the adhesion of EAB and accelerated the biofilm formation on anode surface. The start-up time of the MFC was reduced by 50% and the maximum power density increased by 48%. In addition, nitric acid and ethylenediamine were used to modify the activated carbon fiber felt [91]. The introduction of nitrogenous groups such as ammonium nitrate facilitates the adhesion and growth of bacteria on the electrode surface and promotes the electron transfer from the bacteria to the electrode, thus reducing the start-up time of the MFC and enhancing the cell performance. Feng et al. [92] used acid immersion, heating, and a combination of both treatments to modify the carbon fiber brushes to enhance the MFC power output. The enhancement of MFC performance is thought to be related to the increase of protonated nitrogen and the decrease of oxygen content on the electrode surface. As can be seen above, the introduction of functional groups by chemical modification often requires high energy consumption or the introduction of hazardous chemicals. Therefore it is not commonly applied in the practical application of MFC.

15.3 Microstructure for increasing specific surface area

To maximize the current generation of anode, the microstructure of the porous anode needs to be optimized since both pore size and pore arrangement have a great effect on biofilm establishment [93,94]. Although a smaller pore size of the porous electrode can provide more sites for bacterial adhesion and biofilm formation due to the large specific surface area, it would also result in difficulties in biofilm colonization because of the limited transport of substrates and products [12,13]. In addition, the pore arrangement and distribution of the electrode can also affect the biofilm establishment by influencing the mass transfer process inside the electrode. Compared with the random pore structure of the electrode, the orderly pore structure can facilitate the substrates supply and product removal, leading to an optimal microenvironment for microbial metabolism. This section will discuss the effect of the microstructure of anode materials on biofilm establishment, and therefore the MFC performance.

15.3.1 Electrodes with random microstructure
15.3.1.1 Carbon-based electrodes

In the early days of MFC research, carbon paper, carbon cloth, carbon felt, granular activated carbon, and RVC have been wildly used as the anode due to their high electrical conductivity, good biocompatibility for the microbial colonization, high stability, and durability in MFC relevant conditions, cost-effectiveness, and commercial availability [11,24,95]. Table 15.3 compares the physical and chemical properties of the carbon-based porous materials. Carbon paper is a carbon fiber composite, while carbon cloth is a woven material consisting of hundreds of individual graphite fibers bundled together. Both materials are wildly used in fuel cells and have a random microstructure with a pore size of ~10 um [96,97]. Although carbon paper and carbon cloth have been proven to be suitable for the formation of EAB biofilms, these materials are not suitable for wastewater treatment due to the limited surface area for microbial colonization and the high cost. To reduce the cost of electrode materials, carbon mesh was also used as MFC anode without compromising the anode performance [102]. However, similar to carbon cloth and carbon paper, the surface of carbon mesh cannot effectively utilized by EAB and are susceptible to clogging. As a result, the current density generated by the biofilm established on these materials is generally less than 10 A m^{-2} [96].

Carbon felt is a 3D porous material with a porosity of ~99% that can provide abundant spaces for bacteria attachment and channels for substrate supply and proton removal [99]. However, with the gradual colonization of EABs, a thick biofilm usually develops on the outer surface of the carbon felt, hindering the substrate/proton transfer between the bulk solution and the electrode interior [14]. Therefore most of the carbon felt surface usually remains unutilized. Furthermore, the low electrical

Table 15.3 Comparison of the physical and chemical properties of some common carbon-based porous materials.

Anode material	Pore size	Porosity/specific surface area ($m^2\ m^{-3}$)	Advantages	Disadvantages	References
Carbon paper	~10 um	>70%	Good biocompatibility	Fragile and expensive	[96,97]
Carbon cloth	~10 um	–	Good biocompatibility	Expensive	[96]
Carbon mesh	–	–	Low cost	Limited specific surface area	[98]
Carbon felt	–	99%	High porosity	Low conductivity	[99]
GAC	<50 nm	–	Good biocompatibility	Granular property	[100]
RVC	~ um	95%/3750	High porosity	Expensive	[23]
Carbon brush	um ~ mm scale	18200	High specific surface area	Difficulty in microstructure optimization	[101]

GAC, Granular activated carbon; *RVC*, reticulated vitreous carbon.

conductivity of carbon felt limited its application in bioelectricity recovery from wastewater [100]. Granular activated carbon (GAC) is a porous carbon material that is commonly used in packed bed type anode. Although the specific surface area of GAC is rather high (several hundreds to several thousands $m^2 \, g^{-1}$), the surface area available for effective interaction with EAB is limited since the average pore diameter of activated carbon particles (<50 nm) is much lower than the size of EAB. Besides, the low electrical conductivity resulting from the high porosity is also restricting factor for its application in MFC [100]. RVC is a conductive monolithic carbon material with a high porosity (95%) and a large specific surface area (3750 $m^2 \, m^{-3}$). In contrast to GAC, the pore structure of RVC is in the micron scale, with an average strut thickness and pore size of 100 and 320 um, respectively. Obviously, EABs can be well colonized on the exterior and interior surface of RVC [23,103]. Consequently, the performance of the MFC using RVC anode can reach a maximum power density of 40 $W \, m^{-3}$ [104]. Unfortunately, RVC is expensive and fragile, which prohibits its large-scale application in MFC technology [105].

Carbon brush anode are consider as one of the most suitable electrode for biofilm establishment in MFCs. It has a spiral structure that can easily fabricated by twisting the titanium wire with bundles of carbon fibers [106,107]. As carbon fibers have a small diameter (~ 7 um), the specific surface area of carbon brushes can reach more than $10^4 \, m^2 \, m^{-3}$, which is beneficial for the formation of electroactive biofilms [100,101]. Furthermore, the open pore structure of carbon brushes can facilitate the mass transport between the interior of electrode and the bulk solution, further contributing to the establishment of biofilm and electricity production. Therefore the power density of air-cathode MFC with carbon brush anode can reach 2.4 $W \, m^{-2}$, which is much higher than that of MFC with traditional carbon paper or carbon cloth anode [101]. Further improvement in power generation can be achieved by optimizing the ratio of the length of carbon fiber brush to the loaded fiber mass [106]. However, the optimization of the microstructure of carbon brush electrode is quite difficult because of the random fiber distribution along the titanium core [101].

15.3.1.2 Metal-based electrodes

Compared with carbon materials, the metal-based materials offer a higher mechanical strength, a better machinability and an order of magnitude higher electrical conductivity. Therefore it can be easily fabricated into a wide range of shapes with a significantly low ohmic resistance [108,109]. Although many metals have been used as MFC anode materials in the past decades, some of them are not suitable for electrode fabrication. For example, precious metals such as gold, silver, and platinum have shown to be efficient MFC anode materials (Fig. 15.4A and B), but their high cost limits their application [110,114,115]. It is also noted that although metals are more conductive than carbon materials, some of them can be easily corroded in MFC relevant conditions.

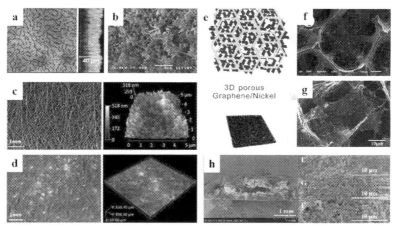

Figure 15.4 Confocal laser scanning microscope images (A) and SEM images (B) of *G. sulfurreducens* biofilms on gold electrodes surface. (C) Light microscope images and AFM micrograph of flame oxidized stainless felt electrodes. (D) Light microscope images and CLSM images of biofilms on flame oxidized stainless felt electrodes. (E) Schematic diagram of the hierarchical porous graphene/nickel composite electrode. (F) SEM images of freeze-dried G/Ni electrodes prepared with GO of 2 mg mL^{-1}. (G) SEM images of *S. putrefaciens* adhered on the G/Ni anode electrode surface. (H) SEM images of biofilm profile in low resolution and the biofilms at outer in high resolution (*AFM*, Atomic force microscopy; *CLSM*, confocal laser scanning microscopy; *GO*, graphene oxide; *SEM*, scanning electron microscopy). *Reprinted with permission from H. Richter, et al. Electricity generation by* Geobacter sulfurreducens *attached to gold electrodes, Langmuir: The ACS Journal of Surfaces and Colloids 24 (2008) 4376−4379, https://doi.org/10.1021/la703469y [110]. Copyright 2008. American Chemical Society. Reprinted with permission from K. Guo, et al. Flame oxidation of stainless steel felt enhances anodic biofilm formation and current output in bioelectrochemical systems, Environmental Science & Technology 48 (2014) 7151−7156, https://doi.org/10.1021/es500720g [111]. Copyright 2014. American Chemical Society. Reprinted with permission from Y. Qiao, X.-S. Wu, C.-X. Ma, H. He, C. M. Li, A hierarchical porous graphene/nickel anode that simultaneously boosts the bio- and electro-catalysis for high-performance microbial fuel cells, RSC Advances 4 (2014) 21788−21793, https://doi.org/10.1039/c4ra03082f [112]. Copyright 2014. The Royal Society of Chemistry. Reprinted with permission from X. B. Ying, et al. Sustainable synthesis of novel carbon microwires for the modification of a Ti mesh anode in bioelectrochemical systems, The Science of the Total Environment 669 (2019) 294-302, https://doi.org/10.1016/j.scitotenv.2019.03.106 [113]. Copyright 2019. Elsevier.*

This phenomenon can not only lead to the increase in electrical conductivity but also resulting in the release of ions toxic to EAB, rendering a decay in performance during operation [115,116]. Therefore the anode made by materials with a good resistance, such as Ni, Ti, and stainless steel, was used in MFC. A summary of different metals-based anodes was listed in Table 15.4.

Pocaznoi et al. [119] used 254SMO stainless steel with a good corrosion resistance as the anode material and produced a high current density of 35 A m^{-2}, which is three times higher that of graphite plate anodes. It was found that a homogeneous biofilm with a thickness of 50~60 um was established on the surface of the stainless steel

Table 15.4 Summary of different metals-based porous materials as MFC anodes

Anode material	Size	Reactor construction	Substrate	Source of inoculation	Pore size (um)	Biofilm thickness (um)	Max power/current density	References
SS plate	20 × 30 cm	SC	Marine sediments	Marine sediments			23 mW m^{-2}	[117]
SS piece	1 × 2.5 × 0.1 cm	Half-cell	Acetate	G. sulfuredaucens			2.4 A m^{-2}	[118]
SS		Half-cell	Acetate	Soil leachate		50—60	35 A m^{-2}	[119]
SS		SC	Acetate	Marine		80—90	4 A m^{-2}	[120]
SS plate	1.5 × 3 × 0.25 cm	DC	Acetate	Effluent of MFC		~70	1500 mW m^{-2}	[121]
SS felt	2 × 1 × 0.07 cm	DC	Acetate	Effluent of BES		~40	19.2 A m^{-2}	[111]
SS felt	1 × 1 × 0.1 cm	DC	Acetate	Effluent of BES		~50	15 A m^{-2}	[122]
Graphene-SS felt	1.8 × 1.8 cm	DC	Acetate	Mixed bacterial culture	15.7		2142 mW m^{-2}	[43]
Carbon black-SS mesh	2.4 × 1.26 × 0.54 cm	DC	Acetate	Wastewater	~500	>10	3215 mW m^{-2}	[35]
SS mesh		Half-cell	Acetate	Wastewater	~300	>20	19.1 A m^{-2}	[108]
PANI-SS wool		SC	Acetate	Landfill leachate			2880 mW m^{-2}	[74]
PPy-SS wool		SC	Acetate	Landfill leachate			1870 mW m^{-2}	[74]
SS foam	0.635 mm of thickness	SC	Acetate	Garden compost leachate	200—800	20	82 A m^{-2}	[123]
Mo mesh	4 × 80 cm	SC	Acetate	Activated sludge			1296 mW m^{-2}	[116]
W mesh	4 × 80 cm	SC	Acetate	Activated sludge			1036 mW m^{-2}	[116]
Nickel	1 × 1 cm	Half-cell	Acetate	Secondary biofilm		77	3.8 A m^{-2}	[115]

Nickel foam	5 × 5 × 0.2 cm	SC	Bad wine	Mixed bacterial culture		8.29 W m^{-3}	[83]
PANI/TC/Chit-Nickel foam	5 × 5 × 0.2 cm	SC	Bad wine	Mixed bacterial culture		18.8 W m^{-3}	[83]
GO-Nickel foam	1 × 2 × 0.18 cm	DC	Yeast extract	S. putrefaciens	0.02–50	3903 mW m^{-2}	[112]
rGO-Nickel foam	1 mm of thickness	DC	Trypticase soy broth	Shewanella oneidensis MR-1	100–500	661 W m^{-3}	[58]
CM-Titanium mesh	2 × 1 cm	DC	Acetate	Effluent of MFC	150	12.19 A m^{-2}	[113]
Copper	1 × 1 cm	Half-cell	Acetate	Secondary biofilm	249	15.2 A m^{-2}	[115]
Sn-Copper mesh		DC	Domestic wastewater	Sludge	1320	271 mW m^{-2}	[124]
Silver	1 × 1 cm	Half-cell	Acetate	Secondary biofilm	154	11.2 A m^{-2}	[115]
Gold	1 × 1 cm	Half-cell	Acetate	Secondary biofilm	127	11.8 A m^{-2}	[115]

SS, Stainless steel; BES, bioelectrochemical system; PANI, polyaniline; TC, titanium carbide; Chit, chitosan; GO, graphene oxide; CM, carbon microwires.

anode, indicating that 254SMO stainless steel can be a good candidate as an highly effective MFC anode. In addition to 254SMO stainless steel, other types of stainless steels such as 304 and 316 have also been proved to be good anode materials [35,122]. It has been revealed that pretreatment by flame oxidation and heat treatment can effectively improve the biocompatibility of stainless steel surfaces and enhance the thickness and density of biofilm covering on the electrode surface (Fig. 15.4C) [111,122]. However, the surface damage caused by heat treatment also makes the stainless steel material more susceptible to corrosion [108]. Although stainless has a good corrosion resistance in the MFC relevant condition, prolonged operation can still lead to the release of toxic heavy metallic ions. To preserve the activity of the EAB biofilm, Peng et al. [108] found that pretreatment of stainless steel electrode by sequential acid etching, carbon black modification and low temperature heat treatment can induce the formation of protective CB/Fe_3O_4 composite layer on the electrode surface, not only protecting the biofilm from the toxic heavy metal ions, but also increasing the EET efficiency at biofilm/electrode interface, and thus leading to a nearly three orders of magnitude higher current density generation than that of the untreated one [108]. To further increase the electrode surface area, many porous electrodes, such as stainless steel felt, mesh, wool, and foam were also used in the MFC anode [35,74,88,108,123].

Yamashita et al. [116] investigated the feasibility of using 14 types of metals and their 31 oxidized forms as MFC anodes. The results showed that four types of metals, Mo, W, Fe, and Sn, or their oxidized forms, can be efficient anodes for MFCs. The metallic Ni was also proven to be a good electrode material in previous studies [83,112]. Qiao et al. [112] prepared graphene/Ni foam anodes with a graded pore structure by using a freeze-drying-assisted self-assembly method. The graded pore structure exhibited a good EAB affinity, allowing the EABs to adhere and interconnect well inside the electrode, producing a complete biofilm network that is substantial to the MFC performance improvement (Fig. 15.4E and F). Ti is a metal with a good corrosion resistance and excellent biocompatibility. Therefore it can be widely found in medical field. In MFCs, Ti is often used as the collector for electrodes. Heijne et al. [125] tried to use Ti sheet as the anode of MFC and found that the power density of the MFC with Ti anode was almost zero because EABs can only colonize on the surface of Ti in small amounts. Additionally, the EET kinetics between EAB and Ti surface is rather slow, resulting in a severe inhibition of the electrons transfer from biofilm to electrode surface [115,126]. Therefore pure metallic Ti seems to be unsuitable as an MFC anode. However, in further studies, it has been demonstrated that the surface modification of Ti with carbon materials and TiO_2 can significantly improve the formation of EAB biofilm and thus enhance MFC power output [113,127]. Ying et al. [113] fabricated Ti-based porous composite electrode with 3D interlaced network structure by carbonizing filamentous fungi on a Ti mesh. It was found that the composite electrode was covered by a dense biofilm with a biomass of 1027 ug cm^{-2} after successful start-up and produced a high current density of 12.19 A m^{-2} (Fig. 15.4H). TiO$_2$ is also a promising material with a good stability and biocompatibility for MFC anodes [82,88,126,127]. However, due to its low electrical conductivity, it is often applied as a modification material with conductive polymers or carbon nanomaterials.

Copper has excellent mechanical properties and it is one of the best metals in terms of electrical conductivity (58 × 10^6 S m^{-1}) [115]. However, the application of copper-based anode in MFC is still under debate. Zhu et al. [128] found that Cu was severely corroded during MFC operation, resulting in the Cu^{2+} overdose in the substrate, which inhibited the biofilm formation on the electrode. Therefore the maximum power density of the MFC with copper anode is as low as 2 mW m^{-2}. Many studies also confirm these results [129–133]. However, it is also reported that EABs can adhere quite well to the copper surface, forming a ~250-um-thick biofilm that is much thicker than that of commonly used anode materials such as graphite (~117 um) and nickel (~77 um). As a results, the current generation from the copper-based anode can reach a value as high as 1.5 mA cm^{-2} [111]. It is known that protecting copper from corrosion can significantly improve the anode performance, but the exact mechanism are still unknown.

15.3.2 Electrode materials with regular microstructure
15.3.2.1 Biomass-derived porous materials
The cost of electrode materials accounts for 20%–50% of the manufacturing cost of MFCs. Therefore reducing the cost of electrodes is an important step to realize the commercial application of MFCs [134–136]. The biomass-derived materials are mostly from waste agricultural and forestry plants, which have a much lower cost compared to traditional carbon-based materials [137]. More importantly, the pore structure of traditional electrode materials is mostly random, which is not beneficial to substrate supply and product removal inside the electrode (Fig. 15.5, Table 15.5). On the other hand, biomass usually contains regular microstructure that is suitable for EAB attachment and biofilm formation [145,146]. Moreover, it is rich in a variety of chemical elements such as oxygen, phosphorus, and nitrogen, which are beneficial for the adhesion of EABs and anode performance improvement [137,145]. However, most of untreated biomass is electrically insulated. Therefore the biomass are usually carbonized in nitrogen atmosphere at a temperature above 800°C to produce heteroatom doped carbon materials [134,152]. During the carbonization process, the inherent regular microstructure can be evolved into interconnected macroporous that favors a reasonable high conductivity and facilitates EAB attachment and biofilm formation.

Zhu et al. [140] obtained MFC anodes with hollow macroporous fiber structure by carbonizing natural kapok at 1100°C. Both the interior and exterior surfaces of the

Figure 15.5 (A) Digital photo of pomelo peel and SEM images of reticulated carbon foam from pomelo peel. (B) Digital photos of reticulated carbon foam electrode (1) before and (2) after biofilm growth. (C) Cross-sectional view images of biofilms in reticulated carbon foam-pomelo peel. (D) Digital photo of natural chestnut before (left) and after (right) carbonization. (E) Digital photo and schematic of a hollow kapok fiber anode. (F) SEM images of the EAB colonizing on the kapok fiber. (G) Digital photo of silk cocoon before carbonization. (H) Digital photo of Chinese characters handwritten using Chinese ink. (I) Digital photos of the loofah sponge before (white) and after (black) application of the ink coating (*EAB*, Electrochemical active bacterium; *SEM*, scanning electron microscopy). Adapted and reprinted with permission from S. Chen, et al., Reticulated carbon foam derived from a sponge-like natural product as a high-performance anode in microbial fuel cells, Journal of Materials Chemistry 22 (2012) 18609–18613 [138]. Copyright 2012. The Royal Society of Chemistry. Reprinted with permission from S. Chen, et al., A hierarchically structured urchin-like anode derived from chestnut shells for microbial energy harvesting, Electrochimica Acta 212 (2016) 883–889, https://doi.org/10.1016/j.electacta.2016.07.077 [139]. Copyright 2016 Elsevier. Reprinted with permission from H. L. Zhu, et al. Lightweight, conductive hollow fibers from nature as sustainable electrode materials for microbial energy harvesting, Nano Energy 10 (2014) 268–276, https://doi.org/10.1016/j.jpowsour.2015.01.065 [140]. Copyright 2014. Elsevier. Reprinted with permission from M. Lu, et al., Nitrogen-enriched pseudographitic anode derived from silk cocoon with tunable flexibility for microbial fuel cells, Nano Energy 32 (2017) 382–388, https://doi.org/10.1016/j.nanoen.2016.12.046 [141]. Copyright 2016. Elsevier. Reprinted with permission from L. H. Zhou, L. H. Sun, P. Fu, C. L. Yang, Y. Yuan, Carbon nanoparticles of Chinese ink-wrapped natural loofah sponge: a low-cost three-dimensional electrode for high-performance microbial energy harvesting, Journal of Materials Chemistry A 5 (2017) 14741–14747, https://doi.org/10.1039/c7ta03869k [142]. Copyright 2017. The Royal Society of Chemistry.

Table 15.5 Summary of different biomass-derived porous materials as MFC anodes (MFC, Microbial fuel cell).

Anode material	Reactor construction	Carbonization temperature (°C)	Residence time (h)	Substrate	Surface area (m² g⁻¹)	Pore size (um)	Biofilm thickness (um)	Max power/ current density	References
Kapok	SC (28 mL)	1100	2	Acetate	820	10–20[a]	≥10	1738.1 mW m⁻²	[140]
Loofah sponge	SC (28 mL)	900		Acetate	445	20–200[a]		701 mW m⁻²	[143]
Chestnut shell	SC (28 mL)	900	1	Acetate	48.12	3–5 (fiber diameter)		759 mW m⁻²	[139]
Chestnut shell	SC (66 mL)	900 (KOH activation followed)	2	Acetate	881			850 mW m⁻²	[137]
King mushroom	Half-cell	1000	1	Acetate		10–120		2.09 mA cm⁻²	[144]
Wild mushroom	Half-cell	1000	1	Acetate		75–200	3–18	3.02 mA cm⁻²	[144]
Corn stem	Half-cell	1000	1	Acetate		2–7	3–18	3.12 mA cm⁻²	[144]
Kenaf	Half-cell	1000		Acetate		25 and 60	>40	3.25 mA cm⁻²	[145]
Bamboo charcoal tube	DC	1000	2	Acetate		10–100		1652 mW m⁻²	[146]
Pomelo peel	Half-cell	900	1	Acetate		>100	>15	4.02 mA cm⁻²	[138]
Silk cocoon	SC (28 mL)	900	3	Acetate	~230			8.6 mW g⁻¹	[140]
CNT-sponge	DC	Dipping and drying		Domestic wastewater	~10⁴ m² m⁻³	300–500		1240 mW m⁻²	[28]
Chinese ink-loofah spong	SC	Dipping and drying		Acetate	182.3	um to mm scale	5	0.82 mW cm⁻³	[142]
Loofah spong	SC	900		Acetate		um to mm scale		0.91 mW cm⁻³	[142]
Almond shell	DC	800	2	Acetate	616.04	2–5		4346 mW m⁻²	[147]
Sewage sludge-coconut shell	SC (28 mL)	900	2	Acetate	36			1069 mW m⁻²	[148]
Coffee waste	SC (40 mL)	900 (After KOH activation)	1	Yeast extract	164.33	Mostly macropore		3927 mW m⁻²	[149]
Corrugated fiberboard	Half-cell	1000	1	Acetate		mm scale	~18	390 A m⁻²	[150]
Eucalyptus leaves	SC (28 mL)	Layer-by-layer assembly		Acetate				33.7 W m⁻³	[47]
Bread	DC	1000	2	Acetate	295.07	0.5–300	2–4	3134 mW m⁻²	[151]

CNT, Carbon nanotube; DC, double chamber; SC, single chamber.
[a]Before carbonization.

hollow fibers provide sites for EAB colonization, which effectively enhances the interaction area between the bacteria and electrode. The macroporous carbon sponge was fabricated by carbonizing loofah, the shrinkage of the material resulted in a large amount of wrinkles on the original smooth electrode surface during carbonization, enhancing the surface roughness and the specific surface area of the electrode, promoting the formation of dense biofilm on electrode surfaces [143]. A sponge-like macroporous reticulated carbon foam was prepared by carbonizing pomelo peel, allowing a dramatic improvement in the availability of EABs on the interior surface of the anode. It was found that biofilm with a thickness of 1.8 um is still existing at a depth of 600 um inside the electrode pores [138], demonstrating the improved substrate supply and product removal within the microstructure of the electrode (Fig. 15.5A–C). In addition, electrode material with an urchin shape can be obtained by carbonizing chestnut shells (Fig. 15.5D). The raised thorn-like structure of the electrode is favorable to increase the specific surface area and enhance the mass transport within the electrode [139]. The king mushroom, wild mushroom and corn stem were carbonized to produce macroporous carbon anodes with pore sizes ranging from 10 to 120 um, 75 to 200 um, and 2 to 7 um, respectively, with porosity of 65%–85%, facilitating the establishment of biofilm within the interior of the electrode [144]. Moreover, the carbonized kenaf exhibits a hierarchically ordered macroporous structure with one pore size of 25 um and the other of 60 um. EAB was able to form biofilm with a thickness >40 um on the electrode surface, resulting in three times higher current generation than the traditional graphite rod anode [145]. Apart from aforementioned materials, many natural biomass (Table 15.5), such as rubber tree sawdust [153], almond shell [147], sewage sludge-coconut shell [148], coffee waste [149], corrugated fiberboard [150], and eucalyptus leaves [47] were also proposed for the fabrication of MFC anode. The natural and recyclable materials provide an alternative way for fabricating cost-effective, highly efficient anodes toward the practical implementation of MFCs.

In addition to carbonization, direct loading of conductive nanomaterials on the surface of natural carbon-based materials is also a method of preparing the high performance MFC anodes, which can greatly reduce the cost of electrode fabrication, compared with the method of carbonization at a high temperature. Xie et al. [28] found that immersing sponge into CNT ink can produce a highly conductive electrode by covering an interconnected layer of CNT skin on sponge. It was found that CNT enhanced the EET efficiency between EAB and electrode surface, promoting the colonization of EABs, and thus achieving a high power density of 1.24 W m^{-2} during domestic wastewater treatment. In another study (Fig. 15.5H), Chinese ink-wrapped natural loofah sponge was served as anode and an electroactive biofilm with a uniform thickness of 5 um was formed on anode surface. Due to the open regular macroporous structure, the blockage of the pores by EAB was not observed after 70 day's operation at 16.3 mA cm^{-3} [142].

15.3.2.2 Electrode with engineered 3D macroporous architectures

Apparently, when using biomass-derived materials as anode, macroporous dominated with a pore size larger than 50 nm are more favorable for bacterial attachment as compared with those with microporous (size less than 2 nm) and mesoporous (size between 2 and 50 nm) structures. However, the microstructure in the biomass-derived materials can be hardly controlled due to the inherent diversity of nature materials, and thus limits their application in MFCs. Therefore to improve the biofilm establishment and anode performance, many researchers have been devoted to constructing orderly macroporous anode materials with controllable and suitable pore size.

To study the effect of electrode microstructure on the biofilm establishment and current generation, Li et al. [154] constructed 3D electrode array by using graphite rod as electrode materials. It was found that the amount of active biomass and current generation gradually decreased from the outer to the inner part of the 3D array anode. This unevenness distribution of biofilm and current generation in 3D porous electrode array is resulted from the restricted mass transport inside the electrodes. To study the biofilm formation inside the electrodes, Moss et al. [12] constructed graphite block model electrodes with different groove structures to simulate macroscopic 3D electrode structures at different scales and they found that the overgrowth of biofilm on the electrode surface gradually blocked the entrance of the channels when the space between graphite blocks was below 400 um, inducing the mass transport limitation within the channels and thus inhibiting the colonization of EAB inside model electrodes. This hypothesis was confirmed by Hu et al [13]. They constructed a 3D orderly porous model array anode with a pore spacing of 500 um by graphite rods and found that the pH in the center of the graphite rod arrays was as low as 5.83, implying that the proton removal is severely inhibited due to the biofouling in the exterior region of the electrode. The proton accumulation in the interior region of the electrode suppressed the metabolic activity of EABs, which resulted in an extra difficulty in biofilm establishment, and therefore an extremely low current generation in the interior electrode unit.

3D printing technology that allows the fabrication of porous electrodes with the desired geometric dimensions and channel structure, have been used in the MFC field (Fig. 15.6) [89,155−158]. Zhou et al. [156] used the Ti and stainless steel to fabricate 3D printing electrodes with a pore size of 1.5 mm and a porosity > 85%. By covering PANI on electrode surfaces, the maximum power densities of the MFCs with Ti and stainless steel anode reached 0.798 and 0.934 W m^{-3}, respectively. Recently, UV curable resins were recognized as the precursors for the ordered microporous electrodes (Fig. 15.6A−D) [155]. After being 3D-printed into a desired porous architecture, an electrically conductive and orderly arranged microstructure can be obtained with a porosity up to 95% by carbonization. Additionally, abundant secondary pores can be formed on the surface of fabricated electrode. Together with the primary pore

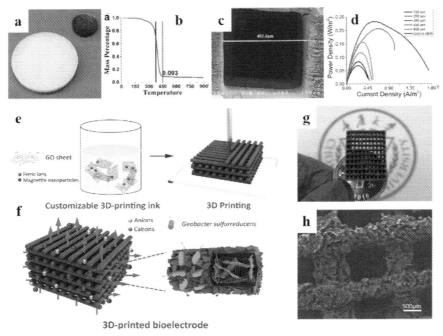

Figure 15.6 (A) Digital photos of 3D-printed porous structures before and after carbonation. (B) Thermogravimetric analysis of the resin as a function of temperature. (C) SEM images of well-printed 3D-printed carbonaceous porous anodes with pore sizes. (D) The MFC performance with 3D-PCP anodes with pore sizes of 100–500 um. (E) The customized GO-based ink was printed into a designed electrode structure. (F) Schematic diagram of the 3D-printed electrode hierarchical pores. (G) Digital photos from side view of the 3D-printed electrodes. (H) SEM images of the hierarchical porous structure (*3D-PCP*, 3D-printed carbonaceous porous; *MFC*, microbial fuel cell; *SEM*, scanning electron microscopy). *Reprinted with permission from X. Ying, et al., Titanium dioxide thin film-modified stainless steel mesh for enhanced current-generation in microbial fuel cells, Chemical Engineering Journal 333 (2018) 260–267, https://doi.org/10.1016/j.cej.2017.09.132 [155]. Copyright 2017. Elsevier. Reprinted with permission from B. Bian, et al., 3D-printed porous carbon anode for enhanced power generation in microbial fuel cell, Nano Energy 44 (2018) 174–180, https://doi.org/10.1016/j.nanoen.2017.11.070 [89]. Copyright 2021. Elsevier.*

structure producing during 3D printing, the hierarchical porous structure greatly enhanced the surface area and roughness of the electrode, providing favorable conditions for EAB attachment and biofilm establishment. As a result, the performance of MFC with the 3D printing anode is three times higher than that of the MFC with the conventional carbon cloth anode. In a more recent study, Freyman et al. [158] reported 3D-printed bacteria structure as a living electrode for the MFC. During the fabrication, EAB was directly printed into electrode and can be directly use to degrade the methyl orange azo dye and to generate electricity without additional inoculation. Recently, a composite anode contain graphene, ferric ions and magnetite nanoparticles has been constructed by 3D printing technology [89]. In the composite anode (Fig. 15.6E–H), graphene can provide a large specific surface area required for bacterial adhesion, and ferric ions and magnetite nanoparticles facilitate the interfacial EET. The composite electrode achieved a high current density of 10,608 A m^{-3}. Apart from 3D printing technology, other technologies such as electrospun and ice-templating technique have also been applied to the fabrication of macroporous anodes. However, although with large specific surface area and high porosity, it is still difficult to precisely regulate the anode structure during the fabrication process.

15.4 Outlook

Although MFC has been intensively studied as a promising technology to recover bioelectricity from wastewater, many barriers need to be overcome toward practical implementation. Among these, fabrication of high-performance and cost-effective anodes with nano/microstructures still poses the most important challenges. Many 3D anode, such as carbon brushes, stainless steel, loofah sponge-derived porous carbon, and 3D printing electrode have been widely used in MFCs. These 3D porous electrodes can achieve a higher performance level than traditional anode materials such as carbon cloth/paper because of the larger electrode surface accessible to EAB. Moreover, surface modification can also effectively increase the specific surface area of electrodes and strengthens the interaction between the EAB and electrode, resulting in an enhancement of the anode performance. However, these materials have drawbacks such as high fabrication cost, being prone to corrosion, a low electrical conductivity and complex preparation processes, which offset their benefits from the performance improvement. Therefore developing a promising anode material with a low cost and simple preparation procedure is always an important research direction. In addition, it is seen from the literatures that the mismatch between the textural properties (such as, pore size, internal channel width, and architecture) of 3D porous electrode and the EAB is the main bottleneck restricting the anode performance due to the serious mass transfer limitation inside the electrodes. Therefore the up-to-date MFC power density is still far below the theoretical value of 53 kW m^{-3} [7]. Thus the development of anodes with an optimized pore structure could be an important direction for future studies.

In addition, most of the achievements on the optimization of anode structure were in bench scale MFCs ranging from microliters to liters. It is well known that the electrical conductivity of the anode and substrate/product concentration gradient in the anolyte will significantly increase in a full scale MFC of several to several hundred m^3, leading to a considerable power loss due to the ohmic and concentration polarization. Moreover, actual domestic wastewater generally contains small particles. The excess suspended particles in wastewater would also block the pathway for EAB adhesion and substrate/product transfer in the electrode, and therefore deteriorating the anode performance. Compared to the artificial wastewater in the laboratory, the complex composition of the actual wastewater is also one of the challenges that needs to be faced in the improvement of MFC performance. Therefore much effort is needed to develop or establish a feasible anode for MFC system that can be operated in a real wastewater treatment plant. In all, MFCs are a carbon neutral technology and they can be used for renewable energy production and wastewater treatment. The practical application of this technology is highly dependent on the full understanding of the effect of micro/nanostructures on the EAB attachment and biofilm formation.

Acknowledgments

This work was supported by the National Natural and Science Foundation of China (Nos. 52076022 and 52021004) and Program for Back-up Talent Development of Chongqing University (No. cqu2018CDHB1A03). We also acknowledge the kind help provided by Wei Yang of Sichuan University, Yang Yang of Chongqing University and Jun Zhang of Zhengzhou University of Light Industry.

References

[1] D. Pant, G. Van Bogaert, L. Diels, K. Vanbroekhoven, A review of the substrates used in microbial fuel cells (MFCs) for sustainable energy production, Bioresource Technology 101 (2010) 1533−1543. Available from: https://doi.org/10.1016/j.biortech.2009.10.017.
[2] A.A. Yaqoob, et al., Outlook on the role of microbial fuel cells in remediation of environmental pollutants with electricity generation, Catalysts 10 (2020). Available from: https://doi.org/10.3390/catal10080819.
[3] G. Palanisamy, et al., A comprehensive review on microbial fuel cell technologies: processes, utilization, and advanced developments in electrodes and membranes, Journal of Cleaner Production 221 (2019) 598−621. Available from: https://doi.org/10.1016/j.jclepro.2019.02.172.
[4] A.J. Slate, K.A. Whitehead, D.A.C. Brownson, C.E. Banks, Microbial fuel cells: an overview of current technology, Renewable and Sustainable Energy Reviews 101 (2019) 60−81. Available from: https://doi.org/10.1016/j.rser.2018.09.044.
[5] B.E. Logan, et al., Microbial fuel cells: methodology and technology, Environmental Science & Technology 40 (2006) 5181−5192. Available from: https://doi.org/10.1021/es0605016.
[6] K. Rabaey, W. Verstraete, Microbial fuel cells: novel biotechnology for energy generation, Trends in Biotechnology 23 (2005) 291−298. Available from: https://doi.org/10.1016/j.tibtech.2005.04.008.
[7] B.E. Logan, Microbial Fuel Cells, John Wiley & Sons, New York, 2008.
[8] K. Scott, E.H. Yu, Microbial Electrochemical and Fuel Cells: Fundamentals and Applications, Woodhead Publishing, Sawston, Cambridge, 2015.
[9] C.C. Goller, T. Romeo, Environmental influences on biofilm development, Current Topics in Microbiology and Immunology 322 (2008) 37−66.
[10] P. Zhang, et al., Nanomaterials for facilitating microbial extracellular electron transfer: recent progress and challenges, Bioelectrochemistry (Amsterdam, Netherlands) 123 (2018) 190−200. Available from: https://doi.org/10.1016/j.bioelechem.2018.05.005.
[11] Y. Liu, X. Zhang, Q. Zhang, C. Li, Microbial fuel cells: nanomaterials based on anode and their application, Energy Technology 8 (9) (2020) 200206. Available from: https://doi.org/10.1002/ente.202000206.

[12] C. Moss, A. Behrens, U. Schroder, The limits of three-dimensionality: systematic assessment of effective anode macrostructure dimensions for mixed-culture electroactive biofilms, ChemSusChem 13 (2020) 582–589. Available from: https://doi.org/10.1002/cssc.201902923.

[13] J. Li, et al., Uneven biofilm and current distribution in three-dimensional macroporous anodes of bio-electrochemical systems composed of graphite electrode arrays, Bioresource Technology 228 (2017) 25–30. Available from: https://doi.org/10.1016/j.biortech.2016.12.092.

[14] T. Cai, et al., Application of advanced anodes in microbial fuel cells for power generation: a review, Chemosphere 248 (2020) 125985. Available from: https://doi.org/10.1016/j.chemosphere.2020.125985.

[15] A.A. Yaqoob, et al., Recent advances in anodes for microbial fuel cells: an overview, Materials (Basel) 13 (9) (2020) 2078. Available from: https://doi.org/10.3390/ma13092078.

[16] B.E. Logan, R. Rossi, A. Ragab, P.E. Saikaly, Electroactive microorganisms in bioelectrochemical systems, Nature Reviews Microbiology 17 (2019) 307–319. Available from: https://doi.org/10.1038/s41579-019-0173-x.

[17] M. Kokko, S. Epple, J. Gescher, S. Kerzenmacher, Effects of wastewater constituents and operational conditions on the composition and dynamics of anodic microbial communities in bioelectrochemical systems, Bioresource Technology 258 (2018) 376–389. Available from: https://doi.org/10.1016/j.biortech.2018.01.090.

[18] M.F. Umar, S.Z. Abbas, M.N. Mohamad Ibrahim, N. Ismail, M. Rafatullah, Insights into advancements and electrons transfer mechanisms of electrogens in benthic microbial fuel cells, Membranes (Basel) 10 (9) (2020) 205. Available from: https://doi.org/10.3390/membranes10090205.

[19] K.S. Aiyer, How does electron transfer occur in microbial fuel cells? World Journal of Microbiology and Biotechnology 36 (2020) 19. Available from: https://doi.org/10.1007/s11274-020-2801-z.

[20] U. Schroder, Anodic electron transfer mechanisms in microbial fuel cells and their energy efficiency, Physical Chemistry Chemical Physics: PCCP 9 (2007) 2619–2629. Available from: https://doi.org/10.1039/b703627m.

[21] B.K. Zaied, et al., A comprehensive review on contaminants removal from pharmaceutical wastewater by electrocoagulation process, The Science of the Total Environment 726 (2020) 138095. Available from: https://doi.org/10.1016/j.scitotenv.2020.138095.

[22] A.S. Mathuriya, J.V. Yakhmi, Microbial fuel cells to recover heavy metals, Environmental Chemistry Letters 12 (2014) 483–494. Available from: https://doi.org/10.1007/s10311-014-0474-2.

[23] Y.Y. Yu, et al., Three-dimensional electrodes for high-performance bioelectrochemical systems, International Journal of Molecular Sciences 18 (1) (2017) 90. Available from: https://doi.org/10.3390/ijms18010090.

[24] S. Li, C. Cheng, A. Thomas, Carbon-based microbial-fuel-cell electrodes: From conductive supports to active catalysts, Advanced Materials 29 (8) (2017) 1602547. Available from: https://doi.org/10.1002/adma.201602547.

[25] D. Nosek, P. Jachimowicz, A. Cydzik-Kwiatkowska, Anode modification as an alternative approach to improve electricity generation in microbial fuel cells, Energies 13 (24) (2020) 6596. Available from: https://doi.org/10.3390/en13246596.

[26] N. Savla, R. Anand, S. Pandit, R. Prasad, Utilization of nanomaterials as anode modifiers for improving microbial fuel cells performance, Journal of Renewable Materials 8 (2020) 1581–1605. Available from: https://doi.org/10.32604/jrm.2020.011803.

[27] J. Zheng, C. Cheng, J. Zhang, X. Wu, Appropriate mechanical strength of carbon black-decorated loofah sponge as anode material in microbial fuel cells, International Journal of Hydrogen Energy 41 (2016) 23156–23163. Available from: https://doi.org/10.1016/j.ijhydene.2016.11.003.

[28] X. Xie, et al., Carbon nanotube-coated macroporous sponge for microbial fuel cell electrodes, Energy & Environmental Science. 5 (2012) 5265–5270. Available from: https://doi.org/10.1039/c1ee02122b.

[29] X. Xie, et al., Three-dimensional carbon nanotube-textile anode for high-performance microbial fuel cells, Nano Letters 11 (2011) 291–296. Available from: https://doi.org/10.1021/nl103905t.

[30] C. Erbay, et al., Control of geometrical properties of carbon nanotube electrodes towards high-performance microbial fuel cells, Journal of Power Sources 280 (2015) 347–354. Available from: https://doi.org/10.1016/j.jpowsour.2015.01.065.

[31] S. Zhou, M. Lin, Z. Zhuang, P. Liu, Z. Chen, Biosynthetic graphene enhanced extracellular electron transfer for high performance anode in microbial fuel cell, Chemosphere 232 (2019) 396−402. Available from: https://doi.org/10.1016/j.chemosphere.2019.05.191.

[32] X. Yang, et al., Eighteen-month assessment of 3D graphene oxide aerogel-modified 3D graphite fiber brush electrode as a high-performance microbial fuel cell anode, Electrochimica Acta 210 (2016) 846−853. Available from: https://doi.org/10.1016/j.electacta.2016.05.215.

[33] J. You, et al., Micro-porous layer (MPL)-based anode for microbial fuel cells, International Journal of Hydrogen Energy 39 (2014) 21811−21818. Available from: https://doi.org/10.1016/j.ijhydene.2014.07.136.

[34] S. Chen, et al., Electrospun and solution blown three-dimensional carbon fiber nonwovens for application as electrodes in microbial fuel cells, Energy & Environmental Science 4 (4) (2011) 1417−1421. Available from: https://doi.org/10.1039/c0ee00446d.

[35] S. Zheng, et al., Binder-free carbon black/stainless steel mesh composite electrode for high-performance anode in microbial fuel cells, Journal of Power Sources 284 (2015) 252−257. Available from: https://doi.org/10.1016/j.jpowsour.2015.03.014.

[36] S.-H. Liu, Y.-C. Lai, C.-W. Lin, Enhancement of power generation by microbial fuel cells in treating toluene-contaminated groundwater: developments of composite anodes with various compositions, Applied Energy 233-234 (2019) 922−929. Available from: https://doi.org/10.1016/j.apenergy.2018.10.105.

[37] S.H. Roh, Layer-by-layer self-assembled carbon nanotube electrode for microbial fuel cells application, Journal of Nanoscience and Nanotechnology 13 (2013) 4158−4161. Available from: https://doi.org/10.1166/jnn.2013.7021.

[38] H. Ren, et al., A high power density miniaturized microbial fuel cell having carbon nanotube anodes, Journal of Power Sources 273 (2015) 823−830. Available from: https://doi.org/10.1016/j.jpowsour.2014.09.165.

[39] P. Zhang, et al., Enhanced performance of microbial fuel cell with a bacteria/multi-walled carbon nanotube hybrid biofilm, Journal of Power Sources 361 (2017) 318−325. Available from: https://doi.org/10.1016/j.jpowsour.2017.06.069.

[40] Y.Z. Zhang, et al., A graphene modified anode to improve the performance of microbial fuel cells, Journal of Power Sources 196 (2011) 5402−5407. Available from: https://doi.org/10.1016/j.jpowsour.2011.02.067.

[41] W. Guo, Y. Cui, H. Song, J. Sun, Layer-by-layer construction of graphene-based microbial fuel cell for improved power generation and methyl orange removal, Bioprocess and Biosystems Engineering 37 (2014) 1749−1758. Available from: https://doi.org/10.1007/s00449-014-1148-y.

[42] J. Hou, Z. Liu, Y. Li, S. Yang, Y. Zhou, A comparative study of graphene-coated stainless steel fiber felt and carbon cloth as anodes in MFCs, Bioprocess and Biosystems Engineering 38 (2015) 881−888. Available from: https://doi.org/10.1007/s00449-014-1332-0.

[43] J. Hou, Z. Liu, S. Yang, Y. Zhou, Three-dimensional macroporous anodes based on stainless steel fiber felt for high-performance microbial fuel cells, Journal of Power Sources 258 (2014) 204−209. Available from: https://doi.org/10.1016/j.jpowsour.2014.02.035.

[44] R. Wang, et al., FeS_2 nanoparticles decorated graphene as microbial-fuel-cell anode achieving high power density, Advanced Materials 30 (2018) e1800618. Available from: https://doi.org/10.1002/adma.201800618.

[45] J.Y. Luo, et al., Compression and aggregation-resistant particles of crumpled soft sheets, ACS Nano 5 (2011) 8943−8949. Available from: https://doi.org/10.1021/nn203115u.

[46] Z. Li, et al., In-situ modified titanium suboxides with polyaniline/graphene as anode to enhance biovoltage production of microbial fuel cell, International Journal of Hydrogen Energy 44 (2019) 6862−6870. Available from: https://doi.org/10.1016/j.ijhydene.2018.12.106.

[47] Y. Cheng, M. Mallavarapu, R. Naidu, Z. Chen, In situ fabrication of green reduced graphene-based biocompatible anode for efficient energy recycle, Chemosphere 193 (2018) 618−624. Available from: https://doi.org/10.1016/j.chemosphere.2017.11.057.

[48] G. Gnana kumar, et al., Synthesis, structural, and morphological characterizations of reduced graphene oxide-supported polypyrrole anode catalysts for improved microbial fuel cell performances, ACS Sustainable Chemistry & Engineering 2 (2014) 2283−2290. Available from: https://doi.org/10.1021/sc500244f.

[49] S.L. Zhao, et al., Three-dimensional graphene/Pt nanoparticle composites as freestanding anode for enhancing performance of microbial fuel cells, Science Advances 1 (2015) 8. Available from: https://doi.org/10.1126/sciadv.1500372.

[50] Y. Zhang, L. Liu, B. Van der Bruggen, F. Yang, Nanocarbon based composite electrodes and their application in microbial fuel cells, Journal of Materials Chemistry A 5 (2017) 12673−12698. Available from: https://doi.org/10.1039/c7ta01511a.

[51] H.-Y. Tsai, C.-C. Wu, C.-Y. Lee, E.P. Shih, Microbial fuel cell performance of multiwall carbon nanotubes on carbon cloth as electrodes, Journal of Power Sources 194 (2009) 199−205. Available from: https://doi.org/10.1016/j.jpowsour.2009.05.018.

[52] H.-F. Cui, L. Du, P.-B. Guo, B. Zhu, J.H.T. Luong, Controlled modification of carbon nanotubes and polyaniline on macroporous graphite felt for high-performance microbial fuel cell anode, Journal of Power Sources 283 (2015) 46−53. Available from: https://doi.org/10.1016/j.jpowsour.2015.02.088.

[53] L. Jourdin, et al., A novel carbon nanotube modified scaffold as an efficient biocathode material for improved microbial electrosynthesis, Journal of Materials Chemistry A 2 (32) (2014) 13093−13102. Available from: https://doi.org/10.1039/c4ta03101f.

[54] A.A. Yazdi, et al., Carbon nanotube modification of microbial fuel cell electrodes, Biosensors & Bioelectronics 85 (2016) 536−552. Available from: https://doi.org/10.1016/j.bios.2016.05.033.

[55] A. Magrez, et al., Cellular toxicity of carbon-based nanomaterials, Nano Letters 6 (2006) 1121−1125. Available from: https://doi.org/10.1021/nl060162e.

[56] T. Cai, M. Huang, Y. Huang, W. Zheng, Enhanced performance of microbial fuel cells by electrospinning carbon nanofibers hybrid carbon nanotubes composite anode, International Journal of Hydrogen Energy 44 (2019) 3088−3098. Available from: https://doi.org/10.1016/j.ijhydene.2018.11.205.

[57] H. Yuan, Z. He, Graphene-modified electrodes for enhancing the performance of microbial fuel cells, Nanoscale 7 (2015) 7022−7029. Available from: https://doi.org/10.1039/c4nr05637j.

[58] H. Wang, et al., High power density microbial fuel cell with flexible 3D graphene-nickel foam as anode, Nanoscale 5 (2013) 10283−10290. Available from: https://doi.org/10.1039/c3nr03487a.

[59] Z. Liu, L. Zhou, Q. Chen, W. Zhou, Y. Liu, Advances in graphene/graphene composite based microbial fuel/electrolysis cells, Electroanalysis 29 (2017) 652−661. Available from: https://doi.org/10.1002/elan.201600502.

[60] M.D. Stoller, S. Park, Y. Zhu, J. An, R.S. Ruoff, Graphene-based ultracapacitors, Nano Letters 8 (2008) 3498−3502. Available from: https://doi.org/10.1021/nl802558y.

[61] J. Liu, et al., Graphene/carbon cloth anode for high-performance mediatorless microbial fuel cells, Bioresource Technology 114 (2012) 275−280. Available from: https://doi.org/10.1016/j.biortech.2012.02.116.

[62] A. ElMekawy, H.M. Hegab, D. Losic, C.P. Saint, D. Pant, Applications of graphene in microbial fuel cells: the gap between promise and reality, Renewable and Sustainable Energy Reviews 72 (2017) 1389−1403. Available from: https://doi.org/10.1016/j.rser.2016.10.044.

[63] W.B. Hu, et al., Graphene-based antibacterial paper, ACS Nano 4 (2010) 4317−4323. Available from: https://doi.org/10.1021/nn101097v.

[64] O. Akhavan, E. Ghaderi, Toxicity of graphene and graphene oxide nanowalls against bacteria, ACS Nano 4 (2010) 5731−5736. Available from: https://doi.org/10.1021/nn101390x.

[65] X. Lu, W. Zhang, C. Wang, T.-C. Wen, Y. Wei, One-dimensional conducting polymer nanocomposites: synthesis, properties and applications, Progress in Polymer Science 36 (2011) 671−712. Available from: https://doi.org/10.1016/j.progpolymsci.2010.07.010.

[66] F. Kazemi, S.M. Naghib, Z. Mohammadpour, Multifunctional micro-/nanoscaled structures based on polyaniline: an overview of modern emerging devices, Materials Today Chemistry 16 (2020) 100249. Available from: https://doi.org/10.1016/j.mtchem.2020.100249.

[67] J.M. Sonawane, S. Al-Saadi, R.K. Singh Raman, P.C. Ghosh, S.B. Adeloju, Exploring the use of polyaniline-modified stainless steel plates as low-cost, high-performance anodes for microbial fuel cells, Electrochimica Acta 268 (2018) 484−493. Available from: https://doi.org/10.1016/j.electacta.2018.01.163.

[68] P. Wang, H.R. Li, Z.W. Du, Polyaniline synthesis by cyclic voltammetry for anodic modification in microbial fuel cells, International Journal of Electrochemical Science 9 (2014) 2038−2046.

[69] K.-B. Pu, et al., Polypyrrole modified stainless steel as high performance anode of microbial fuel cell, Biochemical Engineering Journal 132 (2018) 255–261. Available from: https://doi.org/10.1016/j.bej.2018.01.018.

[70] D.Z. Sun, et al., In-situ growth of graphene/polyaniline for synergistic improvement of extracellular electron transfer in bioelectrochemical systems, Biosensors & Bioelectronics 87 (2017) 195–202. Available from: https://doi.org/10.1016/j.bios.2016.08.037.

[71] Z. Lv, et al., One-step electrosynthesis of polypyrrole/graphene oxide composites for microbial fuel cell application, Electrochimica Acta 111 (2013) 366–373. Available from: https://doi.org/10.1016/j.electacta.2013.08.022.

[72] X. Liu, et al., Facile fabrication of conductive polyaniline nanoflower modified electrode and its application for microbial energy harvesting, Electrochimica Acta 255 (2017) 41–47. Available from: https://doi.org/10.1016/j.electacta.2017.09.153.

[73] Y.C. Yong, X.C. Dong, M.B. Chan-Park, H. Song, P. Chen, Macroporous and monolithic anode based on polyaniline hybridized three-dimensional graphene for high-performance microbial fuel cells, ACS Nano 6 (2012) 2394–2400. Available from: https://doi.org/10.1021/nn204656d.

[74] J.M. Sonawane, S.A. Patil, P.C. Ghosh, S.B. Adeloju, Low-cost stainless-steel wool anodes modified with polyaniline and polypyrrole for high-performance microbial fuel cells, Journal of Power Sources 379 (2018) 103–114. Available from: https://doi.org/10.1016/j.jpowsour.2018.01.001.

[75] C.E. Zhao, J. Wu, S. Kjelleberg, J.S. Loo, Q. Zhang, Employing a flexible and low-cost polypyrrole nanotube membrane as an anode to enhance current generation in microbial fuel cells, Small (Weinheim an der Bergstrasse, Germany) 11 (2015) 3440–3443. Available from: https://doi.org/10.1002/smll.201403328.

[76] B. Li, et al., Modification of PPy-NW anode by carbon dots for high-performance mini-microbial fuel cells, Fuel Cells 20 (2020) 203–211. Available from: https://doi.org/10.1002/fuce.201900210.

[77] Y. Zou, et al., A mediatorless microbial fuel cell using polypyrrole coated carbon nanotubes composite as anode material, International Journal of Hydrogen Energy 33 (2008) 4856–4862. Available from: https://doi.org/10.1016/j.ijhydene.2008.06.061.

[78] Y.L. Kang, S. Ibrahim, S. Pichiah, Synergetic effect of conductive polymer poly(3,4-ethylenedioxythiophene) with different structural configuration of anode for microbial fuel cell application, Bioresource Technology 189 (2015) 364–369. Available from: https://doi.org/10.1016/j.biortech.2015.04.044.

[79] N. Gospodinova, L. Terlemezyan, Conducting polymers prepared by oxidative polymerization: polyaniline, Progress in Polymer Science 23 (1998) 1443–1484. Available from: https://doi.org/10.1016/s0079-6700(98)00008-2.

[80] D. Wang, et al., Surface modification of *Shewanella oneidensis* MR-1 with polypyrrole-dopamine coating for improvement of power generation in microbial fuel cells, Journal of Power Sources 483 (2021). Available from: https://doi.org/10.1016/j.jpowsour.2020.229220.

[81] L. Huang, X. Li, Y. Ren, X. Wang, In-situ modified carbon cloth with polyaniline/graphene as anode to enhance performance of microbial fuel cell, International Journal of Hydrogen Energy 41 (2016) 11369–11379. Available from: https://doi.org/10.1016/j.ijhydene.2016.05.048.

[82] Y. Qiao, et al., Nanostructured polyanifine/titanium dioxide composite anode for microbial fuel cells, ACS Nano 2 (2008) 113–119. Available from: https://doi.org/10.1021/nn700102s.

[83] R. Karthikeyan, et al., Effect of composites based nickel foam anode in microbial fuel cell using *Acetobacter aceti* and *Gluconobacter roseus* as a biocatalysts, Bioresource Technology 217 (2016) 113–120. Available from: https://doi.org/10.1016/j.biortech.2016.02.114.

[84] S. Mwale, M.O. Munyati, J. Nyirenda, Preparation, characterization, and optimization of a porous polyaniline-copper anode microbial fuel cell, Journal of Solid State Electrochemistry 25 (2021) 639–650. Available from: https://doi.org/10.1007/s10008-020-04839-0.

[85] Y.B. Fu, et al., Modified carbon anode by MWCNTs/PANI used in marine sediment microbial fuel cell and its electrochemical performance, Fuel Cells 16 (2016) 377–383. Available from: https://doi.org/10.1002/fuce.201500103.

[86] D.H. Park, J.G. Zeikus, Improved fuel cell and electrode designs for producing electricity from microbial degradation, Biotechnology and Bioengineering 81 (2003) 348–355. Available from: https://doi.org/10.1002/bit.10501.

[87] L. Bouabdalaoui, L. Legrand, D. Feron, A. Chausse, Improved performance of anode with iron/sulfur-modified graphite in microbial fuel cell, Electrochemistry Communications 28 (2013) 1–4. Available from: https://doi.org/10.1016/j.elecom.2012.11.023.

[88] X. Ying, et al., Titanium dioxide thin film-modified stainless steel mesh for enhanced current-generation in microbial fuel cells, Chemical Engineering Journal 333 (2018) 260–267. Available from: https://doi.org/10.1016/j.cej.2017.09.132.

[89] Y.T. He, et al., Customizable design strategies for high-performance bioanodes in bioelectrochemical systems, iScience 24 (2021) 102163. Available from: https://doi.org/10.1016/j.isci.2021.102163.

[90] S.A. Cheng, B.E. Logan, Ammonia treatment of carbon cloth anodes to enhance power generation of microbial fuel cells, Electrochemistry Communications 9 (2007) 492–496. Available from: https://doi.org/10.1016/j.elecom.2006.10.023.

[91] N. Zhu, et al., Improved performance of membrane free single-chamber air-cathode microbial fuel cells with nitric acid and ethylenediamine surface modified activated carbon fiber felt anodes, Bioresource Technology 102 (2011) 422–426. Available from: https://doi.org/10.1016/j.biortech.2010.06.046.

[92] Y.J. Feng, Q. Yang, X. Wang, B.E. Logan, Treatment of carbon fiber brush anodes for improving power generation in air-cathode microbial fuel cells, Journal of Power Sources 195 (2010) 1841–1844. Available from: https://doi.org/10.1016/j.jpowsour.2009.10.030.

[93] P. Chong, B. Erable, A. Bergel, Effect of pore size on the current produced by 3-dimensional porous microbial anodes: a critical review, Bioresource Technology 289 (2019) 121641. Available from: https://doi.org/10.1016/j.biortech.2019.121641.

[94] S. Kerzenmacher, Engineering of microbial electrodes, Advances in Biochemical Engineering/Biotechnology 167 (2019) 135–180. Available from: https://doi.org/10.1007/10_2017_16.

[95] R. Kaur, A. Marwaha, V.A. Chhabra, K.-H. Kim, S.K. Tripathi, Recent developments on functional nanomaterial-based electrodes for microbial fuel cells, Renewable and Sustainable Energy Reviews 119 (2020) 109551. Available from: https://doi.org/10.1016/j.rser.2019.109551.

[96] X. Xie, C. Criddle, Y. Cui, Design and fabrication of bioelectrodes for microbial bioelectrochemical systems, Energy & Environmental Science 8 (2015) 3418–3441. Available from: https://doi.org/10.1039/c5ee01862e.

[97] R.B. Mathur, P.H. Maheshwari, T.L. Dhami, R.K. Sharma, C.P. Sharma, Processing of carbon composite paper as electrode for fuel cell, Journal of Power Sources 161 (2006) 790–798. Available from: https://doi.org/10.1016/j.jpowsour.2006.05.053.

[98] M. Zhou, M. Chi, J. Luo, H. He, T. Jin, An overview of electrode materials in microbial fuel cells, Journal of Power Sources 196 (2011) 4427–4435. Available from: https://doi.org/10.1016/j.jpowsour.2011.01.012.

[99] Y. Ahn, B.E. Logan, Altering anode thickness to improve power production in microbial fuel cells with different electrode distances, Energy & Fuels 27 (2012) 271–276. Available from: https://doi.org/10.1021/ef3015553.

[100] J. Li, et al., Bioreactors for Microbial Biomass and Energy Conversion Green Energy and Technology, Springer, 2018, pp. 391–433). Chapter 10.

[101] B. Logan, S. Cheng, V. Watson, G. Estadt, Graphite fiber brush anodes for increased power production in air-cathode microbial fuel cells, Environmental Science & Technology 41 (2007) 3341–3346. Available from: https://doi.org/10.1021/es062644y.

[102] X. Wang, et al., Use of carbon mesh anodes and the effect of different pretreatment methods on power production in microbial fuel cells, Environmental Science & Technology 43 (2009) 6870–6874. Available from: https://doi.org/10.1021/es900997w.

[103] Mustakeem, Electrode materials for microbial fuel cells: nanomaterial approach, Materials for Renewable and Sustainable Energy 4 (2015) 22. Available from: https://doi.org/10.1007/s40243-015-0063-8.

[104] G. Lepage, F.O. Albernaz, G. Perrier, G. Merlin, Characterization of a microbial fuel cell with reticulated carbon foam electrodes, Bioresource Technology 124 (2012) 199–207. Available from: https://doi.org/10.1016/j.biortech.2012.07.067.

[105] V. Flexer, et al., The nanostructure of three-dimensional scaffolds enhances the current density of microbial bioelectrochemical systems, Energy & Environmental Science 6 (4) (2013) 1291–1298. Available from: https://doi.org/10.1039/c3ee00052d.

[106] C.M. Liu, et al., Effects of brush lengths and fiber loadings on the performance of microbial fuel cells using graphite fiber brush anodes, International Journal of Hydrogen Energy 38 (2013) 15646–15652. Available from: https://doi.org/10.1016/j.ijhydene.2013.03.144.

[107] V. Lanas, Y. Ahn, B.E. Logan, Effects of carbon brush anode size and loading on microbial fuel cell performance in batch and continuous mode, Journal of Power Sources 247 (2014) 228–234. Available from: https://doi.org/10.1016/j.jpowsour.2013.08.110.

[108] X. Peng, S. Chen, L. Liu, S. Zheng, M. Li, Modified stainless steel for high performance and stable anode in microbial fuel cells, Electrochimica Acta 194 (2016) 246–252. Available from: https://doi.org/10.1016/j.electacta.2016.02.127.

[109] Y. Qiao, S.-J. Bao, C.M. Li, Electrocatalysis in microbial fuel cells—from electrode material to direct electrochemistry, Energy & Environmental Science 3 (5) (2010) 544–553. Available from: https://doi.org/10.1039/b923503e.

[110] H. Richter, et al., Electricity generation by *Geobacter sulfurreducens* attached to gold electrodes, Langmuir: the ACS Journal of Surfaces and Colloids 24 (2008) 4376–4379. Available from: https://doi.org/10.1021/la703469y.

[111] K. Guo, et al., Flame oxidation of stainless steel felt enhances anodic biofilm formation and current output in bioelectrochemical systems, Environmental Science & Technology 48 (2014) 7151–7156. Available from: https://doi.org/10.1021/es500720g.

[112] Y. Qiao, X.-S. Wu, C.-X. Ma, H. He, C.M. Li, A hierarchical porous graphene/nickel anode that simultaneously boosts the bio- and electro-catalysis for high-performance microbial fuel cells, RSC Advances 4 (2014) 21788–21793. Available from: https://doi.org/10.1039/c4ra03082f.

[113] X.B. Ying, et al., Sustainable synthesis of novel carbon microwires for the modification of a Ti mesh anode in bioelectrochemical systems, The Science of the Total Environment 669 (2019) 294–302. Available from: https://doi.org/10.1016/j.scitotenv.2019.03.106.

[114] S.R. Crittenden, C.J. Sund, J.J. Sumner, Mediating electron transfer from bacteria to a gold electrode via a self-assembled monolayer, Langmuir: the ACS Journal of Surfaces and Colloids 22 (2006) 9473–9476. Available from: https://doi.org/10.1021/la061869j.

[115] A. Baudler, I. Schmidt, M. Langner, A. Greiner, U. Schröder, Does it have to be carbon? Metal anodes in microbial fuel cells and related bioelectrochemical systems, Energy & Environmental Science 8 (2015) 2048–2055. Available from: https://doi.org/10.1039/c5ee00866b.

[116] T. Yamashita, H. Yokoyama, Molybdenum anode: a novel electrode for enhanced power generation in microbial fuel cells, identified via extensive screening of metal electrodes, Biotechnology for Biofuels 11 (2018) 39. Available from: https://doi.org/10.1186/s13068-018-1046-7.

[117] C. Dumas, et al., Marine microbial fuel cell: use of stainless steel electrodes as anode and cathode materials, Electrochimica Acta 53 (2007) 468–473. Available from: https://doi.org/10.1016/j.electacta.2007.06.069.

[118] C. Dumas, R. Basseguy, A. Bergel, Electrochemical activity of *Geobacter sulfurreducens* biofilms on stainless steel anodes, Electrochimica Acta 53 (2008) 5235–5241. Available from: https://doi.org/10.1016/j.electacta.2008.02.056.

[119] D. Pocaznoi, A. Calmet, L. Etcheverry, B. Erable, A. Bergel, Stainless steel is a promising electrode material for anodes of microbial fuel cells, Energy & Environmental Science 5 (11) (2012) 9645–9652. Available from: https://doi.org/10.1039/c2ee22429a.

[120] B. Erable, A. Bergel, First air-tolerant effective stainless steel microbial anode obtained from a natural marine biofilm, Bioresource Technology 100 (2009) 3302–3307. Available from: https://doi.org/10.1016/j.biortech.2009.02.025.

[121] X. Long, X. Cao, C. Wang, S. Liu, X. Li, Preparation of needle-like Fe_3O_4/Fe_2O_3 nanorods on stainless steel plates to form inexpensive, high-performance bioanodes, Journal of Electroanalytical Chemistry 855 (2019) 113497. Available from: https://doi.org/10.1016/j.jelechem.2019.113497.

[122] K. Guo, et al., Heat-treated stainless steel felt as scalable anode material for bioelectrochemical systems, Bioresource Technology 195 (2015) 46–50. Available from: https://doi.org/10.1016/j.biortech.2015.06.060.

[123] S.F. Ketep, A. Bergel, A. Calmet, B. Erable, Stainless steel foam increases the current produced by microbial bioanodes in bioelectrochemical systems, Energy & Environmental Science 7 (2014) 1633–1637. Available from: https://doi.org/10.1039/c3ee44114h.

[124] E. Taskan, H. Hasar, Comprehensive comparison of a new tin-coated copper mesh and a graphite plate electrode as an anode material in microbial fuel cell, Applied Biochemistry and Biotechnology 175 (2015) 2300–2308. Available from: https://doi.org/10.1007/s12010-014-1439-4.

[125] A. ter Heijne, H.V.M. Hamelers, M. Saakes, C.J.N. Buisman, Performance of non-porous graphite and titanium-based anodes in microbial fuel cells, Electrochimica Acta 53 (2008) 5697–5703. Available from: https://doi.org/10.1016/j.electacta.2008.03.032.

[126] H. Feng, et al., TiO$_2$ nanotube arrays modified titanium: a stable, scalable, and cost-effective bioanode for microbial fuel cells, Environmental Science & Technology Letters 3 (2016) 420–424. Available from: https://doi.org/10.1021/acs.estlett.6b00410.

[127] L. Huang, et al., Effect of heat-treatment atmosphere on the current generation of TiO$_2$ nanotube array electrodes in microbial fuel cells, Electrochimica Acta 257 (2017) 203–209. Available from: https://doi.org/10.1016/j.electacta.2017.10.068.

[128] X. Zhu, B.E. Logan, Copper anode corrosion affects power generation in microbial fuel cells, Journal of Chemical Technology & Biotechnology 89 (2014) 471–474. Available from: https://doi.org/10.1002/jctb.4156.

[129] B. Bian, et al., Application of 3D printed porous copper anode in microbial fuel cells, Frontiers in Energy Research 6 (2018) 50. Available from: https://doi.org/10.3389/fenrg.2018.00050.

[130] E. Taşkan, B. Özkaya, H. Hasar, Comprehensive evaluation of two different inoculums in MFC with a new tin-coated copper mesh anode electrode for producing electricity from a cottonseed oil industry effluent, Environmental Progress & Sustainable Energy 35 (2016) 110–116. Available from: https://doi.org/10.1002/ep.12207.

[131] F. Kargi, S. Eker, Electricity generation with simultaneous wastewater treatment by a microbial fuel cell (MFC) with Cu and Cu–Au electrodes, Journal of Chemical Technology & Biotechnology 82 (2007) 658–662. Available from: https://doi.org/10.1002/jctb.1723.

[132] J. Prasad, R. K. Tripathi. (2017). Maximum electricity generation from low cost sediment microbial fuel cell using copper and zinc electrodes, in: 2017 International Conference on Information, Communication, Instrumentation and Control (ICICIC), Indore, India, August 17–19, 2017.

[133] S. Srikanth, T. Pavani, P.N. Sarma, S. Venkata Mohan, Synergistic interaction of biocatalyst with bio-anode as a function of electrode materials, International Journal of Hydrogen Energy 36 (2011) 2271–2280. Available from: https://doi.org/10.1016/j.ijhydene.2010.11.031.

[134] I. Chakraborty, S.M. Sathe, B.K. Dubey, M.M. Ghangrekar, Waste-derived biochar: Applications and future perspective in microbial fuel cells, Bioresource Technology 312 (2020) 123587. Available from: https://doi.org/10.1016/j.biortech.2020.123587.

[135] T. Huggins, A. Latorre, J. Biffinger, Z. Ren, Biochar sased microbial fuel cell for enhanced wastewater treatment and nutrient recovery, Sustainability 8 (2) (2016) 169. Available from: https://doi.org/10.3390/su8020169.

[136] T. Huggins, H. Wang, J. Kearns, P. Jenkins, Z.J. Ren, Biochar as a sustainable electrode material for electricity production in microbial fuel cells, Bioresource Technology 157 (2014) 114–119. Available from: https://doi.org/10.1016/j.biortech.2014.01.058.

[137] Q. Chen, et al., Activated microporous-mesoporous carbon derived from chestnut shell as a sustainable anode material for high performance microbial fuel cells, Bioresource Technology 249 (2018) 567–573. Available from: https://doi.org/10.1016/j.biortech.2017.09.086.

[138] S. Chen, et al., Reticulated carbon foam derived from a sponge-like natural product as a high-performance anode in microbial fuel cells, Journal of Materials Chemistry 22 (2012) 18609–18613.

[139] S. Chen, et al., A hierarchically structured urchin-like anode derived from chestnut shells for microbial energy harvesting, Electrochimica Acta 212 (2016) 883–889. Available from: https://doi.org/10.1016/j.electacta.2016.07.077.

[140] H.L. Zhu, et al., Lightweight, conductive hollow fibers from nature as sustainable electrode materials for microbial energy harvesting, Nano Energy 10 (2014) 268–276. Available from: https://doi.org/10.1016/j.nanoen.2014.08.014.

[141] M. Lu, et al., Nitrogen-enriched pseudographitic anode derived from silk cocoon with tunable flexibility for microbial fuel cells, Nano Energy 32 (2017) 382–388. Available from: https://doi.org/10.1016/j.nanoen.2016.12.046.

[142] L.H. Zhou, L.H. Sun, P. Fu, C.L. Yang, Y. Yuan, Carbon nanoparticles of Chinese ink-wrapped natural loofah sponge: a low-cost three-dimensional electrode for high-performance microbial energy harvesting, Journal of Materials Chemistry A 5 (2017) 14741–14747. Available from: https://doi.org/10.1039/c7ta03869k.

[143] Y. Yuan, S.G. Zhou, Y. Liu, J.H. Tang, Nanostructured macroporous bioanode based on polyaniline-modified natural loofah sponge for high-performance microbial fuel cells, Environmental Science & Technology 47 (2013) 14525–14532. Available from: https://doi.org/10.1021/es404163g.

[144] R. Karthikeyan, et al., Interfacial electron transfer and bioelectrocatalysis of carbonized plant material as effective anode of microbial fuel cell, Electrochimica Acta 157 (2015) 314–323. Available from: https://doi.org/10.1016/j.electacta.2015.01.029.

[145] S.L. Chen, et al., A three-dimensionally ordered macroporous carbon derived from a natural resource as anode for microbial bioelectrochemical systems, ChemSusChem 5 (2012) 1059–1063. Available from: https://doi.org/10.1002/cssc.201100783.

[146] J. Zhang, et al., Tubular bamboo charcoal for anode in microbial fuel cells, Journal of Power Sources 272 (2014) 277–282. Available from: https://doi.org/10.1016/j.jpowsour.2014.08.115.

[147] M. Li, S. Ci, Y. Ding, Z. Wen, Almond shell derived porous carbon for a high-performance anode of microbial fuel cells, Sustainable Energy & Fuels 3 (2019) 3415–3421. Available from: https://doi.org/10.1039/c9se00659a.

[148] Y. Yuan, T. Liu, P. Fu, J. Tang, S. Zhou, Conversion of sewage sludge into high-performance bifunctional electrode materials for microbial energy harvesting, Journal of Materials Chemistry A 3 (2015) 8475–8482. Available from: https://doi.org/10.1039/c5ta00458f.

[149] Y.-H. Hung, T.-Y. Liu, H.-Y. Chen, Renewable coffee waste-derived porous carbons as anode materials for high-performance sustainable microbial fuel cells, ACS Sustainable Chemistry & Engineering 7 (2019) 16991–16999. Available from: https://doi.org/10.1021/acssuschemeng.9b02405.

[150] S. Chen, et al., Layered corrugated electrode macrostructures boost microbial bioelectrocatalysis, Energy & Environmental Science 5 (12) (2012) 9769–9772. Available from: https://doi.org/10.1039/c2ee23344d.

[151] L. Zhang, et al., Bread-derived 3D macroporous carbon foams as high performance free-standing anode in microbial fuel cells, Biosensors & Bioelectronics 122 (2018) 217–223. Available from: https://doi.org/10.1016/j.bios.2018.09.005.

[152] W. Yang, S. Chen, Biomass-derived carbon for electrode fabrication in microbial fuel cells: a review, Industrial & Engineering Chemistry Research 59 (2020) 6391–6404. Available from: https://doi.org/10.1021/acs.iecr.0c00041.

[153] P. Chaijak, et al., Potential of biochar-anode in a ceramic-separator microbial fuel cell (CMFC) with a laccase-based air cathode, Polish Journal of Environmental Studies 29 (2019) 499–503. Available from: https://doi.org/10.15244/pjoes/99099.

[154] L. Zhang, J. Li, X. Zhu, D.D. Ye, Q. Liao, Anodic current distribution in a liter-scale microbial fuel cell with electrode arrays, Chemical Engineering Journal 223 (2013) 623–631. Available from: https://doi.org/10.1016/j.cej.2013.03.035.

[155] B. Bian, et al., 3D printed porous carbon anode for enhanced power generation in microbial fuel cell, Nano Energy 44 (2018) 174–180. Available from: https://doi.org/10.1016/j.nanoen.2017.11.070.

[156] Y. Zhou, et al., A novel anode fabricated by three-dimensional printing for use in urine-powered microbial fuel cell, Biochemical Engineering Journal 124 (2017) 36–43. Available from: https://doi.org/10.1016/j.bej.2017.04.012.

[157] J. You, R.J. Preen, L. Bull, J. Greenman, I. Ieropoulos, 3D printed components of microbial fuel cells: Towards monolithic microbial fuel cell fabrication using additive layer manufacturing, Sustainable Energy Technologies and Assessments 19 (2017) 94–101. Available from: https://doi.org/10.1016/j.seta.2016.11.006.

[158] M.C. Freyman, T. Kou, S. Wang, Y. Li, 3D printing of living bacteria electrode, Nano Research 13 (2019) 1318–1323. Available from: https://doi.org/10.1007/s12274-019-2534-1.

CHAPTER 16

Nanomaterials in biofuel cells

Sangeetha Dharmalingam, Vaidhegi Kugarajah and John Solomon
Department of Mechanical Engineering, Anna University, Chennai, India

16.1 Introduction

Biofuel cells (BFCs), and more specifically microbial fuel cells (MFCs), are considered as one of the few capable technologies that have the capacity to sustainably treat wastewater and produce electricity simultaneously. BFCs have proved themselves to be a potential solution for the problems of the real world such as energy crisis, wastewater treatment, desalination, and air and water pollution. The applications of this versatile technology are limitless. They have found their way into nutrient recovery [1], bioremediation [2], telemetry [3], wastewater treatment [4], and biosensors [5]. Amidst all the advantages of utilizing BFCs, it is still seen to have limitations: electrode overpotential and energy losses that are caused by various factors such as nature of inoculum, substrate concentration, pH, electrode material, surface area of electrode, polymer electrolyte membrane (PEM), configuration, and so on and so forth, all leading to a reduction in coulombic efficiency, maximum power density, and maximum voltage attained. Out of this, the physical and chemical properties of the three components, viz., PEM, anode, and cathode play a crucial role in determining the overall efficiency of the BFC. Hence the material, structure, and other properties of the three mentioned components are to be better understood and studied before choosing from a variety of choices.

The experimentally attainable maximum voltage in a BFC is lower than that which is attained theoretically. The voltage obtained is the reduction potential of the redox chemical species utilized in BFC, and this output is affected by a number of losses such as activation losses, concentration loss, and ohmic loss [6]. Activation losses are caused by the reduction in electrochemical kinetics of the redox reactions occurring at the surface of the electrodes. To mitigate activation losses suitable catalysts may be used. Concentration losses, typically caused at high current densities, occurs due to decrease in liberation of oxidized chemical species or decrease in the transfer of reduced species toward the anode. The vice versa is applicable in case of cathode concentration loss. The resistance offered by the electrode and ion-exchange membrane to transfer charges is called as ohmic loss. This can be minimized by reducing the interelectrode distance, increasing the surface area of electrode, and prudently selecting the material of electrode and PEM [7,8].

To alleviate the losses that cause detrimental effects in the performance, the three main components of BFC need to be considered. Anode is the electrode that houses the anaerobic, electroactive bacteria, or biocatalysts that aid in the production of electrons; effective electron transfer mechanism is determined by the surface area, electrical conductivity, biocompatibility, and electrocatalytic property of the anode. Generally used anode materials are graphite felt, carbon fiber, carbon cloth, carbon paper, reticulated vitreous carbon, stainless steel mesh, etc. [7]. These anode materials have been known for their extraordinary properties to serve as current collectors but their performance is limited by their high charge-transfer resistance which make them unfit for long term and other practical applications. Similarly, the cathode where the oxygen reduction reaction (ORR) occurs, is generally slothful in its kinetics, hence needs the aid of suitable catalyst to speed up the reaction. Platinum on carbon is the commonly used cathode catalyst owing to its exceptional reducing capability, to aid in ORR. But the utilization of platinum is not appreciated due to its expensiveness and toxic nature [9].

PEM is also considered as the most important component of the BFC, since it is responsible for the transport of ions (anions or cations) through them. The typically used PEM are nonporous and polymer-based ion-exchange membranes. The generally used PEM in fuel cell application is Nafion, a cation-exchange membrane (CEM), whose usage is restricted due to certain disadvantages that limits its full-fledged performance. The fluorinated backbone of this polymer and its hydrophobicity that causes severe biofouling affects the ion transport from the anode to the cathode or vice versa [10].

These drawbacks faced during BFC operation can be overcome by incorporation of nanomaterials. Such composite membranes more sustainable and increases the overall efficiency of BFCs. Development in technology has landed into the discovery of a new domain called the nanotechnology that has rendered us materials with remarkable physical and chemical properties. These nanomaterials are desired due to their extremely high surface area, high functionality, ability to be doped with, chemical stability, catalytic property, electrical and thermal conductivity, and much more. The advent of nanotechnology has shown a radical increase in the development of novel nanocatalysts for ORR and reduction of other electron acceptors in the field of BFCs.

This entire chapter deals with the usage of such nanocomposite as electrode catalysts and in PEMs as membrane fillers in BFCs, and the factors that influences their operation.

16.2 Types of biofuel cell

BFCs are a class of fuel cells aimed for their potential applications in low-powered devices. In a BFC, the chemical energy available in the form of biofuels are converted to electrical energy by the catalytic activity of the enzymes through microbial metabolism rather than high-valued metal catalysts in fuel cells [11]. The attractive features of BFCs are the utilization of wastewater as the substrate with flexibility in fuel selection, the inherent microbes as the catalyst, operation in ambient conditions and neutral pH, and low-cost production and most predominantly simultaneous wastewater treatment with electricity generation. Based on whether the occurrence of catalytic enzymes within or outer surface of the cell, BFCs can be categorized into MFC and enzymatic fuel cell (EFC) [12].

16.2.1 Enzymatic fuel cell

EFC is a type of fuel cell, wherein suitable oxidoreductase enzymes are obtained and purified from the target organisms and utilized as biocatalysts. The salient feature of these isolated enzymes is the ability to conduct electrons to the electrodes of the BFC and generate electricity by the oxidation of target fuels [13]. EFCs can generate high current and power density for microelectronic and minuscule devices [14]. However, the isolation of enzymes requires sensitive chemicals and moreover the process is expensive which requires special protocols for stabilization and application in EFC. Depending on types of fuel, various enzymes such as alcohol dehydrogenase and glucose dehydrogenase are being utilized. For instance, if glucose, the most common fuel in EFCs is utilized, about 24 electrons are produced upon its complete oxidation. However, upon real-time application, glucose is oxidized into gluconic acid which leaves countless electrons within the substrate unutilized. Other enzymes such as laccase, bilirubin oxidase are two other common cathode enzymes that are applicable EFCs [15]. These limitations are overcome by the utilization of whole-cell organelles in EFC [16].

16.2.2 Microbial fuel cell

MFC is the major type of BFC which is a bioelectrochemical system that aims in the production of electricity by the utilization of electrons derived from the biochemical pathways of the microbes available in wastewater [17]. The electrons released by the microbes at the anode are captured through the current collector and transferred to the cathode through an external circuit, whereas the protons travel through a specialized proton-exchange membrane (PEM) specific to enhance and maximize the ion transfer to the cathode. At the cathode terminal, the electrons, protons with a suitable electron acceptor are reduced with water as the only by-product in the process, where the flow of electrons results in electricity generation [18]. The sustained release of the electrons in the anode and its consumption at the cathode is ideal for the effective functioning of an MFC. Here, the attainable metabolic energy for bacteria is related directly to the difference between both the anode potential and the

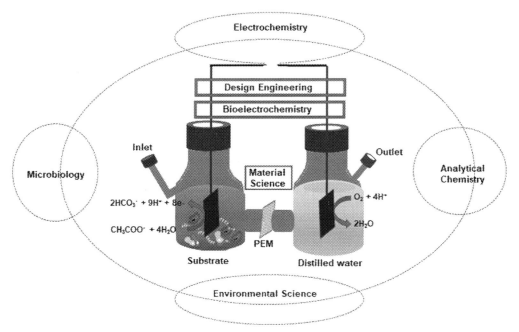

Figure 16.1 Schematic representation of MFC and its fields of application (*MFC*, Microbial fuel cell).

substrate redox potential [17]. Microbes which possess the ability to transfer electrons to the electrode are termed as exoelectrogens, where the electron transfer mechanism is specific to individual strains. While, certain bacteria have the ability to directly transfer the produced electrons through electron shuttles (via flavins, phenazines, and cysteine), conductive pilli, cytochromes, etc., which are categorized under nonmediator types, some bacteria require suitable mediators such as methylene blue, methyl viologen, humic acid, and thionine, which are classified under mediator-less MFC [12]. Enormous research on MFCs is due to its attractive characteristics such as simultaneous electricity generation and wastewater treatment, eco-friendly nature, and portable design. Although, no optimal fabrication is available for MFC, a vast number of fabrications are being reported, including single-chamber microbial fuel cell (SCMFC), H-type, double-chamber MFC, tubular MFC, MFC stacks, etc. [12,17]. The major components in MFC include, an anode, a cathode, and a PEM, each component has a specific role in maximizing the overall output of an MFC (Fig. 16.1).

16.3 Application of nanomaterials in biofuel cells

The defining characteristic of a BFC is based on how well the microbes oxidize and liberate electrons at anode and its consequent reduction at the cathode [8]. To produce electrons and to get those electrons captured, and to reduce the electron acceptors efficiently, the anode and cathode, respectively, need to be modified by suitable methods that make the most yield. Generally, it is said that the anode already has microbes which act as biocatalysts to oxidize the substrate, and do not require much energy; the cathode on the other hand suffers in reducing oxygen ORR, since the reaction is utterly slow and needs the help of a catalyst to speed up the reaction [19,20]. Both the anode and cathode need some kind of a modification for catalysis, to improve the electron transfer mechanism, for greater surface area and to reduce overpotential losses.

Rapid development in technology has led to the utilization of nanomaterials for many applications such as engineering, medicine, pharmaceutical, and forensics. They also found a way to be utilized in bioelectrochemical systems (BES) as catalysts for anode and cathode. Nanomaterials can be in a variety of forms such as nanorods, nanocomposites, nanotubes, nanosphere, nanocapsule, and nanohybrids, but broadly they are classified into organic and inorganic nanoparticles. Particles behave in a completely different manner in the nanoscale when compared with the larger ones, which makes the nanoparticles so peculiar in their physical and chemical aspects. With the right scientific knowledge and effort, nanoparticles can be orchestrated according to the application. BES has already in many different ways used the nanoparticles as nanocatalysts, nanofillers in membranes and electrode materials. Further research is going on to attain greater efficiencies and upscaling for real-time applications of BES. The following part of this chapter will deal with various applications of nanoparticles in MFC and other BESs.

16.3.1 Nanomaterials in electrodes

Modification of electrodes with viable nanoparticles or its other forms to increase the projected surface area and to improve the catalytic activity seems to be a very good approach in ameliorating the performance of any BFC [21].

16.3.1.1 Anode modification

Anode is a vital member of a BFC since it affects the bacterial biofilm formation, electron transport mechanism, and organic matter degradation, all of which are important in proper functioning of the system [22]. Electron transport efficiency from the bacterial biofilm to the anode electrode of a BES stands as a deciding factor in the overall power output [23]. Many researches focus on altering the cathode, to develop a material that behaves best in terms of ORR catalysis which is a huge limitation of any BES, hence the attention required for anode is very minimal. The detailed chemical reaction occurring at the anode of a BFC is depicted in Fig. 16.2.

Since, bacterial biofilm plays a significant part in the performance of a BES (Fig. 16.3), modifications of anode with respect to its surface area, biocompatibility,

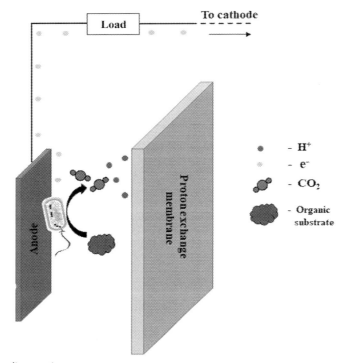

Figure 16.2 Anodic reaction.

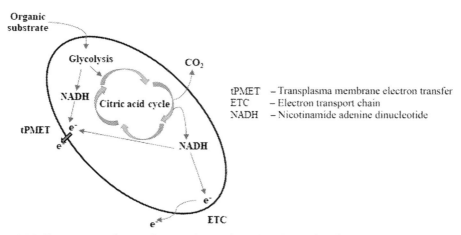

Figure 16.3 Electron transfer mechanism [24]. *Adapted and modified from O. Schaetzle, F. Barrière, K. Baronian, Bacteria and yeasts as catalysts in microbial fuel cells: electron transfer from microorganisms to electrodes for green electricity, Energy & Environmental Science 1 (6) (2008) 607–620.*

electron conductivity, and chemical stability are a few to be considered to improve the bacterial adherence on the anodic surface. The electron transfer can be improved either by (1) physically altering the surface properties of the anode by incorporating carbon-based electrodes for better biocompatibility, metal or metal oxide particles [25] for improved electron transport [26], nanoparticles or nanocomposites for increased surface area for enhanced microbial attachment and better electron conductivity, or (2) by introducing redox chemical species (both of natural or artificial origin) on the anode compartment as electron shuttle or mediator to aid in the overall transport of electrons from the bacterial cell to the anode surface [27,28]. The physical modification of the anode electrode with nanomaterials is discussed in detail in the successive parts of this chapter.

The physical and chemical properties of anode play a major role in deciding the efficiency of electron transfer mechanism. The conductivity of electrons and the biocatalytic activity of the bacteria from the biofilm are important parameters to be considered for better electron transfer [22]. The mechanism of microbial electron transfer is depicted in Fig. 16.3. Conventionally, since the start of BFC technology, carbonaceous electrodes such as carbon paper, carbon veil, graphite felt, graphite rod, and much more, have caught the attention of various researchers due to their exceptional biocompatibility, chemical stability, large projected surface area, and low cost. Many studies have extensively made use of these carbon-based electrodes with various modifications in applications such as MFC, constructed wetland MFC, sediment MFCs, and microbial electrolysis cell. Although being highly versatile for the usage in BFCs, carbon-based electrodes do not possess significant electrocatalytic activity toward the redox reactions in the *exo*-electrogenic bacteria [29]. Carbon-based electrodes are also hydrophobic in nature which hiders the microbial attachment on the anode surface, which ultimately increases the resistance of electron transfer from anode to cathode that results in decreased performance of a BFC system [30].

16.3.1.1.1 Transition metals-based nanoparticles and nanocomposites as anode catalyst

The electrocatalytic activity of the carbon-based anodes are generally improved by incorporating nanoparticles or nanocomposites as surface coatings. Oxides, hydroxides, and carbides of various transition metals have found to be better catalysts on anode surface than just bare electrode. Incorporating such transition metal-based nanoparticle coating over carbonaceous electrode have found to give more power density in BFC applications, since they improve the wettability, decrease the internal resistance, and enhance the biocompatibility of the anode [31]. Nanoparticles of transition metals have an inimitable ability of reaction selectivity and reactivity that makes them potent as catalysts [32]. The catalytic activity of the nanocatalyst is exhibited by lowering the activation energy of the oxidation of organic substrates by improving the electrocatalytic activity toward the redox reaction of the microbes. Moreover, the transition metal oxides should be chosen in such a way that they do not exhibit antibacterial activity to the microbes at the anode compartment instead they should aid in the enrichment of the biofilm over the anode surface for efficient electron transfer.

Iron oxide (FeO) nanoparticles-coated anode showed a maximum open circuit voltage (OCV) of 765 mV and the maximum power density was observed to be 140.5 mW m^{-2} [33] in MFC. On a previous relatable study made in MFC, the same FeO nanocatalyst on its application on anode electrode under alkaline condition performed well by reaching a maximum current density of 159.6 mA m^{-2} and a maximum power density of 63.8 mW m^{-2} [34]. A carbon-felt anode was coated with ruthenium oxide (RuO$_2$) nanoparticles by electrodeposition method to produce a maximum power density of 3080 mW m^{-2} on its application in MFC. The high performance of RuO$_2$ nanoparticles was attributed to its increased projected surface area and low resistance that actively helped in the anode electron transfer mechanism [25]. Electrodeposited iron (Fe) and cobalt (Co) nanoparticles on carbon paper anode in SCMFC exhibited a maximum power density of 117.8 and 165.6 mW m^{-2}, respectively, where the former was 50% and the latter 111% higher than the pristine carbon paper anode when mixed bacterial culture were used as inoculum [35]. Biosynthesized palladium nanoparticles, on its application on carbon cloth enhanced the MFC performance by achieving a voltage of 585 mV, at a current density of 1.1 A m^{-2}, maximum power density of 605 mW m^{-2}, and coulombic efficiency of 40.5%. The improved performance was contributed by the palladium nanoparticles which increased the electrical conductivity by providing ample surface area for microbial attachment which in turn contributed to better electron transfer from the electroactive biofilm [36].

Transition metal nanomaterials/nanocomposites being good electrocatalysts is one point, but the most important question is that whether they are biocompatible, especially to be used in a bio-based system such as BFCs? A catalyst composite of nanomolybdenum carbide (MoC$_2$) with carbon nanotube (CNT) were coated on a carbon-based anode for MFC application and tested for its catalyst activity and biocompatibility where, the cyclic voltammogram showed notable and distinct oxidation and reduction peaks that confirmed the enriched biofilm formed over the nanocomposite coated anode which makes the nanocomposite biocompatible. Also, the same showed an OCV of 0.702 V and maximum power density of 1.05 W m^{-2}, both of which were very close to those obtained from Pt/C catalyst, a common and commercially available one, and proved its good electrocatalytic performance [37]. Another novel study made an approach to synthesize nano-Fe$_3$C@PGC (nanoiron carbide dispersed in porous graphitized carbon) incorporated on anode electrode aiming to improve the electrocatalytic activity for better electron transfer. It was observed from the impedance spectra obtained for the further mentioned nanocomposite coated anode that the ohmic and charge-transfer resistance

Table 16.1 Different nanocomposite-based anode catalysts.

Chamber type	Anode modification	Output	References
Membrane-less single-chambered MFC	Nano-Fe$_3$O$_4$ on activated carbon coated stainless steel mesh	809 mW m^{-2}	[39]
Dual	Titanium dioxide nanosheets/carbon paper	690 mW m^{-3}	[40]
Dual	NiFe$_2$O$_4$-MXene/carbon felt	1385 mW m^{-2}	[41]
Dual	Fe$_2$O$_3$-polyaniline-dopamine/carbon felt	3184.4 mW m^{-2}	[42]
Single	Biosynthesized Fe/Fe$_2$O$_3$/carbon felt	970 mW m^{-2}	[31]
Dual	CoFe$_2$O$_4$ microspheres/carbon cloth	1964 mW m^{-2}	[43]

MFC, Microbial fuel cell.

was the least of 11.91 and 29.5 Ω, respectively, when compared with pristine carbon felt anode (R$_{ohmic}$ = 33.15 Ω and R$_{charge\ transfer}$ = 47.58 Ω) which proved better electrocatalytic activity [38] (Table 16.1).

16.3.1.1.2 Carbon-based nanomaterials and nanocomposites as anode catalysts

It is a well-known fact that the material used as electrode has a direct relation to the overall efficiency from any BFC. Especially an ideal anode should possess properties such as increased surface area for microbial adhesion, excellent biocompatibility, conductivity, longevity, durability toward EPS (extra-cellular polymeric substances) and other microbial excretions, ease in fabrication, and affordability [44]. Carbon-based anode material has found its profound application in BFCs owing to their excellent physical and chemical properties. However, the poor electrical conductivity of carbon-based electrodes has made them less suitable to be utilized in BFC applications. Therefore to overcome this limitation so as to achieve the best results out of the system, carbon-based electrodes are coated with carbon-based coatings or modifications to catalyze the reactions and at the same time to support microbial growth. Carbon materials such as CNTs, multiwalled carbon nanotubes (MWCNTs), chitosan, [45], reduced graphene oxide [46], cellulose [47], polydopamine [48], and so much more are used extensively in combination with the already existing carbon-based electrodes to enhance performance of the BFC system in both research purposes and upscaling. Nitrogen-doped CNT shaped like bamboo shoots used as anode catalyst have been proved to have low internal resistance and better biocompatibility than the plain anode, reason being the availability of active sites introduced by the bamboo-like joints on their microstructures. This quality resulted in the highest power density of 1.04 W m^{-2} during its application in MFC [49].

Table 16.2 Different carbon-based nanomaterials as anode catalysts.

Chamber type	Anode modification	Output	References
Dual	Mesoporous carbon nanofiber aerogel	1747 mW m^{-2}	[53]
Dual	Polyaniline/reduced graphene oxide/Pt	2059 mW m^{-2}	[54]
Dual	Graphene/stainless steel mesh	2668 mW m^{-2}	[55]
Single	Activated carbon/carbon fibers	1626 mW m^{-2}	[56]
–	Carbon nanoweb/reticulated vitreous carbon (RVC)	6.8 mA cm^{-2}	[57]
Dual	N-doped graphene nanosheets/carbon cloth	1008 mW m^{-2}	[58]

A study made with nitrogen-doped carbon nanomaterial proved to facilitate direct electron transfer from c-cytochrome since the catalyst improved the adsorption of flavin (flavin and c-cytochrome plays vital role in electron transfer) which was secreted by *Shewanella oneidensis* [50]. Yet another study made with functionalized activated carbon-polyaniline introduced functional groups at the anode surface for better electron capture which reduced the electron transfer resistance ultimately resulting in better overall functioning of MFCs [51]. Carbon cloth coated with poly-4-hydroxyalkanoates/carbon nanotubes, when used as anode, showed better electron transfer, caused by low internal resistance and better redox reactions due to abundance in the electrochemically active mediators for electron transfer given out by the microbes [52]. All these studies point toward a common output when carbon-based catalyst was used in anode of BFCs, that is, decreased internal resistance due to reduction in charge-transfer resistance, improved surface area for microbial attachment, excellent biocompatibility, introduction of sites for electron transfer from the cell membrane of the microbes to the anode, all leading to improved power generation. Other advantages of using carbon-based catalysts in anode is that they are inert to microbial exudates or other chemicals present on wastewater, easily accessible, and high electrical conductivity make them unique. These properties are in confirmation to that of a perfect and ideal catalyst for BFC application (Table 16.2).

16.3.1.1.3 Polymer-based anodes

Polymers that are conductive have great potential to be used in BFC anodes to improve electron transfer. Due to their one-dimensional chemical structure, polymers are known for their exceptional chemical and physical properties [59]. The main role of polymer-coated or modified anode is that they enhance microbial cell adherence since the long-chained polymers have the ability to adsorb proteins and other microbial products [60]. Electroactive polymers also possess low electrical resistance to aid in better flow of electrons through the electrode resulting in enhanced BFC performance. For, example,

Table 16.3 Different conductive polymer-based anode catalysts.

Chamber type	Anode modification	Output	References
Dual	Reduced graphene oxide @ polydopamine/carbon cloth	988.1 mW m^{-2}	[48]
Single	Bacterial cellulose-CNT-polyaniline	450 mW m^{-2}	[67]
Dual	Polyacrylamide/graphite nanopowder functionalized polyvinyl formaldehyde sponge	8 W m^{-2}	[68]
Dual	Polypyrrole-co-polyaniline encapsulated manganite nanoparticles	4.9 W m^{-3}	[69]
H-shaped dual	1-ethyl-3-vinylimidazolium tetrafluoroborate [VEIm]BF$_4$/carbon felt	4400 mW m^{-2}	[70]
Dual	Poly(3,4-ethylenedioxythiophene)/graphite felt	3.5 A m^{-2}	[71]

CNT, Carbon nanotube.

polyaniline, a conductive polymer has exceptional electrochemical properties due to its unique physicochemical properties, chemical stability, and is easy to work with. Polyaniline is a highly versatile polymer that can be formed composites with other carbon-based materials such as carbon cloth, graphite felt, and carbon nanotubes, to make them more conductive and to develop porous structure for increased surface area [61,62]. Polyaniline, when used in conjunction with other carbon-based nanomaterials shows exceptional conductive properties. For instance, polyaniline with carbon nanotube on graphite nanoribbons displayed an increase in surface area for better bacterial adherence and *exo*-cellular transfer of electrons [63]. A natural macroporous loofah sponge modified with polyaniline used as anode showed a maximum power density of 1090 mW m^{-2} which was 11.68% higher than graphene-coated electrodes. Here, the improvement in power density is owed to the macroporous structure that gave enough space for microbial biofilm growth and the nitrogen-rich carbon coating that enhanced the electron transfer [64]. Polypyrrole, another electrically conductive polymer, when used as anode coating on carbon-base showed high energy output due to high surface area, surface biocompatibility, chemical stability, and most importantly its positively charged surface, attracting electrons as they come in its vicinity serving in electron transfer at the anode [65]. Carbon nanotubes along with poly *N*-isopropylacrylamide hydrogels used in anode was observed to show improved electrical conductivity than that of plain electrode. This enhancement is due to increase in the number of functional groups, surface area, and exposure of the catalyst layer to the anolyte [66] (Table 16.3).

16.3.1.2 Cathode modification

It is a well-known fact for decades that the ORR occurring at the cathode of a BFC is tardy in kinetics, following the same, a myriad of research works was carried on to get to a better and efficient solution. Electrons generated as the end product of anaerobic microbial metabolism are captured by the anode and then transferred via the external circuit to reach the cathode where the electrons and protons (transferred through PEM) combine with a final electron acceptor to reduce the chemical species (Fig. 16.4). Typically, oxygen is preferred to be the final electron acceptor at the cathode due to its availability in abundance and its high reduction potential ($E_o = 1.229$ V) [8]. To accelerate the ORR kinetics, cathode catalysts are used. Platinum (Pt) and platinum-based catalysts are deemed to be in terms of ORR catalysis since it readily provides active sites for oxygen and it very much reduces the activation energy of the ORR than any other catalysts so far [59]. But the main limitation of using platinum is that it is highly expensive and it is not suitable for long-term applications especially in BFCs where replacements may be required. Hence alternative electrode catalysts for

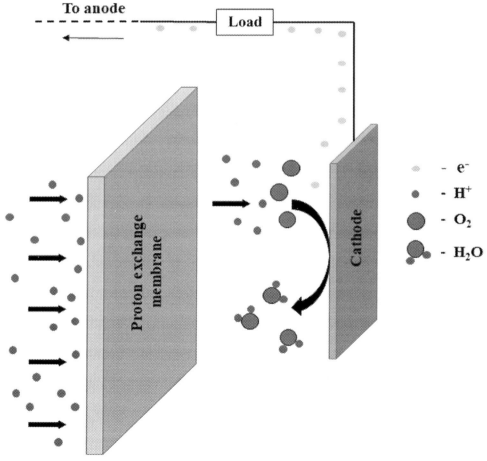

Figure 16.4 Cathodic reaction.

cathode are in demand to solve this issue. This has led to the synthesis of other nonnoble metal cathode catalysts such as metal oxides, metal carbides, metal nanocomposites, doped metal oxides and carbides, and metal-free nanocomposites, and has proved to be efficient in catalytic activity when compared with commercial Pt/C [72]. The next part of this chapter will deal with alternative cathode catalysts especially for ORR.

16.3.1.2.1 Metal-free cathode catalysts

The very first thought of metal-free catalysts leads to carbon-based cathodes, a very eco-friendly material when compared with any other metal-based catalysts. Metal-free cathode catalysts include carbon doped with heteroatoms (B, F, N, S, etc.), graphene, CNT, and other polymer-based cathodes. A study revealed that vertically aligned N-doped CNT as cathode catalyst possessed three times higher catalytic activity than platinum (20%) on carbon. The N-doping creates a net-positive charge on the surface that readily attracts oxygen in its vicinity facilitating an efficient ORR [73]. It was later on found that heteroatom-doped carbon nanomaterials depict good electrocatalytic activity and they can also function as catalysts for purposes not only like ORR but also oxygen evolution reaction and hydrogen evolution reaction [74]. Doping with atoms other than carbon changes the charge distribution and electrical property based on the electronegativity and size of the newly introduced atom. Hence this completely alters the structure of the carbon nanomaterial and makes them more active chemically. This also increases the electrocatalysis toward ORR [75]. Carbon black doped with fluorine and nitrogen under ammonium atmosphere along with polytetrafluoroethylene resulted in an efficient ORR catalyst that was used in SCMFC. The results of the study depict increased catalytic performance of the as-synthesized catalyst (672 mA cm^{-2}) than Pt/C (572 mA cm^{-2}) [76].

The discovery of graphene in 2004 has made an impact in almost all fields of research and real-time applications all owing to the unique properties of graphene. The tremendously large surface area (2630 m^2 g^{-1}), remarkable thermal conductivity (5000 W m^{-1} K^{-1}), and outstanding electrical conductivity (10^6 S cm^{-1}) [77], make it a suitable material to be used in fuel cells, especially BFCs. A two chamber MFC with cathode catalyst as N-doped GO/CoNi dispersed on CNT has observed a maximum power density of 2000 mW m^{-2} [78].

Polypyrrole with carbon black composite, when used as cathode catalyst showed a power density of 401.8 mW m^{-2} which was higher than just carbon black. Also, the power per unit cost of polypyrrole with carbon black was 15 times greater than the commercial platinum-coated cathode in MFC [79]. A novel study utilized the web of the spider Pholcus opilionoides and pyrolyzed them under N$_2$ atmosphere to produce carbon nanofibers (CNFs) to be incorporated as catalyst for ORR in MFC. The resulted catalyst showed a maximum power density of 1800 mW m^{-2}, which was 1.5 times greater than platinum-based cathode. The reason for such high performance

Table 16.4 Various metal-free cathode catalysts.

Final electron acceptor	Cathode catalyst	Output	References
Oxygen	Nitrogen-doped carbon nanodots	196 mW m^{-2}	[81]
Oxygen	Activated carbon nanofiber	61.3 mW m^{-2}	[82]
Oxygen	Nanobiocomposite polypyrrole/kappa-carrageenan	72.1 mW m^{-2}	[83]
Oxygen	Nitrogen and fluorine codoped carbon black	672 mA cm^{-2}	[76]
Oxygen	Mesoporous N-carbon	979 mW m^{-2}	[84]
Oxygen	N-graphene/C$_3$N$_4$	1618 mW m^{-2}	[85]

in spider web-derived CNF is due to its large surface area and a plethora of active sites resulted due to the mesoporous structure and the presence of electronegative sulfur and nitrogen atoms on the carbon lattice, respectively [80]. The possibilities to synthesize and fabricate metal-free/carbon-based cathodes to be utilized in BFCs are truly endless, owing to the properties of carbon-based nanomaterials and composites which make them unique (Table 16.4).

16.3.1.2.2 Metal-based carbon cathode catalysts

Carbon-based cathode catalysts are desirable in fuel cell applications due to their ease in synthesis, handling, and unique molecular structure that enable them to withstand thermal and mechanical stress and also due to their good electrocatalytic properties [86]. But, on the other hand, to furthermore increase the performance, especially in terms of ORR, metal dispersed or doped carbon catalysts are needed. The performance of MFC that utilized carbon-based cathode catalysts is approximately twofolds to threefolds higher than platinum/carbon catalyst, but the performance increased to greater extents when m metal-based carbon cathode catalysts come into play.

Cathode coated with samarium-doped nanoceria dispersed on carbon powder showed an energy generation of 113 mWh m^{-2} (31 mWh m^{-2} for control) when sodium acetate was used as sugar source in a membrane-less SCMFC. The increased performance could be the result of enhanced ORR at the cathode where the nanoceria acted as a reservoir of oxygen for the aerobic bacteria to be used. Also, the enhancement in ORR may be associated to the Ce$^{4+/3+}$ redox reactions [87]. Copper oxide (Cu$_2$O) doped into activated carbon cathodes produced a maximum power density of 1390 mW m^{-2}, which was approximately 59% higher than that of bare cathode MFC. The study also depicted decreased charge-transfer resistance, high electron transfer kinetics, and an exchange current density of 1.03×10^{-3} A cm^{-2}. This improved performance was due to efficient ORR aided by the increased active sites on the copper oxide nanoparticles [88]. Nickel oxide, a nonprecious metal oxide was used as cathode catalyst along with CNT to enhance ORR efficiency. Cyclic voltammogram taken at the scan rate of 50 mV s^{-1} showed high reduction peaks at 0.2 V at NiO concentration of 77%. Rotational disk electrode analysis at 1600 portrayed and confirmed a four-electron pathway, an efficient pathway to ORR to produce water instead of hydrogen peroxide (acidic) as in two-electron pathway [72,89] (schematic of two- and four-electron pathway to ORR is displayed in Fig. 16.5). The cost analysis of catalysts showed that Pt/C 10% was 42 \$ g^{-1} but the cost of nickel oxide dispersed CNT was just 0.3 \$ g^{-1}, which was cost effective ([91]). Activated carbon dispersed with Co_3O_4/$NiCo_2O_4$ nanocage had large surface area of 112.9 m^2 g^{-1} and showed a maximum power density of 1810 mW m^{-2}. There exist redox couples for both Co (Co^{3+}/Co^{2+}) and Ni (Ni^{3+}/Ni^{2+}), which when combined into one nanocomposite cage like structure enhances the ORR, which ultimately improves the performance of the MFC [92].

Since, the cathodic reduction is the rate determining step in the functioning of BFCs, the modification of the cathode with novel catalysts or utilizing suitable electrode material to achieve maximum reduction is highly encouraged when it comes to extend this technology to large-scale applications (Table 16.5).

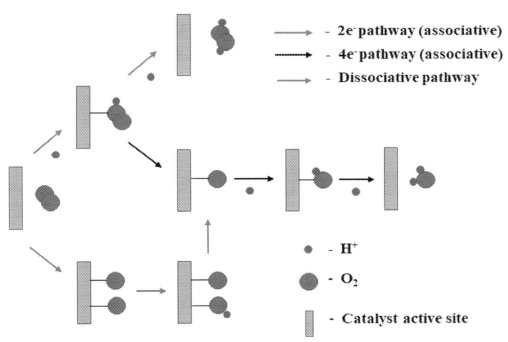

Figure 16.5 ORR through two- and three-electron pathways (*ORR*, Oxygen reduction reaction) [90]. *Adopted and modified from J. Han, J. Bian, C. Sun, Recent advances in single-atom electrocatalysts for oxygen reduction reaction, Research 2020 (2020) 1−51.*

Table 16.5 Various metal-based cathode catalysts.

Final electron acceptor	Cathode catalyst	Output	References
Oxygen	Manganese dioxide	1554 mW m^{-2}	[91]
Oxygen	Titanium dioxide/activated carbon	494 mW m^{-2}	[93]
Oxygen	Silver-platinum	1000 mW m^{-2}	[94]
Oxygen	Iron-manganese	1940 mW m^{-2}	[95]
Oxygen	CoMn$_2$O$_4$/rGO	361 mW m^{-2}	[96]
Oxygen	Hollow-spherical Co/N-C	2514 mW m^{-2}	[97]

rGO, Reduced graphene oxide.

16.3.2 Nanomaterials in membranes

One of the essential components in MFC functioning include the membrane, despite the research available on membrane-less MFC [98–100] which has a prominent role of isolating the anode (fuel consuming) from the cathode (reduction) [101]. Although there are major investigations on membranes, its rapid and continuous functionality becomes uncertain in MFC operation during ion transfer, as the membranes are adversely susceptible to limitations such as substrate or oxygen crossover [102,103]. Further, one of the major problems associated in commercializing MFC is due to drawbacks in long term operation particularly affected through biofouling ([104,105]). The effect of biofouling is rapid when the substrate permeates into the cathode or oxygen from cathode permeates into the anode which abate the oxidation reduction reaction kinetics (ORR) since the oxygen entering the anode disrupts the anaerobic environment maintained and causes substate loss leading to predominant growth of species, including facultative organisms [106–108]. Thus to combat on the drawbacks and for the development of a low cost, highly selective and proton conducting membrane, investigations on membranes are widely explored.

16.3.2.1 Ion-exchange membranes

Ion-exchange membranes widely utilized in MFC are categorized into CEM, anion-exchange membrane, and bipolar membrane. Among the same, CEMs are being widely reported in MFC due to their high affinity toward proton transfer [109]. Numerous CEMs such as Nafion [110], Zirfon [111], Hyflon [112], and Ultrex [113] have been widely reported as separators due to the ability in ease of proton transfer from the anode to cathode. Among the same, Nafion, a product of Dupont, is the most widely reported CEM which has better proton conductivity due to its negatively charged sulfonic hydrophilic group attached to the hydrophobic fluorocarbon thereby promoting ion transfer [114]. The ion conduction in Nafion is through the microphase and nanophase separation between the hydrophobic/hydrophilic moieties predominantly by water molecules. The microstructure of Nafion comprises of a hydrophobic carbon backbone and the hydrophilic region containing water, protons and hydrated protons [115]. For instance, Kim et al., studied the ion transport of Nafion using an electron strain microscopy. When a bias voltage or positive charges are applied, the hydrated protons travel toward the bottom whereas the hydrophobic region remains intact as shown in Fig. 16.6A and B whereas when negative voltage is applied, the hydrated protons move to the top as seen in Fig. 16.6C [115]. Here, ion conduction can be classified into vehicular mechanism, where the ions travel by mobility of water molecules or hopping (Grotthuss) mechanism, where the ion transport is through fixed ion-exchangeable groups. Proton conduction is convoyed with activation energies; that is, $0.1 < E_{act} < 0.4$ eV by hopping (Grotthuss) and $E_{act} > 0.5$ eV by vehicular mechanism [116]. Ion conduction in Nafion is predominantly through vehicular mechanism, wherein the sulfonate groups attached enhances the ionic conductivity (10^{-2} S cm^{-1}) thereby promoting its commercial application worldwide.

There are several limitations however upon Nafion utilization which include, less specificity toward cation transport, oxygen leakage into anode and substrate loss to cathode, ion accumulation, and biofouling. For instance, the oxygen entering the anode compartment decreases the bacterial electron production due to the introduction of facultative and aerobic microbes as it is well known that oxygen is the most preferred electron acceptor for major microbial metabolic pathways. Thus a higher oxygen mass transfer coefficient of 2.80×10^{-4} cm s^{-1} of Nafion membrane hindered the long-term application. Additionally, the cost of Nafion approximately 1500 \$ m^{-2} imparts the hunt for alternative membranes. In this regard, a wide range of polymers and sulfonated form of polymers are being rapidly investigated for combating the drawbacks faced by Nafion. Some of the most common employed CEMs include polyvinylidene fluoride (PVDF), polyvinyl alcohol (PVA), polybenzimidazole, polyvinylidene fluoride grafted polystyrene sulfonic acid (PVDF-g-PSSA), sulfonated polyether ether ketone (SPEEK), sulfonated

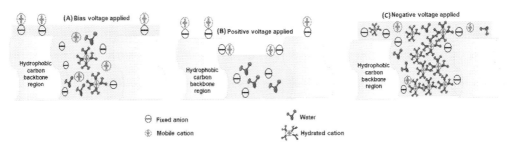

Figure 16.6 Schematic representation of protons, fixed anions and water near the surface of Nafion membrane (A) bias voltage, (B) positive, and (C) negative charge applied. *Adopted and modified from S. Kim, K. No, S. Hong, Visualization of ion transport in Nafion using electrochemical strain microscopy. Chemical Communications 52 (4) (2016) 831−834.*

polyeether ether sulfone, sulfonated polystyrene ethylene butylene polystyrene (SPSEBS), sulfonated polyether sulfone, polyvinyl chloride (PVC), sulfonated polysulfone, sulfonated polystyrene, sulfonated polyphenylene oxide, sulfonated poly(phenylenebenzophenone), etc. [12]. Some of the most common polymers reported are tabulated in Table 16.1. The advent of sulfonated polymers exhibited an upgraded MFC performance which can be related to the improved hydrophilicity, ion transfer through the added SO_3H groups which in case on sulfonated polymers are highly advantages as these SO_3H groups are directly attached to the polymer ring unlike Nafion (Table 16.6).

Table 16.6 Common polymers used in MFC.

Polymer	Structure	References
Nafion		[117]
SPEEK		[118]
SPSU		[119]
SPSEBS		[12]
SPS		[120]
PBI		[121]
SPEES		[122]

(*Continued*)

Table 16.6 (Continued)

Polymer	Structure	References
SPPO		[123]
SPES		[124]
SPPBP		[125]

MFC, Microbial fuel cell; *PBI*, polybenzimidazole; *SPEEK*, sulfonated polyether ether ketone; *SPEES*, sulfonated polyeether ether sulfone; *SPES*, sulfonated polyether sulfone; *SPPBP*, sulfonated poly(phenylenebenzophenone); *SPPO*, sulfonated polyphenylene oxide; *SPS*, sulfonated polystyrene; *SPSEBS*, sulfonated polystyrene ethylene butylene polystyrene; *SPSU*, sulfonated polysulfone.

16.3.2.1.1 Nanocomposite membrane (hybrid organic–inorganic)

Although the polymer composites exhibited better characteristics such as improved hydrophilicity, higher proton transfer, higher water uptake, and sufficient mechanical-chemical stability, the maximum power density was not remarkably higher and was found to be similar to that of Nafion membrane. Thus inorganic fillers either as such or in its sulfonated form have been utilized as an alternative to improve the entire MFC performance [103,126,127]. Organic–inorganic composite membranes are categorized into two main types: (1) membranes comprising proton conductive polymer and nonconductive inorganic particles and (2) membranes comprising proton conductive polymers and less-proton conductive inorganic particles [128]. The introduction of these inorganic nanoparticles into the polymer backbone has a very strong influence on the characteristics of the polymer as their inclusion improves the mechanical stability and the water management. Further, they also aid in suppressing the fuel crossover by creating transport pathways [117]. However, the preparation of the composite membrane containing the organic polymers and proton-conducting inorganic particles are vital as it will affect the microstructure of the membrane. The inherent characteristics of the particles such as surface area, size, shape and their mode of interaction within the polymer matrix are particularly essential and have a major contribution in

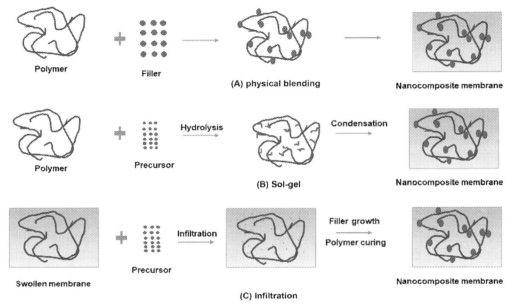

Figure 16.7 Schematic representation for the preparation of nanocomposite membranes: (A) physical blending; (B) sol-gel and (C) infiltration *Adapted and modified from K. Pourzare, Y. Mansourpanah, S. Farhadi, Advanced nanocomposite membranes for fuel cell applications: a comprehensive review. Biofuel Research Journal 3 (4) (2016) 496−513.*

MFC performance [129,120]. Nanomaterials are incorporated in the membrane to improve the surface area, hydrophilicity, and mechanical strength, which are prepared through various techniques such as sol-gel, blending, and infiltration methods [130,131], as shown in Fig. 16.7. Here, blending is the most common and simplest protocol applied to incorporate the required inorganic fillers into the polymer matrix, where it is possible through solution blending or melt blending. However, the major concern during blending is the filler agglomeration which can be reduced by modifications on the surface of the inorganic components. Sol-gel, a low-temperature method, is widely applicable to synthesize organic-inorganic nanocomposites. The process is normally adopted by the hydrolysis and condensation of the metal alkoxides inside the polymer by dissolution in aqueous and nonaqueous solutions [123]. In situ or infiltration techniques are also used to prepare organic − inorganic nanocomposite membranes, where the inorganic fillers infiltrate into hydrogel or swollen like polymer matrix and then the nanocomposite membranes are obtained by removing impurities, polymer curing. Addition of these metal particles in nanoforms improved the essential membrane characteristics by the increased stability, specificity during ion transfer, increased hydrophilicity, and reducing the biofouling [132]. These fillers are further,

sulfonated to improve the MFC performance. The advent of sulfonated fillers improved the MFC performance by:
1. Increasing the mechanical stability of the membrane for efficient long-term operation
2. Introduction of an extra charge carrier by the added (SO_3H) groups thereby enabling proton channeling through Grotthuss mechanism
3. Improving the water uptake for enhanced proton transport through vehicular mechanism

Various nanocomposite membranes have been reported utilizing sulfonated forms of various fillers such as SiO_2, TiO_2, polyhedral oligomeric silsesquioxane (POSS), TNT, GO, and Fe_3O_4 [133,134,18,103,135], wherein the addition of the respective filler in the polymers enhanced the performance compared to plain polymer and polymer + unfunctionalized filler.

16.3.2.1.2 Nanocomposite membranes with Fe_3O_4

Among the common transition metals, iron (Fe) is most dominant metal accessible on earth. To improve the membrane characteristics, variations in the form of nanoparticles or oxides to increase the conductivity and reactivity are being proposed creating it as a pyrophoric material [136]. For instance, Rahimnejad et al., reported the utilization of Fe_3O_4 prepared by the coprecipitation of iron chloride and sodium hydroxide was incorporated in various concentrations (5, 10, 15, and 20 wt.%) in polyether sulfone (PES) for as an effective alternative for maximizing the MFC performance. From the experimental analysis, it was observed that the upon incorporation of 15 wt.% of Fe_3O_4, a maximum power density of 20 mW m^{-2} was attained compared to 15 mW m^{-2} (Nafion) [137]. Another report by Prabhu and Sangeetha portrays the utilization of Fe_3O_4 (2.5, 5, 7.5, and 10 wt.%) in SPEEK in a single-chamber MFC, where a maximum performance of 104 mW m^{-2} was achieved by SPEEK/Fe_3O_4 (7.5 wt.%) compared to the power density of Nafion (47 mW m^{-2}). The obtained power density was attained due to the selectivity during proton transfer, reduced oxygen crossover and the vehicular mechanism of ion transfer [138]. Fe_3O_4 nanoparticles were sulfonated and incorporated in PES to further improve the essential membrane properties. From the results it was observed that 10 wt.% of sulfonated Fe_3O_4 exhibited the maximum performance (65.24 mW m^{-2}) amongst the tested concentrations and Nafion [134]. The nanocomposite membranes exhibited better ion conduction, drastic reduction in fuel crossover and higher affinity toward ion transfer further enhancing its application in MFC [130].

16.3.2.1.3 Nanocomposite membranes with TiO_2

TiO_2 is yet another widely reported filler incorporated in many polymer membranes due to its characteristics such as antibacterial activity and hygroscopic property prepared by solution casting or sol-gel methods. A report by Venkatesan and Sangeetha shows the incorporation of TiO$_2$ (rutile) in SPEEK where the addition of TiO$_2$ reduced the oxygen crossover thereby improving power production up to 98.1 mW m^{-2} [139]. As per a report by Kumar et al., a polymer blend comprised of PVDF-HFP (polyvinylidene fluoride-hexafluoropropylene), Nafion, and TiO$_2$ was prepared. A maximum power density of 552.12 mW m^{-2} with better characteristics was observed compared to Nafion [140]. Further TiO$_2$ (anatase) was sulfonated and incorporated into SPEEK and SPSEBS polymer backbones. The addition of the sulfonic acid group increased the affinity toward better proton transfer and enhanced the power generation. For instance, SPSEBS-S-TiO$_2$ 7.5% exhibited a maximum performance of 1345 mW m^{-2} compared to 695 and 835 mW m^{-2} attained by plain SPSEBS and SPSEBS-TiO$_2$ ([141]). Similarly, SPEEK-S-TiO$_2$ portrayed a maximum power density of 1202 mW m^{-2} compared to 676 and 766 mW m^{-2} obtained by SPEEK and SPEEK-TiO$_2$ [126]. TiO$_2$ (anatase) was converted to titanium nanotubes (TNTs) for improving hydrophilicity, water uptake and surface area. The conversion of TiO$_2$ to TNT improved essential membrane characteristics, which further upon sulfonation the proton transfer increased by both vehicular and hopping mechanism. The incorporation of 7.5 wt.% S-TiO$_2$ improved the proton conductivity from 1.18×10^{-2} S cm^{-1} to 1.37×10^{-2} S cm^{-1} upon utilization of 7.5 wt.% S-TNT thus showing the improved tendency of S-TNT due to the improved surface area [18].

16.3.2.1.4 Nanocomposite membranes with SiO$_2$

Another metallic compound reported in membrane preparation is silica which has been incorporated in various polymers such as Nafion, SPEEK, and SPSEBS due to its inherent attractive features such as thermal and mechanical stability, and large surface area [142,143,144]. Metal oxide polymers undergo proton conductivity through vehicular mechanism through the OH ions inherent in the metal oxide frame, where the rate of ion transfer increases in case of a hydrated membrane. Thus the power production and overall stability was improved by addition of these fillers [145]. The incorporation of functional group such as sulfonic acid in the polymer and filler, create negative ion conducting channels which serve as an extra charge carrier promoting hopping mechanism thereby further promoting ion transfer [118]. A report by Wang et al. shows the introduction of bifunctionalized silica in Nafion (NBS) prepared by sol-gel method where it was observed that the ion conductivity increased with the addition of functionalized silica. The incidence of bifunctional silica increased the ion-exchange capacity up to 1.82 meq g^{-1} with 7.5 wt.% of filler loading. The hydrophilic -OH and proton conductive (SO$_3$H) group proved to improve both water uptake and proton conductivity of the membranes [144]. A report by Ayyaru and Dharmalingam showed the effect of sulfonated SiO$_2$ as an effective additive in SPSEBS by varying the filler concentration from 2.5 to 10 wt.% where a maximum power density of 1209 mW m^{-2} [142]. For instance, Sugumar and Dharmalingam reported the utilization of sulfonated polyhedral oligomeric silsesquioxane (SPOSS) in SPSEBS, where POSS is a new class of nanocaged silica structure with explicit properties such as ease of functionality, hydrophilicity and improved mechanical strength, where 6% SPOSS showed a maximum power density of 126 mW m^{-2} [135]. Similarly, a recent study by Kugarajah and Dharmalingam showed the effect of SPOSS upon incorporation into SPEEK for enhancing the MFC performance. From the experimental deductions, it was observed at a concentration of 5 wt.% SPOSS the prepared nanocomposite membrane exhibited a maximum MFC performance of 162 mW m^{-2} upon three weeks of operation [103] displaying the advantage of SPOSS as an effective filler for improving the performance.

16.3.2.1.5 Nanocomposite membranes with carbon materials

Graphene oxide (GO), an amphiphilic material, has a two-dimensional laminated structure containing epoxy and hydroxy groups available on the base plane and the carboxylic acid groups at the sheet edges. The presence of the oxygen functionalized groups enables in better hydration of GO thereby enhancing the water uptake which in turn promotes the proton conductivity. Owing to these attractive features and other properties such as large surface area, chemical stability, inherent mechanical strength, GO is one of the prominent utilized filler in fuel cell research [146–148]. Rudra et al. showed the effect of GO loaded in situ polymerized sulfonated polystyrene and glutaraldehyde, cross-linked PVA. GO in three molar concentrations of 0.2, 0.4, and 0.6 molar concentrations were utilized upon which $GO_{0.4}$ exhibited a maximum power density of 193.6 mW m^{-2} [149]. A report by Liew et al. shows the incorporation of silver graphene oxide (Ag-GO) and graphene oxide (GO) nanoparticles in SPEEK. Initially, due to the renowned antibacterial activity of silver, GO-SPEEK generated a maximum power density of 134 mW m^{-2}; however, the addition of silver aided in the reduced biofilm and for the long-term operation. Thus upon the 100th day of operation, AgGO-SPPEK attained the maximum power density of 896 mW m^{-2} [102]. Recently, Ayyaru et al. [133] reported that the addition of 2 wt.% of sulfonated GO in SPEEK exhibited a maximum power density of 1028 mW m^{-2}, which were higher than both neat SPEEK (570 mW m^{-2}) and SPEEK-GO (704 mW m^{-2}). The obtained enhanced results were correlated to the addition of the sulfonic acid groups which facilitated the higher water uptake, proton conductivity and IEC of the prepared membranes [133]. A report by Ghasemi et al. exhibited the utilization of CNF, activated carbon nanofiber (ACNF) in Nafion. The utilization of these nanocarbon materials which are highly conductive, improved tendency due to the Van der Waal interaction. ACNF/Nafion exhibited a better coulombic efficiency and power density [150,151]. Thus the addition of nanomaterials in membranes improved the membrane properties by increasing the surface area, stability, and promoting better ion conduction. Venkatesan and Dharmalingam studied the effect of MWCNT-incorporated chitosan (CHIT) in an SCMFC. It was found that CHIT-MWCNT exhibited a lower oxygen mass transfer coefficient (2.66×10^{-4} cm s^{-1}) which was one order lesser than CHIT (2.28×10^{-3} cm s^{-1}) [138].

16.3.2.1.6 Nanocomposite membranes with miscellaneous materials

Zeolites are a class of porous fillers with high surface area, crystalline structure, and hydrated microporous minerals. Zeolite possess a 3D structure by the coordination bonding of aluminate tetrahedral (AlO$_4$) and silicate (SiO$_4$) that are linked at the corners inducing the thermal, mechanical, and structural stability of the membrane with its stability suitable to resist extreme environments [124,152]. A report by Nagar et al. showed the incorporation of 15 wt.% of zeolite in hydrophobic PVC to produce a maximum power density of 250 mW m^{-2} [153]. Further, to facilitate both ion transfer and reduce biofouling, silver has been introduced [102,109]. For instance, a recent report by Kugarajah et al. showed the incorporation of silver in SPEEK to produce a maximum power density of 156 mW m^{-2} upon 7.5 wt.% of filler loading with reduced biofouling [109]. Various nanocomposite membranes are being continuously emerging to combat the obstacles faced by Nafion and to improve the entire MFC performance.

16.4 Conclusion

There are a lot of factors that affect the performance of BFC such as overpotential, substrate crossover, oxygen crossover, and pH splitting. All of these are dependent on three major components, viz., anode, cathode, and PEM. The advancement in nanotechnology has made the application of materials cost effective, sustainable, and with unique electrocatalytic properties. The increased surface area of nanoparticles, size, applicability in minimal quantities, and ease toward modifications has led to wide explorations of nanoparticles in BFCs. However, the major hurdles, including cost of nanoparticles and ease toward synthesis are to be considered for improving the overall efficiency of BFC. Modification of anode with nanocomposites increased the projected surface area available for higher microbial adherence, increased the electron transfer mechanism thus decreasing the charge-transfer resistance and increased the biocompatibility. When cathode is modified with nanomaterials it increased the active sites for ORR catalytic activity and reduced the activation energy for quicker and efficient ORR. The nanocatalysts mostly paves way for four-electron ORR which is deemed as the preferable kind of ORR in BFCs.

Furthermore, blending the polymer electrolytes with inorganic or organic nanomaterials or its composites positively affects various parameters that are essential for better operation of a BFC such as, improved proton conductivity through both Grotthuss and vehicular mechanisms, increased the hydrophilicity to aid in water uptake and for enhanced antifouling properties, and increased mechanical, thermal, and chemical stability. Certain nanoparticles inherently possess antibacterial or antimicrobial or antifungal properties which helps in reducing biofilm formation over the PEM. With all the aforementioned merits of utilizing nanomaterials in various ways on a BFC, the output measured in terms of power density, current density, COD/BOD removal, and coulombic efficiency were seen to rise drastically higher than their corresponding controls. This arena of application of nanomaterials in BFC or any fuel cell for that matter has great potential for improvement. Studies relating to optimization of nanocatalyst and nanofiller loading rate on electrodes and PEM, respectively may be performed. Also, other studies regarding the reaction kinetics of ORR or any other compound/ion reduction can be performed in the presence of novel cathode nanocatalyst. Thus nanomaterials and their composites still have room for improvement in the near future where they could be used to their fullest extent to produce BFCs with extreme performance.

References

[1] V. Kiran Kumar, K. Manmohan, S.P. Manangath, P. Mamju, S. Gajalakshmi, Resource recovery from paddy field using plant microbial fuel cell, Process Biochemistry 99 (2020) 270–281. Available from: https://doi.org/10.1016/j.procbio.2020.09.015.

[2] N. Pous, S. Puig, M. Coma, M.D. Balaguer, J. Colprim, Bioremediation of nitrate-polluted groundwater in a microbial fuel cell, Journal of Chemical Technology & Biotechnology 88 (9) (2013) 1690–1696.

[3] A. Schievano, A. Colombo, M. Grattieri, S.P. Trasatti, A. Liberale, P. Tremolada, et al., Floating microbial fuel cells as energy harvesters for signal transmission from natural water bodies, Journal of Power Sources 340 (2017) 80–88. Available from: https://doi.org/10.1016/j.jpowsour.2016.11.037.

[4] K. Gunaseelan, D.A. Jadhav, S. Gajalakshmi, D. Pant, Blending of microbial inocula: an effective strategy for performance enhancement of clayware biophotovoltaics microbial fuel cells, Bioresource Technology 323 (2021) 124564. Available from: https://doi.org/10.1016/j.biortech.2020.124564.

[5] A. Kaur, J.R. Kim, I. Michie, R.M. Dinsdale, A.J. Guwy, G.C. Premier, et al., Microbial fuel cell type biosensor for specific volatile fatty acids using acclimated bacterial communities, Biosensors and Bioelectronics 47 (2013) 50–55.

[6] K. Gunaseelan, S. Gajalakshmi, S.-K. Kamaraj, J. Solomon, D.A. Jadhav, Electrochemical losses and its role in power generation of microbial fuel cells, Bioelectrochemical Systems, Springer, Singapore, 2020, pp. 81–118.

[7] A. Dumitru, K. Scott, Anode materials for microbial fuel cells, In Microbial Electrochemical and Fuel Cells: Fundamentals and Applications, Elsevier Ltd., Cambridge, 2016. Available from: https://doi.org/10.1016/B978-1-78242-375-1.00004-6.

[8] B.E. Logan, B. Hamelers, R. Rozendal, U. Schröder, J. Keller, S. Freguia, et al., Microbial fuel cells: methodology and technology, Environmental Science and Technology 40 (17) (2006) 5181–5192. Available from: https://doi.org/10.1021/es0605016.

[9] G.D. Bhowmick, S. Das, H.K. Verma, B. Neethu, M.M. Ghangrekar, Improved performance of microbial fuel cell by using conductive ink printed cathode containing Co_3O_4 or Fe_3O_4, Electrochimica Acta 310 (2019) 173–183. Available from: https://doi.org/10.1016/j.electacta.2019.04.127.

[10] S. Dharmalingam, V. Kugarajah, M. Sugumar, Membranes for microbial fuel cells, Biomass, Biofuels, Biochemicals: Microbial Electrochemical Technology: Sustainable Platform for Fuels, Chemicals and Remediation, Elsevier, San Diego, 2018. Available from: https://doi.org/10.1016/B978-0-444-64052-9.00007-8.

[11] F. Mashayekhi Mazar, M. Alijanianzadeh, Z. Jamshidy Nia, A. Molaei Rad, Introduction to biofuel cells: a biological source of energy, Energy Sources, Part A: Recovery, Utilization, and Environmental Effects 39 (4) (2017) 419–425. Available from: https://doi.org/10.1080/15567036.2016.1219794.

[12] S. Dharmalingam, V. Kugarajah, M. Sugumar, Membranes for microbial fuel cells, Microbial Electrochemical Technology, Elsevier, Amsterdam, 2019, pp. 143–194.

[13] E. Katz, I. Willner, A biofuel cell with electrochemically switchable and tunable power output, Journal of the American Chemical Society 125 (22) (2003) 6803–6813.

[14] S. Calabrese Barton, J. Gallaway, P. Atanassov, Enzymatic biofuel cells for implantable and microscale devices, Chemical Reviews 104 (10) (2004) 4867–4886.

[15] S. Ha, Y. Wee, J. Kim, Nanobiocatalysis for enzymatic biofuel cells, Topics in Catalysis 55 (16–18) (2012) 1181–1200.

[16] S.V. Mohan, S.V. Raghavulu, P.N. Sarma, Biochemical evaluation of bioelectricity production process from anaerobic wastewater treatment in a single chambered microbial fuel cell (MFC) employing glass wool membrane, Biosensors and Bioelectronics 23 (9) (2008) 1326–1332.

[17] M.R. Beylier, M.D. Balaguer, J. Colprim, C. Pellicer-Nàcher, B.-J. Ni, B.F. Smets, et al., Biological nitrogen removal from domestic wastewater, in: M. Moo-Young (Ed.), Comprehensive Biotechnology, 3rd ed., Pergamon, Oxford, 2011, pp. 285–296. Available from: https://doi.org/10.1016/B978-0-444-64046-8.00360-8.

[18] V. Kugarajah, S. Dharmalingam, Investigation of a cation exchange membrane comprising sulphonated poly ether ether ketone and sulphonated titanium nanotubes in microbial fuel cell and preliminary insights on microbial adhesion, Chemical Engineering Journal 398 (2020) 125558. Available from: https://doi.org/10.1016/j.cej.2020.125558.

[19] T. Cai, Y. Huang, M. Huang, Y. Xi, D. Pang, W. Zhang, Enhancing oxygen reduction reaction of supercapacitor microbial fuel cells with electrospun carbon nanofibers composite cathode, Chemical Engineering Journal 371 (2019) 544–553. Available from: https://doi.org/10.1016/j.cej.2019.04.025.

[20] W. Yang, J.E. Lu, Y. Zhang, Y. Peng, R. Mercado, J. Li, et al., Cobalt oxides nanoparticles supported on nitrogen-doped carbon nanotubes as high-efficiency cathode catalysts for microbial fuel cells, Inorganic Chemistry Communications 105 (2019) 69–75. Available from: https://doi.org/10.1016/j.inoche.2019.04.036.

[21] H. Rismani-Yazdi, S.M. Carver, A.D. Christy, O.H. Tuovinen, Cathodic limitations in microbial fuel cells: an overview, Journal of Power Sources 180 (2) (2008) 683–694. Available from: https://doi.org/10.1016/j.jpowsour.2008.02.074.

[22] X. Xie, C. Criddle, Y. Cui, Design and fabrication of bioelectrodes for microbial bioelectrochemical systems, Energy and Environmental Science 8 (12) (2015) 3418–3441. Available from: https://doi.org/10.1039/c5ee01862e.

[23] U. Schröder, Anodic electron transfer mechanisms in microbial fuel cells and their energy efficiency, Physical Chemistry Chemical Physics 9 (21) (2007) 2619–2629. Available from: https://doi.org/10.1039/b703627m.

[24] O. Schaetzle, F. Barrière, K. Baronian, Bacteria and yeasts as catalysts in microbial fuel cells: electron transfer from micro-organisms to electrodes for green electricity, Energy & Environmental Science 1 (6) (2008) 607–620.

[25] Z. Lv, D. Xie, X. Yue, C. Feng, C. Wei, Ruthenium oxide-coated carbon felt electrode: a highly active anode for microbial fuel cell applications, Journal of Power Sources 210 (2012) 26–31. Available from: https://doi.org/10.1016/j.jpowsour.2012.02.109.

[26] A. Mehdinia, E. Ziaei, A. Jabbari, Multi-walled carbon nanotube/SnO_2 nanocomposite: A novel anode material for microbial fuel cells, Electrochimica Acta 130 (2014) 512–518. Available from: https://doi.org/10.1016/j.electacta.2014.03.011.

[27] Y. Tekle, A. Demeke, Review on microbial fuel cell, Basic Research Journal of Microbiology 2 (1) (2015) 5.

[28] P. Bhunia, K. Dutta, Biochemistry and electrochemistry at the electrodes of microbial fuel cells, Progress and Recent Trends in Microbial Fuel Cells, Elsevier, Cambridge, MA, 2018. Available from: https://doi.org/10.1016/b978-0-444-64017-8.00016-6.

[29] J. Xiong, M. Hu, X. Li, H. Li, X. Li, X. Liu, et al., Porous graphite: a facile synthesis from ferrous gluconate and excellent performance as anode electrocatalyst of microbial fuel cell, Biosensors and Bioelectronics 109 (2018) 116−122. Available from: https://doi.org/10.1016/j.bios.2018.03.001.
[30] J. Wei, P. Liang, X. Huang, Recent progress in electrodes for microbial fuel cells, Bioresource Technology 102 (20) (2011) 9335−9344. Available from: https://doi.org/10.1016/j.biortech.2011.07.019.
[31] H.O. Mohamed, M. Obaid, K.M. Poo, M. Ali Abdelkareem, S.A. Talas, O.A. Fadali, et al., Fe/Fe$_2$O$_3$ nanoparticles as anode catalyst for exclusive power generation and degradation of organic compounds using microbial fuel cell, Chemical Engineering Journal 349 (2018) 800−807. Available from: https://doi.org/10.1016/j.cej.2018.05.138.
[32] G. Ji, S. Silver, Bacterial resistance mechanisms for heavy metals of environmental concern, Journal of Industrial Microbiology 14 (2) (1995) 61−75. Available from: https://doi.org/10.1007/BF01569887.
[33] M. Harshiny, N. Samsudeen, R.J. Kameswara, M. Matheswaran, Biosynthesized FeO nanoparticles coated carbon anode for improving the performance of microbial fuel cell, International Journal of Hydrogen Energy 42 (42) (2017) 26488−26495. Available from: https://doi.org/10.1016/j.ijhydene.2017.07.084.
[34] N. Samsudeen, T.K. Radhakrishnan, M. Matheswaran, Bioelectricity production from microbial fuel cell using mixed bacterial culture isolated from distillery wastewater, Bioresource Technology 195 (2015) 242−247. Available from: https://doi.org/10.1016/j.biortech.2015.07.023.
[35] H.O. Mohamed, E.T. Sayed, M. Obaid, Y.J. Choi, S.G. Park, S. Al-Qaradawi, et al., Transition metal nanoparticles doped carbon paper as a cost-effective anode in a microbial fuel cell powered by pure and mixed biocatalyst cultures, International Journal of Hydrogen Energy 43 (46) (2018) 21560−21571. Available from: https://doi.org/10.1016/j.ijhydene.2018.09.199.
[36] X. Quan, B. Sun, H. Xu, Anode decoration with biogenic Pd nanoparticles improved power generation in microbial fuel cells, Electrochimica Acta 182 (2015) 815−820. Available from: https://doi.org/10.1016/j.electacta.2015.09.157.
[37] Y. Wang, B. Li, D. Cui, X. Xiang, W. Li, Nano-molybdenum carbide/carbon nanotubes composite as bifunctional anode catalyst for high-performance Escherichia coli-based microbial fuel cell, Biosensors and Bioelectronics 51 (2014) 349−355. Available from: https://doi.org/10.1016/j.bios.2013.07.069.
[38] M. Hu, X. Li, J. Xiong, L. Zeng, Y. Huang, Y. Wu, et al., Nano-Fe$_3$C@PGC as a novel low-cost anode electrocatalyst for superior performance microbial fuel cells, Biosensors and Bioelectronics 142 (2019) 111594. Available from: https://doi.org/10.1016/j.bios.2019.111594.
[39] X. Peng, H. Yu, X. Wang, N. Gao, L. Geng, L. Ai, Enhanced anode performance of microbial fuel cells by adding nanosemiconductor goethite, Journal of Power Sources 223 (2013) 94−99.
[40] T. Yin, Z. Lin, L. Su, C. Yuan, D. Fu, Preparation of vertically oriented TiO$_2$ nanosheets modified carbon paper electrode and its enhancement to the performance of MFCs, ACS Applied Materials & Interfaces 7 (1) (2015) 400−408.
[41] K. Tahir, W. Miran, J. Jang, N. Maile, A. Shahzad, M. Moztahida, et al., Nickel ferrite/MXene-coated carbon felt anodes for enhanced microbial fuel cell performance, Chemosphere 268 (2021) 128784.
[42] M. Jian, P. Xue, K. Shi, R. Li, L. Ma, P. Li, Efficient degradation of indole by microbial fuel cell based Fe$_2$O$_3$-polyaniline-dopamine hybrid composite modified carbon felt anode, Journal of Hazardous Materials 388 (2020) 122123.
[43] M. Rethinasabapathy, A.T.E. Vilian, S.K. Hwang, S.-M. Kang, Y. Cho, Y.-K. Han, et al., Cobalt ferrite microspheres as a biocompatible anode for higher power generation in microbial fuel cells, Journal of Power Sources 483 (2021) 229170.
[44] B. Gutarowska, A. Michalski, Microbial degradation of woven fabrics and protection against biodegradation, Woven Fabrics, IntechOpen, 2012. Available from: https://doi.org/10.5772/38412.
[45] K. Katuri, M.L. Ferrer, M.C. Gutiérrez, R. Jiménez, F. Del Monte, D. Leech, Three-dimensional microchanelled electrodes in flow-through configuration for bioanode formation and current generation, Energy and Environmental Science 4 (10) (2011) 4201−4210. Available from: https://doi.org/10.1039/c1ee01477c.

[46] G.G. Kumar, C.J. Kirubaharan, S. Udhayakumar, C. Karthikeyan, K.S. Nahm, Conductive polymer/graphene supported platinum nanoparticles as anode catalysts for the extended power generation of microbial fuel cells, Industrial and Engineering Chemistry Research 53 (43) (2014) 16883−16893. Available from: https://doi.org/10.1021/ie502399y.

[47] M. Christwardana, A.S. Handayani, R. Yudianti, Joelianingsih, Cellulose—carrageenan coated carbon felt as potential anode structure for yeast microbial fuel cell, International Journal of Hydrogen Energy 46 (8) (2021) 6076−6086. Available from: https://doi.org/10.1016/j.ijhydene.2020.05.265.

[48] H. Liu, Z. Zhang, Y. Xu, X. Tan, Z. Yue, K. Ma, et al., Reduced graphene oxide@polydopamine decorated carbon cloth as an anode for a high-performance microbial fuel cell in Congo red/saline wastewater removal, Bioelectrochemistry 137 (2021) 107675. Available from: https://doi.org/10.1016/j.bioelechem.2020.107675.

[49] S. Ci, Z. Wen, J. Chen, Z. He, Decorating anode with bamboo-like nitrogen-doped carbon nanotubes for microbial fuel cells, Electrochemistry Communications 14 (1) (2012) 71−74. Available from: https://doi.org/10.1016/j.elecom.2011.11.006.

[50] Y.Y. Yu, C.X. Guo, Y.C. Yong, C.M. Li, H. Song, Nitrogen doped carbon nanoparticles enhanced extracellular electron transfer for high-performance microbial fuel cells anode, Chemosphere 140 (2015) 26−33. Available from: https://doi.org/10.1016/j.chemosphere.2014.09.070.

[51] M. Yellappa, J. Annie Modestra, Y.V. Rami Reddy, S. Venkata Mohan, Functionalized conductive activated carbon-polyaniline composite anode for augmented energy recovery in microbial fuel cells, Bioresource Technology 320 (Part B) (2021) 124340. Available from: https://doi.org/10.1016/j.biortech.2020.124340.

[52] Y. Hindatu, M.S.M. Annuar, R. Subramaniam, A.M. Gumel, Medium-chain-length poly-3-hydroxyalkanoates-carbon nanotubes composite anode enhances the performance of microbial fuel cell, Bioprocess and Biosystems Engineering 40 (6) (2017) 919−928.

[53] L. Zou, Y. Qiao, Z. Wu, X. Wu, J. Xie, S. Yu, et al., Tailoring unique mesopores of hierarchically porous structures for fast direct electrochemistry in microbial fuel cells, Advanced Energy Materials 6 (4) (2016) 1501535.

[54] G.G. Kumar, C.J. Kirubaharan, S. Udhayakumar, C. Karthikeyan, K.S. Nahm, Conductive polymer/graphene supported platinum nanoparticles as anode catalysts for the extended power generation of microbial fuel cells, Industrial and Engineering Chemistry Research 53 (43) (2014) 16883−16893. Available from: https://doi.org/10.1021/ie502399y.

[55] Y. Zhang, G. Mo, X. Li, W. Zhang, J. Zhang, J. Ye, et al., A graphene modified anode to improve the performance of microbial fuel cells, Journal of Power Sources 196 (13) (2011) 5402−5407.

[56] I. Gajda, J. You, C. Santoro, J. Greenman, I.A. Ieropoulos, A new method for urine electrofiltration and long term power enhancement using surface modified anodes with activated carbon in ceramic microbial fuel cells, Electrochimica Acta 353 (2020) 136388.

[57] V. Flexer, J. Chen, B.C. Donose, P. Sherrell, G.G. Wallace, J. Keller, The nanostructure of three-dimensional scaffolds enhances the current density of microbial bioelectrochemical systems, Energy & Environmental Science 6 (4) (2013) 1291−1298.

[58] Y.-C. Yong, X.-C. Dong, M.B. Chan-Park, H. Song, P. Chen, Macroporous and monolithic anode based on polyaniline hybridized three-dimensional graphene for high-performance microbial fuel cells, ACS Nano 6 (3) (2012) 2394−2400.

[59] S. Li, C. Cheng, A. Thomas, Carbon-based microbial-fuel-cell electrodes: from conductive supports to active catalysts, Advanced Materials 29 (8) (2017) 1602547.

[60] Y. Hindatu, M.S.M. Annuar, A.M. Gumel, Mini-review: anode modification for improved performance of microbial fuel cell, Renewable and Sustainable Energy Reviews 73 (2017) 236−248.

[61] Z. Huang, E. Liu, H. Shen, X. Xiang, Y. Tian, C. Xiao, et al., Preparation of polyaniline nanotubes by a template-free self-assembly method, Materials Letters 65 (13) (2011) 2015−2018.

[62] P. Wang, H. Li, Z. Du, Polyaniline synthesis by cyclic voltammetry for anodic modification in microbial fuel cells, International Journal of Electrochemical Science 9 (2014) 2038−2046.

[63] C. Zhao, P. Gai, C. Liu, X. Wang, H. Xu, J. Zhang, et al., Polyaniline networks grown on graphene nanoribbons-coated carbon paper with a synergistic effect for high-performance microbial fuel cells, Journal of Materials Chemistry A 1 (40) (2013) 12587−12594.

[64] Y. Yuan, S. Zhou, Y. Liu, J. Tang, Nanostructured macroporous bioanode based on polyaniline-modified natural loofah sponge for high-performance microbial fuel cells, Environmental Science & Technology 47 (24) (2013) 14525−14532.

[65] Y. Yuan, S.-H. Kim, Polypyrrole-coated reticulated vitreous carbon as anode in microbial fuel cell for higher energy output, Bulletin of the Korean Chemical Society 29 (1) (2008) 168−172.

[66] X.-W. Liu, Y.-X. Huang, X.-F. Sun, G.-P. Sheng, F. Zhao, S.-G. Wang, et al., Conductive carbon nanotube hydrogel as a bioanode for enhanced microbial electrocatalysis, ACS Applied Materials & Interfaces 6 (11) (2014) 8158−8164.

[67] M. Mashkour, M. Rahimnejad, M. Mashkour, F. Soavi, Electro-polymerized polyaniline modified conductive bacterial cellulose anode for supercapacitive microbial fuel cells and studying the role of anodic biofilm in the capacitive behavior, Journal of Power Sources 478 (2020) 228822.

[68] P. Mukherjee, P. Saravanan, Graphite nanopowder functionalized 3-D acrylamide polymeric anode for enhanced performance of microbial fuel cell, International Journal of Hydrogen Energy 45 (43) (2020) 23411−23421.

[69] M.K. Sarma, M.G.A. Quadir, R. Bhaduri, S. Kaushik, P. Goswami, Composite polymer coated magnetic nanoparticles based anode enhances dye degradation and power production in microbial fuel cells, Biosensors and Bioelectronics 119 (2018) 94−102.

[70] L. Yang, W. Deng, Y. Zhang, Y. Tan, M. Ma, Q. Xie, Boosting current generation in microbial fuel cells by an order of magnitude by coating an ionic liquid polymer on carbon anodes, Biosensors and Bioelectronics 91 (2017) 644−649.

[71] Y.L. Kang, S. Ibrahim, S. Pichiah, Synergetic effect of conductive polymer poly (3, 4-ethylenediox-ythiophene) with different structural configuration of anode for microbial fuel cell application, Bioresource Technology 189 (2015) 364−369.

[72] L. Dai, Y. Xue, L. Qu, H.-J. Choi, J.-B. Baek, Metal-free catalysts for oxygen reduction reaction, Chemical Reviews 115 (11) (2015) 4823−4892.

[73] K. Gong, F. Du, Z. Xia, M. Durstock, L. Dai, Nitrogen-doped carbon nanotube arrays with high electrocatalytic activity for oxygen reduction, Science 323 (5915) (2009) 760−764.

[74] L. Dai, D.W. Chang, J. Baek, W. Lu, Carbon nanomaterials for advanced energy conversion and storage, Small (Weinheim an der Bergstrasse, Germany) 8 (8) (2012) 1130−1166.

[75] P. Avouris, Z. Chen, V. Perebeinos, Carbon-based electronics, Nanoscience and Technology: A Collection of Reviews From Nature Journals, World Scientific, Singapore, 2010, pp. 174−184.

[76] K. Meng, Q. Liu, Y. Huang, Y. Wang, Facile synthesis of nitrogen and fluorine co-doped carbon materials as efficient electrocatalysts for oxygen reduction reactions in air-cathode microbial fuel cells, Journal of Materials Chemistry A 3 (13) (2015) 6873−6877.

[77] A.K. Geim, K.S. Novoselov, The rise of graphene, Nanoscience and Technology: A Collection of Reviews from Nature Journals, World Scientific, Singapore, 2010, pp. 11−19.

[78] Y. Hou, H. Yuan, Z. Wen, S. Cui, X. Guo, Z. He, et al., Nitrogen-doped graphene/CoNi alloy encased within bamboo-like carbon nanotube hybrids as cathode catalysts in microbial fuel cells, Journal of Power Sources 307 (2016) 561−568. Available from: https://doi.org/10.1016/j.jpowsour.2016.01.018.

[79] Y. Yuan, S. Zhou, L. Zhuang, Polypyrrole/carbon black composite as a novel oxygen reduction catalyst for microbial fuel cells, Journal of Power Sources 195 (11) (2010) 3490−3493.

[80] L. Zhou, P. Fu, X. Cai, S. Zhou, Y. Yuan, Naturally derived carbon nanofibers as sustainable electrocatalysts for microbial energy harvesting: a new application of spider silk, Applied Catalysis B: Environmental 188 (2016) 31−38.

[81] A. Loukanov, A. Angelov, Y. Takahashi, I. Nikolov, S. Nakabayashi, Carbon nanodots chelated with metal ions as efficient electrocatalysts for enhancing performance of microbial fuel cell based on sulfate reducing bacteria, Colloids and Surfaces A: Physicochemical and Engineering Aspects 574 (2019) 52−61. Available from: https://doi.org/10.1016/j.colsurfa.2019.04.067.

[82] M. Ghasemi, S. Shahgaldi, M. Ismail, B.H. Kim, Z. Yaakob, W.R.W. Daud, Activated carbon nanofibers as an alternative cathode catalyst to platinum in a two-chamber microbial fuel cell, International Journal of Hydrogen Energy 36 (21) (2011) 13746−13752.

[83] C. Esmaeili, M. Ghasemi, L.Y. Heng, S.H.A. Hassan, M.M. Abdi, W.R.W. Daud, et al., Synthesis and application of polypyrrole/carrageenan nano-bio composite as a cathode catalyst in microbial fuel cells, Carbohydrate Polymers 114 (2014) 253−259.

[84] Y. Ahn, I. Ivanov, T.C. Nagaiah, A. Bordoloi, B.E. Logan, Mesoporous nitrogen-rich carbon materials as cathode catalysts in microbial fuel cells, Journal of Power Sources 269 (2014) 212−215.

[85] L. Feng, L. Yang, Z. Huang, J. Luo, M. Li, D. Wang, et al., Enhancing electrocatalytic oxygen reduction on nitrogen-doped graphene by active sites implantation, Scientific Reports 3 (1) (2013) 1−8.

[86] T. Zhang, H. Nie, T.S. Bain, H. Lu, M. Cui, O.L. Snoeyenbos-West, et al., Improved cathode materials for microbial electrosynthesis, Energy & Environmental Science 6 (1) (2013) 217−224.

[87] S. Marzorati, P. Cristiani, M. Longhi, S.P. Trasatti, E. Traversa, Nanoceria acting as oxygen reservoir for biocathodes in microbial fuel cells, Electrochimica Acta 325 (2019) 134954. Available from: https://doi.org/10.1016/j.electacta.2019.134954.

[88] X. Zhang, K. Li, P. Yan, Z. Liu, L. Pu, N-type Cu_2O doped activated carbon as catalyst for improving power generation of air cathode microbial fuel cells, Bioresource Technology 187 (2015) 299−304. Available from: https://doi.org/10.1016/j.biortech.2015.03.131.

[89] J. Huang, N. Zhu, T. Yang, T. Zhang, P. Wu, Z. Dang, Nickel oxide and carbon nanotube composite (NiO/CNT) as a novel cathode non-precious metal catalyst in microbial fuel cells, Biosensors and Bioelectronics 72 (2015) 332−339. Available from: https://doi.org/10.1016/j.bios.2015.05.035.

[90] J. Han, J. Bian, C. Sun, Recent advances in single-atom electrocatalysts for oxygen reduction reaction, Research 2020 (2020) 1−51.

[91] P. Zhang, K. Li, X. Liu, Carnation-like MnO_2 modified activated carbon air cathode improve power generation in microbial fuel cells, Journal of Power Sources 264 (2014) 248−253. Available from: https://doi.org/10.1016/j.jpowsour.2014.04.098.

[92] S. Zhang, W. Su, K. Li, D. Liu, J. Wang, P. Tian, Metal organic framework-derived Co_3O_4/$NiCo_2O_4$ double-shelled nanocage modified activated carbon air-cathode for improving power generation in microbial fuel cell, Journal of Power Sources 396 (2018) 355−362. Available from: https://doi.org/10.1016/j.jpowsour.2018.06.057.

[93] G.D. Bhowmick, I. Chakraborty, M.M. Ghangrekar, A. Mitra, TiO_2/Activated carbon photo cathode catalyst exposed to ultraviolet radiation to enhance the efficacy of integrated microbial fuel cell-membrane bioreactor, Bioresource Technology Reports 7 (2019) 100303. Available from: https://doi.org/10.1016/j.biteb.2019.100303.

[94] M.T. Noori, B.R. Tiwari, C.K. Mukherjee, M.M. Ghangrekar, Enhancing the performance of microbial fuel cell using Ag−Pt bimetallic alloy as cathode catalyst and anti-biofouling agent, International Journal of Hydrogen Energy 43 (42) (2018) 19650−19660. Available from: https://doi.org/10.1016/j.ijhydene.2018.08.120.

[95] X. Guo, J. Jia, H. Dong, Q. Wang, T. Xu, B. Fu, et al., Hydrothermal synthesis of Fe−Mn bimetallic nanocatalysts as high-efficiency cathode catalysts for microbial fuel cells, Journal of Power Sources 414 (2019) 444−452. Available from: https://doi.org/10.1016/j.jpowsour.2019.01.024.

[96] Z. Hu, X. Zhou, Y. Lu, R. Jv, Y. Liu, N. Li, et al., $CoMn_2O_4$ doped reduced graphene oxide as an effective cathodic electrocatalyst for ORR in microbial fuel cells, Electrochimica Acta 296 (2019) 214−223. Available from: https://doi.org/10.1016/j.electacta.2018.11.004.

[97] T. Yang, K. Li, L. Pu, Z. Liu, B. Ge, Y. Pan, et al., Hollow-spherical Co/N-C nanoparticle as an efficient electrocatalyst used in air cathode microbial fuel cell, Biosensors and Bioelectronics 86 (2016) 129−134. Available from: https://doi.org/10.1016/j.bios.2016.06.032.

[98] S. Singh, C. Dwivedi, A. Pandey, Electricity generation in membrane-less single chambered microbial fuel cell, in: 2016 International Conference on Control, Computing, Communication and Materials (ICCCCM), Allahabad, India, October 21−22, 2016.

[99] P. Tanikkul, N. Pisutpaisal, Membrane-less MFC based biosensor for monitoring wastewater quality, International Journal of Hydrogen Energy 43 (1) (2018) 483−489. Available from: https://doi.org/10.1016/j.ijhydene.2017.10.065.

[100] M.M. Ghangrekar, V.B. Shinde, Performance of membrane-less microbial fuel cell treating wastewater and effect of electrode distance and area on electricity production, Bioresource Technology 98 (15) (2007) 2879−2885.

[101] K. Scott, Membranes and separators for microbial fuel cells, in: K. Scott, E.H. Yu (Eds.), Microbial Electrochemical and Fuel Cells, Woodhead Publishing, Cambridge, 2016, pp. 153−178. Available from: https://doi.org/10.1016/B978-1-78242-375-1.00005-8.

[102] K. Ben Liew, J.X. Leong, W.R. Wan Daud, A. Ahmad, J.J. Hwang, W. Wu, Incorporation of silver graphene oxide and graphene oxide nanoparticles in sulfonated polyether ether ketone membrane for power generation in microbial fuel cell, Journal of Power Sources 449 (2020) 227490. Available from: https://doi.org/10.1016/j.jpowsour.2019.227490.

[103] V. Kugarajah, S. Dharmalingam, Sulphonated polyhedral oligomeric silsesquioxane/sulphonated poly ether ether ketone nanocomposite membranes for microbial fuel cell: Insights to the miniatures involved, Chemosphere 260 (2020) 127593. Available from: https://doi.org/10.1016/j.chemosphere.2020.127593.

[104] B. Erable, N.M. Duțeanu, M.M. Ghangrekar, C. Dumas, K. Scott, Application of electro-active biofilms, Biofouling 26 (1) (2010) 57−71.

[105] F. Meng, S. Zhang, Y. Oh, Z. Zhou, H.-S. Shin, S.-R. Chae, Fouling in membrane bioreactors: an updated review, Water Research 114 (2017) 151−180.

[106] M. Miskan, M. Ismail, M. Ghasemi, J. Md Jahim, D. Nordin, M.H. Abu Bakar, Characterization of membrane biofouling and its effect on the performance of microbial fuel cell, International Journal of Hydrogen Energy 41 (1) (2016) 543−552. Available from: https://doi.org/10.1016/j.ijhydene.2015.09.037.

[107] M.T. Noori, M.M. Ghangrekar, C.K. Mukherjee, B. Min, Biofouling effects on the performance of microbial fuel cells and recent advances in biotechnological and chemical strategies for mitigation, Biotechnology Advances 37 (8) (2019) 107420. Available from: https://doi.org/10.1016/j.biotechadv.2019.107420.

[108] M.T. Noori, S.C. Jain, M.M. Ghangrekar, C.K. Mukherjee, Biofouling inhibition and enhancing performance of microbial fuel cell using silver nano-particles as fungicide and cathode catalyst, Bioresource Technology 220 (2016) 183−189.

[109] V. Kugarajah, S. Dharmalingam, Effect of silver incorporated sulphonated poly ether ether ketone membranes on microbial fuel cell performance and microbial community analysis, Chemical Engineering Journal 415 (2021) 128961. Available from: https://doi.org/10.1016/j.cej.2021.128961.

[110] S.K. Chaudhuri, D.R. Lovley, Electricity generation by direct oxidation of glucose in mediatorless microbial fuel cells, Nature Biotechnology 21 (10) (2003) 1229−1232.

[111] D. Pant, G. Van Bogaert, M. De Smet, L. Diels, K. Vanbroekhoven, Use of novel permeable membrane and air cathodes in acetate microbial fuel cells, Electrochimica Acta 55 (26) (2010) 7710−7716.

[112] I. Ieropoulos, J. Greenman, C. Melhuish, Improved energy output levels from small-scale microbial fuel cells, Bioelectrochemistry 78 (1) (2010) 44−50.

[113] C.-Y. Lee, K.-L. Ho, D.-J. Lee, A. Su, J.-S. Chang, RETRACTED: Electricity harvest from wastewaters using microbial fuel cell with sulfide as sole electron donor, International Journal of Hydrogen Energy 37 (20) (2012) 15787−15791. Available from: https://doi.org/10.1016/j.ijhydene.2012.03.114.

[114] K.A. Mauritz, R.B. Moore, State of understanding of Nafion, Chemical Reviews 104 (10) (2004) 4535−4586.

[115] S. Kim, K. No, S. Hong, Visualization of ion transport in Nafion using electrochemical strain microscopy, Chemical Communications 52 (4) (2016) 831−834. Available from: https://doi.org/10.1039/C5CC07412F.

[116] Y. Ren, G.H. Chia, Z. Gao, Metal−organic frameworks in fuel cell technologies, Nano Today 8 (6) (2013) 577−597.

[117] S.J. Peighambardoust, S. Rowshanzamir, M. Amjadi, Review of the proton exchange membranes for fuel cell applications, International Journal of Hydrogen Energy 35 (17) (2010) 9349−9384. Available from: https://doi.org/10.1016/j.ijhydene.2010.05.017.

[118] V. Kugarajah, M. Sugumar, S. Dharmalingam, Nanocomposite membrane and microbial community analysis for improved performance in microbial fuel cell, Enzyme and Microbial Technology 140 (2020) 3–10. Available from: https://doi.org/10.1016/j.enzmictec.2020.109606.

[119] N. Awang, A.F. Ismail, J. Jaafar, T. Matsuura, H. Junoh, M.H.D. Othman, et al., Functionalization of polymeric materials as a high performance membrane for direct methanol fuel cell: A review, Reactive and Functional Polymers 86 (2015) 248–258. Available from: https://doi.org/10.1016/j.reactfunctpolym.2014.09.019.

[120] S.-L. Li, Q. Xu, Metal–organic frameworks as platforms for clean energy, Energy & Environmental Science 6 (6) (2013) 1656–1683. Available from: https://doi.org/10.1039/C3EE40507A.

[121] O.Z. Sharaf, M.F. Orhan, An overview of fuel cell technology: fundamentals and applications, Renewable and Sustainable Energy Reviews 32 (2014) 810–853. Available from: https://doi.org/10.1016/j.rser.2014.01.012.

[122] S. Shenvi, A.F. Ismail, A.M. Isloor, Enhanced permeation performance of cellulose acetate ultrafiltration membranes by incorporation of sulfonated poly (1, 4-phenylene ether ether sulfone) and poly (styrene-co-maleic anhydride), Industrial & Engineering Chemistry Research 53 (35) (2014) 13820–13827.

[123] A.D. Pomogailo, Polymer sol-gel synthesis of hybrid nanocomposites, Colloid Journal 67 (6) (2005) 658–677.

[124] B.P. Tripathi, V.K. Shahi, Organic–inorganic nanocomposite polymer electrolyte membranes for fuel cell applications, Progress in Polymer Science 36 (7) (2011) 945–979.

[125] A.-C. Dupuis, Proton exchange membranes for fuel cells operated at medium temperatures: materials and experimental techniques, Progress in Materials Science 56 (3) (2011) 289–327. Available from: https://doi.org/10.1016/j.pmatsci.2010.11.001.

[126] S. Ayyaru, S. Dharmalingam, Improved performance of microbial fuel cells using sulfonated polyether ether ketone (SPEEK) TiO_2–SOH nanocomposite membrane, RSC Advances 3 (47) (2013) 25243–25251.

[127] H. Muthukumar, S.N. Mohammed, N. Chandrasekaran, A.D. Sekar, A. Pugazhendhi, M. Matheswaran, Effect of iron doped zinc oxide nanoparticles coating in the anode on current generation in microbial electrochemical cells, International Journal of Hydrogen Energy 44 (4) (2019) 2407–2416.

[128] L. Zhang, S.-R. Chae, Z. Hendren, J.-S. Park, M.R. Wiesner, Recent advances in proton exchange membranes for fuel cell applications, Chemical Engineering Journal 204–206 (2012) 87–97. Available from: https://doi.org/10.1016/j.cej.2012.07.103.

[129] C. Laberty-Robert, K. Vallé, F. Pereira, C. Sanchez, Design and properties of functional hybrid organic–inorganic membranes for fuel cells, Chemical Society reviews 40 (2) (2011) 961–1005. Available from: https://doi.org/10.1039/C0CS00144A.

[130] K. Pourzare, Y. Mansourpanah, S. Farhadi, Advanced nanocomposite membranes for fuel cell applications: a comprehensive review, Biofuel Research Journal 3 (4) (2016) 496–513.

[131] F.S. Victor, V. Kugarajah, M. Bangaru, S. Ranjan, S. Dharmalingam, Electrospun nanofibers of polyvinylidene fluoride incorporated with titanium nanotubes for purifying air with bacterial contamination, Environmental Science and Pollution Research 28 (2021) 37520–37533. Available from: https://doi.org/10.1007/s11356-021-13202-3.

[132] E. Antolini, Composite materials for polymer electrolyte membrane microbial fuel cells, Biosensors and Bioelectronics 69 (2015) 54–70.

[133] S. Ayyaru, Y.-H. Ahn, Enhanced performance of sulfonated GO in SPEEK proton-exchange membrane for microbial fuel-cell application, Journal of Environmental Engineering 147 (2) (2021) 4020153.

[134] I. Bavasso, M.P. Bracciale, F. Sbardella, D. Puglia, F. Dominici, L. Torre, et al., Sulfonated Fe_3O_4/PES nanocomposites as efficient separators in microbial fuel cells, Journal of Membrane Science 620 (2021) 118967.

[135] M. Sugumar, S. Dharmalingam, Statistical optimization of process parameters in microbial fuel cell for enhanced power production using sulphonated polyhedral oligomeric silsesquioxane

[136] L.Y. Ng, A.W. Mohammad, C.P. Leo, N. Hilal, Polymeric membranes incorporated with metal/metal oxide nanoparticles: a comprehensive review, Desalination 308 (2013) 15−33.
[137] M. Rahimnejad, M. Ghasemi, G.D. Najafpour, M. Ismail, A.W. Mohammad, A.A. Ghoreyshi, et al., Synthesis, characterization and application studies of self-made Fe3O4/PES nanocomposite membranes in microbial fuel cell, Electrochimica Acta 85 (2012) 700−706. Available from: https://doi.org/10.1016/j.electacta.2011.08.036.
[138] P. Narayanaswamy Venkatesan, S. Dharmalingam, Characterization and performance study on chitosan-functionalized multi walled carbon nano tube as separator in microbial fuel cell, Journal of Membrane Science 435 (2013) 92−98. Available from: https://doi.org/10.1016/j.memsci.2013.01.064.
[139] P. Narayanaswamy Venkatesan, S. Dharmalingam, Effect of cation transport of SPEEK—Rutile TiO$_2$ electrolyte on microbial fuel cell performance, Journal of Membrane Science 492 (2015) 518−527. Available from: https://doi.org/10.1016/j.memsci.2015.06.025.
[140] P. Kumar, R.P. Bharti, Nanocomposite polymer electrolyte membrane for high performance microbial fuel cell: Synthesis, characterization and application, Journal of the Electrochemical Society 166 (15) (2019) F1190.
[141] S. Ayyaru, S. Dharmalingam, A study of influence on nanocomposite membrane of sulfonated TiO$_2$ and sulfonated polystyrene-ethylene-butylene-polystyrene for microbial fuel cell application, Energy 88 (2015) 202−208.
[142] A. Sivasankaran, D. Sangeetha, Y.-H. Ahn, Nanocomposite membranes based on sulfonated polystyrene ethylene butylene polystyrene (SSEBS) and sulfonated SiO$_2$ for microbial fuel cell application, Chemical Engineering Journal 289 (2016) 442−451.
[143] D. Tang, R. Yuan, Y. Chai, Magnetic control of an electrochemical microfluidic device with an arrayed immunosensor for simultaneous multiple immunoassays, Clinical Chemistry 53 (7) (2007) 1323−1329.
[144] H. Wang, B.A. Holmberg, L. Huang, Z. Wang, A. Mitra, J.M. Norbeck, et al., Nafion-bifunctional silica composite proton conductive membranes, Journal of Materials Chemistry 12 (4) (2002) 834−837.
[145] S. Dharmalingam, V. Kugarajah, M. Sugumar, Chapter 1.7—Membranes for microbial fuel cells, in: S.V. Mohan, S. Varjani, A. Pandey (Eds.), Microbial Electrochemical Technology, Elsevier, 2019, pp. 143−194. Available from: https://doi.org/10.1016/B978-0-444-64052-9.00007-8.
[146] H.-C. Chien, L.-D. Tsai, C.-P. Huang, C. Kang, J.-N. Lin, F.-C. Chang, Sulfonated graphene oxide/Nafion composite membranes for high-performance direct methanol fuel cells, International Journal of Hydrogen Energy 38 (31) (2013) 13792−13801. Available from: https://doi.org/10.1016/j.ijhydene.2013.08.036.
[147] S.J. Lue, Y.-L. Pai, C.-M. Shih, M.-C. Wu, S.-M. Lai, Novel bilayer well-aligned Nafion/graphene oxide composite membranes prepared using spin coating method for direct liquid fuel cells, Journal of Membrane Science 493 (2015) 212−223. Available from: https://doi.org/10.1016/j.memsci.2015.07.007.
[148] T. Bayer, S.R. Bishop, M. Nishihara, K. Sasaki, S.M. Lyth, Characterization of a graphene oxide membrane fuel cell, Journal of Power Sources 272 (2014) 239−247. Available from: https://doi.org/10.1016/j.jpowsour.2014.08.071.
[149] R. Rudra, V. Kumar, N. Pramanik, P.P. Kundu, Graphite oxide incorporated crosslinked polyvinyl alcohol and sulfonated styrene nanocomposite membrane as separating barrier in single chambered microbial fuel cell, Journal of Power Sources 341 (2017) 285−293. Available from: https://doi.org/10.1016/j.jpowsour.2016.12.028.
[150] A. Tofighi, M. Rahimnejad, Synthetic polymer-based membranes for microbial fuel cells, Synthetic Polymeric Membranes for Advanced Water Treatment, Gas Separation, and Energy Sustainability, Elsevier, 2020, pp. 309−335.

[151] M. Ghasemi, S. Shahgaldi, M. Ismail, Z. Yaakob, W.R.W. Daud, New generation of carbon nanocomposite proton exchange membranes in microbial fuel cell systems, Chemical Engineering Journal 184 (2012) 82–89.

[152] H. Sudhakar, C. Venkata Prasad, K. Sunitha, K. Chowdoji Rao, M.C.S. Subha, S. Sridhar, Pervaporation separation of IPA-water mixtures through 4A zeolite-filled sodium alginate membranes, Journal of Applied Polymer Science 121 (5) (2011) 2717–2725.

[153] H. Nagar, N. Badhrachalam, V.V.B. Rao, S. Sridhar, A novel microbial fuel cell incorporated with polyvinylchloride/4A zeolite composite membrane for kitchen wastewater reclamation and power generation, Materials Chemistry and Physics 224 (2019) 175–185. Available from: https://doi.org/10.1016/j.matchemphys.2018.12.023.

Index

Note: Page numbers followed by "*f*" and "*t*" refer to figures and tables, respectively.

A

Activated carbon (AC), 261–262
Activated carbon fiber (ACF), 158
Activated carbon nanofiber nonwoven (ACNFN), 144
Activated carbon nanofibers (ACNFs), 265
Activation losses (η_{act}), 87
Activity enhancement, 354–355
Air-breathing counter-flow formic acid MFC, 19*f*, 24
Air-breathing immersed fuel micro-jet MFC, 19*f*
Alcohol solvent process, 370
Alkaline fuel cell (AFC), 123, 201–202
Amino groups, 305–306, 306*f*
Anode catalyst
  carbon-based nanomaterials and nanocomposites, 419–420, 420*t*
  conductive polymer-based, 421*t*
  transition metals-based nanoparticles and nanocomposites, 417–419
Anodic reaction, 416*f*

B

Bioelectrochemical systems (BES), 139, 415–417
Biofuel cells (BFCs)
  enzymatic fuel cell, 413
  microbial fuel cell, 413–414
Biomass-derived porous materials, 395–398
Bipolar plates (BPPs), 80–81, 84–85
Bleach microfluidic fuel cell, 17–18, 18*f*
Boron-doped diamond (BDD), 266
Bottom-up approach, 141
Buckminsterfullerene (C_{60}), 206–207

C

CAD software, 17
Carbon, 201, 203–204
  different allotropes of, 204*f*
Carbon-based electrodes, 388–390, 389*t*
Carbon-based nanomaterials, 263–266, 264*f*, 265*f*
Carbon brush anode, 390
Carbon cloth (CC), 261–262, 264–265
Carbon dots (CDs), 142, 209–210, 212*f*
Carbon felt, 388–390
Carbon fibers (CFs), 33–35, 34*f*
Carbon monoxide (CO) poisoning, 175–176
Carbon nanofibers (CNFs), 144, 423–424
Carbon nanofibrils (CNFs), 272–273, 273*t*, 323–325, 424–425
Carbon nanomaterials decoration, 377–383
Carbon nanotubes (CNTs), 263–265, 272–273, 273*t*, 294, 326–331, 333–335, 364, 381, 398
Carbon nanotube-supported active materials, 215–216
Carbon nanotubes-based 1D nanostructures, 213–216, 213*f*
  carbon nanotube-supported active materials, 215–216
  heteroatom-doped, 214–215
Carbon nitride (CN_x) nanofibers, 25
Carbon–nitrogen-based approach, 244–245
Catalyst-coated membrane, electrochemical dissolution of, 369*f*
Catalyst layers (CLs), 80–81, 83
Cathode modification, 421–425
Cathodic reaction, 422*f*
Cation-exchange membrane (CEM), 159, 412
Ceramatec-Advanced Ionic Technologies, 4–5
Chemically reduced graphene oxide (CGO), 146
Clay/polymer nanocomposites, 322
Clean energy, advantages, 173
Co catalyst, 353–354
COF-derived SAC-based materials, 239–241, 240*f*
Commercial fuel cell, 41–42
Computer Numerical Control (CNC), 19–20
Co-N-C catalysts, 223*f*
Concentration/mass transport losses (η_{mass}), 87
Conducting polymers, 274–275, 274*t*, 275*t*
Conductive polyaniline nanofibers (PANInf), 156
Conductive polymer coating, 383–387, 384*f*

445

Conductive polymers (CP), 266–267
Confocal laser scanning microscope, 391f
Cross-linked full aromatic sulfonated polyamide (Cr-FA-SPA), 312–313
Cross-linking membranes, 310–315
Current density, 205
Cyclic and spiral nanostructures, 144–147
Cylindrical and flat water structures, ion charge distribution in, 292f

D

Darcy's viscous resistance, 108
Densely sulfonated units, 306–310, 307f, 309f, 310f
Diamond, 266
Different bonding-based approaches, 244–245, 244f
Direct Alcohol Fuel Cells (DAFCs), fuels for, 41–42
Direct alkaline fuel cell, 131–132
Direct glycerol fuel cells (DGFCs), 32
Direct methanol fuel cell (DMFC), 6–7, 77–79, 123, 201–203, 208–209
  operating mechanism of, 203f
Direct numerical simulation (DNS), 90–91, 92f
Dispersion method, effect of, 256, 256f
Dissolved oxygen, 42
Double-chamber microbial fuel cell, 262f

E

Electrochemical active bacteria (EAB), 376–377, 383–401
Electrodes
  engineered 3D macroporous architectures, 423–424
  nanomaterials in, 415–425
    anode modification, 415–421, 416f
Electron transfer mechanism, 416f
Electrospinning, 144
Electrospun nanofibers, 144
ELSAM, 4–5
EnergieNed, 4–5
Enzymatic fuel cell (EFC), 413
Ethanol (EtOH), 41–42
Ethanol microfluidic fuel cells, nanoparticle anodes and cathodes for, 56–59, 57f, 59f
Ethylene glycol (EgOH), 41–42
Ethylene glycol microfluidic fuel cells, nanoparticle anodes and cathodes for, 60
Ethylenediamine-nitrogen-containing carbon nanotube (NCNT), 214–215, 215f
Extracellular electron transfer (EET), 376–382, 384–387

F

Fe catalyst, 353–354
Flow-over mode MFC, 27f
Formic acid MFC, 24, 25f
  scaled-up array arrangement of, 26f
4-PAEK-xx, 301–304, 302f
Fuel cell on-a-chip, 42–45
Fuel cell vehicles (FCVs), 77–79
Fuel cells (FCs), 5–6, 41, 77, 123, 174
  application, nano-ink preparation and utilization in, 254–258
  benefits of microfluidics in, 12f
  chemistry, 230–231
  COF-based electrocatalysts, 242–245
    carbon–nitrogen-based approach, 244–245
    imide-based approach (PICOFs), 245
    thin film and nanofiber-based COFs, 243–244
    3D-based COFs, 242
    triazine-based approach (CTFs), 245
    2D-based COFs, 242–243
  COFs
    chemical stability of, 235
    interlayer stacking in, 235
    macrocycle-incorporated, 237–239, 238f
    nanocarbon-supported, 241–242
    porosity of, 234
    pristine metal-free, 236–237, 237f
  comparison of technologies, 78t
  components of, 3
  covalent organic frameworks, chemistry of, 232–235
  history of, 3–5
  mathematical formulation in, 124–134
    challenges of, 134–135
  metal-organic frameworks for, 183–189
  molecular building blocks linking and crystallization, 232–234
  nanomaterials. See Nanomaterials
  nanotechnology and, 6–7
  ORR, COF-based electrocatalysts for, 235–242
    COF-derived SAC-based materials, 239–241
  process design of, 124–134

schematic representation of, 125f
simulation and modeling, 124
types of, 123, 174, 201—202
  direct alkaline fuel cell, 131—132
  microbial, 132—134, 133f
  molten carbonate, 126—128
  proton exchange membrane, 128—131, 129f
  solid oxide, 125, 126t
Fuel cells applications, shape-controlled metal nanoparticles for
  Co catalyst, 353—354
  Fe catalyst, 353—354
  Ni catalyst, 353—354
  platinum catalyst, 350—352
  shape-controlled catalyst, 350
  shape-selected nanoparticles, various functionality of, 354—357
    activity enhancement, 354—355
    optical behavior, 356—357
    selectivity enhancement, 355—356
Fuel cells recycling
  microbial fuel cell recycling mechanism, 370—371
  nanomaterials used in, 362—366
    molten carbonate fuel cells, 364—365
    phosphoric acid fuel cell, 365
    polymer electrolyte membrane/proton-exchange membrane, 363—364
    solid oxide fuel cells, 365—366
  recycling processes, 366—370
    alcohol solvent process, 370
    hydrometallurgical process, 367
    hydrothermal process, 368
    membrane and Pt-recovery acid process, 369
    potential alteration, transient dissolution through, 369
    pyrohydrometallurgical process, 367—368
    selective electrochemical dissolution, 368—369
Fuel cells, nano-inks for, 252f, 253f
  fuel cell application, nano-ink preparation and utilization in, 254—258
    dispersion method, effect of, 256
    ink composition, effect of, 256—258
    solvents and additives, effect of, 254—255
  ink rheological parameters/influencing factors, 253—254
Fuel cells, nanomembranes in
  description of, 285—288
    characteristic parameters of, 286—287
    classification of, 287—288
    proton-exchange membranes, 285—286
  hybrid membranes containing nanofillers, 316—337
    effect of nanofillers, 316—317
    nanocarbon materials, 323—337
    nanoparticles, 317—323
  polymer-based nanomembranes, 295—315
    cross-linking membranes, 310—315
    functional side chains, polymer membranes with, 299—304
    special groups, polymer membranes with, 304—310
    tailored main chains, polymer membranes with, 295—299
  proton transport channels, 288—294
    proton transfer mechanism, 288—289
    proton transport channel models, 289—292
    proton transport channel, modification strategies of, 292—294
Fuels, types of, 5—6
Fuji Electric Corporate Research and Development, Ltd, 4—5
Fullerenes, 206, 333—337, 335f, 336f, 337t
  -based electrocatalysts, 207—209
  for catalyzing ORR, 208
Fulleropyrrolidine (PyC$_{60}$), 209
Fully cross-linking, 312—315, 313f, 314f, 315f, 315t
Functional groups, 180—181

G

Gas diffusion layers (GDLs), 80—81, 83—84
Geobacter sulfurreducens, 264—265
Glassy carbon power (GCP), 208
Global Thermoelectric's Fuel Cell Division, 4—5
Glycerol (GlOH), 41—42
Glycerol microfluidic fuel cells, nanoparticle anodes and cathodes for, 17—18, 18f, 61—65, 64f
Granular activated carbon (GAC), 388—390
Graphene, 152—153, 382—383
Graphene-based nanomaterial, 271—272, 272t
Graphene-based two-dimensional carbonaceous materials, 216—220
  graphene-supported carbonaceous materials, 219—220

Graphene-based two-dimensional carbonaceous materials (*Continued*)
  heteroatom-doped graphene materials, 217–219, 217f, 218f
Graphene-modified electrodes, 146–147
Graphene oxide (GO), 144–145, 145f, 277, 294, 330–335, 433–434
Graphene-supported carbonaceous materials, 219–220
Green energy cells, 123
Grotthuss mechanism, 288–289
Grove cell, 3

H
Heteroatom-doped carbon nanomaterials, 158
Heteroatom-doped CNTs, 214–215, 214f
Heteroatom-doped graphene materials, 217–219, 217f, 218f
High resolution transmission electron microscopy (HRTEM), 61f
High temperature proton-transferring MOFs, 186–189
Histamine-based MOF, 188
Hybrid continuum/pore-scale modeling, 92–94
Hybrid modeling, 93f
Hybrid organic–inorganic membrane, 429–431
Hydrated Nafion morphology, cluster network model of, 290f
Hydrogen, 362t
Hydrogen oxidation reaction (HOR), 30–32, 80–81, 230–232
Hydrogen PEMFCs, 77–79
Hydrometallurgical process, 367
Hydrophilic side chains, 299–301, 300f, 301f
Hydrophobic side chains, 301–304
Hydrothermal process, 368

I
Imidazole-based MOF, 187
Imide-based approach (PICOFs), 245
Influencing factors, 253–254
Ink composition, effect of, 256–258, 257f
Ink rheological parameters
Inkjet printing method, 251–252
Inter-diffusion zone, 14–15, 15f
International Fuel Cells, 3–4
Ion-exchange capacity (IEC), 294–308, 311

Ion-exchange membranes (IEM), 275–278, 426–434

L
Laminar flow-based fuel cells (μFCs), 42
  advantage of, 45
  mechanics of, 44
Large and rigid groups, 304–305
Liquid-feed DMFCs, 77–79
Loofah sponge, Digital photo of, 378f
Low-temperature proton-transferring MOFs, 185–186, 187f

M
Macroscopic continuum modeling, in PEMFCs, 85f, 86f, 88–89
Membrane and Pt-recovery acid process, 369, 370f
Membrane electrode assembly (MEA), 11, 80–81, 82f, 285–286, 363–364, 367–370
Meso-micro-Fe-N-CNTs, 216f
Meso-micro-PoPD catalyst, 221f
Mesopore-carbon microbeads (MCMBs), 208
Metal oxide-based carbon nanocomposite cathode, 158–159
Metal polymer nanocomposite (M-P cathode), 159
Metal sites, in MOFs, 179
Metal sulfides, 195
Metal/nitrogen@carbon (MNC) framework, 192–194
Metal-based carbon cathode catalysts, 424–425
Metal-based carbon nanocomposites cathode, 157–158
Metal-based electrodes, 390–395
Metal-based nanomaterials, 268–269, 270f, 270t
Metal-carbon nanocomposites (M-C anode), 153–155
  metal-based carbon nanocomposite anodes, 153–154
  metal oxide-based carbon nanocomposite anode, 154–155
  metal-polymer nanocomposite, 155–156
Metal-carbon nanocomposites (M-C cathode), 157–159
  metal-based carbon nanocomposites cathode, 157–158
  metal oxide-based carbon nanocomposite cathode, 158–159
Metal-free cathode catalysts, 423–424, 424t, 426t

Metal-organic framework-derived porous carbon materials, 222
Metal-organic frameworks (MOFs), 175−176, 222, 294, 320−322
  for fuel cells applications, 183−189
    proton-conducting, 184−189
  as oxygen reduction reaction catalyst, 189−195
  structure of, 176−183, 177f
    derivative, design of, 183, 184f
    development of porous structure, 181−183
    functional groups, 180−181
    open metal sites, 179
    pores, 179−180
    secondary building units, 178−179, 178f
Metal-polymer nanocomposite (M-P anode), 155−156
Methanol (MeOH), 41−42
Methanol microfluidic fuel cells, nanoparticle anodes and cathodes for, 55−56
Methanol oxidation reaction (MOR), 30−32
μDEgFC, 47−49, 47f
Micro/nanostructures, microbial fuel cells
  bacteria/electrode interaction, 377−387
    carbon nanomaterials decoration, 377−383
    conductive polymer coating, 383−387
    modifications, types of, 387
  increasing specific surface area, microstructure for, 388−401
    microstructure, electrode materials with, 395−401
    random microstructure, electrodes with, 388−395
Microbial fuel cells (MFCs), 132−134, 133f, 139, 270t, 370−371, 375−377, 376f, 379t, 382−383, 385t, 387−390, 392t, 395−401, 397t, 411, 413−414, 414f, 417−418, 420, 423−426, 428t, 429−431
  anodes, nanomaterial structures in, 263−269
    carbon-based nanomaterials, 263−266
    metal-based nanomaterials, 268−269
    nanoconducting polymer, 266−268
  cathodes, nanomaterial structures in, 270−275, 271f
    carbon nanotubes and nanofibers, 272−273
    conducting polymer nanomaterials, 274−275
    graphene-based nanomaterial, 271−272
    transition metal oxide nanomaterials, 273−274
  classification, 140f
  components of, 139−140
  ion-exchange membranes, 275−278
    nonfluorinated and nonsulfonated membranes, 277−278
    perfluorinated membranes, 276
    sulfonated membranes, 276−277
  nanocomposite materials. *See* Nanocomposite materials
  nanostructured materials in, 140−148
Microfluidic direct alcohol fuel cells (μDAFCs), 42−43
  architecture used in, 43f
  general view of, 45−50, 46f, 47f, 48f
  nanotechnology in, 54−65
    nanoparticle anodes and cathodes for ethanol, 56−59, 57f, 59f
    nanoparticle anodes and cathodes for ethylene, 60
    nanoparticle anodes and cathodes for glycerol, 61−65
    nanoparticle anodes and cathodes for methanol, 55−56
  testing, 50−54, 51f, 52f
Microfluidic direct ethanol fuel cells (μDEtFCs), 46−49, 47f
Microfluidic direct glycerol fuel cells (μDGFCs), 46, 47f
Microfluidic direct methanol fuel cells (μDMFCs), 46−48, 47f
Microfluidic fuel cell (MFC), 12−13
  with carbon fibers, 33−35, 34f
  characteristics and key findings, 22t
  electrochemical reaction rate in, 16
  fabrication and design of, 16−21
  fuel utilization in, 28−29
  glycerol/bleach, 17−18, 18f
  microchannels, fabrication of, 33−35
  performance evaluation of, 21−35
  theory, 13−16
Microporous layer (MPL), 85
Mineral materials, 322−323, 322f
Mixed potential losses (η_{mp}), 87
Mixing region. *See* Inter-diffusion zone
Molecular building blocks linking/crystallization, 232−234, 234f
Molten carbonate fuel cells (MCFCs), 123, 126−128, 201−202, 364−365
  current density and cell voltage, 128
  electrochemical reactions, 127

450 Index

Molten carbonate fuel cells (MCFCs) (*Continued*)
  energy balance, 127–128
  mass balance, 127
Montmorillonite (MMT), 322–323
Multiwalled carbon nanotubes (MWCNTs), 56, 142–143, 264

N

N- and S-doped carbon nanotubes (NSCNTs), 60
Nafion membrane, 81–83, 289–294
  parallel water-channel model of, 291f
Nafion PEM, 184–185
Nanobelts, 144–147
Nanocarbon materials, 323–337
Nanocarbons
  approaches to synthesize, 206–224
    graphene-based two-dimensional carbonaceous materials, 216–220
    one-dimensional carbonaceous materials, 211–216
    three-dimensional carbon materials, 220–224
    zero-dimensional carbonaceous materials, 206–211
  classification of, 207f
Nanocarbon-supported COFs, 241–242, 241f
Nanocomposite materials, in MFCs, 148–162
  anode materials, 149–156, 150t
    metal-carbon nanocomposites, 153–155
    polymer carbon nanocomposites, 149–153
  cathode materials, 156–159, 160t
    metal polymer nanocomposite, 159
    metal-carbon nanocomposites, 157–159
    polymer-carbon nanocomposites, 156
  membranes, 159–162
    polymer-carbon nanocomposites, 161
    polymer-metal nanocomposites, 162
    polymer-polymer nanocomposites, 161
Nanocomposite membrane, 429–431, 430f
  carbon materials, 433–434
  Fe_3O_4, 431
  miscellaneous materials, 434
  SiO_2, 432–433
  TiO_2, 431–432
Nanocomposite-based anode catalysts, 419t
Nanoconducting polymer, 266–268
Nanofibers, 142–144, 324–325, 325f, 326f
Nanofillers, effect of, 316–317
Nanofuel cells, 6–7

current research on, 7–8
Nanomaterials
  electrode materials, 141–148
    carbon-based structures, 141–148
  in microbial fuel cells, 140–148
Nanomaterials, biofuel cells
  application of, 414–434
    carbon-based nanomaterials and nanocomposites, anode catalyst, 419–420
    in electrodes, 415–425
    metal-based carbon cathode catalysts, 424–425
    metal-free cathode catalysts, 423–424
    polymer-based anodes, 420–421
    transition metals-based nanoparticles and nanocomposites, anode catalyst, 417–419
  in membranes, 426–434
    carbon materials, nanocomposite membranes with, 433–434
  Fe_3O_4
    nanocomposite membranes with, 431
    ion-exchange membranes, 426–434
    miscellaneous materials, nanocomposite membranes with, 434
    nanocomposite membrane, 429–431
  SiO_2
    nanocomposite membranes with, 432–433
  TiO_2
    nanocomposite membranes with, 431–432
  types of, 412–414
    enzymatic fuel cell, 413
    microbial fuel cell, 413–414
Nanomembranes
  characteristic parameters of, 286–287
  classification of, 287–288
Nanometal oxides, 317–320, 318f
Nanoparticles, 142
Nanoporous carbon materials (NPCs), 191–192
Nanorods, 142–144
Nanosheets, 144–147
Nanospheres, 142
Nanotechnology
  and fuel cells, 6–7
  in direct alcohol microfluidic fuel cells, 54–65
    nanoparticle anodes and cathodes for ethanol, 56–59, 57f, 59f
    nanoparticle anodes and cathodes for ethylene, 60

nanoparticle anodes and cathodes for glycerol, 61–65
nanoparticle anodes and cathodes for methanol, 55–56
Nanotubes, 142–144, 326–329, 327f, 328f, 329f, 330f
Nanowires, 142–144
National Aeronautics and Space Administration (NASA), 3–4, 361–362
National Fuel Cell Research Centre, 4–5
N-doped carbon foam (NCF), 147
N-doped CDs, 211f
Nernst equation, 375–376
Nernst–Monod relationship, 134
Ni catalyst, 353–354
Nitrogen, as doping agent, 208
Nonfluorinated membranes, 277–278, 278t
Nonsulfonated membranes, 277–278, 278t
Northern Research and Engineering Corporation, 4–5
NOVEM cooperate, 4–5

O

Ohmic losses (η_{ohm}), 87, 411
One-dimensional (1D) carbon nanostructures, 142–144
One-dimensional carbonaceous materials, 211–216
  carbon nanotubes-based 1D nanostructures, 213–216
Optical behavior, 356–357
Oxygen reduction reaction (ORR), 80–81, 229–231, 234–242, 271–272, 365, 412, 414–415, 421–426, 425f
  characterization notions, 204–206
    current density, 205
    electron transfer number, 205
    HO_2^- percentage, 205
    onset potential (E_{ons}), 204–205
    Tafel slope, 205
    turnover frequency, 206
  metal-organic frameworks as, 189–195
Oxygen reduction, 48–49

P

PANI/MWCNT composite, 156
Perfluorinated membranes, 276
Perfluorinated sulfonic acid (PFSA), 287, 290f, 291
Perfluorosulfonic acid polymers, 159–161

Phosphoric acid fuel cell (PAFC), 123, 201–202, 365
Plain carbon brush (PCB), 146
Platinum (Pt), 158–159
Platinum catalyst, 350–352
PLUXter technique, 181
Poly(2-acrylamido-2-methylpropane sulfonic acid) (PAMPS)-g-TiO_2, 317–320
Poly(3,4-ethylene dioxythiophene) (PEDOT), 383
Poly(vinyl alcohol) (PVA), 320–322
Polyacrylonitrile (PAN), 144
Polyaniline (PANI), 266–267, 317–320, 382–387
Polybenzimidazoles, structural formula of, 296f
Polydimethylsiloxane (PDMS), 17
Polyethylene terephthalate glycol (PETG), 20–21
Polylactic acid (PLA), 20–21
Polymer carbon nanocomposites (P-C anode), 149–153
Polymer electrolyte membrane (PEM) fuel cells, 77–79, 201–203, 363–364, 366t, 367–368, 370, 411–412. See also Proton exchange membrane fuel cells (PEMFCs)
  diagrammatic illustration of, 202f
  operating principle for, 202
Polymer membranes
  cross-linking membranes, 310–315
    fully cross-linking, 312–315
    semiinterpenetrating, 311
  functional side chains, 299–304
    hydrophilic side chains, 299–301
    hydrophobic side chains, 301–304
  special groups, 304–310
    amino groups, 305–306
    densely sulfonated units, 306–310
    large and rigid groups, 304–305
  tailored main chains, 295–299
    flexible and rigid chain segments, 295–297
    regular and atactic chain segments, 297–299, 298f, 299f
Polymer-based anodes, 420–421
Polymer-carbon nanocomposites (P-C cathode/membrane), 156, 161
Polymer-metal nanocomposites (P-M membrane), 162
Polymer-polymer nanocomposites (P-P membrane), 161
Polymethyl methacrylate (PMMA), 18–19
Polypyrrole (Ppy), 149–152, 266–267

Polyvinyl alcohol (PVA), 277
Pomelo peel and SEM, 396f
Pore network modeling (PNM), 90−91, 90f
Pores, in MOFs, 179−180
Pore-scale modeling, 90−92
Porous nitrogen-doped carbon nanosheet (PNCN), 271−272
Potential alteration, transient dissolution through, 369
Proton conductivity, 288−289
Proton exchange membrane fuel cells (PEMFCs), 80−87, 123, 128−131, 363−364
  cross-section of, 98f
  components
    bipolar plates, 80−81, 84−85
    catalyst layers, 80−81, 83
    gas diffusion layers, 80−81, 83−84
    microporous layer, 85
    proton exchange/polymer electrolyte membrane, 81−83
  electrochemical reaction, 80−81
  electron and proton transport, 81
  macroscopic modeling, 94−112
    assumptions, 88f, 97−98
    boundary conditions, 110−112
    conservations equations, 89f, 90f, 92f, 93f, 98−103
    source terms, 104−109
  numerical modeling, 88
    hybrid continuum/pore-scale modeling, 92−94
    macroscopic continuum, 85f, 86f, 88−89
    pore-scale modeling, 90−92
  operation principle, 81, 81f
  product removal, 81
  reactant transport, 80
  schematic representation of, 129f
  X-ray computed tomography images, 84f
Proton jumping mechanism, 289
Proton transfer mechanism, 288−289, 289f
Proton transport channel models, 289−292, 303f
  modification strategies of, 292−294
Proton-conducting metal-organic frameworks, 184−189
Proton-exchange membrane (PEM), 261, 275, 277, 285−289, 292−299, 304−308, 315t, 326−327, 330−331, 333−335, 337t, 413−414

Proton-exchange membrane fuel cells (PEMFCs), 3−4, 174, 285−286, 293−295, 332−333, 363, 367
  advantages of, 175
  H_2 as fuel, 174−175
  schematic diagram of, 175f
Proton-exchange membrane water electrolysis (PEMWE), 363−364, 367
Pt-Ru/PAni-C_{60} electrocatalyst, 209f
Pyrohydrometallurgical Pt-recovery process, 367−368, 368f

R
Recycling processes, 366−370
Reduced graphene oxide (rGO), 144−145, 190−191
Reduced graphene oxide-based carbon brush (RGO-CB), 146
Reticulated vitreous carbon (RVC), 381, 388−390
Reynolds number, 13
rGO-based cathodes, 146

S
SDF-PAEK, 295−297, 301−304, 302f, 303f
Secondary building units (SBUs), 176−179, 178f
Selective electrochemical dissolution, 368−369
Selectivity enhancement, 355−356
Semiinterpenetrating, 311, 312f
SFPAE-A-70 and SFPAE-S-70, 306−308
Shape-controlled catalyst, 350, 351t
Shewanella oneidensis, 420
Siemens and Sulzer, 4−5
Single atoms catalysts (SACs), 229, 239
Single-chamber microbial fuel cell (SCMFC), 162
Single-wall carbon nanohorns (SWCNHs), 265
Single-walled carbon nanotubes (SWCNTs), 142−143, 264
Soft lithography technique, 17−18
Solid oxide fuel cell (SOFC), 3−5, 77−79, 123, 125, 126t, 201−202, 365−366
Solid-state lithium-ion batteries, 11
Solvents and additives, effect of, 254−255
Standard lithography technique, 17
Sulfonated carbon nanofibers (SCNFs), 287−288
Sulfonated membranes, 276−277, 277f
Sulfonated poly(ether ether ketone) (SPEEK), 162, 276, 287−288
Sulfonated polyethersulfone (SPAES), 317−320

T

Tafel slope, 205
Tartaric acid-doped polyaniline (PANI), 152–153
Thin film and nanofiber-based COFs, 243–244, 243f
2D-based COFs, 242–243
3D graphene nanosheets (3D-GNS), 147–148
3D printing technology, 20–21, 399–401, 400f
  for MFCs, 35
3D-based COFs, 242
3D-printed glycerol microfluidic fuel cell, 64f
3D-printed μFC, 63–65
Three-dimensional carbon materials, 220–224
  heteroatom-doped porous carbon nanomaterials, 220–222
  metal-organic framework-derived porous carbon materials, 222
  three-dimensional porous metal-nitrogen-carbon materials, 222–224
Three-dimensional carbon nanostructures, 147–148
Three-dimensional ordered porous carbon (3DOM-C), 155
Three-dimensional porous metal-nitrogen-carbon materials, 222–224
Titania nanowire structures TiO_2 (TiO_2-NWs), 154
Titanium dioxide (TiO_2), 154, 162
Tokyo Electric Power Co., 4–5
Top-down approach, 141
Transition metal oxides (TMOs), 273–274
Triazine-based approach (CTFs), 245
Triazole, 188
Two-cell stack air-breathing MFC, 29–30, 30f
Two-dimensional (2D) carbon nanostructures, 144–147
Two-phase flow H_2O_2 MFC, 32–33, 33f

V

Vapor feed methanol MFC, 28–29, 29f
Vapor-feed DMFCs, 77–79

W

Westinghouse Electrical Cooperation, 4–5

Y

Yttria-stabilized zirconia (YSZ), 365–366, 368, 368f

Z

Zeolites, 434
Zero-dimensional (0D) carbon nanostructures, 142
Zero-dimensional carbonaceous materials, 206–211
  carbon dots-based electrocatalysts, 209–211, 212f
  fullerene-based electrocatalysts, 207–209
Zirconium-sulphoterephthalate (ZST) MOFs, 186
Zn-Fe-zeolitic imidazolate framework (ZIF), 214

Printed in the United States
by Baker & Taylor Publisher Services